Numerical Modelling Handbook

Numerical Modelling Handbook

Edited by **Gregory Rago**

CLANRYE INTERNATIONAL

New Jersey

Published by Clanrye International,
55 Van Reypen Street,
Jersey City, NJ 07306, USA
www.clanryeinternational.com

Numerical Modelling Handbook
Edited by Gregory Rago

International Standard Book Number: 978-1-63240-397-1 (Hardback)

Printed in China.

Contents

Preface IX

Part 1 **Fluid and Gas Dynamics** 1

Chapter 1 **Simulation of the Scavenging Process in Two-Stroke Engines** 3
María Isabel Lamas Galdo and Carlos G. Rodríguez Vidal

Chapter 2 **3D Numerical Modelling of Mould Filling of a Coat Hanger Distributer and Rectangular Cavity** 21
Rekha R. Rao, Lisa A. Mondy, David R. Noble, Matthew M. Hopkins, Carlton F. Brooks and Thomas A. Baer

Chapter 3 **3D Multiphase Numerical Modelling for Turbidity Current Flows** 45
A. Georgoulas, P. Angelidis, K. Kopasakis and N. Kotsovinos

Chapter 4 **Numerical Simulation of the Unsteady Shock Interaction of Blunt Body Flows** 73
Leonid Bazyma, Vasyl Rashkovan and Vladimir Golovanevskiy

Chapter 5 **Modelling Dynamics of Valley Glaciers** 91
Surendra Adhikari and Shawn J. Marshall

Chapter 6 **Numerical Modelling of Heavy Metals Transport Processes in Riverine Basins** 119
Seyed Mahmood Kashefipour and Ali Roshanfekr

Chapter 7 **Numerical Modelling of Dynamic Nitrogen at Atmospheric Pressure in a Negative DC Corona Discharge** 143
A.K. Ferouani, B. Liani and M. Lemerini

Part 2 **Maxwell's Equations** 157

Chapter 8 **Numerical Study of Diffusion of Interacting Particles in a Magnetic Fluid Layer** 159
Olga Lavrova, Viktor Polevikov and Lutz Tobiska

Chapter 9 **Numerical Modelling and Design of an Eddy Current Sensor** 179
Philip May and Erping Zhou

Chapter 10 **Finite Element Method Applied to the Modelling and Analysis of Induction Motors** 203
M'hemed Rachek and Tarik Merzouki

Part 3 **Mechanics and Materials** 227

Chapter 11 **Numerical Modelling of Steel Deformation at Extra-High Temperatures** 229
Marcin Hojny and Miroslaw Glowacki

Chapter 12 **Numerical Evaluation of Product Development Processes** 251
Nadia Bhuiyan

Chapter 13 **Distinct Element Method Applied on Old Masonry Structures** 277
Marwan Al-Heib

Chapter 14 **Inverse Analysis Applied to Mushy Steel Rheological Properties Testing Using Hybrid Numerical-Analytical Model** 303
Miroslaw Glowacki

Chapter 15 **Numerical Schemes for Fractional Ordinary Differential Equations** 329
Weihua Deng and Can Li

Chapter 16 **Phenomenological Modelling of Cyclic Plasticity** 349
Radim Halama, Josef Sedlák and Michal Šofer

Chapter 17 **Biorthogonal Decomposition for Wide-Area Wave Motion Monitoring Using Statistical Models** 375
P. Esquivel, D. Cabuto, V. Sanchez and F. Chan

Permissions

List of Contributors

Preface

A descriptive account based on numerical modelling has been presented in this book. It documents case studies and applications developed by veterans for students and professionals associated with the fields of engineering, decision making management, technology, materials and other industries in which mathematical modelling plays a vital role. It elucidates various examples ranging from well-known standards to novelty applications. The term "model" is primarily comprehended as a group of equations which present the variables and interconnections of physical system or procedure. Models are developed and analyzed extensively, and authors cautiously consider the process for formulating a mathematical alternative to a phenomenon under consideration. For most cases, this results in partial differential equations, for the solution of which numerical methodologies need to be applied. Advancements in computer technology and associated software have presented several tools of enhancing power for specialists in mathematical modelling. This book will serve as a valuable reference for those interested in the field.

This book is a result of research of several months to collate the most relevant data in the field.

When I was approached with the idea of this book and the proposal to edit it, I was overwhelmed. It gave me an opportunity to reach out to all those who share a common interest with me in this field. I had 3 main parameters for editing this text:

1. Accuracy - The data and information provided in this book should be up-to-date and valuable to the readers.
2. Structure - The data must be presented in a structured format for easy understanding and better grasping of the readers.
3. Universal Approach - This book not only targets students but also experts and innovators in the field, thus my aim was to present topics which are of use to all.

Thus, it took me a couple of months to finish the editing of this book.

I would like to make a special mention of my publisher who considered me worthy of this opportunity and also supported me throughout the editing process. I would also like to thank the editing team at the back-end who extended their help whenever required.

Editor

Part 1

Fluid and Gas Dynamics

1

Simulation of the Scavenging Process in Two-Stroke Engines

María Isabel Lamas Galdo and Carlos G. Rodríguez Vidal
Universidade da Coruña
Spain

1. Introduction

It is widely known that the scavenging process plays a very important role in the performance and efficiency of two-stroke engines. Briefly, scavenging is the process by which the fresh charge displaces the burnt gas from the cylinder. Due to the difficulties associated with the measurement techniques, CFD (Computational Fluid Dynamics) is a very helpful tool to analyze the flow pattern inside the cylinder. CFD simulations can provide more detailed information than experimental studies. For this reason, this chapter focuses on a numerical analysis to simulate the fluid flow and heat transfer inside the cylinder at the scavenging process.

This chapter is a continuation and extension of previous works (Lamas-Galdo *et al.*, 2011; Lamas & Rodriguez, 2012), in which CFD models were developed and validated with experimental results. The content is organized as follows. A brief description of two-stroke engines is given in Section 2. The mathematical model, i.e., the governing equations are presented in Section 3 and the numerical model is discussed in Section 4. After that, the results are shown in Section 5 and the conclusions of this chapter are discussed in Section 6.

2. Introduction to the two-stroke engine

Although the focus of this chapter is the numerical treatment of the scavenging process, it is important to introduce certain introductory aspects about the performance of two-stroke engines. This will facilitate the reader's understanding of the chapter.

2.1 Mechanical aspects

A two-stroke engine is an internal combustion engine that completes the process cycle in one revolution of the crankshaft or two strokes of the piston: an up stroke and a down stroke. Both spark ignition and compression ignition engines exist today. Spark ignition engines are employed in light applications (chainsaws, motorcycles, outboard motors, etc) due to its low cost and simplicity. On the other hand, diesel compression ignition engines are mainly employed in large and weight applications, such as large industrial and marine engines, heavy machinery, locomotives, etc. Fig. 1 (a) shows a spark ignition engine installed on a motorbike and Fig. 1 (b) shows a large compression ignition engine, the MAN B&W 7S50MC, typically used in marine propulsion and industrial plants.

(a) (b)

Fig. 1. (a) Spark ignition gasoline engine installed on a motorcycle. (b) Compression ignition diesel engine MAN B&W 7S50MC installed on a ship.

There are several mechanical details which vary from one engine to another. For example, Fig. 2 (a) shows a cross section of the spark ignition engine shown in Fig. 1 (a). In this engine, the charge is introduced to the cylinder by ports. The opening and closing of the ports is controlled by the sides of the piston covering and uncovering them as it moves up and down in the cylinder. As can be seen in the bottom part of Fig. 2 (a), this engine has a crankcase. This is a separate charging cylinder which employs the volume below the piston as a charging pump. On the other hand, Fig. 2 (b) shows a cross section of the compression ignition engine illustrated on Fig. 1 (a). This engine has one exhaust valve and several intake ports. In this case, the external air is introduced directly in the cylinder instead of being pumped from the crankcase.

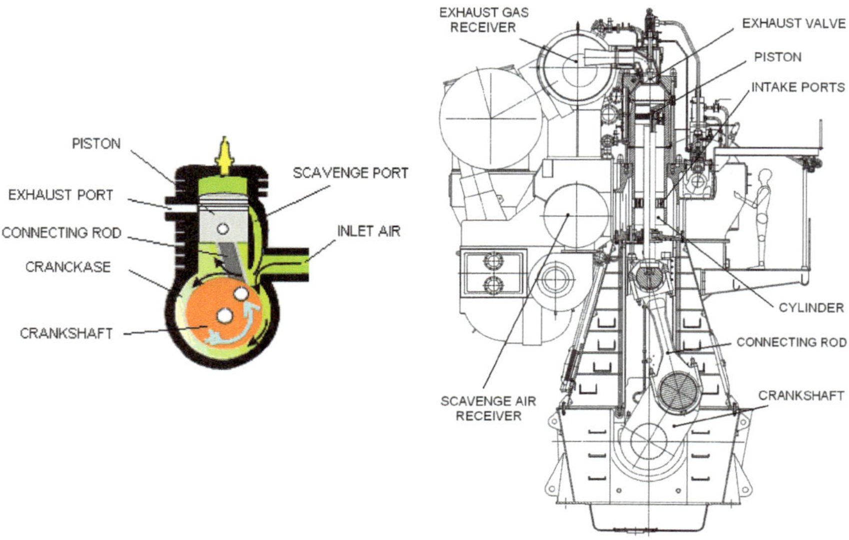

(a) (b)

Fig. 2. (a) Cross section of a spark ignition engine. (b) Cross section of a compression ignition engine, the MAN B&W 7S50MC.

2.2 The scavenging process

Before discussing the scavenging process, it is useful to describe the operation cycle of the two-stroke engine with direct injection. For this purpose, an engine with scavenge and exhaust ports instead valves will be considered. At the beginning of the cycle, when fuel injection and ignition have just taken place, the piston is at the TDC (top dead center). The temperature and pressure rise and consequently the piston is driven down, Fig. 3 (a) (note that the arrows indicate the direction of the piston). Along the power stroke, the exhaust ports are uncovered (opened) and, consequently, the burnt gases begin to flow out, Fig. 3 (b). The piston continues down. When the piston pasts over (and consequently opens) the scavenge ports, pressurized air enters and drives out the remaining exhaust gases, Fig. 3 (c). This process of introducing air and expelling burnt gases is called scavenging. The incoming air is used to clean out or scavenge the exhaust gases and then to fill or charge the space with fresh air. After reaching BDC (bottom dead center), the piston moves upward on its return stroke. The scavenge ports and then the exhaust ports are closed, Fig. 3 (d), and the air is then compressed as the piston moves to the top of its stroke. Soon before the piston reaches TDC, the injectors spray the fuel, the spark plug ignite the mixture and the cycle starts again.

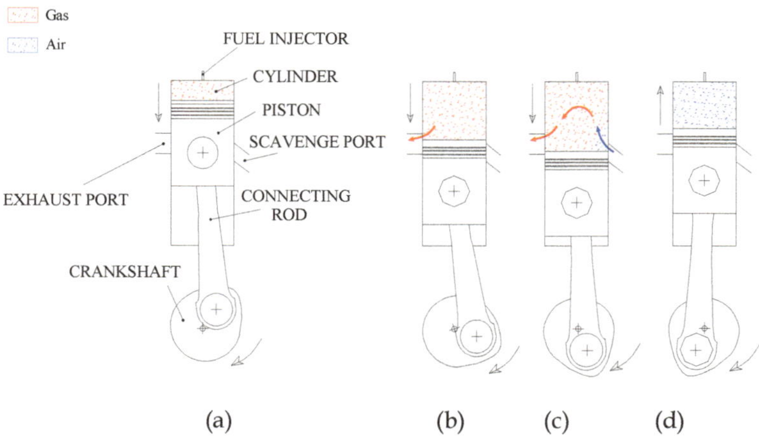

Fig. 3. Basic engine operation. (a) Injection; (b) exhaust; (c) scavenge; (d) compression.

A drawback which has a decisive influence, not only on consumption but also on power and pollution, is the process of displacing the burnt gases from the cylinder and replacing them by the fresh-air charge, known as scavenging. In ideal scavenging, the entering scavenge air acts as a wedge in pushing the burnt gases out of the cylinder without mixing with them. Unfortunately, the real scavenging process is characterized by two problems common to two-stroke engines in general: *short-circuiting losses* and *mixing*. *Short-circuiting* consists on expelling some of the fresh-air charge directly to the exhaust and *mixing* consists on the fact that there is a small amount of residual gases which remain trapped without being expelled, being mixed with some of the new air charge.

The main difficulty involved in designing an effective scavenging system is that there are too involving variables: piston chamber geometry, intake and exhaust ports design, opening and closing timings, compression ratio, fuel composition, inlet and exhaust pressures, etc, being necessary a detailed study to embrace all this factors. For years, the study of the fluid flow inside engines has been mainly supported by experimental tests such as PIV (Particle Image Velocity), LDA (Laser Doppler Anemometry), ICCD cameras, etc. However, these experimental tests are very laborious and expensive. As an alternative solution to experimental techniques, CFD has recently become a useful tool to study the fluid flow inside engines. In the field of engines, CFD is especially useful to design complex components such as combustion chambers, manifolds, injectors and other parameters. The first numerical simulations of engines appeared in the eighties (Sher, 1980; Carpenter & Ramos, 1986; Sweeny *et al.*, 1985; Ahmadi-Befrui *et al.*, 1989) but, unfortunately, these first numerical studies only provided, with poor accuracy, information about the general configuration of the flow field inside the cylinder. Besides, at that time it was very difficult to carry out a three-dimensional analysis. After these first numerical studies, a lot of works appeared in the nineties and in recent years. The number of CFD codes has also increased noticeably, appearing studies using KIVA (Epstein *et al.*, 1991; Amsden *et al.*, 1992), STAR-CD (Raghunathan & Kenny, 1997; Yu *et al.* 1997; Zahn *et al.*, 2000; Hariharan *et al.*, 2009), FIRE (Hori *et al.*, 1995; Laimböck *et al.*, 1998) Fluent (Pitta & Kuderu, 2008; Lamas-Galdo *et al.*, 2011), CFX (Albanesi *et al.*, 2009), etc.

3. Mathematical model

Once the basic performance of two-stroke engines was described, the methodology to simulate the scavenging process will be treated in this section.

3.1 Governing equations

The governing equations of the flow inside the cylinder are the Navier-Stokes ones. The energy equation is also needed to compute the thermal problem. Finally, as there are two components (air and burnt gases), one more equation must be added to characterize the propagating interface. These equations are briefly described in what follows.

In Cartesian tensor form, the continuity equation is given by:

$$\frac{\partial \rho}{\partial t} + \frac{\partial}{\partial x_i}(\rho u_i) = 0 \tag{1}$$

where ρ is the density and u the velocity. It is very common to consider the flows as ideal gasses, so the density can be calculated as follows:

$$\rho = \frac{p}{RT} \tag{2}$$

where p is the pressure, T the temperature and R the ideal gas constant. The momentum conservation equation is given by:

$$\frac{\partial}{\partial t}(\rho u_i)+\frac{\partial}{\partial x_j}(\rho u_i u_j)=-\frac{\partial p}{\partial x_i}+\frac{\partial \tau_{ij}}{\partial x_j} \tag{3}$$

where τ_{ij} is the stress tensor. If the fluid is treated as Newtonian, the stress tensor components are given by:

$$\tau_{ij}=\mu\left(\frac{\partial u_i}{\partial x_j}+\frac{\partial u_j}{\partial x_i}-\frac{2}{3}\delta_{ij}\frac{\partial u_k}{\partial x_k}\right) \tag{4}$$

As only the scavenging process and not the combustion is treated on this chapter, only two components need to be computed: burnt gas and unburnt gas (air). In order to characterize the propagating interface, the following equation is solved:

$$\frac{\partial(\rho Y_{air})}{\partial t}+\nabla\cdot(Y_{air}\rho\vec{V})=0 \tag{5}$$

where Y_{air} is the mass fraction of the air. The mass fraction of the burnt gases, Y_{gas} , is given by the restriction that the total mass fraction must sum to unity:

$$Y_{gas}=1-Y_{air} \tag{6}$$

3.2 Turbulence

Today's standard in engine simulation are Reynolds Averaged Navier-Stokes (RANS) methods. Another approach are Large Eddy Simulation (LES) techniques. LES and RANS techniques differ in the way they address the present impossibility to resolve all the scales present in engine flows. RANS simulations are based on a statistical averaging to solve only the mean flow. This implies that modelling concerns the whole spectrum of scales. In LES, a spatial or temporal filtering is used to represent the large turbulent scales of the flow, which are directly resolved, while the small scales are modeled. In LES, modeling thus concerns a much smaller part of the spectrum, which leads to an improvement of predictivity as compared to RANS. LES inherently allows to address large scale unsteady phenomena, and thus has a good potential to predict engine unsteadiness. The problem is that LES would lead to a CPU time that is way beyond reach of present supercomputers. Therefore, the use o LES is not very common.

In the field of RANS methods, the two-equation model standard k-ε is the most used to simulate engines. The RNG k-ε model is also widely employed, specially in the cases of swirling flows.

The momentum conservation equation for a turbulent flow is given by:

$$\frac{\partial}{\partial t}(\rho u_i)+\frac{\partial}{\partial x_j}(\rho u_i u_j)=-\frac{\partial p}{\partial x_i}+\frac{\partial \tau_{ij}}{\partial x_j}+\frac{\partial}{\partial x_j}(-\rho\overline{u_i^{'}u_j^{'}}) \tag{7}$$

A common method to model the Reynolds stresses, $-\rho\overline{u_i^{'}u_j^{'}}$, is the Boussinesq hypothesis to relate the Reynolds stresses to the mean velocity gradients:

$$-\rho\overline{u_i'u_j'} = \mu_t\left(\frac{\partial u_i}{\partial x_j}+\frac{\partial u_j}{\partial x_i}\right)-\frac{2}{3}\left(\rho k+\mu_t\frac{\partial u_k}{\partial x_k}\right)\delta_{ij} \tag{8}$$

where δ_{ij} is the Kronecker delta (δ_{ij}=1 if $i=j$ and δ_{ij}=0 if $i\neq j$), which is included to make the formula applicable to the normal Reynolds stresses for which $i=j$ (Versteeg and Malalasekera, 2007) and μ_t is the turbulent viscosity. The k-ε model includes two differential equations, corresponding to the turbulent kinetic energy (k), and its dissipation rate (ε), given by Ecs. (9) and (10) respectively.

$$\frac{\partial}{\partial t}(\rho k)+\frac{\partial}{\partial x_i}(\rho k u_i)=\frac{\partial}{\partial x_j}\left[\alpha_k\mu_t\frac{\partial k}{\partial x_j}\right]+G_k+G_b-\rho\varepsilon-Y_M \tag{9}$$

$$\frac{\partial}{\partial t}(\rho\varepsilon)+\frac{\partial}{\partial x_i}(\rho\varepsilon u_i)=\frac{\partial}{\partial x_j}\left[\alpha_\varepsilon\mu_t\frac{\partial \varepsilon}{\partial x_j}\right]+C_{1\varepsilon}\frac{\varepsilon}{k}\left(G_k+G_{3\varepsilon}G_b\right)-C_{2\varepsilon}\rho\frac{\varepsilon^2}{k} \tag{10}$$

In the above equations, G_k represents the generation of turbulence kinetic energy due to the mean velocity gradients; G_b is the generation of turbulence kinetic energy due to buoyancy; Y_M represents the contribution of the fluctuating dilatation in compressible turbulence to the overall dissipation rate. C_μ, $C_{1\varepsilon}$, $C_{2\varepsilon}$, $C_{3\varepsilon}$, σ_k and σ_ε are constants and the terms a_k and a_ε represent the inverse effective Prandtl numbers for k and ε respectively. These quantities were obtained by a RNG modified method which accounts for the effects of swirl or rotation. Details of the procedure are given elsewhere, (Fluent Inc., 2006).

The turbulent viscosity, μ_t, is computed by combining k and ε as follows:

$$\mu_t=\rho C_\mu\frac{k^2}{\varepsilon} \tag{11}$$

Concerning the heat transfer problem, turbulent heat transport can be modeled using the concept of Reynolds' analogy to turbulent momentum transfer. The energy equation is thus given by the following:

$$\frac{\partial}{\partial t}(\rho E)+\frac{\partial}{\partial x_i}\left[u_i(\rho E+p)\right]=\frac{\partial}{\partial x_j}\left[\left(k_t+\frac{C_p\mu_t}{\Pr}\right)\frac{\partial T}{\partial x_j}+u_i\tau_{ij}\right] \tag{12}$$

where E is the total energy.

4. Numerical procedure

In this section, the generation of the mesh and other numerical details will be described. Particularly, this section focuses on the engine studied in Lamas-Galdo *et al.* (2011), which is shown in Fig. 1 (a) and Fig. 2 (a). This is a single cylinder two-stroke engine. The geometry and distribution diagram are shown in Fig. 4, and other technical specifications are summarized in Table 1.

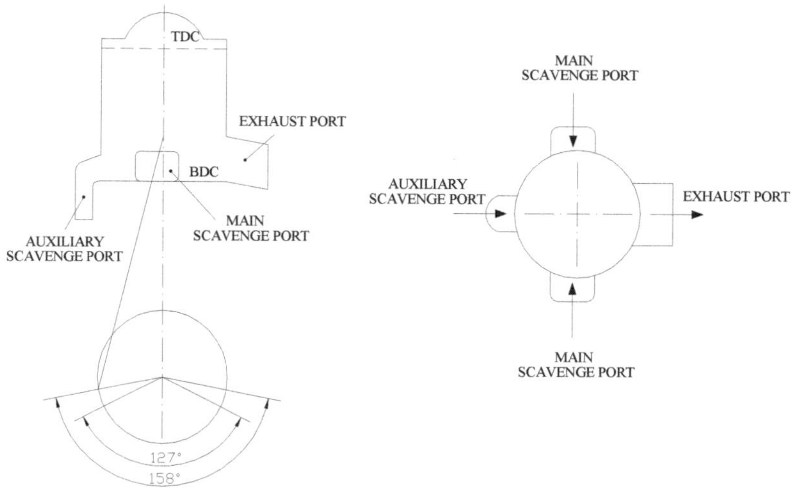

(a) (b)

Fig. 4. Cylinder geometry and distribution diagram. (a) Lateral view; (b) Plant view. Lamas-Galdo *et al.* (2011).

Parameter	Value
Type of engine	Two-stroke, Otto
Displacement (cm^3)	127.3
Compression rate	9.86:1
Bore (mm)	53.8
Stroke (mm)	56
Connecting rod length (mm)	110
Scavenging system	Loop scavenge
Fuel system	Direct injection
Power (W)	7500
Speed (rpm)	6000

Table 1. Technical specifications.

At maximum continuum rating, the in-cylinder, exhaust and intake pressures were measured experimentally. Piezoresistive sensors were employed to measure the exhaust and intake pressures, while a piezoelectric sensor was employed to measure the in-cylinder pressure. These sensors were connected to its corresponding charge amplifier and data acquisition system. The data were analyzed using the software *LabVIEW SignalExpress LE*. The in-cylinder pressure is shown in Fig. 5 and the intake and exhaust pressures are shown in Fig. 6. Note that, in this work, the crank angles were chosen with reference to TDC.

Concerning the temperatures, unfortunately, the in-cylinder temperatures can not be measured experimentally because a temperature sensor is not fast enough to accurate capture the in-cylinder temperature along the whole cycle.

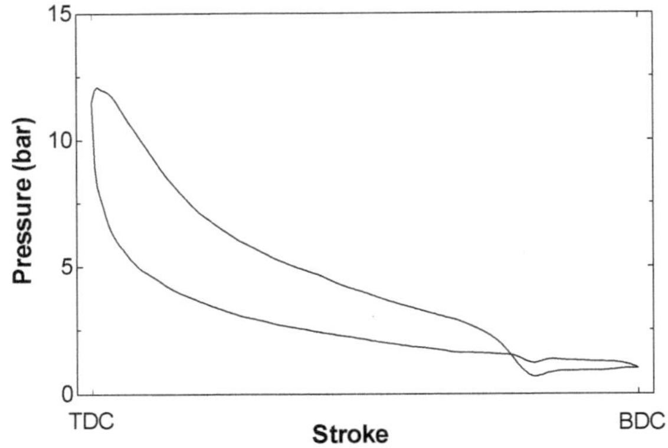

Fig. 5. Evolution of the in-cylinder pressure.

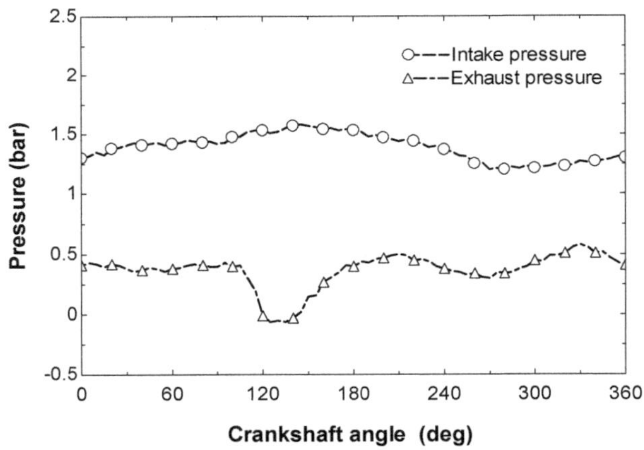

Fig. 6. Evolution of the exhaust and intake pressures.

4.1 Mesh generation

The principle of operation of CFD codes is subdividing the domain into a number of smaller, non-overlapping sub-domains. The result is a grid (or mesh) of cells (or elements). In this work, a grid generation program, Gambit 2.4.6, was used to generate the mesh. In order to implement the movement of the piston, a moving mesh must be used. Figure 7 shows the mesh at several crankshaft angles. The computational domain includes the scavenge ports, exhaust port, cylinder and cylinder head.

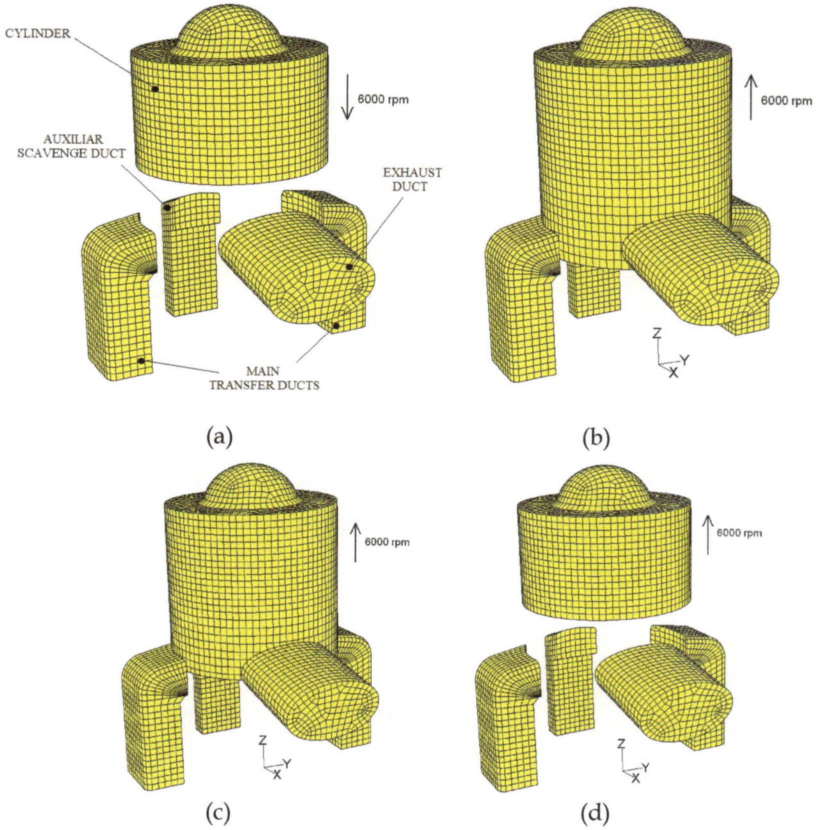

Fig. 7. Computational mesh. (a) 92° crank angle; (b) 190°; (c) 215°; (d) 270° crank angle. Lamas-Galdo *et al.* (2011).

Hexahedral elements provide better accuracy and stability, so a structured hexahedral mesh was adopted. The numerical algorithm implemented automatically updates the mesh after each time step relative to the piston motion using a meshing tool called "dynamic layering", which consists on adding or removing layers of cells adjacent to a moving boundary based on the height of the layer adjacent to the moving surface. The procedure is shown in Fig. 8.

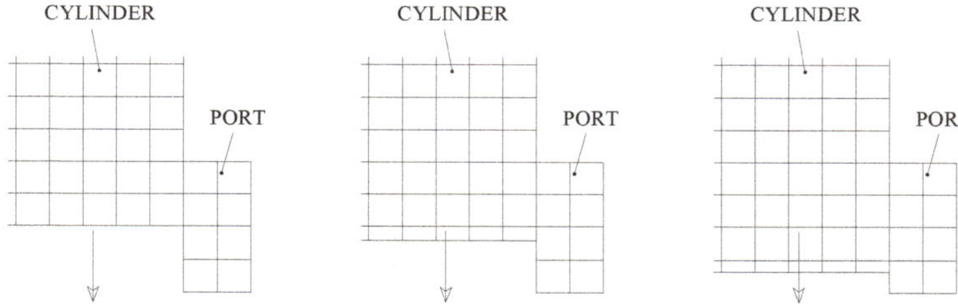

Fig. 8. Layering procedure.

Sometimes it is not possible to employ hexahedral elements in the totality of the control volume. For example, the engine studied in Lamas & Rodríguez (2012), Fig. 1 (b) and Fig. 2 (b), has an exhaust valve in every cylinder. Due to the complex geometry of the valve and duct, tetrahedral elements were employed in that region. Besides, it was necessary to refine the region closed to the valve in order to capture the complex characteristics of the flow. The result is shown in Fig. 9.

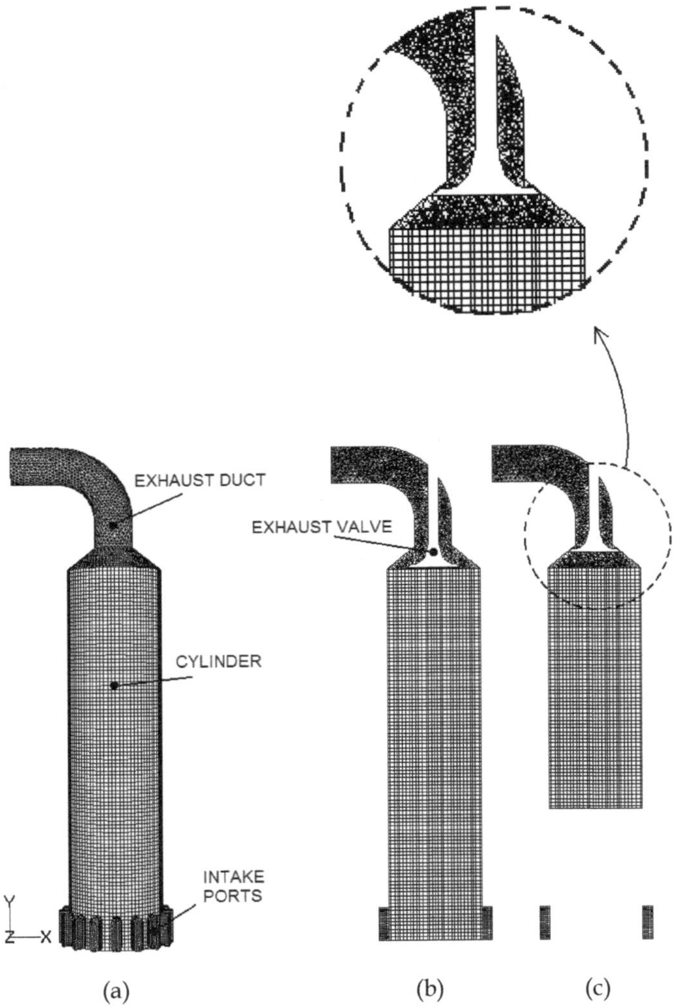

Fig. 9. (a) Tri-dimensional mesh, 180° crankshaft angle. (b) Cross-section mesh, 180° crankshaft angle; (c) Cross-section mesh, 270° crankshaft angle. Lamas & Rodríguez (2012).

It is very important to include the ports and ducts in the computational grid because they notably influence the movement of gases inside the cylinder and therefore the characteristics of the scavenging. For example, in the engine of Fig. 1 (b) and Fig. 2 (b), the intake ports and ducts are inclinated respect to the cylinder axis. Consequently, a swirling motion is promoted by the tangential velocities around the cilinder axis. This phenomena is shown in Fig. 10, which represents the velocity field in a tri-dimensional view, Fig. 10 (a), and in a transversal section at the base of the cylinder, Fig. 10 (b).

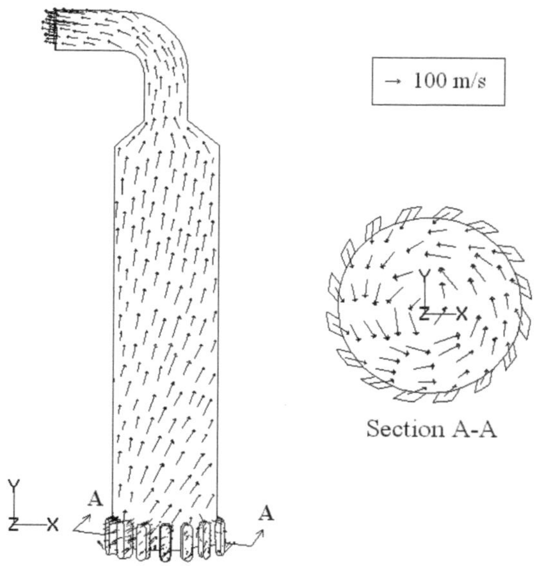

(a) (b)

Fig. 10. Velocity field [m/s] for 150° crankshaft angle. (a) Tri-dimensional view. (b) Transversal section A-A, at the base of the cylinder. Lamas & Rodríguez (2012).

Obviously, not all the engines are so sensible to the inlet ports and ducts geometry, but it is recommended to include them in the mesh instead a surface in which a boundary condition is imposed.

4.2 Boundary and initial conditions

All CFD models require initial and boundary conditions. Concerning the pressures, the experimentally values mentioned in the beginning of section 4 were employed as initial and boundary conditions.

As the in-cylinder temperature can not be measured experimentally, the initial temperature must be estimated from an adaptation of the ideal Otto cycle, Fig. 11 (a) and 11 (b). Details of the procedure can be found in most undergraduate textbooks on internal combustion engines or thermodynamics, so they are not repeated here. As can be seen in Fig. 6 (b), the temperature at 90° crankshaft angle is 1027 K.

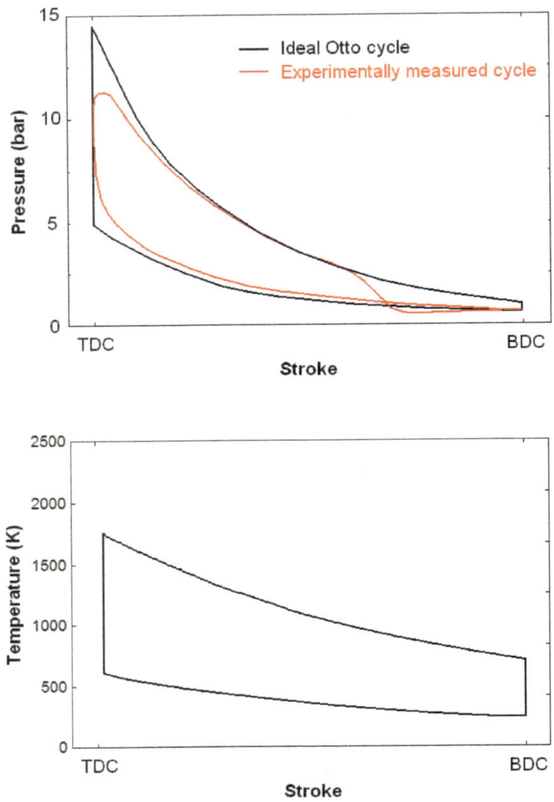

Fig. 11. (a) In-cylinder pressure experimentally measured and obtained from the ideal Otto cycle. (b) In-cylinder temperature obtained from the ideal Otto cycle.

4.3 Resolution of the equations

In this case, the software ANSYS Fluent 6.3 was employed. This is based on the finite volume method. Concerning the time discretization, an implicit method was chosen, with a constant timestep equivalent to 0.1° crankshaft angle. An explicit method could also have been chosen, but implicit methods are unconditionally stables and allow greater time steps. Concerning the pressure-velocity coupling, the PISO algorithm was employed because it is more recommended for transient calculations than the SIMPLE algorithm (Versteeg, 1995). A second order scheme was chosen for discretization of the continuity, momentum, energy and mass fraction equations.

Both the grid and time step sensibility were studied and it was verified that the size of the computational mesh and time increment are adequate to obtain results that are insensitive to further refinement of numerical parameters. In order to ensure this grid independence, several calculations with different mesh sizes and time step sizes were compared.

5. Results

5.1 Pressure field and validation of the code

In order to ensure that the CFD model is accurate enough, numerical results were compared to experimental ones. Particularly, the in-cylinder gauge pressure was validated. For the interval of time studied, from 90° to 270° crankshaft angles, the numerical and experimental results are shown in Fig. 12. Note that an acceptable concordance is obtained between CFD and experimental results.

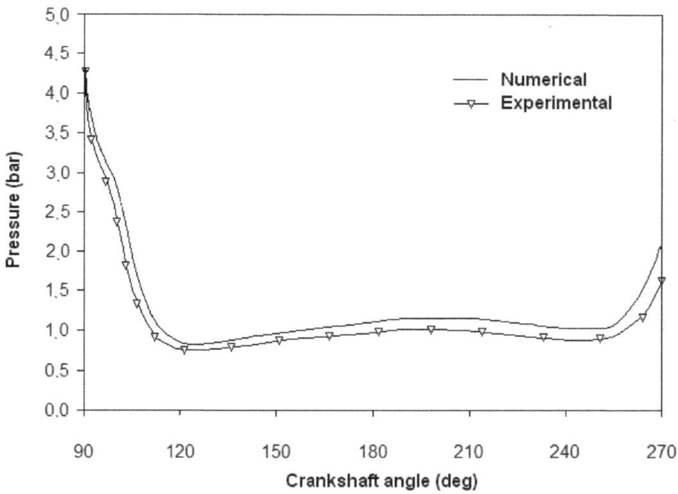

Fig. 12. In-cylinder pressure numerically and experimentally obtained.

Figure 13 shows the gauge pressure field at several crank angles. As can be seen, the initial in-cylinder pressure, Fig. 13 (a), is 4.26 bar. As mentioned before, the intake and exhaust pressures are variable, imposed as boundary conditions at the intake and exhaust ports. At the beginning of the simulation, the pressure descends drastically due to the expansion of the piston (note that the arrows indicate the direction of the piston). When the ports are opened, Fig. 13 (b) and (c), the in-cylinder pressure is slightly superior to the exhaust pressure and slightly inferior to the intake pressure, therefore burnt gasses are expelled through the exhaust port and fresh air enters through the scavenge ports. Finally, when all ports are closed, Fig. 13 (d), the piston is ascending and the gasses are compressed, Fig. 13 (d).

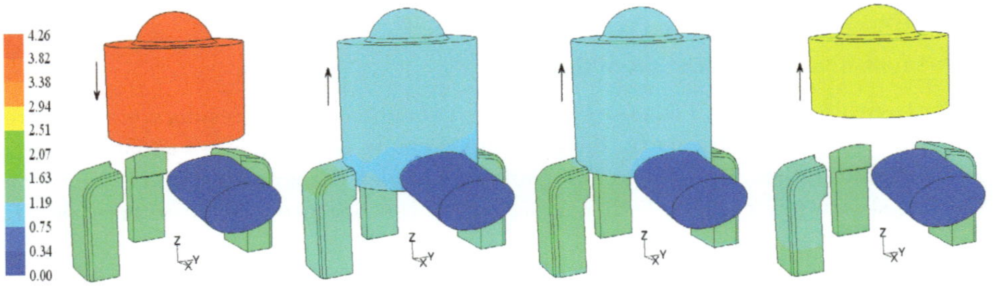

(a) (b) (c) (d)

Fig. 13. Pressure field [bar]. (a) 92° crank angle; (b) 190° crank angle; (c) 215° crank angle; (d) 270° crank angle. Lamas-Galdo *et al.* (2011).

5.2 Mass fraction field

The mass fraction field is shown in Fig. 14. Four positions were represented, 92.5°, 190°, 215° and 270° crank angles. Initially, the cylinder is full of burned gases (blue color), Fig. 14 (a). When the scavenging process begins, the fresh air charge (red color) throws away the burned gases out the cylinder, Fig. 14 (b) and (c). At the end of the process, Fig. 14 (d), the cylinder is full of fresh air charge.

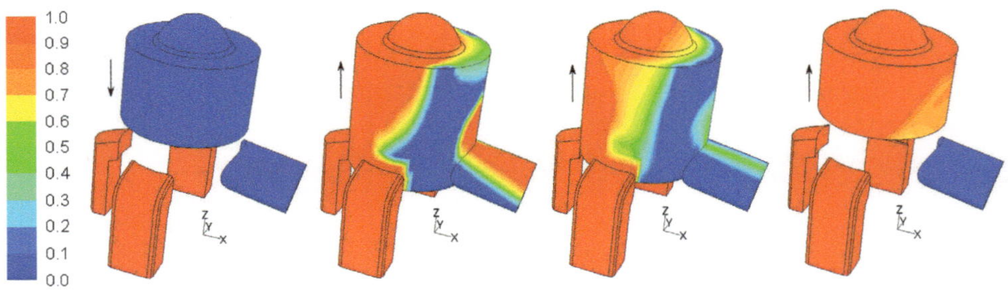

(a) (b) (c) (d)

Fig. 14. Mass fraction field [-]. (a) 92° crank angle; (b) 190° crank angle; (c) 215° crank angle; (d) 270° crank angle. Lamas-Galdo *et al.* (2011).

A very important advantage of CFD codes over experimental setups is that it is very easy to compute the portion of burnt gases which could not be expelled. In this work, it was quantified by means of the scavenging efficiency. This indicates the mass of delivered air that was trapped by comparison with the total mass of air and fresh charge that was retained at exhaust closure, Ec. (13), and its value was 82.5 for the parameters studied.

$$\eta = \frac{mass\ \ of\ \ delivered\ \ air\ \ retained}{mass\ \ of\ \ mixture\ \ in\ \ the\ \ cylinder} \tag{13}$$

The mass fraction field of air of the engine described in Fig. 1 (b) and Fig. 2 (b) is shown in Fig. 15. As can be seen, fresh air (red color) enters through the inlet ports situated at the bottom part of the cylinder and burnt gases (blue color) are expelled through the exhaust valve situated at the top part of the cylinder.

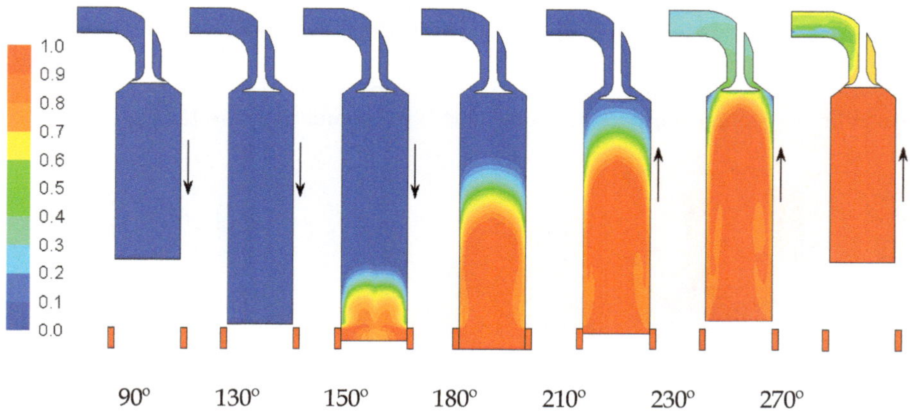

Fig. 15. Mass fraction field of air for several crankshaft positions. Lamas & Rodríguez (2012).

5.3 Velocity field

Fig. 16 shows the velocity field at 92.5° and 190°crankshaft angles. It is represented in a cross plane containing the auxiliary transfer port and the exhaust port. As the intake and exhaust ports are opened, fresh charge flows to the cylinder through the scavenge ports and exhaust gasses are expelled thought the exhaust port.

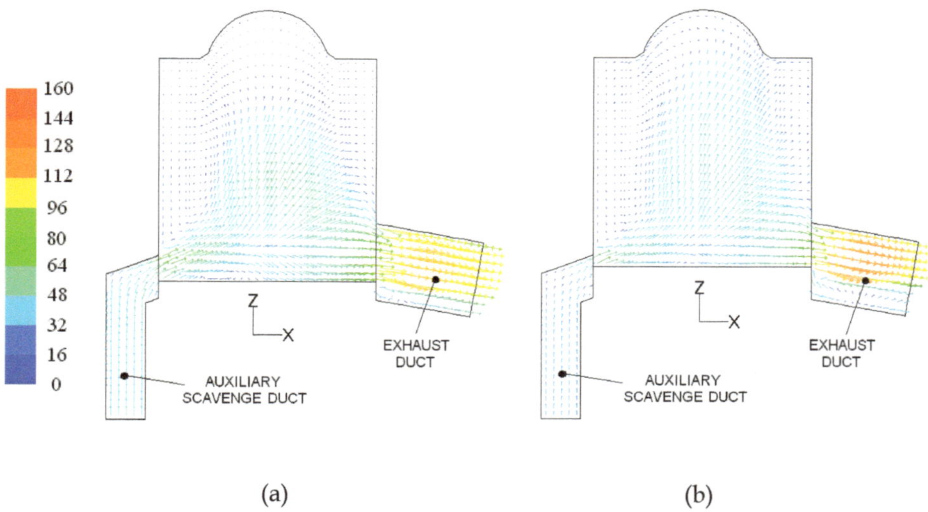

Fig. 16. Velocity field (m/s). (a) 92° crankshaft angle; (b) 190° crankshaft angle.

5.4 Temperature field

The temperature field at various crank angles is given in Fig 17. As mentioned above, the initial temperature, obtained from the ideal thermodynamic Otto cycle, was imposed as 1027 K, Fig. 17 (a). At the beginning of the simulation, the in-cylinder temperature descends due to the expansion of the piston. When the ports are opened, Fig. 17 (b) and (c), the temperature descends again because fresh air at 300 K enters through the scavenge ports and hot exhaust gases are expelled. At the end of the simulation all the ports are closed and the piston is rising. The compression of the piston makes the temperature increase. Finally, the in-cylinder average temperature at the end of the simulation, Fig. 17 (d), is 677 K.

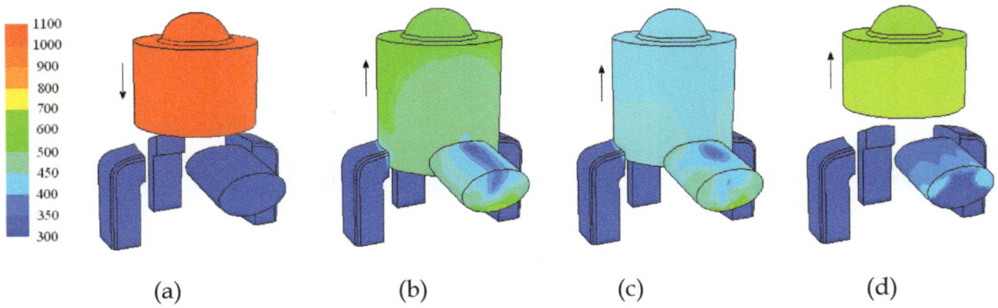

Fig. 17. Temperature field. (a) 92.5° crank angle; (b) 190° crank angle; (c) 215° crank angle; (d) 270° crank angle.

The in-cylinder average temperature and heat transfer from 90° to 270° crankshaft angles is shown in Fig. 18.

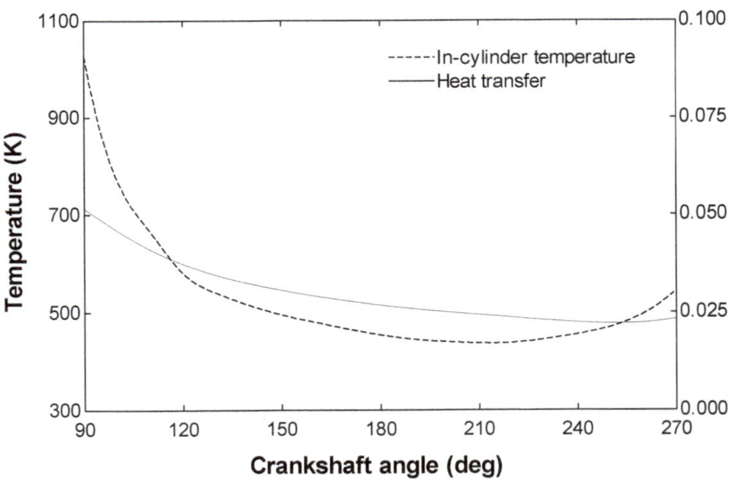

Fig. 18. In-cylinder average temperatura and heat transfer.

6. Conclusions

In the present chapter, a CFD analysis was carried out to study the scavenging process of two-stroke engines. The results were satisfactory compared to experimental data. In general, this study shows that CFD predictions yield reasonably accurate results that allow improving the knowledge of the fluid flow characteristics.

This model is very useful to design the scavenging system of new two-stroke engines. The pressure field is useful for identifying areas where the gas flow is inefficient and should be corrected. The velocity field is useful for locating areas with too high, too low or inadequate orientation velocities. Finally, the mass fraction field is useful for checking the filling of fresh gases into the cylinder and detecting problems of short circuiting and gas drag.

Finally, it is very important to mention the disadvantages of CFD. First of all, a 3D CFD model is very tedious due to the large computational resources. Besides, the moving mesh required to simulate the movement of the piston is too computationally expensive to solve. Other disadvantage is that it must not be applied blindly as it has the capability to produce non-physical results due to erroneous modeling. The process of verification and validation of a CFD model is necessary to ensure the numerical model accurately captures the physical phenomena present. By comparing numerically obtained results with experimental results, confidence in the numerical model is achieved. Once thoroughly validated, a numerical model may be used to accurately predict the effect of design changes and experimentally unobservable phenomena.

7. References

Ahmadi-Befrui, B.; Brandstatter, W.; Kratochwill, H. (1989). Multidimensional calculation of the flow processes in a loop-scavenged two-stroke cycle engine. *SAE Paper 890841.*

Albanesi A., Destefanis C, Zanotti A. (2009) Intake port shape optimization in a four-valve high performance engine. *Mecánica Computacional.* Vol. 28, pp. 1355-1370.

Amsden, A. A.; O'Rourke, P. J.; Butler, T. D.; Meintjes, K. and Fansler, T. D. Comparisons of computed and measured three-dimensional velocity fields in a motored two-stroke engine. SAE Paper 920418, 1992.

Blair G.P. (1996). *Design and Simulation of Two-Stroke Engines.* SAE International. ISBN 978-1-56091-685-7, USA.

Carpenter, M. H.; Ramos, J. I. (1986). Modelling a gasoline-injected two-stroke cycle engine. *SAE Paper 860167.*

Creaven J.P., Kenny K.G., Cunningham G. (2001). A computational and experimental study of the scavenging flow in the transfer duct of a motored two-stroke cycle engine. *Proc Instn Mech Engrs.* Vol.215-D.

Epstein, P. H.; Reitz, R. D. and Foster, D. E. (1991). Computations of two-stroke cylinder and port scavenging. *SAE Paper 919672.*

Fluent 6.3 Documentation, 2006. Fluent Inc.

Hariharan Ramamoorthy, Mahalakshmi N. V., Krishnamoorthy Jeyachandran. (2009). Setting up a comprehensive CFD model of a small two stroke engine for simulation. *International Journal of Applied Engineering Research.* Vol. 4-11.

Hori, H.; Ogawa, T. and Toshihiko, K. (1985). CFD in-cylinder flow simulation of an engine and flow visualization. *SAE Paper 950288.*

Kato S., Nakagawa H., Kawahara Y., Adachi T., Nakashima M. (1991) Numerical analysis of the scavenging flow in a two stroke- cycle gasoline engine. *JSME International Journal*. Vol. 34-3, pp. 385-390.

Laimböck, F. J.; Meist, G. and Grilc, S. (1998). CFD application in compact engine development. *SAE Paper 982016*.

Lamas-Galdo, M.; Rodríguez-Vidal, C.; Rodríguez-García, J.; Fernández-Quintás, M. (2011). Modelo de Mecánica de Fluidos Computacional para el proceso de barrido en un motor Otto de dos tiempos. DYNA Ingeniería e Industria, vol. 86-2, pp. 165-172.

Lamas, M. I.; Rodríguez, C. G. (2012) CFD analysis of the scavenging process in the MAN B&W 7S50MC two-stroke diesel marine engine. Submitted to Journal of Ship Research.

Payri F., Benajes J., Margot X. et al. (2004). CFD modeling of the in-cylinder flow in direct-injection diesel engines. *Computers & Fluids*. Vol.33 p.995-1021.

Pitta S. R., Kuderu R. (2008). A computational fluid dynamics analysis on stratified scavenging system of medium capacity two-stroke internal combustion engines. *Thermal Science*. Vol. 12-1, pp. 33-42.

Raghunathan, B. D. and Kenny, R. G. (1997). CFD simulation and validation of the flow within a motored two-stroke engine. *SAE Paper 970359*.

Rahman M.M., Hamada K.I., Noor M.M. et al. (2010) Heat transfer characteristics of intake port for spark ignition engine: A comparative study. *Journal of applied sciences*. Vol.10-18, pp. 2019-2026.

Sher, E. (1989). An improved gas dynamic model simulating the scavenging process in a two-stroke cycle engine. *SAE Paper 800037*.

Sweeny, M. E. G.; Kenny, R. G.; Swann, G. B. G. and Blair, G. P. (1985). Computational fluid dynamics applied to two-stroke engine scavenging. *SAE Paper 851519*.

Yu, L.; Campbell, T. and Pollock, W. (1997). A simulation model for direct-fuel-injection of two-stroke gasoline engines. *SAE Paper 970367*.

Zahn, W.; Rosskamp, H.; Raffenberg, M. and Klimmek, A. (2000). Analysis of a stratified charging concept for high-performance two-stroke engine. *SAE Paper 2000-01-0900*.

Zancanaro F.V., Vielmo H.A. (2010) Numerical analysis of the fluid flow in a high swirled diesel engine. *Proceedings of the 7th International Conference on Heat Transfer, Fluid Mechanics and Thermodynamics*. Antalya-Turkey, 19-21 July 2010, pp. 387-392.

2

3D Numerical Modelling of Mould Filling of a Coat Hanger Distributer and Rectangular Cavity

Rekha R. Rao[1], Lisa A. Mondy[1], David R. Noble[1], Matthew M. Hopkins[1], Carlton F. Brooks[1] and Thomas A. Baer[2]
[1]*Sandia National Laboratories*
[2]*Procter and Gamble Company*
USA

1. Introduction

Filling processes occur in a wide range of industries, ranging from packaging of consumer products to manufacturing processes for making polymeric, metal and ceramic components. These processes involve the complex interplay of extrusion of a viscous liquid into a mould or container where it displaces a gas phase. Numerical modelling based on computational fluid dynamics can be useful for understanding the filling process. However, complexity arises in that the fluid dynamics in both the viscous liquid and gas phase must be resolved while concurrently determining the location of the fluid-gas interface and the interaction of this interface with the solid surface, *i.e.*, the wetting behaviour. Determining the free surface location and wetting behaviour is an integral part of the numerical method.

Numerical methods have been applied to bottle and container filling for consumer products where the rheology can include shear thinning and viscoelastic effects and instabilities such as buckling and coiling may be prevalent [Tome, et al., 2001; Oishi et al., 2008; Roberts & Rao, 2011; Ville et al., 2011]. In metal casting simulations, the fluids are generally Newtonian, but complexity arises from the high injection rates leading to turbulent flow and temperature-dependent behaviour such as solidification [Ilinca & Hetu, 2000; Cross et al, 2006]. For injection moulding of polymers, time- and temperature-dependent effects are seen in conjunction with non-Newtonian rheology [Ilinca & Hetu, 2001; Kumar & Ghoshdastidar, 2002]. In powder injection moulding for ceramic and metal forming, a suspension of particles is injected into a mould to create a green part, which later sees further processing steps to produce the final part [Hwang & Kwon, 2002; Ilinca & Hetu, 2008]. Numerical methods for these problems range from finite difference, to finite volume, and finite element.

General classes of algorithms for determining the location of the free surface include Eulerian, Lagrangian and arbitrary Lagrangian-Eulerian (ALE) descriptions. Eulerian methods use a fixed-grid with an interface capturing technique such as the volume of fluid [Hirt & Nichols, 1981] or level-set method [Sethian, 1999] to determine the location of the free surface. Traditional Lagrangian methods use a moving mesh as a material interface that advects with the fluid. These methods often require multiple remeshing steps to avoid mesh distortion and tangling [see for instance, Bach & Hassager, 1985; R. Radovitzky & M. Ortiz,

1998; Zhang & Khayat, 2001]. To avoid these problems, Lagrangian mesh-free methods have been developed using smooth particle hydrodynamics or the material point method to avoid meshing issues [Kulasegaram et al, 2006; Kauzlaric et a;., 2011; Love & Sulsky, 2006]. These methods work well for being able to capture a moving interface with topological changes, but have difficulty in accurately solving the base physics, *e.g.* viscosity, and applying boundary conditions such as surface tension. ALE methods are hybrid techniques that seek to exploit the benefits of both the Eulerian and Lagrangian description in a hybrid manner, to determine the location of the interface [i.e. Sackinger et al, 1996; Lewis et al., 1998; Nithiarasu, 2005].

Injection loading of a ceramic paste is a high-rate process used to create green ceramic parts, which subsequently experience binder burnout and sintering to produce the final ceramic part. In the process of interest, the mould is a rectangular cavity, with an inflow from a coat hanger die that should distribute the flow evenly across the mould inflow. The cavity is small with dimensions of 1.3 cm by 3.6 cm in plane and a height of 0.4 cm. The inflow tube to the distributer has a diameter of 0.5cm. Figure 1 shows short shots (or incomplete filling of a mould) for the injection loading process, illustrating some of the defects that can occur when a fluid with complex shear and temperature-dependent rheology interacts with a high rate process.

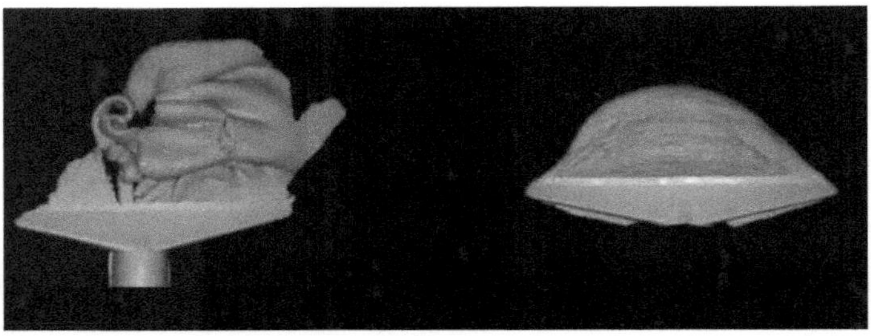

Fig. 1. Short shots of injection loading of a ceramic paste in a part are shown [Rao et al, 2006]. The photo on the left is injection loading using a slow injection speed, while the photo on the right uses a higher injection speed. At low filling speed, the paste acts like a solid material. Even at high filling rates, when the paste begins to act as a fluid, pooling at the centre of the mould is seen and the desired mould shape is not achieved.

Temperature control and high-rate processing can limit the folding instability seen on the left of Figure 1. However, the pooling phenomenon shown on the right of Figure 1 occurs when optimal processing conditions are used, indicating that the design of the distributer may be responsible for material building up in the centre of the die. The complex shear-rate and temperature-dependent rheology of the ceramic paste was determined to follow a power-law dependence on shear rate and a Williams-Landau-Ferry temperature dependence [Rao et al., 2006]. The material shear thinned quickly to a constant viscosity value at moderate shear rates. Thus, because of the high shear rates in the injection loader, it was determined that the rheology was essentially Newtonian as long as the temperature remained constant at the processing temperature. Therefore, a study was undertaken to better understand the filling dynamics and reduce pooling by changing the distributer design using a Newtonian fluid.

1.1 Dynamic wetting models and mould filling

Understanding dynamic wetting, or the interaction between the free surface and the mould walls, has been the subject of numerous experimental and theoretical studies and is still an outstanding research topic [see for instance Blake, 2006; Ren et al, 2011]. The difficulty arises from the contradiction of a moving contact line at the fluid-gas interface and the no slip boundary conditions traditionally applied at solid surfaces. How can the contact line advance when the velocity vanishes at the solid surface? Highly viscous materials such as polymers and particle suspensions have large capillary numbers (a measure of the ratio of viscous forces to capillary forces) and are often hypothesized to obey a rolling motion condition with an 180° dynamic contact angle. Numerically, this approach is difficult to apply and other methods have been proposed. The simplest models use a Navier slip condition to allow either slip on the entire solid surface, slip for the gas phase only, or slip only at the dynamic contact line. These models are *ad hoc* and ignore any thermodynamics considerations such as the static contact angle and surface energy. More advanced wetting models generally give dynamic contact angle as a function of the local Ca, the static contact angle and other material properties of the fluid and the solid surface [see Schunk et al., 2006 for a brief review of this work]. Hoffman [1975] used experimental measurements to develop a universal correlation for the dynamic contact angle as a function of static contact angle and a local capillary number, while Cox [1986] developed a competing model using asymptotic analysis. Kistler [1983] used a linear model that was easy to implement in numerical computer codes. Shikhmurzaev [1994] used hydrodynamic theory and included a surface phase as part of the wetting model. Blake [1969] developed a molecular kinetic theory that reduces to a linear model for small contact angles. Blake noted that the advancing angle is a monotonically increasing function of Ca. The degree of velocity dependence will however increase steeply as viscosity increases or surface tension decreases.

To model dynamic contact for the filling process, the dynamic angle is tied to the balance of forces at the advancing wetting line, namely the tangential wetting line force, liquid-gas surface tension force, and fluid viscous force. Here, a version of the Blake model is used that is straightforward to populate with experiments. The dynamic contact angle is measured in the laboratory as a function of the velocity of the wetting line for the fluids and surface used in our experiments as input to the wetting model. The Blake model is also easy to implement numerically, since the wetting speed can be written as a function of the dynamic and static contact angles. The performance of the Blake model at the high capillary number limit may be suspect, since it exhibits an unbounded dynamic contact angle while more physical models such as Cox and Hoffman reach a limit of 180° for large capillary numbers [Schunk et al, 2006].

1.2 Mould design

Because the initial design of the mould exhibited pooling of the fluid in the centre of the cavity and this would lead to poor filling (see Figure 1), we tried two minor redesigns to the distributor to see if we could improve the flow into the mould and reduce pooling. Ideally, we would like to see a flat profile coming out of the distributor and a more one-dimensional front shape. Figure 2 shows the original mesh, Mesh 1, and a variation, Mesh 2, with a longer distributor, and Mesh 3 with a longer-taller distributor. The idea behind Mesh 2 was to give a longer length for the flow to develop a flat profile and fill up the distributor before entering the main cavity. Mesh 3 kept this longer distributor and made it wider on inflow to ease the fluid entering the cavity. These ideas were inspired by discussions in Sartor [1990] about die design. The meshes themselves all have the same cavity size and a similar amount of refinement, though Mesh 2 and Mesh 3 have more elements and unknowns due to the longer distributor.

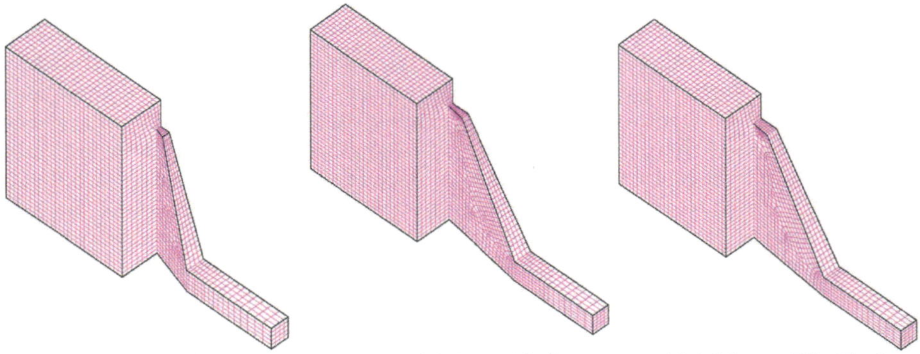

Fig. 2. Geometries to be investigated: Mesh 1 is the original design, Mesh 2 incorporates a longer distributer, and Mesh 3 incorporates a longer distributer with a wider inflow to the cavity.

1.3 Chapter organization

In this chapter, we investigate the design of an injection loading mould using flow visualization experiments and numerical models. The finite element method is used to understand the interaction of the inflowing viscous liquid with the geometry and the displaced gas phase. Filling dynamics are determined with a diffuse-interface implementation of the level set method [Sethian 1999]. The Blake wetting model is used to represent the interaction of the free surface with the mould surface at the dynamic contact line [Blake & Haynes, 1969; Blake, 2006]. The flow visualization experiments are carried out under isothermal conditions using an acrylic mould and a viscous Newtonian fluid. Three different moulds are examined in two different orientations with gravity. Simulation results are given for these six cases.

The chapter is organized in the following manner. First, the equations and numerical method are presented. Next, the experimental methods used to provide input parameters for the models and flow visualization studies to better understand the filling dynamics and provide confidence in the numerical method are discussed. In the subsequent section, the results for injection moulding process are given for a Newtonian fluid into a coat hanger die distributer and a rectangular cavity, where the 3D level set simulations are compared to experiments. We conclude by summarizing the results and discussing future efforts.

2. Equations and method

2.1 Equations of motion

We can write the equations of motion for a single-phase fluid and then generalize them for our multiphase flow problem, where a viscous fluid displaces a gas. The fluids of interest are assumed to be incompressible and have a constant density, meaning that the velocity field, *u*, will be solenoidal and the continuity equation contains no density or pressure variables. (Note, this is a good assumption for the viscous liquid but a simplification for the gas phase, which is actually compressible.)

$$\nabla \cdot u = 0 \tag{1}$$

Conservation of momentum for a Newtonian fluid takes into account gradients in the fluid stress tensor, T, defined as the product of the viscosity μ and the shear rate tensor, $\nabla u + (\nabla u)^t$, gradients in the pressure, p, as well as gravitational effects and inertial terms that can be dependent on time, t, and the fluid density, ρ. Note that gravity, g, can be an important body force in filling processes and most filling processes fill counter to gravity.

$$\rho(\frac{\partial u}{\partial t} + u \cdot \nabla u) = \mu \nabla^2 u - \nabla p + \rho g \tag{2}$$

Because we have a viscous fluid displacing a gas phase, the location of the interface between fluids is unknown *a priori*. To determine the location of the free surface as it evolves in time, we use an Eulerian interface-capturing scheme based on the level set method, the details of which are included in the following section.

2.2 Interface capturing

We use the level set method of Sethian [1999] to determine the evolution of the interface with time. The level set is a scalar distance function, the zero of which coincides with the free surface or fluid-gas interface, *e.g.*

$$\phi(x, y, z) = 0 \tag{3}$$

We initialize this function to have a zero value at the fluid-gas interface, with negative distances residing in the fluid phase and positive distances in the gas phase. An advection equation is then used to determine the location of the interface over time.

$$\frac{\partial \phi}{\partial t} + v \cdot \nabla \phi = 0 \tag{4}$$

Derivatives of the level set function can give us surface normal, *n*, and curvature, *κ*, at the interface, which is useful for applying boundary conditions.

$$n = \frac{\nabla \phi}{|\nabla \phi|}$$
$$\kappa = \frac{-\nabla^2 \phi}{|\nabla \phi|} \tag{5}$$

We use the equations of motion described in the previous section, but vary the material properties across the phase interface. This variation is handled using a smooth Heaviside function that modulates material properties to account for the change in phase.

$$H_{gas}(\phi) = \frac{1}{2}(1 + \frac{\phi}{\alpha} + \frac{1}{\pi}\sin(\frac{\pi\phi}{\alpha})); -\alpha \le \phi \le \alpha$$
$$H_{fluid}(\phi) = 1 - H_{gas}(\phi) \tag{6}$$

This is a diffuse interface implementation of the level set method, which allows for an interfacial zone of length 2α. This zone is usually chosen to be four to six elements wide. Equation averaging is done using a Heaviside for the gas and viscous fluid equations. Because the properties are linear, this process results in Heaviside-averaged properties in a single momentum equation and an unchanged continuity equation:

$$\rho_{average}\left(\frac{\partial u}{\partial t}+u\bullet\nabla u\right)=\mu_{average}\nabla^2 u-\nabla p+\rho_{average}g \qquad (7)$$
$$\nabla\bullet u=0$$

The properties have fluid properties in some regions and gas properties in other regions as modulated by the numerical Heaviside. In the diffuse interface region, the properties are averaged between fluid and gas values. Figure 3 shows a schematic representation of the Heaviside function.

$$\rho_{average}=H_{gas}\rho_{gas}+H_{fluid}\rho_{fluid} \qquad (8)$$
$$\mu_{average}=H_{gas}\mu_{gas}+H_{fluid}\mu_{fluid}$$

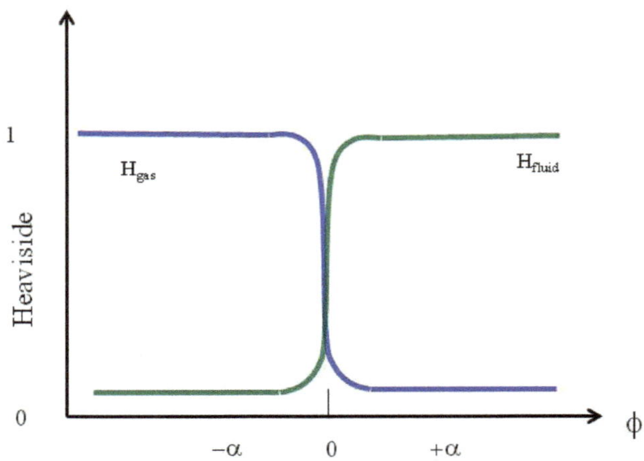

Fig. 3. Numerical Heaviside for averaging material properties.

The regularized Dirac delta function, which is defined as

$$\delta_\alpha(\phi)=\frac{dH(\phi)}{d\phi}=\frac{|\phi|}{2\alpha}\left(1+\cos\left(\frac{\pi\phi}{\alpha}\right)\right)' \qquad (9)$$

is used to apply surface tension and capillary boundary conditions via a continuous surface force approach [Brackbill et al, 1992]. This applies surface tension as a volumetric body force on the momentum equation, which is distributed throughout the interfacial zone region through the regularized Dirac delta function.

$$(\mu_{gas}-\mu_{fluid})n\bullet(\nabla u+(\nabla u)^t)=2\sigma\delta_\alpha(\phi)\kappa n \qquad (10)$$

2.3 Finite element implementation

The equations of motion (7) and the level set advection (4) were solved using a finite element method as implemented in ARIA [Notz et al., 2006]. Bilinear shape functions were used for the three velocity components, pressure, and level set. The LBB requirement on the velocity and pressure space was circumvented using Dohrmann-Bochev pressure stabilized pressure-projection (PSPP) [Dohrmann and Bochev, 2004] to allow for this equal order, bilinear, interpolation of all variables. (LBB compliant elements have the velocity space higher than the pressure space [Hughes, 2000].) The velocity vector and pressure unknowns were solved in the same matrix, while the level set equation was solved in a separate matrix, but at the same time step intervals. The level set equation was stabilized using a Taylor-Galerkin method [Donea, 1985]. The PSPP stabilization method greatly improved the condition number of the discretized matrix equations when compared to LBB elements or other stabilization methods, allowing for the use of an ILU preconditioner with a BiCGStab Krylov iterative solver. Further details of the modelling approach and equations, the numerical methods used and the finite element implementation can be found in Rao *et al.* [2011].

2.4 Contact-line wetting model

Boundary conditions for the dynamic contact line where the free surface and wall intersects are handled with a Blake wetting condition [Blake and Haynes, 1969; Blake, 2006]. Parameters for the model are informed by goniometer experiments that determine the wetting speed, v_{wet} as a function of the dynamic contact angle, θ, and the static contact angle, θ_s for the various fluids and surfaces of interest [Mondy *et al.*, 2007].

$$v_{wet} = v_o \sinh(\bar{\gamma}(\cos\theta_s - \cos\theta)) - \tau \frac{\partial v_{wet}}{\partial t} \tag{11}$$

The dynamic contact angle can be calculated from the level set function.

$$\cos\theta = n_w \cdot \nabla\phi / |\nabla\phi| \tag{12}$$

When we integrate the stress, T, by parts in the finite element implementation of the momentum equation, a surface term is created as a natural boundary condition. We exploit the surface stress term and add on a Navier slip condition that includes the wetting speed from the Blake model.

$$\begin{aligned} \vec{n}\cdot T &= -\frac{1}{\beta}\left(-\tau\frac{\partial\vec{v}}{\partial t} + \vec{v} - f_\alpha(\phi) v_{wet}\vec{t}\right) \\ f_\alpha(\phi) &= 1 - \frac{|\phi|}{\alpha}, -\alpha \le \phi < \alpha; 0, |\phi| > \alpha \end{aligned} \tag{13}$$

The value of the tangential wall velocity ramps from zero at a level set length scale away from the contact line to v_{wet} from equation (11) at the contact line. Away from the contact line, we revert to no slip for the tangential wall velocity boundary condition. The normal velocity is enforced as no penetration everywhere. The shape of this ramp must be smooth in order to get a realistic wetting line that shows a smooth transition from the contact line to the bulk flow. For a sharp transition from slip to Blake, we ended up with unphysical looking cusps in the interface shape near the wall. The transient terms introduce dynamics into the wetting line motion and allow smooth movement of the contact line. The τ parameter is taken to be approximately the time step. The β parameter is generally small so that equation (13) functions almost like a Dirichlet requirement on the fluid velocity.

3. Experiments

It was decided to visually record the flow of a simple, single phase, Newtonian liquid through transparent moulds to build confidence in the front tracking and wetting models used in the computations. Details of the geometries, experimental conditions, and properties of the materials used in each test will be given in the following sections.

3.1 Geometries

For these tests, we built transparent acrylic moulds identical to the three mesh geometries (Figure 2). The strength of the acrylic material making up the transparent moulds dictated that we inject at a lower injection pressure and lower operating temperature than the actual injection loading process. To mimic the actual injection process, we used a pressure driven syringe held at a constant pressure of 29.95 ± 0.10 psig during injection. The syringe of liquid was degassed in a vacuum chamber prior to the experiment. The syringe was modified to prevent leakage around the plunger and subsequent bubble formation. However, this resulted in more friction and a flow rate that varied somewhat from experiment to experiment. Hence, the time to fill the moulds varied from test to test, but was determined and recorded for each test. Reported below is the median and spread of the measured filling time for four repeated experiments in each geometry.

The tests were conducted at a room temperature that ranged from 23 to 24°C. To minimize the effect of temperature on the viscosity, the liquid was held in a water bath set to 23.5^{o}C after degassing and before being used in an experiment. These conditions resulted in an injection rate that was approximately ten times slower than the actual process. However, the Reynolds number for the actual process and the validation experiments were similar and in the Stokes regime of much smaller than one. For an average fill time of 20 s, a cavity volume of 1.87 cm^3, and characteristic length scale from the inflow port of 0.5cm, the Reynolds number was 0.0006 and the capillary number was 4.5.

3.2 Materials and properties

We chose a liquid, UCON 75-H-90,000 (oxyethylene/oxypropylenes from Dow Chemical) as our model fluid. This UCON was chosen since its viscosity is an order of magnitude lower than the ceramic paste at processing conditions. Combined with an order of magnitude decrease in the injection rate compared to the real system, this gives a similar Reynolds number. The liquid surface tension and wetting properties of the UCON lubricant were determined at 23.5°C. The viscosity of the lubricant was measured with a Rheologica™ constant stress rheometer and was equal to 390 Poise over a shear rate ranging from 0.1 to 10 sec^{-1}, indicating Newtonian rheological behaviour. The density of the liquid was measured with a densitometer to be 1.09 g/cm^3. The surface tension measured with a Du Noüy ring (mean circumference of 5.935 cm) was 42.4±0.1 dyne/cm.

The dynamic contact angle on acrylic was measured with a feed-through goniometer [Mondy *et al.*, 2007], in which liquid can be continuously injected to achieve "high" velocities, or the sessile drop can be allowed to relax to obtain "low" velocities. Figure 4 shows a schematic of the experimental apparatus and the results of this wetting test. During each experiment, the angle of contact and the location of the triple point of contact was recorded. The wetting line speed was then determined from the location of the triple point with time. The data fit to the Blake model (equation 9) gives a wetting constant v_o of 0.00130cm/s and scale factor γ of 2.29 with a static contact angle θ of 37.3°.

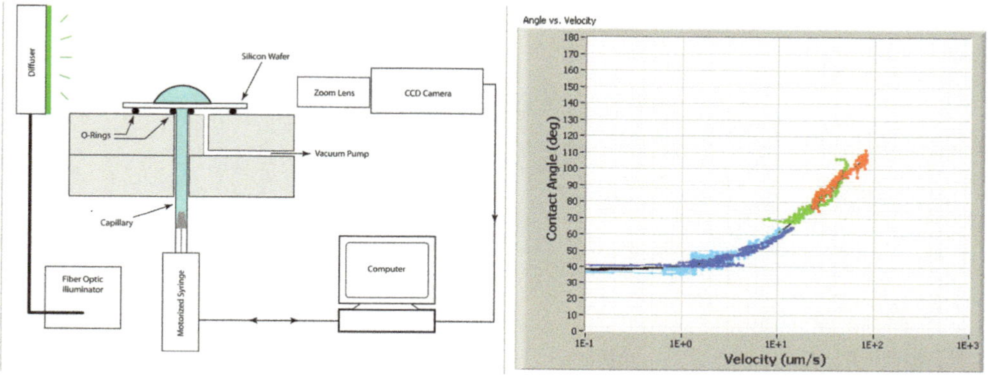

Fig. 4. Sketch of apparatus for wetting parameter measurements (left). Dynamic wetting measurements of contact angle vs. velocity for 75-H-90000 UCON on acrylic (right).

3.3 Flow tests

Figures 5 through 7 are representative frames from the video recordings of the fill process using a vertical alignment of the mould as in the proposed ceramic injection process. The time it took to completely fill the original geometry, Mesh 1, (defined by the time when the liquid began to exit along the entire length of the vent) was 24.6 ± 1.2 s, to fill the geometry of Mesh 2 was 26.9 ± 2.0 s, and to fill the geometry of Mesh 3 was 24.6 ± 2.3 s. Because the fill rates varied, the time is shown in these figures in a nondimensional form. The initial time is taken to be when the front passes a line in the square entry channel 0.16 cm from the entrance of the distributor. Here, one can see the effects of changing the distributor geometry. In Figure 5 the liquid has just filled the distributor. All of the modified geometries help flatten the leading front, especially Mesh 2. Mesh 2 fills the distributor with the fastest relative time.

Mesh 1	**Mesh 2**	**Mesh 3**
Time/total time=0.32	Time/total time=0.13	Time/total time=0.24

Fig. 5. Comparison of the effect of distributor geometry on the shape of the fluid front entering the mould for vertical mould orientation.

Figure 6 shows the times at which the leading front hits the wall opposite the injection port. In this case, the front in Mesh 2 takes the longest relative time to reach this stage. The other geometries once again follow the same pattern as before, with Mesh 2 giving the flattest flow profile and the original mesh the displaying the most curvature of the interface. At this point, the fluid has wetted more of the side walls with the Mesh 2 geometry and the front is flatter. Because Mesh 2 has a flatter profile, it takes the longest time to reach the top wall compared to Mesh 1 and Mesh 3.

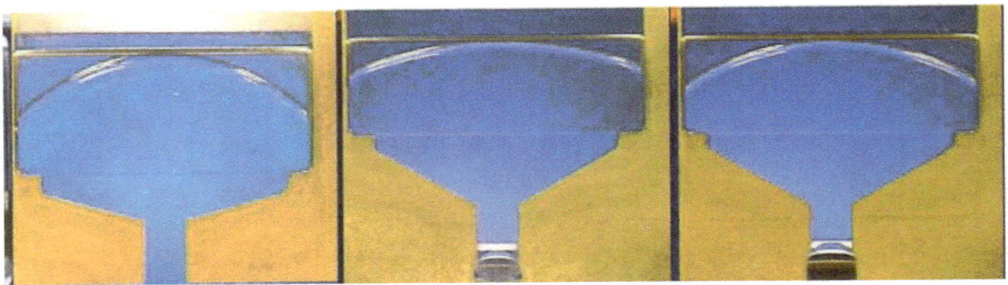

Mesh 1	Mesh 2	Mesh 3
Time/total time=0.80	Time/total time=0.86	Time/total time=0.82

Fig. 6. Comparison of the effect of distributor geometry on the time it takes to reach the wall farthest from the injection port for vertical mould orientation.

Figure 7 shows the locations of the voids remaining in the mould once filled but without any over pressure (a relative time of 1). Small bubbles away from the corners are artefacts of the syringe loading process. The voids in geometries with redesigned distributors remain in the same locations as those seen in the original geometry. The relative areas of the bubbles on the images, which reflect the volume of air trapped, were determined and compared quantitatively in Table 1. One pixel resolution represents about 1×10^{-5} cm^2. All redesigned distributors result in smaller bubbles in the upper corners than those in the original mesh. The lower bubbles of Mesh 2 are also smaller, whereas those in Mesh 3 are approximately the same as those in the original geometry.

Mesh 1 Mesh 2 Mesh 3

Fig. 7. Voids remain in the front upper and lower corners of each geometry for the vertical mould orientation. Each frame consists of two views of the mould (top view and side view). These voids can be seen more easily from the side view.

Geometry	Area Upper Corners (cm^2)	Area Lower Corners (cm^2)
Original	0.0153 ± 0.0012	0.0020 ± 0.0005
Mesh 2	0.0086 ± 0.0007	0.0012 ± 0.0002
Mesh 3	0.0128 ± 0.0005	0.0021 ± 0.0001

Table 1. Bubble sizes remaining in the corners for vertical mould orientation.

Next, we experimented with the orientation of the mould. Moulds were turned so that the distributor was perpendicular to gravity on the lower surface and the vent was moved to be on the upper surface. In other words, the flow direction and gravity are now perpendicular, with gravity acting in the thinnest cavity direction. Results for the original mesh showed that orientation with respect to gravity had a large impact on the likelihood of voids remaining in the corners of the mould. Figure 8 shows representative video frames at three stages of the fill: 1) when the distributor was completely filled, 2) when the front hit the far wall, and 3) when the liquid began to exit through the vent and the sides had completely wetted. Because the fill rates varied, the time is shown in a nondimensional form. The time it took to completely fill the original geometry in the horizontal position was 23.3 ± 0.9 s, to fill the geometry of Mesh 2 was 28.9 ± 1.5 s, and to fill the geometry of Mesh 3 was 27.3 ± 1.8 s.

The top views of the horizontal orientation show that the liquid wets the top later than the bottom. The oval area of the middle image in Figure 8 indicates where the liquid has wetted the upper surface. The side views are more difficult to interpret because of the lighting challenges accompanying trying to see through the thickest section of the mould. However, the dark areas of the side view are where the liquid has wetted the sides.

It is interesting to note that the horizontal alignment causes the front to hit the far wall before even wetting the sides at all. In other words, the front was less "flat" entering the mould from the distributor. Nevertheless, the results also show that the horizontal orientation, with the vent on top and the distributor entrance on bottom, resulted in bubbles only in the upper corners nearest the distributor. When oriented vertically, bubbles are trapped in corners both opposite the distributor and opposite the vent. The bubbles nearest the distributor with the horizontal orientation are approximately the same size as the bubbles trapped in the upper corners opposite the vent in the vertical orientation.

The effect of distributor geometry on the filling in the horizontal orientation is shown in Figure 9 and Figure 10, which correspond to Figure 5 and Figure 6 in the vertical orientation. Again, the geometry of Mesh 2 results in the flattest front entering the mould from the distributor (Figure 9). The front reaches the back wall at approximately the same time using the distributors of Mesh 2 and 3. By the time the front reaches the back wall both modified geometries result in more liquid filling the mould than with the original geometry; however, Mesh 2 results in somewhat more than Mesh 3 (Figure 10).

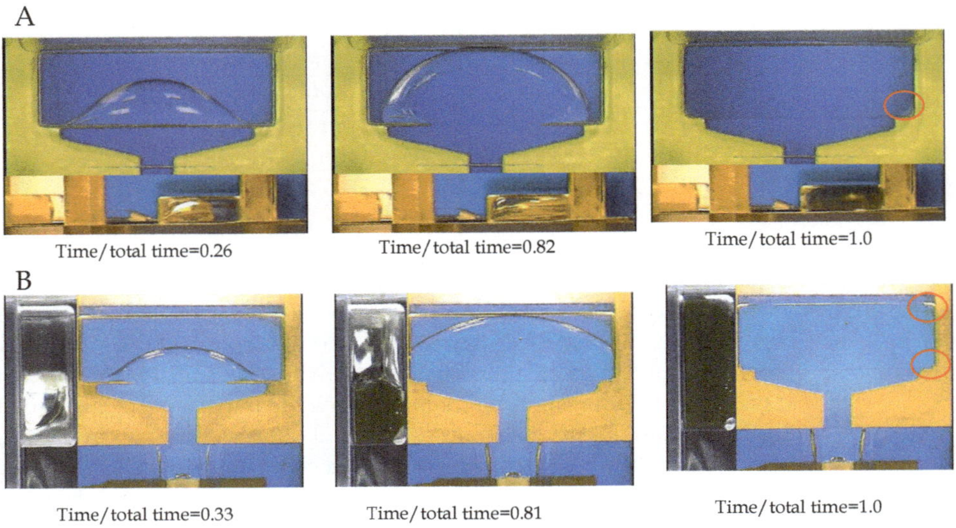

Fig. 8. Original geometry (Mesh 1) oriented horizontally (A) and vertically (B). Images on the left show when the distributor fills completely, middle images show when the front first hits the back wall, and images on the right show when the part is filled to the point that the fluid leaves the vent area through the entire length of the vent. Bubbles left in the corners are circled in red.

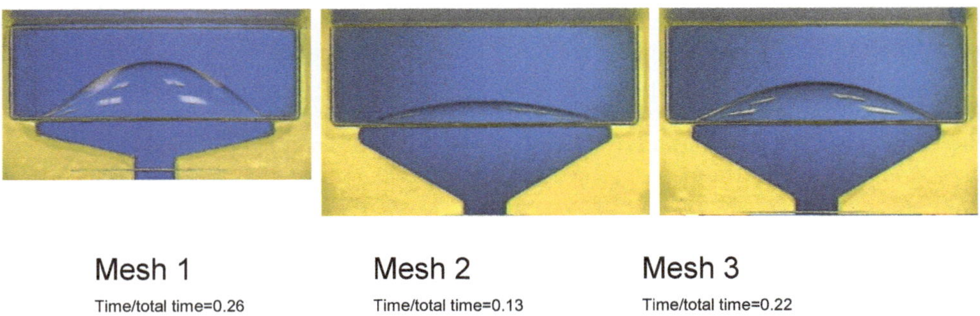

Fig. 9. Comparison of the effect of distributor geometry on the shape of the fluid front entering the mould in a horizontal orientation. The flow direction and gravity are perpendicular, with gravity acting in the thinnest cavity direction.

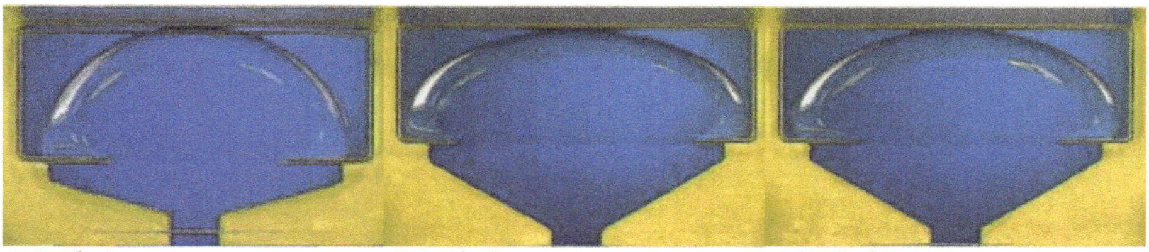

Mesh 1 Time/total time=0.82 **Mesh 2** Time/total time=0.71 **Mesh 3** Time/total time=0.70

Fig. 10. Comparison of the effect of distributor geometry on the time it takes to reach the wall farthest from the injection port in a horizontal orientation.

Sizes of the bubbles left in the various geometries in the horizontal orientation are listed in Table 2. Comparing these values with the measurements in Table 1, one can see that the amount of gas left in the vertical orientation is roughly the same as that in the horizontal orientation, although in the horizontal orientation there are fewer bubbles. Bubbles observed in other locations than the corners are almost always artefacts of the syringe loading process.

Geometry	Area (cm^2)
Original	0.0189 ± 0.0002
Mesh 2	0.0111 ± 0.0002
Mesh 3	0.0123 ± 0.0017

Table 2. Bubble sizes remaining in horizontal orientation.

4. Simulations

The simulations runs were made on 64 processors of Thunderbird (Sandia National Laboratories capacity computing platform) and ran in less than 3.5 hours. This allowed us to do real time design and sensitivity calculations for parameters such as wetting speed, second phase viscosity, level set length scale and inflow pressure. Properties for the liquid used for the validation simulations were the measured values discussed in the experimental section above. The density and viscosity of the displaced gas phase were taken as fictitious values of one thousand times smaller than the liquid phase density and viscosity. These values are summarized in Table 3.

Material Property	Value
Density of liquid	1.09 g/cm^3
Viscosity of liquid	390 Poise
Density of gas	0.0011 g/cm^3
Viscosity of gas	0.39 Poise
Wetting speed, v_o	0.0013 cm/s
Blake scale factor, γ	2.29
Static contact angle	37.3^o
Surface tension	42.4 dyne/cm
Inflow pressure	$1.0x10^6$ $dyne/cm^2$

Table 3. Material properties used for validation simulations.

At 25°C, the viscosity of air is 2.0×10^{-4} Poise and the density of air is 0.0012 g/cm^3, thus our second phase properties are very close for density, but three orders of magnitude too high for viscosity. The numerical method fails to converge for values of the liquid/gas viscosity ratio of more than 1000 for a diffuse interface implementation of the level set equations, so this is a necessary expedient requiring us to adhere to this fictitious viscosity value.

For the inflow condition, we used a constant inflow pressure of 1.0×10^6 $dyne/cm^2$. This value was chosen to match the horizontal fill time of 23 seconds in the original mesh and then used for all other meshes and geometries. A shooting method was used, where different values of the inflow pressure were used and the solution was examined to see if it filled in the correct time. This required many simulations to be run. It is believed that the actual boundary condition for the experiments is somewhere between a constant velocity and constant pressure condition, but this is hard to replicate numerically. From the simulations, we found that the velocity changed quickly in the beginning for the pressure inflow boundary condition and subsequently reached a steady value.

The initial 3D mesh and boundary conditions are given Figure 11 for the mould filling simulations. We assume symmetry about the centreline and only solve half the problem to improve the computational efficiency. The mesh contains 6744 8-Node hexahedral elements giving 41300 total degrees of freedom for bilinear velocity/bilinear pressure interpolation. This mesh was shown to be adequate, as a more refined version of this mesh gave the same fill times and meniscus shapes [Rao et al., 2006].

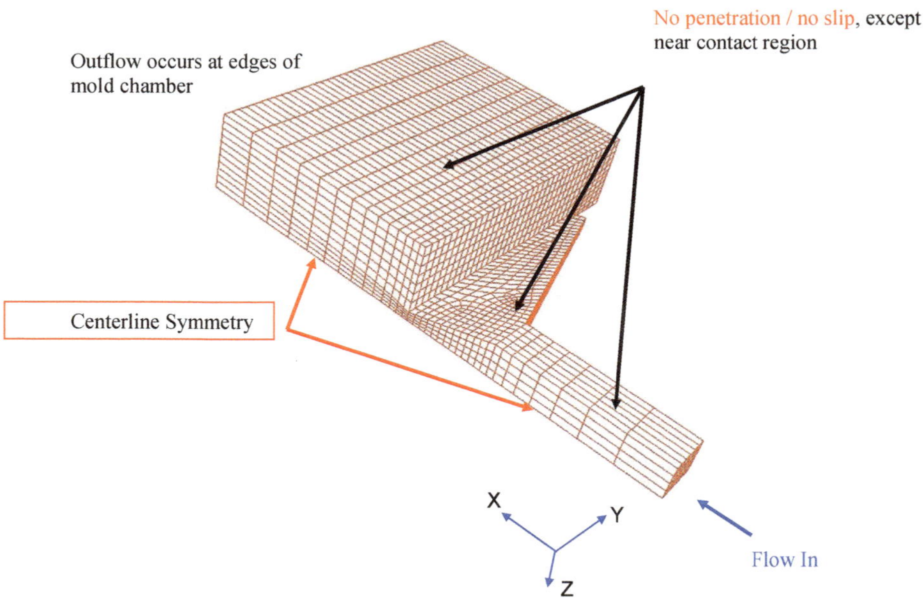

Fig. 11. Initial mesh and boundary conditions for 3D level set simulations. The outflow vent is on the same side as the distributor for vertical simulations and opposite the distributor for horizontal simulations.

The time to fill the mould for the vertical orientation for Mesh 1, Mesh 2, and Mesh 3 are 15.2s, 17.5s, and 13.2s, which was much faster than the experimental values of 24.6s, 26.9s, and 24.6s. We defined our fill time as when the vent had filled completely to a distance of the half the level-set length-scale or two elements. Unfortunately, the numerical fill times are difficult to obtain accurately as the gas phase viscosity makes a large difference in how fast a simulation will fill for the same value of pressure. For instance, when we reduced the second phase viscosity by a factor of 10 we got an increase in the inflow velocity when keeping all other parameters constant. Thus, the fact that the gas phase is harder to push out than it is for the experiments, adds a great deal of uncertainty to the fill times. However, we hope to be able to predict trends. Figure 12 shows the effect of the distributor geometry on the meniscus shape for Mesh 1, Mesh 2, and Mesh 3 in a vertical orientation.

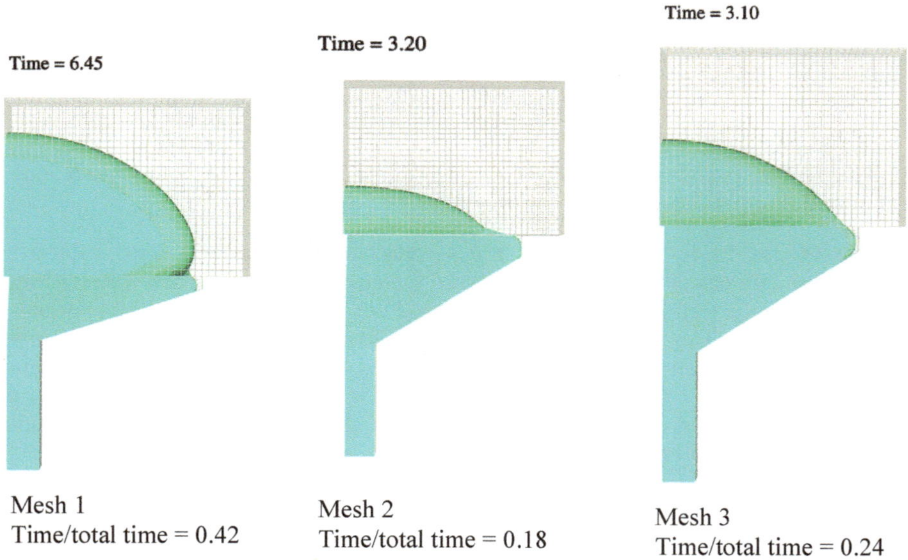

Fig. 12. Free surface profile after filling the distributor for Mesh 1, Mesh 2, and Mesh 3 for vertical mould orientation.

Comparing Figure 12 and Figure 5, the numerical and experimental version of this profile, we can see that the simulations are exhibiting the physically correct trends. The original mesh takes the longest fractional time to fill the distributor, 42%, and gives the most bulging front shape. Mesh 2 is an improvement, taking 18% of the time to fill the distributor, while Mesh 3 is somewhere in between at 24%. The values for the experimental distributor dimensionless fill times are 32%, 13%, and 24%, so the simulations are also capturing the correct trends for fill time though they are not quantitative. The shape of the meniscus for Mesh 1 has more of a bulge at the edge of the distributor than the experimental meniscus, which looks as if the front is pinned at the distributor.

We can also look at the profiles and dimensionless time to hit the back wall. These results are given in Figure 13.

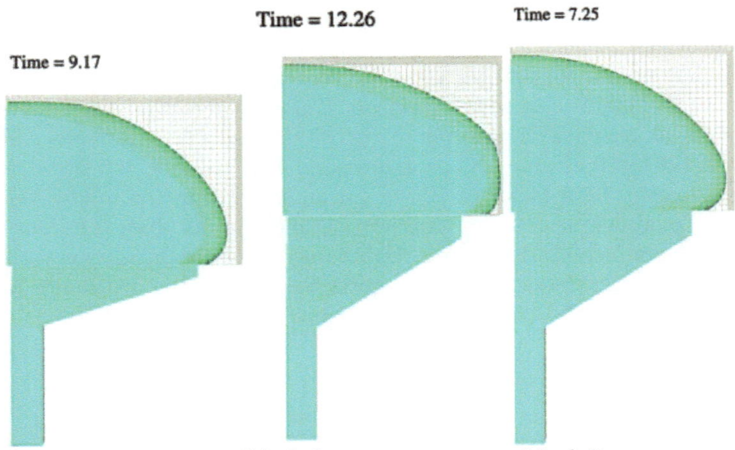

Mesh 1 Time/total time = .60 **Mesh 2** Time/total time = 0.70 **Mesh 3** Time/total time = 0.55

Fig. 13. Free surface profile after hitting the back wall for Mesh 1, Mesh 2, and Mesh 3 for vertical mould orientation.

Comparing Figure 13 to Figure 6, for the simulation versus experiment, we can see differences in the meniscus shape. The numerical interface reaches the back wall for Mesh 1 and Mesh 3, before it wets the sidewall and Mesh 2 has a flatter profile in the experiments than the simulations. The percentage time to reach the back wall for the simulations on Mesh 1, Mesh 2, and Mesh 3 are 60%, 70%, and 55% compared to 80%, 86% and 82% for the experiments. Again, we capture the correct trends, but are still unable to match the data quantitatively. Figure 14 shows the full meniscus shape and void locations for the simulations on Mesh 1, Mesh 2, and Mesh 3.

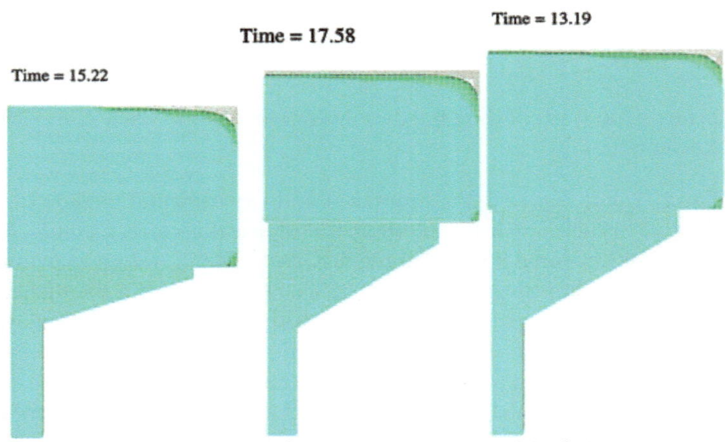

Mesh 1 Time/total time = 1.0 **Mesh 2** Time/total time = 1.0 **Mesh 3** Time/total time = 1.0

Fig. 14. Final void location and front profile for Mesh 1, Mesh 2, and Mesh 3.

The void in the corner near the distributor does eventually fill in, since its size is less than the level-set length scale and we are using a diffuse interface method. The larger void at the vent never fills in as the viscous gas phase is trapped away from the vent by the fluid. For the numerical solutions it is hard to make any predictions about void size, though we can say that for similar values of the dimensionless time the voids for Mesh 2 will be smaller than Mesh 1, with Mesh 3 being somewhere in between, which does follow the experimental trend.

We also examined the horizontal orientation numerically, which gave fill times for Mesh 1, Mesh 2, and Mesh 3 of 21.4s, 23.1s, and 22.2s compared to 23.3s, 28.9s, and 27.3s for the experiments. Again, we follow the trends of the experiment, but do not match quantitatively. Mesh 2 seems to take a longer time to fill for the experiment than one would predict numerically. Figure 15 shows a comparison of the profiles for Mesh 1 in a vertical and horizontal orientation.

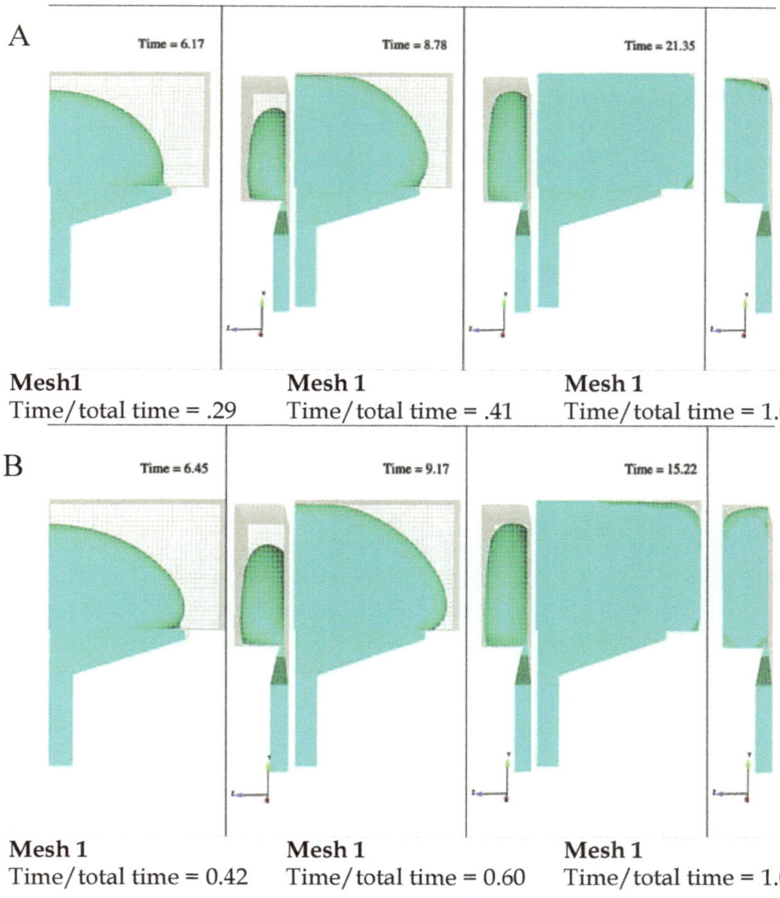

Fig. 15. Mesh 1 oriented horizontally (A) and vertically (B). Leftmost pictures show profiles when the distributor is filled, middle shows profile when the fluid hits the back wall, and rightmost pictures show final profile. Both front and side views are given to highlight void location.

Comparing Figure 15 to the experimental equivalent, Figure 8, we can see that we have quantitative differences but do match some trends. The numerical meniscus shape leaving the distributor looks similar to the experimental profile as it bulges more in the centre for the horizontal orientation, though the simulation is less dramatic. The numerical profiles when the fluid first hits the back wall are flatter for the vertical orientation than the horizontal, though the vertical should be even flatter to match the data. The numerical solutions predict two voids for each orientation, though the experiments do not show a second void for the horizontal orientation near the vent. However, this void may just be difficult to see experimentally. Conversely, the horizontal void at the outflow may be an artefact of the numerical method as we have a difficult balance at the outflow between wetting forces, gravity, the gas phase viscosity, and the material flowing out the vent. Also, our numerical vent is not identical to the experimental one and exhibits a slightly different area and shape.

Figure 16 shows the meniscus profiles for Mesh 1, Mesh 2, and Mesh 3 after filling the distributor for the horizontal orientation.

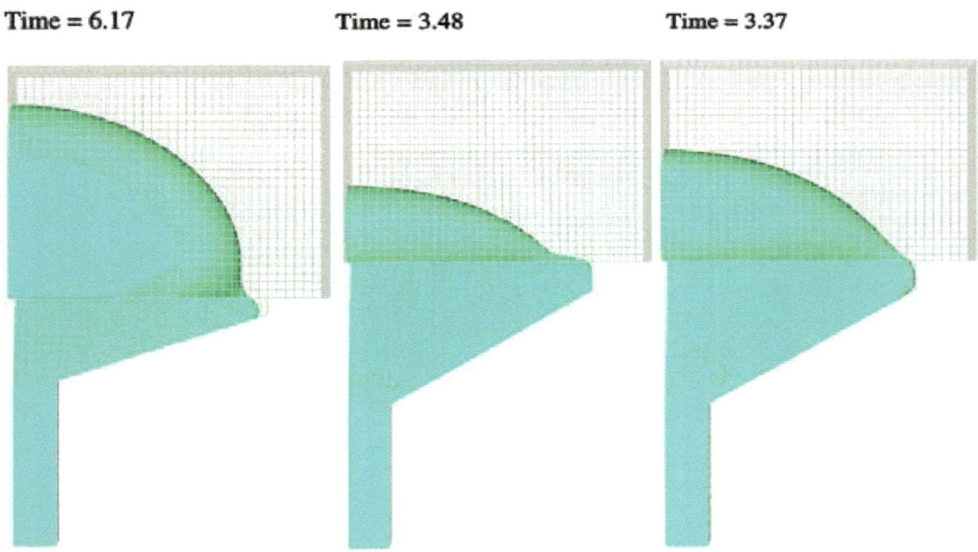

Mesh 1
Time/total time = .29

Mesh 2
Time/total time = .15

Mesh 3
Time/total time = .21

Fig. 16. Free surface profile after filling the distributor for Mesh 1, Mesh 2, and Mesh 3 for horizontal mould orientation.

Comparing Figure 16 and Figure 9, the numerical and experimental version of this profile, we can see that the simulations are again exhibiting the correct trends of the physical situation. Mesh 1 shows the most pooling at the centre of the mould, Mesh 2 has a flatter profile as does Mesh 3. The experiments predict filling times for Mesh 3 to be in between Mesh 1 and Mesh 2, and the simulations follow this trend. The dimensionless times to fill the distributor numerically for Mesh 1, Mesh 2, and Mesh 3 are 29%, 15%, and 21% compared to 26%, 13% and 22% for the experiments. In general, the simulations predict the vertical filling to be faster overall than the horizontal by several seconds for each of the geometries, whereas the experiments are faster in the vertical for Mesh 2 and 3, but slower for Mesh 1. This could have resulted from some experimental errors or from the uncertainty in injection rates and flow profiles from the experiments to the simulations.

Figure 17 shows the free surface profile as the fluid hits the back wall for Mesh 1, Mesh 2, and Mesh 3 in the horizontal orientation. The dimensionless times it takes to hit the back wall for Mesh 1, Mesh 2, and Mesh 3 are 41%, 55%, and 52% compared to values of 82%, 71%, and 70% for the experiments seen in Figure 10.

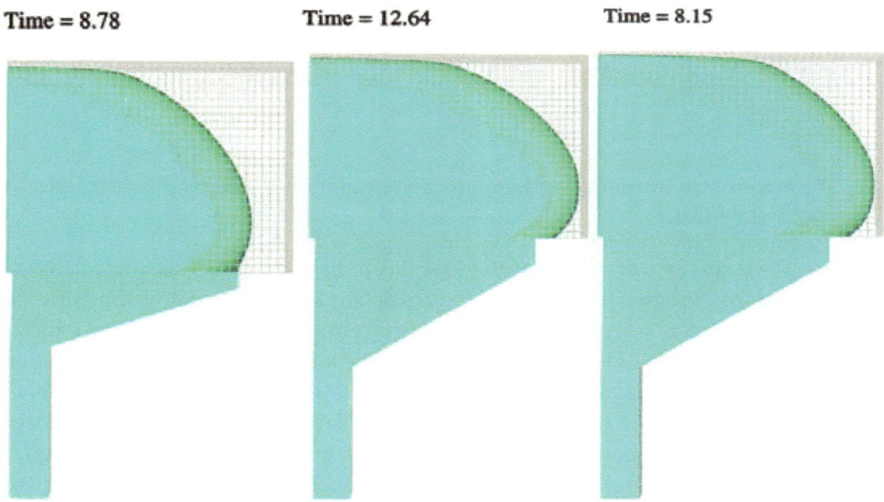

Mesh 1
Time/total time = .41

Mesh 2
Time/total time = .55

Mesh 3
Time/total time = .52

Fig. 17. Free surface profile after hitting the back wall for Mesh 1, Mesh 2, and Mesh 3 for horizontal mould orientation.

For the simulations, Mesh 1 hits the back wall for the smallest dimensionless time while Mesh 2 and Mesh 3 take about the same time. For the experiments, Mesh 1 takes the longest time, while Mesh 2 and Mesh 3 do take about the same time. Thus for this case, we are capturing one trend, but not the differences between Mesh 1 and Mesh 2.

Table 4 summarizes the fill times to reach the distributor, back wall, and complete mould for the simulations and experiments for vertical and horizontal orientations on all three meshes.

Mesh	Orientation	Expt. Time - Full	Expt. % Time - Dist	Expt. % Time - Wall	Sim. Time- Full	Sim. % Time - Dist	Sim. % Time - Wall
1	Vertical	24.6s	32%	80%	15.2s	42%	60%
2	Vertical	26.9s	13%	86%	17.5s	18%	70%
3	Vertical	24.6s	24%	82%	13.2s	24%	55%
1	Horizontal	23.3s	26%	82%	21.4s	29%	41%
2	Horizontal	28.9s	13%	71%	23.1s	15%	55%
3	Horizontal	27.3s	22%	71%	15.6s	21%	52%

Table 4. Summary of fill times for experiment and simulations to reach the distributor, the wall and completely full.

From Table 4 we can see that we capture some of the correct trends, especially for the vertical orientation, though some features elude us like the time to fill to the back wall for Mesh 1 in the horizontal orientation. Sources of uncertainty in the simulations include: 1) Lack of clarity of inflow conditions from the experiment to the simulation, since it is somewhere in between constant pressure and constant velocity, 2) Possible poor performance of the Blake wetting model at moderate capillary number and large dynamic contact angles, 3) Possible poor performance of the wetting model at wetting speed higher than experiments used to populate the model. Sources of error in the simulations include: 1) Numerically expedient of high gas phase viscosity, 2) Lack of compressibility for the gas dynamics, 3) The use of a diffuse interface model that smears out material property jumps, allowing viscous bleed through of the liquid phase into the gas phase. For future work, we will try to reduce these uncertainties and errors by using some of the advanced features in ARIA, which should be available soon, to allow for smaller values of the gas phase viscosity, such as sharp integration and a compressible gas phase. The optimal choice of wetting model for moderate to high capillary numbers continues to be an ongoing focus of our research.

5. Conclusion

A diffuse interface finite element/level-set algorithm has been used to investigate filling behaviour for injection loading using a Blake wetting model. The modelling has been successful in matching experimental data qualitatively, but quantitative agreement is still lacking especially for the wetting dynamics and meniscus shape. For future work, we will investigate an advanced version of the level set method termed the conformal decomposition finite element method (CDFEM). CDFEM is a hybrid moving boundary algorithm, which uses a level set field to determine the location of the fluid-fluid interface and then dynamically adds mesh on the interface to facilitate the resolution of discontinuous material properties and fields, as well as the application of boundary conditions such as capillarity. This is a sharp interface method, where it is possible to apply jumps in material properties, material models, and field variables [Noble et al, 2010]. We believe this algorithm, which should be available soon, will lead to better agreement with experiments and should allow for straightforward inclusion of a compressible gas phase.

6. Acknowledgments

This work was supported by the Laboratory Directed Research and Development program at Sandia National Laboratories. We would like to thank Dr. Pin Yang, the project principal investigator, for his support of this study. We would also like to thank our Sandia reviewer Daniel Guildenbecher and P. Randall Schunk for their insightful editorial comments.

Sandia National Laboratories is a multi-program laboratory managed and operated by Sandia Corporation, a wholly owned subsidiary of Lockheed Martin Corporation, for the U.S. Department of Energy's National Nuclear Security Administration under contract DE-AC04-94AL85000.

7. References

Bach, P. & Hassager, O. (1985). "An algorithm for the use of the Lagrangian specification in Newtonian fluid mechanics and applications to free surface flow," *J. Fluid Mech.*, 152, 173-190.

Blake, T. D. & Haynes, J. M. (1969). "Kinetics of liquid/liquid displacement," *J. Colloid Interface Sci.*, 30, 421.

Blake, T. D. & De Coninck, J. (2002). "The Influence of Solid-Liquid Interactions on Dynamic Wetting," *Adv. Colloid & Int. Sci.*, 96, 21-36.

Blake, T. D. (2006). "The physics of moving wetting lines," *J. Colloid Interface Sci.*, 299, 1-13.

Brackbill, J. U.; Kothe, D. B. & Zemach, C. (1992). "A continuum method for modelling surface-tension," *J. Comp. Phys.*, 100, 335-354.

Brooks, C. F.; Grillet, A. M. & Emerson, J. A. (2006). "Experimental investigation of the spontaneous wetting of polymers and polymer blends," *Langmuir*, 22, 9928-9941.

Cox, R. G. (1986). "The dynamics of the spreading of liquids on a solid surface. Part 1. Viscous flow," *J. Fluid Mech.*, 168, 169-194.

Dohrmann, C. R. & Bochev, P. B. (2004). "A stabilized finite element method for the Stokes problem based on polynomial pressure projections," *Int. J. Num. Meth. Fluids*, 46, 183–201.

Donea, J. (1984). "A Taylor-Galerkin Method for Convective Transport Problems," *Int. J. Num. Meth. Engn.*, 20, 101-119.

Hirt, C. W. & Nichols, B. D. (1981). "Volume of fluid (VOF) method for the dynamics of free boundaries," *J. Comp. Phys.*, 39, 201-225.

Hoffman, R. L. (1975). "A study of the advancing interface. I. Interface shape in liquid-gas systems," *J. Colloid Inter. Sci.*, 50, 228-241.

Hwang, C. J. & Kwon, T. H. (2002). "A full 3D finite element analysis of the powder injection moulding filling process including slip phenomena," *Poly. Eng. Sci*, 42, 33-50.

Hughes, T. J. R. (2000). The Finite Element Method. Dover Publications, New York, USA.

Kauzlaric, D.; Pastewka, L.; Meyer, H.; et al., (2011). "Smoothed particle hydrodynamics simulation of shear-induced powder migration in injection moulding," *Phil. Trans. Roy. Soc. A Math. Phys. Engn. Sci.*, 369, 2320-2328.

Kistler, S. F. (1983). "The fluid mechanics of curtain coating and related viscous free surface flows with contact lines," Ph.D. Thesis, University of Minnesota, Minneapolis.

Kulasegaram, S.; Bonet, J.; Lewis, R. W. & Profit, M., (2003). "High pressure die casting simulation using a Lagrangian particle method," *Comm. Appl. Numer. Meth. Engn.*, 19, 679-687.

Kumar, A. & Ghoshdastidar, P. S. (2002). "Numerical Simulation of Polymer Flow in a Cylindrical Cavity," *J. Fluids Engn.*, 124, 251-261.

Lewis, R. W.; Navti, S. E. & Taylor, C. (1997). "A mixed Lagrangian-Eulerian approach to modelling fluid flow during mould filling," *Int. J. Num. Meth. Fluids*, 25, 931-952.

Love, E. & Sulsky, D. L. (2006). "An energy-consistent material-point method for dynamic finite deformation plasticity," *Int. J. Num. Meth. Engn.*, 65, 1608-1638.

Ilinca, F. & Hetu, J. -F. (2000). "Finite element solution of three-dimensional turbulent flows applied to mould-filling problems," *Int. J. Num. Meth. Fluids*, 34, 729-750.

Ilinca, F. & Hetu, J.-F. (2001). "Three-dimensional filling and post-filling simulation of polymer injection moulding," *Int. Poly. Proc.*, 16, 291-301.

Ilinca, F. & Hetu, J.-F. (2008). "Three-dimensional free surface flow simulation of segregating dense suspensions ," *Int. J. Num. Meth. Fluids*, 58, 451-472.

Mondy, L.A.; Rao, R. R.; Brooks, C. F.; Noble, D. R.; *et al.*, (June 2007). "Wetting and free surface flow modelling for potting and encapsulation," SAND2007-3316, Sandia National Laboratories, Albuquerque, NM.

Nithiarasu, P. (2005). "An arbitrary Lagrangian Eulerian (ALE) formulation for free surface flows suing a characteristic-base split (CBS) scheme," *Int. J. Num. Meth. Fluids*, 48, 1415-1428.

Noble, D. R.; Newren, E. & Lechman, J. B. (2010). "A conformal decomposition finite element method for modelling stationary fluid interface problems", *Int. J. Num. Meth. Fluids*, 63, 725-742.

Notz, P. K.; Subia, S. R.; Hopkins, M. M.; Moffat, H. K. & Noble, D. R. (April 2007). "ARIA Manual Aria 1.5: User's Manual," SAND2007-2734, Sandia National Laboratories, Albuquerque, NM.

Oishi, C. M.; Tome, M. F.; Cuminato, J. A. & McKee, S. (2008). "An implicit technique for solving 3D low Reynolds number moving free surface flows, *J. Comp. Phys.*, 227, 7446-7468.

Radovitzky, R. & Ortiz, M. (1998). "Lagrangian finite element analysis of Newtonian fluid flows, *Int. J. Num. Meth. Engn.*, 43, 607-617.

Rao, R. R.; Mondy, L. A.; Noble, D. R.; Hopkins, M. M.; Notz, P. K.; Baer, T. A.; Halbleib, L.; Yang, P.; Burns, G.; Grillet, A. M.; Brooks, C.; Cote, R. O. & Castaneda, J. N. (September 2006). "Modeling Injection Molding of Net-Shape Active Ceramic Components," SAND2006-6786, Sandia National Laboratories, Albuquerque, NM.

Rao, R. R.; Mondy, L. A.; Noble, D. R.; Moffat, H. K; Adolf, D. B. & Notz, P.K. (2011). "A Level Set Method to Study Foam Processing: A Validation Study," *Int. J. Num. Meth. Fluids*, early view.

Ren, W.; Hu, D. & E, W. (2010). "Continuum models for the contact line problem," *Phys. Fluids*, 22.

Roberts, S. A. & Rao, R. R. (2011). "Entraining flow of a shear-thinning jet impinging in a container: A finite element approach," *J. Non-Newtonian Fluid Mech.*, 166, 1100–1115.

Sackinger, P. A.; Schunk, P. R. & Rao, R. R. (1996). "A Newton-Raphson pseudo-solid domain mapping technique for free and moving boundary problems: a finite element implementation," *J. Comp. Phys.*, 125, 83-103.

Sartor, L. (1990). "Slot Coating: Fluid Mechanics and Die Design," Ph.D. Thesis, University of Minnesota, Minneapolis.

Schunk, P. R.; Noble, D. R.; Baer, T. A.; Secor, R. B. & Jendoubi, S. (September 2006). "Implementation and performance of published wetting models in level-set and ALE-based algorithms for free and moving boundary problems," Poster Presentation, 13th International Coating Science and Technology Symposium, Denver, Colorado.

Sethian, J. A. (1999). *Level Set Methods and Fast Marching Methods*, Volume 3 of Cambridge Monographs on Applied and Computational Mathematics. Cambridge University Press, New York, USA, 2nd edition.

Shikhmurzaev, Y. D. (1994). "Mathematical modelling of wetting hydrodynamics," *Fluid Dynamics Res.*, 13, 45-64.

Tome, M. F.; Filho, A. C.; Cuminato, J. A.; Mangiavacchi, N. & McKee, S. (2001). "GENSMAC3D: a numerical method for solving unsteady three-dimensional free surface flows," *Int. J. Num. Meth. Fluids*, 37, 747-796.

Ville, L.; Silva, L.; & Coupez, T. (2011) "Convected level set method for the numerical simulation of fluid buckling," *Int. J. Num. Meth. Fluids*, 66, 324-344.

Voinov, O. V. (1976). "Hydrodynamics of wetting," *Fluid Dynamics*, 11, 714-721.

Zhang, J. & Khayat, R. E. (2001). "A Lagrangian boundary element approach to transient three-dimensional free surface flow in thin cavities, *Int. J. Num. Meth. Fluids*, 37, 399-418.

3

3D Multiphase Numerical Modelling for Turbidity Current Flows

A. Georgoulas, P. Angelidis, K. Kopasakis and N. Kotsovinos
Democritus University of Thrace, Xanthi
Greece

1. Introduction

Gravity or density currents constitute a large class of natural flows that are generated and driven by the density difference between two or even more fluids. The density difference between two fluids usually arises due to differences in temperature or salinity, but it can also arise due to the presence of suspended solid particles. These particulate currents, in the case of sediment laden water that enters a water basin, are classified according to the density difference with the ambient fluid into three major categories: a) hypopycnal currents, when the density of the sediment laden water is lower than that of the receiving water basin, b) homopycnal currents, when the density of the sediment laden water is almost equal to that of the receiving water basin, and c) hyperpycnal currents when their density is much greater than that of the receiving water body (Mulder & Alexander, 2001). In the case of floods, the suspended sediment concentration of river water rises to a great extent. Hence, the river plunges to the bottom of the receiving basin and forms a hyperpycnal plume which is also known as turbidity current. Such flows are usually formed at river mouths in oceans, lakes or reservoirs, and can travel remarkable distances transferring, eroding and depositing large amounts of suspended sediments (Mulder & Alexander, 2001).

Turbidity currents are very difficult to be observed and studied in the field. This is due to their rare and unexpected occurrence nature, as they are usually formed during floods. Therefore, field investigations are usually limited to the study of the deposits originating from such currents. The anatomy of deposits originating from turbidity currents can be studied on a large scale, in order to identify the various depositional elements such as lobes, levees and submarine channels (Janbu et al., 2009). Furthermore, considerable research on the morphology of turbiditic systems and general deep-marine depositions is being increasingly done with the use of 3D seismic sections (Posamentier & Kolla, 2003; Saller et al., 2006).

On the other hand, scaled laboratory experiments constitute an alternative and widely used method for simulating and studying the dynamics of turbidity currents. Many researchers have been focused in the study of the flow dynamics, depositional and erosional characteristics of laboratory turbidity currents, using scaled experimental models (Britter & Linden, 1980; Lovell, 1971; Garcia & Parker, 1989; Simpson & Britter, 1979). Advances in experimental technology in the last decades have increased the existing knowledge from macroscopic and qualitative descriptions of turbidity current behaviour and deposits, to detailed, quantitative results relating to the actual flow characteristics, such as the velocity, concentration as well as the turbulence structure of such flows (Baas et al., 2004; Garcia, 1994; Gladstone et al., 1998; Kneller et al., 1997).

Mathematical and numerical models when properly designed and tested against field or laboratory data, can provide significant knowledge for turbidity current dynamics as well as for erosional and depositional characteristics. Up to present, there are various numerical investigations dealing with turbidity current dynamics and flow characteristics, providing valuable results regarding these complex phenomena. The characteristics of a gravity-current head have been studied by (Hartel et al., 2000), using 3D Direct Numerical Simulations (DNS) of flow fronts in the lock-exchange configuration. (Kassem & Imran, 2001) present a 2D numerical approach for investigating the transformation of a plunging river flow into a turbidity current. In the work of (Heimsund et al., 2002) a computational, 3D, fluid-dynamics model for sediment transport, erosion and deposition by turbidity currents has been constructed using the CFD (Computational Fluid Dynamics) software Flow-3D. Another 3D numerical model using the CFX-4 code was developed, in order to simulate turbidity currents in Lake Lugano (Switzerland), in the work of (Lavelli et al., 2002). (Necker et al., 2002) presented 2D and 3D Direct Numerical Simulations of particle-driven gravity currents, placing special emphasis on the sedimentation of particles, and the influence of particle settling on the flow dynamics. (Cantero et al., 2003) present two and three-dimensional CFD simulations of a discontinuous density current, using a stabilized equal-order finite element method. A comparative study on the convergence of CFD commercial codes, when simulating dense underflows is presented by (Bombardelli et al., 2004). Two codes are used for the proposed simulations: the first one is a comprehensive finite-element platform, whereas the other one is a commercial code. The lateral development of density-driven flow in a subaqueous channel is studied using a 3D numerical model, in the work of (Imran et al., 2004). The conditions under which turbidity currents may become self-sustaining through particle entrainment are investigated in the work of (Blanchette et al., 2005), using 2D Direct Numerical Simulations of resuspending gravity currents. A numerical model of turbidity currents with a deforming bottom boundary, that predicts the vertical structure of the flow velocity and concentration as well as the change in the bed level, due to erosion and deposition of suspended sediment, is developed in the work of (Huang et al., 2005). Lock-exchange gravity current flows, produced by the instantaneous release of a heavy fluid, are investigated by means of 2D Large-Eddy Simulation (LES) in the work of (Ooi et al., 2007). A numerical simulation of turbidity current using the $\overline{v^2}-f$ turbulence model is carried out in the work of (Mehdizadeh et al., 2008). (Cantero et al., 2008a), perform 2D Direct Numerical Simulations in order to investigate the effect of particle inertia on the dynamics of particulate gravity currents. They introduce an Eulerian-Eulerian formulation for gravity currents driven by inertial particles. 3D Direct Numerical Simulations of planar gravity currents have been conducted with the objective of identifying, visualizing and describing turbulent structures and their influence on flow dynamics, in the work of (Cantero et al., 2008b). The investigation of the effect of initial aspect ratio on the flow characteristics of suspension gravity currents as well as the diffusion of the turbidity under the presence of a turbidity fence is carried out in the work of (Singh, 2008), using 3D Large Eddy Simulations.

Most of these previous CFD-based investigations treat turbidity currents with a quasi-single-phase approach, since the transport of sediment particles is taken into account through an advection-diffusion equation for sediment concentration. The present chapter aims to present the validity, usefulness and applicability of a three-dimensional, "uncommon", CFD-based, multiphase numerical approach for the simulation and study of the hydrodynamic and depositional characteristics of turbidity currents that are usually formed at river outflows in the sea, lakes and reservoirs. The numerical model is based in a multiphase modification of the Reynolds Averaged Navier-Stokes Equations (RANS). Turbulence closure is achieved through the application of the RNG (Renormalization-Group) k-ε turbulence model. The calculations of the model are performed using the robust CFD solver FLUENT. The proposed numerical model for the simulation of turbidity current hydrodynamics was firstly introduced in the work of (Georgoulas et al., 2010).

In the present section of the chapter (Section 1) a brief introduction on turbidity currents and a literature review on field, experimental and numerical studies are conducted while the main aim of the chapter is also stated. In Section 2 the theoretical background of the proposed numerical approach is presented and discussed in detail, while in Section 3 some main validation results are presented (Georgoulas et al., 2010). Section 4 presents the results of a laboratory-scale (Georgoulas, 2010) and a field scale (Georgoulas et al, 2009) application of the numerical approach. Finally, in Section 5 the main concussions that are withdrawn from the present chapter are summarized.

2. Numerical model description

2.1 Overview

Turbidity current flows can be characterized as multiphase flow systems, since they consist of a primary fluid phase (water) and secondary granular phases (suspended sediment classes) dispersed into the primary phase. Therefore, turbidity currents can be modeled through the application of suitable multiphase numerical models. Since, the particulate loading of turbidity currents may vary from small to considerably large values, an Eulerian-Eulerian multiphase numerical approach is considered to be more appropriate, as it can handle a wider range of particle volume fractions than an Eulerian-Lagrangian approach (maximum particles volume fraction of 10-12%). FLUENT provides various multiphase models that are based in the Eulerian-Eulerian approach. The "Eulerian" model that has been selected for the numerical approach that is presented in the present chapter, may require more computational effort, but it can handle a wider range of particulate loading values and is more accurate than the other available multiphase models in FLUENT. In this multiphase model, the different phases are treated mathematically as interpenetrating continua and therefore the concept of phasic volume fraction is introduced, where the volume fraction of each phase is assumed to be a continuous function of space and time. The sum of the volume fractions of the various phases is equal to unity. An accordingly modified set of momentum and continuity equations for each phase is solved. Pressure and inter-phase exchange coefficients are used in order to achieve coupling for these equations (Georgoulas et al., 2010).

The motion of the suspended sediment particles within a turbidity current as well as the motion generated in the ambient fluid are of highly turbulent nature. In order to account for the effect of turbulence in the numerical simulations of the present investigation, the instantaneous governing equations are not applied directly but they are ensemble-averaged, converting turbulent fluctuations into Reynolds stresses, which represent the effects of turbulence. This averaging procedure for the numerical simulation of turbulent flows is known as RANS (Reynolds-averaged Navier-Stokes equations). The averaged governing equations contain additional unknown variables, and turbulence models are needed to determine these variables in terms of known quantities. Therefore, with this averaging approach the turbulence is modeled and only the unsteady, mean flow structures that are primarily larger than the turbulent eddies are resolved. This is the main difference with the other two widely used numerical approaches for turbulent flows, known as DNS (Direct Numerical Simulation) and LES (Large Eddy Simulation). In DNS, the Navier-Stokes equations are applied and solved directly without the application of a turbulence model, resolving the whole range of turbulent eddies. In LES on the other hand, large eddies are resolved directly, while small eddies are modeled. DNS and LES may provide detailed information on turbidity current flows but their major disadvantage is that their application is limited due to large computational requirements. On the other hand, RANS may not provide detailed information from a microscopic point of view, but is quite accurate and attractive for modeling large scale, three dimensional flows of practical engineering interest due to the relatively low computational cost (Georgoulas et al., 2010).

In the numerical approach presented here, the Renormalization-group (RNG) k-ε model is applied for turbulence closure. This model was derived using a rigorous statistical technique, the renormalization group theory. The basic form of the RNG k-ε model is similar to the standard k-ε model, but it includes a number of refinements, rendering it more appropriate for the case of turbidity currents, as it is more accurate for swirling flows and rapidly strained flows and also accounts for low Reynolds number effects. Moreover, it provides an analytical formula for the calculation of the turbulent Prandtl numbers. At this point it should be mentioned that in the present numerical approach, the RNG k-ε model is also modified accordingly in order to simultaneously account for the primary (continuous) phase and the secondary (dispersed) phases of the simulated flows. This modification in FLUENT is based on a number of assumptions. In more detail, turbulent predictions for the continuous phase are obtained using the RNG k-ε model, supplemented with extra terms that include the interphase turbulent momentum transfer. Predictions for turbulence quantities for the dispersed phases are obtained using the Tchen theory of dispersion of discrete particles by homogeneous turbulence. Interphase turbulent momentum transfer is also assumed, in order to take into account the dispersion of the secondary phases transported by the turbulent fluid motion. Finally, a phase-weighted averaging process is assumed, so that no volume fraction fluctuations are introduced into the continuity equations (Georgoulas et al., 2010).

2.2 Governing equations

The volume of phase q, V_q is defined by the following relationship (ANSYS FLUENT Documentation, 2010):

$$V_q = \int_V \alpha_q dV \tag{1}$$

where,

$$\sum_{q=1}^{n} \alpha_q = 1 \tag{2}$$

and α_q is the volume fraction of phase q.

The effective density of phase q is:

$$\hat{\rho}_q = \alpha_q \rho_q \tag{3}$$

where ρ_q is the physical density of phase q.

The continuity, the fluid-fluid, and fluid-solid momentum equations that are actually solved by the model are described by equations (4), (5) and (6) respectively, for the general case of a n-phase flow consisting of granular and non-granular secondary phases (ANSYS FLUENT Documentation, 2010):

$$\frac{1}{\rho_{rq}}\left(\frac{\partial}{\partial t}(\alpha_q \rho_q) + \nabla \cdot (\alpha_q \rho_q \vec{\upsilon}_q) = 0\right) \tag{4}$$

$$\begin{gathered}\frac{\partial}{\partial t}(\alpha_q \rho_q \vec{\upsilon}_q) + \nabla \cdot (\alpha_q \rho_q \vec{\upsilon}_q \vec{\upsilon}_q) = -\alpha_q \nabla p + \nabla \cdot \overline{\overline{\tau}}_q + \alpha_q \rho_q \vec{g} + \\ \sum_{p=1}^{n} (K_{pq}(\vec{\upsilon}_p - \vec{\upsilon}_q)) + (\vec{F}_q + \vec{F}_{lift,q} + \vec{F}_{vm,q})\end{gathered} \tag{5}$$

$$\begin{gathered}\frac{\partial}{\partial t}(\alpha_s \rho_s \vec{\upsilon}_s) + \nabla \cdot (\alpha_s \rho_s \vec{\upsilon}_s \vec{\upsilon}_s) = -\alpha_s \nabla p - \nabla p_s + \nabla \cdot \overline{\overline{\tau}}_s + \alpha_s \rho_s \vec{g} + \\ \sum_{l=1}^{N} (K_{ls}(\vec{\upsilon}_l - \vec{\upsilon}_s)) + (\vec{F}_s + \vec{F}_{lift,s} + \vec{F}_{vm,s})\end{gathered} \tag{6}$$

where ρ_{rq} is the phase reference density, or the volume averaged density of the qth phase in the solution domain, $\vec{\upsilon}_q$ is the velocity of phase q, $\vec{\upsilon}_p$ is the velocity of phase p, p is the pressure shared by all phases, $\overline{\overline{\tau}}_q$ is the qth phase stress-strain tensor, $\vec{g}$ is the gravitational acceleration, K_{pq} is the interphase momentum exchange coefficient, $\vec{F}_q$ is an external body force, $\vec{F}_{lift,q}$ is a lift force and $\vec{F}_{vm,q}$ is a virtual mass force. $K_{ls} = K_{sl}$ is the momentum exchange coefficient between fluid phase l and solid phase s and N is the total number of phases. The stress-strain tensors $\overline{\overline{\tau}}_q$ and $\overline{\overline{\tau}}_s$ are calculated by the following relationships:

$$\overline{\overline{\tau}}_q = \alpha_q \mu_q \left(\nabla \vec{\upsilon}_q + \nabla \vec{\upsilon}_q^{\,T}\right) + \alpha_q \left(\lambda_q - \frac{2}{3}\mu_q\right) \nabla \cdot \vec{\upsilon}_q \overline{\overline{I}} \tag{7}$$

$$\bar{\bar{\tau}}_s = \alpha_s \mu_s \left(\nabla \vec{\upsilon}_s + \nabla \vec{\upsilon}_s^{\,T} \right) + \alpha_s \left(\lambda_s - \frac{2}{3} \mu_s \right) \nabla \cdot \vec{\upsilon}_s \bar{\bar{I}} \tag{8}$$

where, μ_q and μ_s are the shear viscosities of phases q and s, λ_q and λ_s are the bulk viscosities of phases q and s, and $\bar{\bar{I}}$ is the identity tensor.

The momentum exchange between the various phases involved in a multiphase flow is based in the value of the interphase exchange coefficients. Therefore, these coefficients are very important for the simulation of granular multiphase flows, as turbidity currents. In the numerical approach presented here, the fluid-solid momentum exchange coefficient, between the ambient water (primary phase) and the suspended sediment particles (secondary phase) is calculated using the Syamlal-O'Brien model, which is based on measurements of the terminal velocities of particles in fluidized or settling beds. This model was selected, as a series of trial numerical runs indicated that this gives the best results, in comparison with corresponding experimental measurements, for the case of turbidity currents (Georgoulas et al., 2010). As it can be seen from Equation (6), in the case of granular flows, in the regime where the solids volume fraction is less than its maximum allowed value, a solids pressure is calculated independently and used for the pressure gradient term (∇p_s), in the fluid-solid momentum equation. This solids pressure is composed of a kinetic term as well as a second term due to particle collisions and is calculated using the following relationship (ANSYS FLUENT Documentation, 2010):

$$p_s = \alpha_s \rho_s \Theta_s + 2\rho_s (1 + e_{ss}) \alpha_s^2 g_{0,ss} \Theta_s \tag{9}$$

where e_{ss} is the coefficient of restitution for particle collisions, $g_{0,ss}$ is the radial distribution function, and Θ_s is the granular temperature which is proportional to the kinetic energy of the fluctuating particle motion. Trial numerical simulations indicated that the solids pressure is significant at various regions and stages of turbidity current flows (Georgoulas et al., 2010).

The effect of lift forces in the secondary phase solid particles is also taken into account. These lift forces act on particles mainly due to velocity gradients in the primary-phase flow field. The lift force will be more significant for larger particles. A main assumption is that the particle diameter is much smaller than the interparticle spacing. Hence, the inclusion of lift forces is not appropriate for closely packed particles or for very small particles. The lift force acting on a secondary phase p in a primary phase q is calculated in FLUENT, using the following equation (ANSYS FLUENT Documentation, 2010):

$$\vec{F}_{lift} = -C_L \rho_q \alpha_p (\vec{\upsilon}_q - \vec{\upsilon}_p) \times (\nabla \times \vec{\upsilon}_q) \tag{10}$$

where C_L is the lift coefficient. For the turbidity current cases that are presented in the present chapter, values of the lift coefficient ranging from 0.1 to 0.5 give the best results in comparison with corresponding experimental measurements (Georgoulas et al., 2010).

The virtual mass force is usually significant, in cases where the secondary phase density is much smaller than the primary phase density (ANSYS FLUENT Documentation, 2010). For example, the virtual mass force would be significant in the case of air bubbles moving through water, as in this case the density of air is much smaller than the density of the ambient water and the added mass (by the surrounding water) in the air bubbles would be much larger than their own mass. In all of the turbidity current cases considered in the present chapter, the secondary phase density (solid particles) is larger than the primary phase density (fresh water) and therefore the virtual mass force is not taken into consideration.

The general transport equations for the turbulence kinetic energy k and the turbulence dissipation rate ε, of the RNG k-ε turbulence model, can be described by equations (11) and (12) respectively (ANSYS FLUENT Documentation, 2010):

$$\frac{\partial}{\partial t}(\rho k)+\frac{\partial}{\partial x_i}(\rho k u_i)=\frac{\partial}{\partial x_j}\left(\alpha_k \mu_{eff}\frac{\partial k}{\partial x_j}\right)+G_k+G_b-\rho\varepsilon \tag{11}$$

$$\begin{aligned}\frac{\partial}{\partial t}(\rho \varepsilon)+\frac{\partial}{\partial x_i}(\rho \varepsilon u_i)=\frac{\partial}{\partial x_j}\left(\alpha_\varepsilon \mu_{eff}\frac{\partial \varepsilon}{\partial x_j}\right)+\\ C_{1\varepsilon}\frac{\varepsilon}{k}(G_k+C_{3\varepsilon}G_b)-C_{2\varepsilon}\rho\frac{\varepsilon^2}{k}-R_\varepsilon\end{aligned} \tag{12}$$

where u represents velocity, ρ is the local mixture density, G_k is the generation of turbulence kinetic energy due to mean velocity gradients, G_b is the generation of turbulence kinetic energy due to buoyancy, α_k and α_ε are the inverse effective Prandtl numbers for k and ε respectively, μ_{eff} is the effective viscosity and $C_{1\varepsilon}$, $C_{2\varepsilon}$ and $C_{3\varepsilon}$ are turbulence model constants. The term R_ε in the ε equation accounts for the effects of rapid strain and streamline curvature (ANSYS FLUENT Documentation, 2010).

2.3 Solution procedure

The governing equations in the proposed multiphase numerical approach are solved sequentially, using the control-volume method. Hence, the equations are integrated about each control-volume, yielding discrete equations for the conservation of each quantity. An implicit formulation is used, in order for the discretized equations to be converted to linear equations for the dependent variables in every computational cell. Further details regarding the solution procedure can be found at (ANSYS FLUENT Documentation, 2010).

2.4 Boundary conditions

At inlets, a velocity-inlet boundary condition is used. With this type of boundary condition, a uniform distribution of all the dependent variables is prescribed at the face representing the sediment laden water inflow. In more detail, the velocity magnitudes of the primary and secondary phases with directions normal to the inlet face are specified, assuming constant, uniform values. Moreover, the volume fractions of the secondary phases at the inlet are also specified.

For the outlets, a pressure-outlet boundary condition is applied. Using this type of boundary condition, all flow quantities at the outlets are extrapolated from the flow in the interior domain. A set of "backflow" conditions can be also specified, allowing reverse direction flow at the pressure outlet boundary during the solution process. In other words, this type of outlet condition serves as an open flow boundary, allowing the flow to freely exit or enter the computational domain during the calculations.

At the free ambient water surfaces, a symmetry boundary condition is used, which is typically well above the generated turbidity currents. Thus, there are neither convective nor diffusive fluxes across the top surface. This type of free surface boundary condition has also been used by other researchers in literature (Imran et al., 2004; Huang et al., 2005) for the case of turbidity currents. (Farrell & Stefan, 1986) have found that for a plunging reservoir flow, the relative error that can be introduced by this approximation of the free surface, is of the order of 10^{-3} and does not influence the velocity field.

The solid boundaries are specified as stationary walls with a no-slip shear condition. Turbulent flows are significantly affected by the presence of walls. Very close to the wall, viscous damping and kinematic blocking reduce the tangential and normal velocity fluctuations respectively. However, in the outer part of the near wall region, the turbulence is rapidly augmented by the production of turbulent kinetic energy due to the relatively large gradients in mean velocity. In FLUENT, there are two different approaches for modeling the near wall region. In the first approach, the viscous sub-layer and the buffer sub-layer are not resolved. Instead, semi-empirical formulas known as "wall functions" (e.g. "standard wall functions") are used in order to link the viscosity affected sub-layers between the wall and the fully-turbulent region. In the second approach, known as "near-wall modeling" approach (e.g. "enhanced wall treatment), the turbulence models are modified in order for the viscosity affected near-wall regions to be resolved with a computational mesh all the way to the wall. However, the computational mesh must be significantly fine in these regions. This approach may require more computational effort, but it gives more accurate predictions at the near-wall region of the computational domain. Therefore, wall functions should only be used in cases where the complexity and size of the computational domain as well as the available computational resources, do not allow the construction of very fine meshes at the near-wall regions (ANSYS FLUENT Documentation, 2010).

3. Numerical model validation

A detailed verification of the proposed numerical model is conducted in the work of (Georgoulas et al., 2010), where two different series of published laboratory experiments on turbidity currents, conducted by (Gladstone et al., 1998) and (Baas et al., 2004) are reproduced numerically, and the results are compared aiming to evaluate how realistic and reliable the numerical simulations of the proposed model are. The first series of laboratory experiments (Gladstone et al., 1998) consist of fixed-volume lock-gate releases of dilute mixtures containing two different sizes of suspended silicon carbide particles, in various initial proportions, within a rectangular flume (Run A - Run G). The second series of laboratory experiments (Baas et al., 2004) consist of high-density sediment-water mixtures released with a steady rate, through a small inflow gate, into an inclined channel which is connected to a tank, were an expansion table covered with loose sediment is positioned. The mixtures consist of either fine sand, very fine sand or coarse silt. Apart from the suspended sediment grain size, the initial suspended sediment volume fraction, the water-sediment mixture discharge and the channel slope angle and bed roughness, are varied among the experimental runs (Run 1 - Run 14).

Details regarding the above mentioned laboratory experiments (experimental set-up, initial conditions) and their numerical reproduction (computational geometry, computational mesh, boundary conditions, etc.) can be found in the work of (Georgoulas et al., 2010). However, for the purposes of the present chapter, the key quantitative results that prove that the proposed numerical model predictions are realistic and reliable are presented and discussed in subsections 3.1 and 3.2 that follow, for the cases of the fixed-volume releases (Gladstone et al., 1998) and the steady-state releases (Baas et al., 2004), respectively.

3.1 Fixed-volume releases

Front speed is one of the most studied parameters for lock-exchange turbidity currents. Figure 1 compares the simulated and observed current front position versus time for all the lock-gate cases considered in the work of (Georgoulas et al., 2010). As it can be seen, in general the numerical simulations show a good match with the experimental data, adequately predicting the differences in the flow front advance among the generated currents with respect to the different relative proportions of coarse (%C in the legend) and fine particles (%F in the legend) that were used in the initial suspensions. The observed divergence between the experimental and the numerical curves at various times, might be partially attributed to possible over-estimation or under-estimation of the flow front position in the particular laboratory runs, due to the difficulty in the visual definition of the exact flow front position, since these laboratory difficulties are stated in the work of (Gladstone et al., 1998). Another possible reason for the observed divergence might be the overall assumptions in the numerical simulations (e.g. uniform grain size in each particle class).

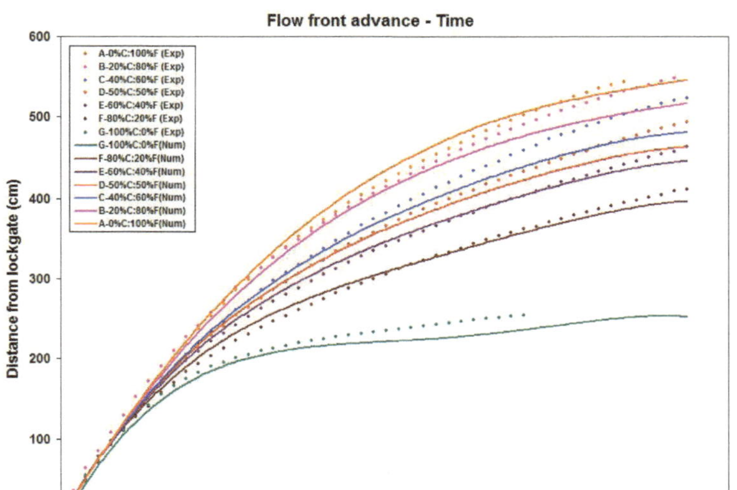

Fig. 1. Comparison of numerical (Georgoulas et al., 2010) and experimental (Gladstone et al., 1998) results, of flow front advance with respect to time.

In order to also examine the validity of the vertical structure of the simulated lock-gate cases, the non-dimensional vertical profiles of the stream-wise velocity component for numerical runs A and D are constructed and compared with analogous dimensionless experimental data from the laboratory work of (Garcia, 1994). The numerical profiles and the corresponding experimental data are compared in Figure 2. As it can be seen, the numerically predicted dimensionless profiles (Georgoulas et al., 2010) fall within the general scatter range of the dimensionless data for subcritical currents that resulted from the laboratory experiments of (Garcia, 1994). Therefore, it can be concluded that the proposed numerical model gives fairly reasonable predictions regarding the vertical structure of the simulated currents.

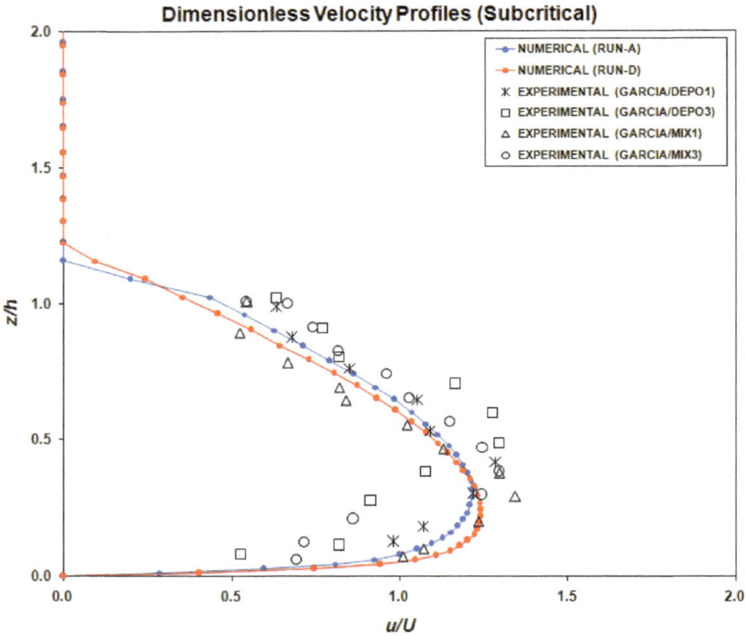

Fig. 2. Comparison of numerical dimensionless velocity profiles (Georgoulas et al., 2010) with analogous experimental data (Garcia, 1994), for numerical Runs A, and D that reproduce the experiments of (Gladstone et al., 1998).

3.2 Steady-state releases

The relationship between head velocity and initial suspended sediment concentration for fine-sand, very-fine sand and coarse silt laden turbidity currents is depicted in Figure 3, both for the numerical (Georgoulas et al., 2010) and the corresponding experimental runs (Runs 1, 3, 4, 7, 8, 13 and 14) (Bass et al., 2004). Once again, the numerical values are very close to the corresponding experimental values. Moreover, it is evident that the numerical model captures the same trend in the head velocity variation with respect to the increase of the initial suspended sediment concentration, in comply with the experimental runs.

In order to examine the validity of the vertical structure of the simulated steady-state releases, the non-dimensional vertical profiles of the streamwise velocity component for numerical runs 1, 7 and 14 (Georgoulas et al., 2010) are constructed and compared with corresponding dimensionless experimental data from the laboratory work of (Garcia, 1994). The numerical profiles and the corresponding experimental data are illustrated in Figure 4. As it can be seen, the numerically predicted dimensionless data fall within the scatter range of the dimensionless data for supercritical currents that resulted from the laboratory experiments of (Garcia, 1994). However, at the near-wall region of the numerical profiles, a sharp change is observed in relation to the experimental values. This sharp change at the near-wall region could be attributed to the 3cm mesh resolution that was used in the steady-state release runs and the application of the standard wall functions that do not resolve but instead link the viscosity affected near-wall region with the fully turbulent outer region, though the use of empirically derived formulas. Since, this sharp change is not presented in the lock-gate cases (Figure 2), it can be concluded that the application of the "enhanced wall treatment" that was used in the numerical reproduction of lock-gate releases should be preferable at the bottom wall boundaries, in cases that the complexity and size of the computational domain geometry as well as the available computational resources, allow the construction of high-resolution meshes at the near-wall regions, since this provides more accurate and detailed predictions in the vicinity of the bottom wall boundaries.

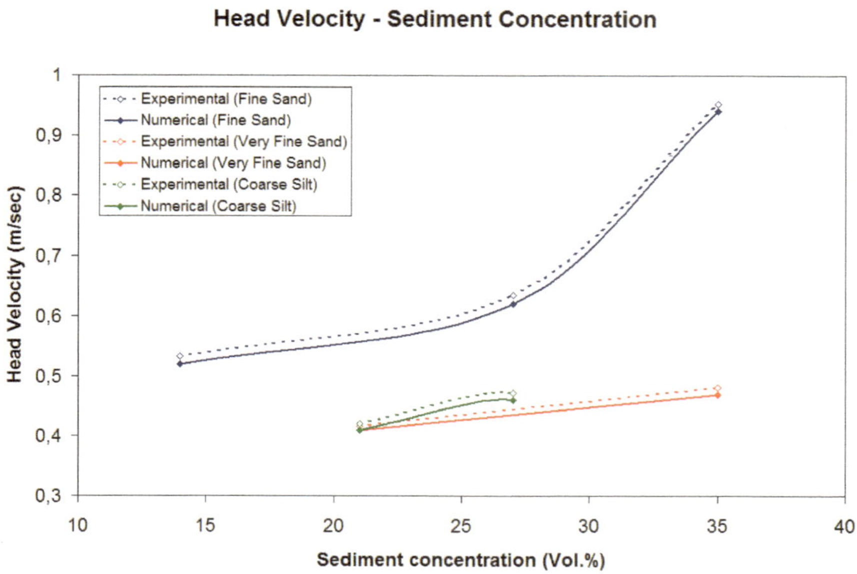

Fig. 3. Head velocity variation with respect to the initial suspended sediment concentration, for turbidity currents laden with fine sand, very fine sand and coarse silt. Comparison of numerical (Georgoulas et al., 2010) and experimental results (Baas et al., 2004).

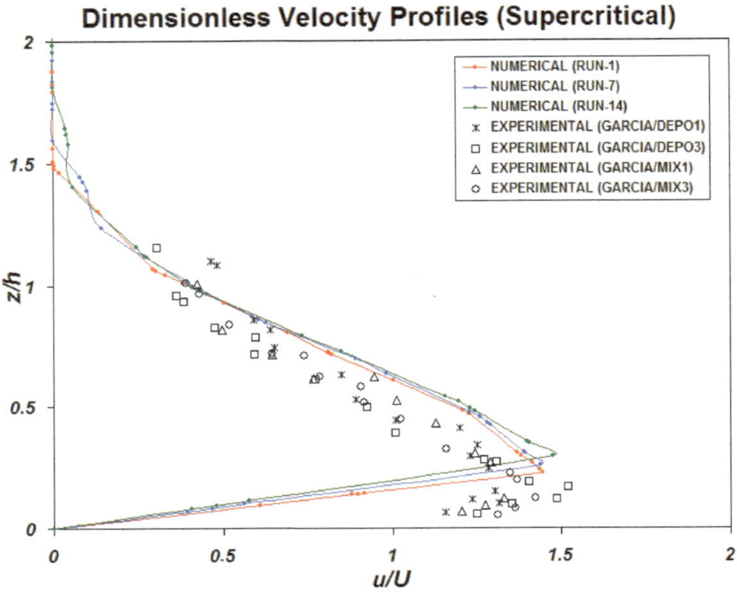

Fig. 4. Comparison of numerical dimensionless velocity profiles (Georgoulas et al., 2010) with analogous experimental data (Garcia, 1994), for numerical Runs 1, 7 and 14 that reproduce the experiments of (Baas et al., 2004).

4. Applications of numerical model

4.1 Laboratory scale application

The present subsection of the chapter describes a laboratory scale application of the proposed numerical model that aims to identify the effect of various flow controlling parameters (bed slope, bed roughness, initial suspended sediment concentration and suspended sediment diameter) in the hydrodynamic and depositional characteristics of continuous, high density turbidity currents (Georgoulas, 2010). For this purpose, four different series of parametric numerical experiments are conducted, using a laboratory scale experimental set-up, similar to the one used in the laboratory experiments of (Baas et al., 2004). In each series of numerical experiments, the initial value of only one of the above mentioned controlling parameters is varied, while the initial values of the rest parameters are kept constant.

The geometry and the general conditions of the physical problem under investigation are depicted in Figure 5. As it can be seen, the physical problem consists of turbidity currents that are generated during the continuous inflow of fresh water - suspended sediment mixtures (through an inflow gate, of height h_{gate}=0.035 m, width w_{gate}=0.18 m and cross-sectional area of A_{gate}=0.0063 m), into an inclined channel connected to a horizontal bottomed tank at its downstream end. The turbidity current flow within the inclined channel is laterally confined (confined turbidity current), while after its exit from the inclined channel into the tank, the turbidity current is free to expand in all directions (unconfined turbidity current). The proposed laboratory scale configuration, serves as a simplified experimental analog of natural, hyperpycnal turbidity currents that initially travel, laterally confined within a subaqueous canyon with a sloped bottom and then, after they exit from the downstream end of the canyon, they spread out laterally unconstrained in the horizontal or mild sloped bottom of the receiving basin (sea, lake or reservoir).

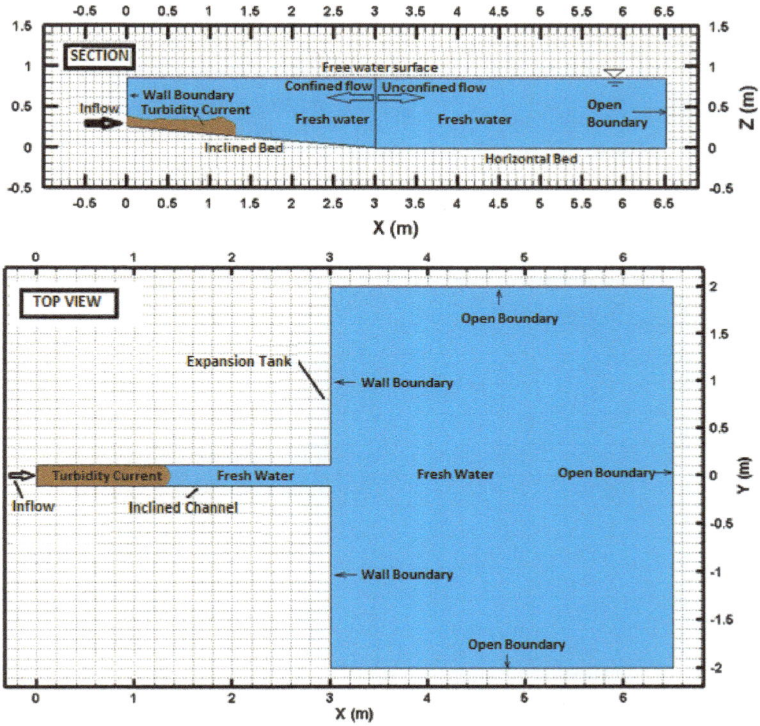

Fig. 5. General configuration of investigated physical problem.

The symbols and the explanations of the controlling flow parameters that are investigated (varied) in each series of numerical experiments, in the present application, are summarized in Table 1. Each series of numerical experiments consists of four runs. The initial conditions of these runs are summarized in Table 2. The numerical experiments in each case are named accordingly to the varied parameter and its corresponding value in each numerical experiment. It should also be mentioned that in each series of numerical experiments (A, B, C and D) there is a common Reference Numerical Experiment (R.N.E.), which for ease purposes in the analysis of the results is named as S5, C25, D150 and R0 for Series A, B, C and D, respectively. Finally, it should be mentioned that the inflow discharge of the incoming fresh water - suspended sediment mixtures is continuous and steady, with a value of Q_{inflow}=0.0078 m^3/sec (that corresponds in an inflow velocity value of V_{inflow}=1.24 m/sec) in all series of numerical experiments.

Series of Numerical Experiments	Investigated/Varied Parameter	Symbol	Explanation
A	Channel slope	S_i	"inclination angle of channel bed"
B	Suspended sediment concentration	C_i	"Initial, volumetric concentration of suspended sediment particles in the inflow mixture"
C	Grain diameter	D_i	"Grain diameter of suspended sediment particles in the inflow mixture"
D	Bed roughness	R_i	"Roughness of channel and tank bed expressed as equivalent roughness of uniformly distributed suspended sediment particles of specific grain size"

Table 1. Investigated, fundamental controlling parameters, of turbidity current flows.

Series of Numerical Experiments	Numerical Experiment Name	S_i (°)	C_i (% vol.)	D_i (μm)	R_i (μm)
A	S1	1	25	150	0
A	S5	5	25	150	0
A	S10	10	25	150	0
A	S20	20	25	150	0
B	C5	5	5	150	0
B	C10	5	10	150	0
B	C15	5	15	150	0
B	C25	5	25	150	0
C	D80	5	25	80	0
C	D100	5	25	100	0
C	D120	5	25	120	0
C	D150	5	25	150	0
D	R0	5	25	150	0
D	R80	5	25	150	80
D	R235	5	25	150	235
D	R500	5	25	150	500

Table 2. Numerical experiments initial conditions.

As it can be seen from Table 2, the overall channel slope values that were used in the numerical experiments are 1°, 5°, 10° and 20°. Therefore, in order to conduct the numerical experiments of Series A, four different computational geometries, one for each channel slope, where constructed. In all the rest series of numerical experiments (B, C and D) the geometry with 5° channel slope is used. The computational geometry, computational mesh and boundary conditions, which were used in the numerical simulations are illustrated in Figure 6, for the case of the 5° channel slope that also corresponds to the R.N.E.. For the rest configurations these characteristics are similar and therefore are not illustrated schematically. In the computational geometries, that correspond to a channel slope of 1°, 5°, 10° and 20°, the computational meshes consist of a total number of cells (control volumes) of 51770, 58398,

69370 and 93487, respectively. In all situations the same mesh characteristics (cell size, cell clustering growth rates, cell layers in the vicinity of the bottom boundary etc.) are used. As it can be seen from Figure 6, the largest part of the computational mesh consists of tetrahedral cells of varying size, that are locally refined at regions where more computational accuracy is required (regions of sudden changes in the calculated quantities), such as the region in the vicinity of the inflow boundary and the downstream end of the inclined channel.

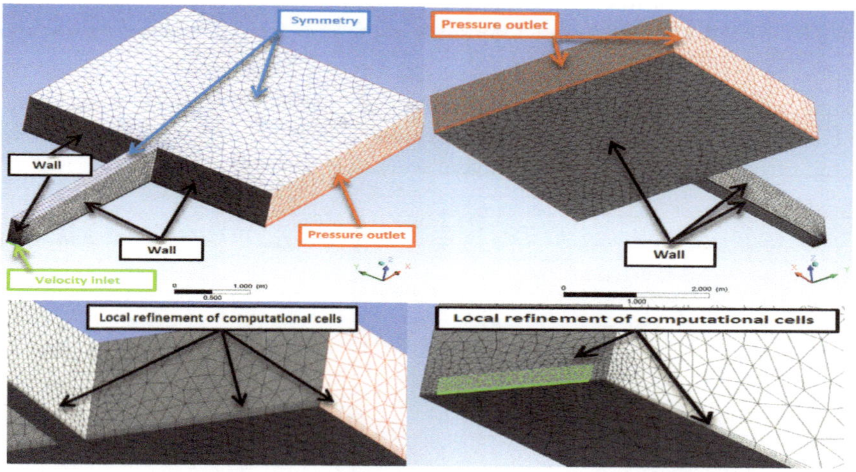

Fig. 6. Computational geometry, mesh and boundary conditions (R.N.E.).

In order to ensure that the numerical solutions presented are mesh independent, sensitivity tests were performed with computational meshes of different total cell number. Figure 7 illustrates indicatively, the flow front position of the generated turbidity current with respect

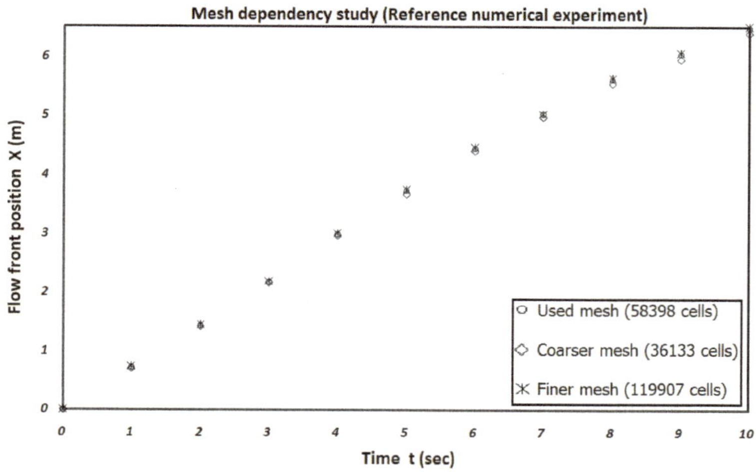

Fig. 7. Mesh size sensitivity test, on turbidity current front position with respect to time (R.N.E.).

to time, for three different computational meshes in the case of the R.N.E. The first computational mesh is the one used in the simulations (58,398 computational cells), the second one is a coarser mesh (36,133 computational cells) and the third one is a finer mesh (119,907 computational cells). It is obvious (Figure 7) that the resulting curves in each case show a good degree of convergence and therefore the solution can be considered to be mesh independent. In more detail, comparing the results of the coarser mesh with the corresponding results of the finer mesh, it is concluded that increasing the total number of cells by a factor of 3.33, the average differences of the flow front position values with respect to time is only 1.85%.

In order to visualize the flow of the generated turbidity currents in the simulations of the present investigation, the three-dimensional time evolution of the interface, between the generated turbidity current and the ambient water, for the case of the R.N.E., is depicted in Figure 8.

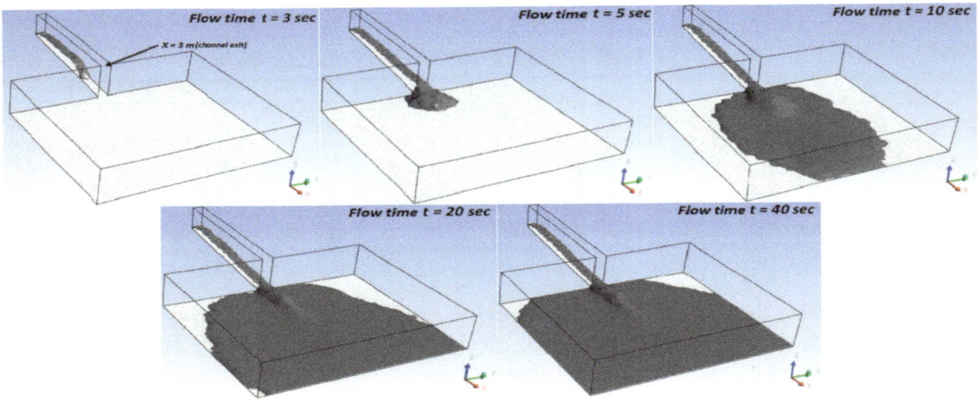

Fig. 8. Three dimensional time evolution of the interface (grey surface) between the generated turbidity current and the ambient water (R.N.E.).

It is obvious that 3sec after the inflow of the fresh water - suspended sediment mixture the generated turbidity current, flows within the inclined channel (laterally confined part of the flow). At t=5 sec, the turbidity current head has already exited from the downstream part of the channel and has started to expand radically in the horizontal bed of the tank (unconfined part of the flow). At t=10 sec the head of the current has just reached the downstream open boundary of the computational domain, while at t=20 sec it has already exited the computational domain from the downstream as well as the left and right side open boundaries. Finally, at t=40sec the evolution of the current within the computational domain has already reached a quasi-steady state.

In Figure 9 the resulting curves of the generated turbidity current flow front position with respect to time, are illustrated in dimensionless form, for the numerical experiments of Series A, B, C and D respectively. For comparison purposes, the varied parameter in each series of numerical experiments is normalized with its lowest value ($S_1=1°$ for Series A, $C_5=5\%$ by vol. for Series B, $D_{80}=80$ μm for Series C and $R_{80}=80$ μm for Series D), the horizontal distance X of the flow front from the inflow gate is normalized with the width of the inclined channel (b=0.22 m) and the flow time t is normalized with the time needed for the slowest of the generated turbidity currents (in each series of numerical experiments) to exit from the downstream boundary of the expansion tank ($t_{exit(S1)}$=12 sec for Series A, $t_{exit(C5)}$=18 sec for Series B, $t_{exit(D80)}$=10 sec for Series C and $t_{exit(R500)}$=13 sec for Series D). As it can be seen the resulting curves in each numerical experiment of Series A, B, C and D have a similar form, consisting of three distinct parts. In the first part the flow front velocity of the generated turbidity currents is almost constant, in the second part a gradual acceleration of the flow front is observed and in the third part, a gradual deceleration of the flow front is evident. In the first part, the flow of the generated turbidity currents is primarily controlled by their initial momentum, due to the continuous and steady discharge of the inflowing fresh water - suspended sediment mixtures from the inflow gate and therefore the flow front velocity remains constant. In the second part that the flow front of the currents has already traveled almost half the length of the inclined channel, the observed acceleration of the front is due to the continuous increase of the gravitational force effect, since the currents are flowing over an inclined bottom boundary. At the third part, the turbidity currents have already entered the expansion tank and their flow is laterally unconfined, expanding radically in all directions over the horizontal bottom boundary of the tank. Therefore, the continuous reduction of their excess density, due to the continuous entrainment of the ambient water of the tank and the consequent gradual deposition of suspended sediment particles, causes a gradual dissipation and deceleration of the generated turbidity current flows.

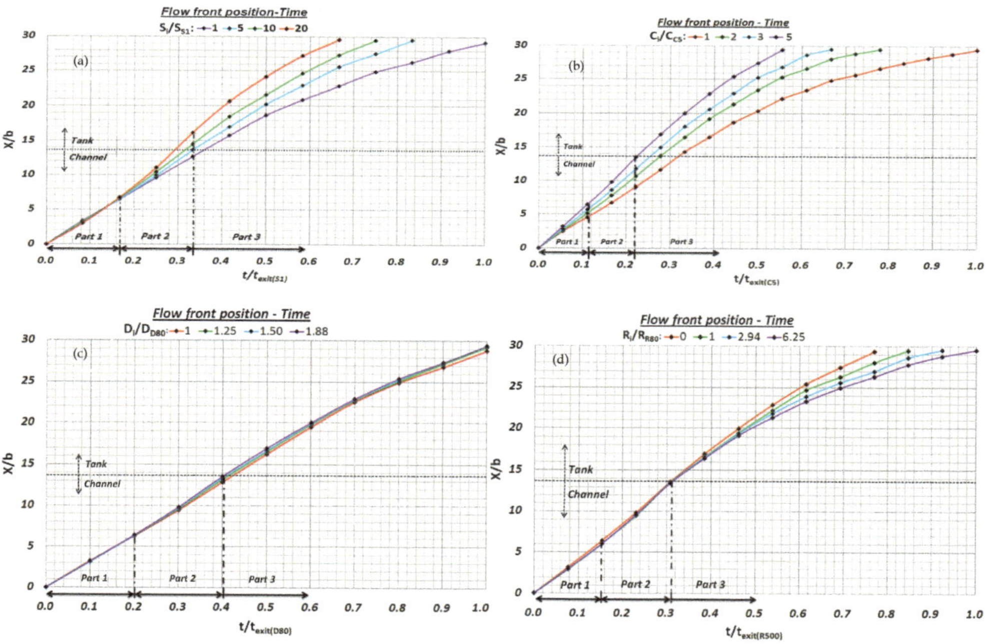

Fig. 9. Dimensionless flow front position with respect to dimensionless time for, (a) Series A numerical experiments, (b) Series B numerical experiments, (c) Series C numerical experiments and (d) Series D numerical experiments.

In order to investigate the exact quantitative effect of the varied controlling parameters to the depositional characteristics of the generated turbidity currents, in Figure 10 the suspended sediment volumetric concentration at the bottom boundary of the domain is plotted against the horizontal distance from the inflow gate, for flow time t=40 sec, where the flow of generated turbidity currents within the computational domain have reached a quasi-steady state. For comparison purposes, the varied parameter in each series of numerical experiments is normalized with its lowest value (S_1=1° for Series A, C_5=5% by vol. for Series B, D_{80}=80 μm for Series C and R_{80}=80 μm for Series D), the horizontal distance X of the flow front from the inflow gate is normalized with the width of the inclined channel (b=0.22 m) and the suspended sediment volume fraction at the bottom boundary C_{vol} is normalized with the values, C_{S1}=0.25 for Series A, C_{C5}=0.05 for Series B, C_{D80}=0.25 for Series C and C_{R80}=0.25 for Series D numerical experiments. It should be mentioned that the suspended sediment volume fraction values at the bottom boundary of the computational domain are taken at the central axis of the generated flows. It is obvious that in each case the resulting curves have a similar form. In more detail, in the laterally constrained and sloped bottom part of the flow (channel), the suspended sediment volumetric concentration at the bottom boundary increases rapidly with the longitudinal distance from the inflow gate, up

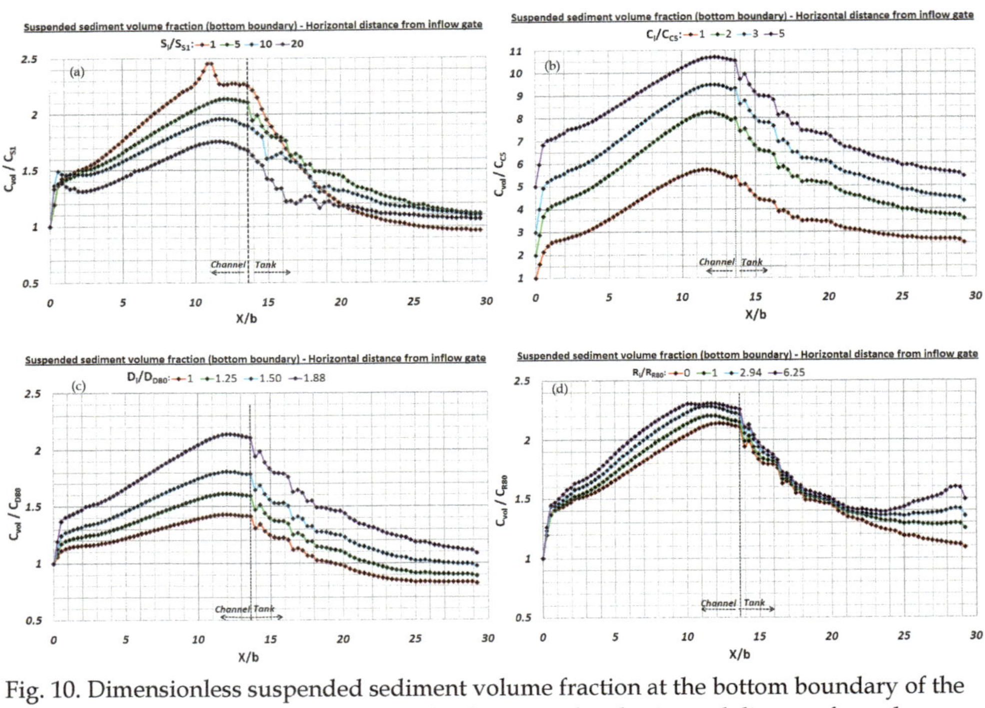

Fig. 10. Dimensionless suspended sediment volume fraction at the bottom boundary of the computational domain, with respect to the dimensionless horizontal distance from the inflow gate for, (a) Series A numerical experiments, (b) Series B numerical experiments, (c) Series C numerical experiments and (d) Series D numerical experiments, 40 sec after the beginning of the inflow of the fresh water – suspended sediment mixtures.

to a distance of X/b=1 and then follows a less rapid increase up to a maximum value, at a distance of X/b=11 that is close to the downstream end of the channel (X/b=13.6). The rapid increase of the volume fraction values in the vicinity of the inflow point (X/b=0 to 1) is probably due to the local increase of the volume fraction value of the inflowing mixtures, as a result of the resistance that is exerted from the ambient fluid. In the unconstrained and horizontal bottom part of the flow (tank), the suspended sediment volumetric concentration follows an irregular decrease with respect to the longitudinal distance, reaching an almost constant minimum value in the vicinity of the downstream boundary of the computational domain. The fact that in all cases, the maximum value of the suspended sediment volumetric concentration at the bottom boundary of the computational domain is found near the downstream end of the channel, is probably due to the sudden reduction in the velocity of the generated turbidity currents which is a result of the flow transition from the laterally constrained (channel) to the unconstrained (tank) part of the computational domain. This sudden drop of velocity is reasonable to cause intense particle deposition just upstream of the channel exit to the expansion tank.

Examining separately the effect of each controlling parameter in the flow front advance velocity and in the deposit density of the current at the bottom boundary, it can be concluded that in general, the increase of the channel slope causes an increase in the flow front advance velocity and a reduction in the deposit density. The increase of the initial suspended sediment concentration causes an increase both in the flow front advance velocity and in the deposit density. The increase of the suspended sediment grain diameter causes an increase both in the flow front advance velocity and in the deposit density. Finally, the increase of the bed roughness causes a reduction in the flow front advance velocity and an increase in the deposit density.

From the presentation and the analysis of the above results so far, it is evident that the investigated controlling parameters affect with a different way and in a comparably different degree the dynamic and depositional characteristics of turbidity currents. Therefore in order to compare the relative percentage effect of the varied controlling parameters in the main flow characteristics of the generated turbidity currents, Figure 11 presents diagrams of the relative percentage change of the maximum flow front advance velocity (Figure 11 a) and the maximum value of suspended sediment volume fraction at the bottom boundary (Figure 11 b), in relation to the relative percentage change of the varied controlling parameters. It should be mentioned that for comparison purposes, the relative percentage change in each case is calculated using absolute differences. It should also be mentioned that in the case of Series D numerical experiments, only the experiments R80, R235, and R500 are taken into consideration, where the values of the bottom boundary roughness are greater than zero. It is obvious that the variation of the initial suspended sediment concentration as well as the suspended sediment grain diameter have the biggest effect in the flow of the generated turbidity currents. This can be probably attributed to the direct effect of the proposed controlling parameters in the main driving force of turbidity currents, which is the excess density of the current in relation to the ambient water density. The variation of the bed roughness has the smallest effect, while the variation of the channel slope causes a moderate effect in the turbidity current flows, in relation to the rest controlling parameters.

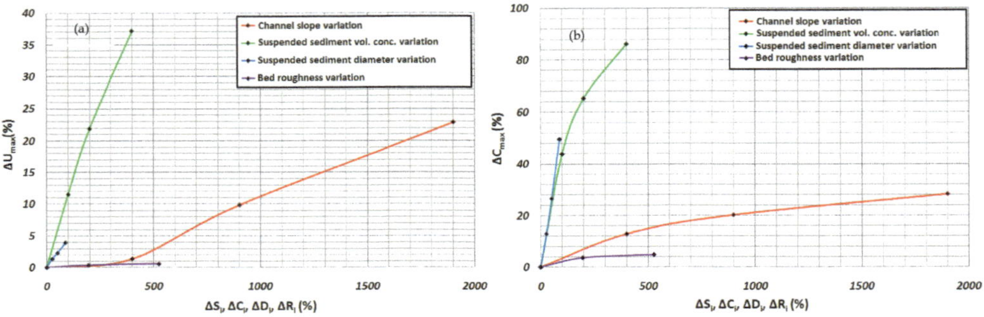

Fig. 11. Dependence of the maximum flow front velocity (a) and the maximum suspended sediment volume fraction at the bottom boundary of the computational domain (b), from the investigated flow controlling parameters (expressed as relative percentage change).

Summarizing, the overall results of the present numerical investigation contribute considerably in the understanding of the dependence of the suspended sediment transport and deposition mechanism, from fundamental flow controlling parameters of natural, continuous, high-density turbidity currents that are usually formed during flood discharges at river outflows in the sea, lakes and reservoirs.

4.2 Field scale application

The present subsection of the chapter describes a field scale application of the proposed numerical model that aims to identify the dynamic behavior and the main flow characteristics of turbidity currents, which are potentially formed at Evros river mouth (Georgoulas et al., 2009). More specifically, the numerical model is applied at Evros river mouth (Greece), in order to simulate the river's suspended sediment transport and dispersal into the North Aegean Sea, in the case of a flood discharge, where the suspended sediment concentration of the river water is considerably high, in order for a turbidity current to be formed. It should be mentioned that the effects of the bed morphology and the Coriolis force are taken into account, during the numerical simulation.

The flow examined, is a flood discharge of Evros River that is based in existing flood data. It is treated numerically as a multiphase flow, with saline water (North Aegean Sea) being the primary phase and fresh water and suspended sediment particles (Evros River) being the secondary phases. For the present numerical application, two separate phases of suspended sediment particles are assumed. The first consists of fine sand particles of 0.235 mm diameter and the second consists of very fine sand particles of 0.069mm diameter.

The geometry used in the numerical simulation, has been extracted from a 3D digitized bottom relief model of the North Aegean Sea, which is illustrated in Figure 12 below. The region, denoted by number 1 in the digitized bottom relief model, is the wider region of Evros river mouth, while the hatched area in the sub-region denoted by number 2, is the part that was selected for the numerical simulation.

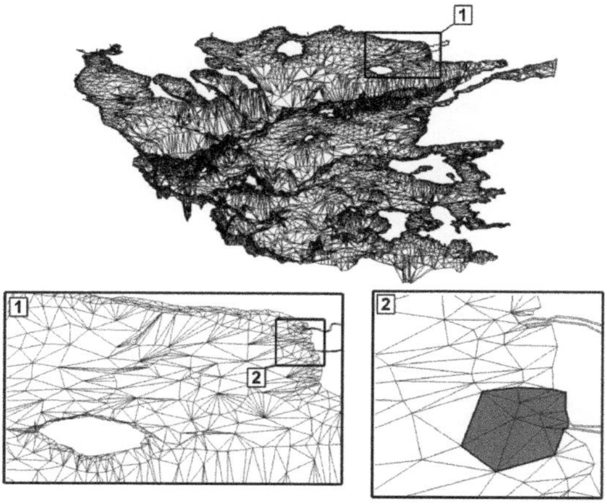

Fig. 12. 3D digitized bottom relief model of the North Aegean Sea, region of Evros River Outflow (region 1) and region of numerical simulation (hatched region within Sub-region 2).

The resulting numerical geometry, the computational mesh and the boundary conditions that are used for the simulation of the present paper, are illustrated in Figure 13. The computational mesh consists of a total number of 56,720 hexahedral cells (Figure 13 a). For the inflow (Figure 13 b) a "velocity inlet" boundary condition is applied. For the shoreline east and west of Evros River outflow (Figure 13 c), a "wall" boundary condition is applied. For the open sea boundary of the flow field (Figure 13 d) a "pressure outlet" boundary condition is applied. For the bottom boundary a "wall" boundary condition is used (Figure 13 e), while for the free water surface of the ambient sea water a "symmetry" boundary condition is used (Figure 13 f). At this point, it should be mentioned that during the simulation, the entire flow field is rotated with respect to the Z-axis (vertical axis), with a rotational speed that corresponds to the rotational frequency (Coriolis parameter) of the North Aegean Sea region (latitude $\varphi = 40°N$), in order for the Coriolis force effect to be taken into account.

The initial conditions applied for the simulation, are summarized in Table 3. As it can be seen the numerical simulation was conducted with relatively simplified conditions, in order to investigate more clearly the effects of the bottom topography and the Coriolis force, in the results of the studied flow. Therefore, the inflow discharge from Evros River was assumed to be steady, and the potential effects of other parallel to the shore, subaqueous and/ or surface currents were not taken into consideration. The value of the Evros river discharge, which was indicatively used in the simulation (4,555 m^3/sec), corresponds in a big flood discharge of the proposed river. Since there are not any available data for the maximum sediment discharge at Evros River mouth, the initial suspended sediment concentration that was used for the numerical simulation, was estimated, taking into consideration the sediment discharge measurements, upstream of the river mouth, in the work of (Gergov, 1996). The initial condition for the ambient water was assumed to be constant for the entire flow field, with a salinity of 38.6 ppt and a temperature of 15.0 °C that correspond to a density value of 1028.75 (kg/m3). The width of Evros River at the inflow position into the North Aegean Sea was assumed to be 450 m, while the corresponding depth was assumed to be 2.5 m.

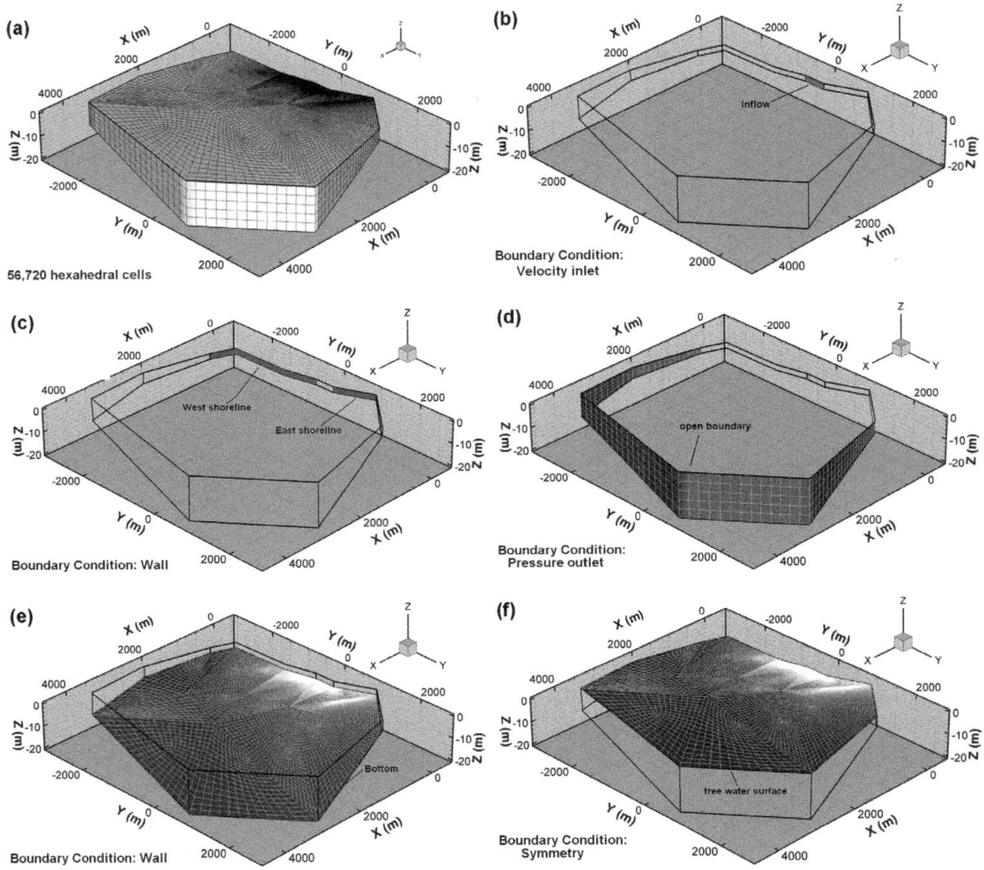

Fig. 13. Numerical simulation geometry, computational mesh and boundary conditions.

Inflow discharge (m^3/sec)	4,555
Fine sand particle diameter (mm)	0.235
Very fine sand particle diameter (mm)	0.069
Fine sand concentration (vol. %)	0.7%
Very fine sand concentration (vol. %)	0.7%
Saline water density (kg/m^3)	1028,75
Fresh water density (kg/m^3)	998,2
Sand particle density (kg/m^3)	2,650
Coriolis parameter f (Hz)	9.35x10-5

Table 3. Numerical simulation initial conditions.

The 3D, time evolution of the root and dispersal of the suspended sediment - fresh water mixture that enters the flow field, is illustrated in Figure 14. It is observed that a part of the suspended sediment – fresh water mixture, that enters the flow field, is spreading at the free water surface, while most of the mixture plunges underneath the free sea water surface, forming a turbidity current, which continuous to flow along the bottom. This subaqueous current, initially spreads out radically in all directions, with an irregular shape, due to the mixing of the current with the ambient fluid. After the second hour of real flow, a major part of the turbidity current clearly deviates to the left of the main inflow axis, due to the general slope gradient of the bottom relief in this direction. However, it is obvious that a smaller part of the turbidity current deviates to the right of the main inflow axis, due to the effect of the Coriolis force, encountering even negative slope gradients. The effect of the Coriolis force is also evident in the part of the suspended sediment – fresh water mixture that is spreading in the free surface. The surface that is enclosed by the dashed line in Figure 14, constitutes the so called "plunge region" of the generated turbidity current. After this line the mixture plunges underneath the free water surface and continuous to flow along the bottom of the receiving basin.

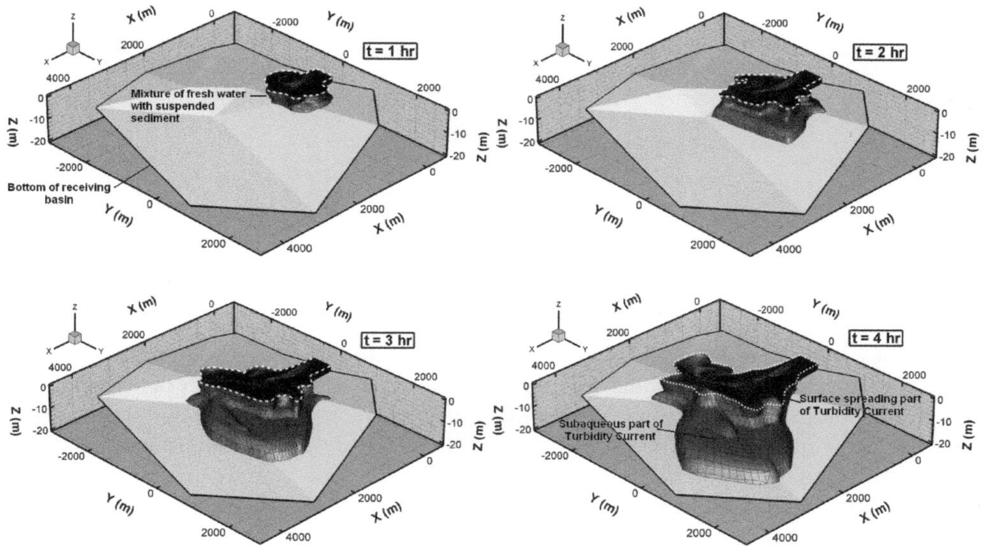

Fig. 14. Time evolution of the root and dispersal of the suspended sediment – fresh water mixture that enters the flow field.

In order to investigate the different concentration distributions, for the two separate suspended sediment phases, Figure 16 is plotted, which illustrates the corresponding volume fraction contours for each of these phases, in two (perpendicular to each other) vertical sections within the flow field, 4 hours (real flow time) after the beginning of the simulation. It is reminded that at the river inflow into the sea, the initial concentration values for the fine sand and the very fine sand phases are 0.7% by volume (table 3). The position of each of the proposed sections, within the flow field, is shown in Figure 15.

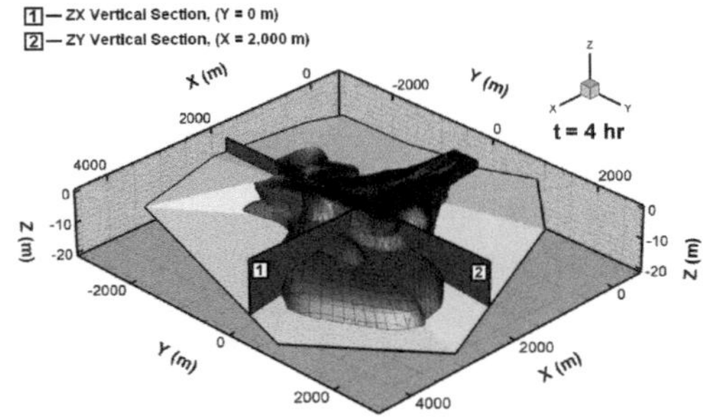

Fig. 15. Position of vertical sections within the flow field (Flow time, t = 4hr).

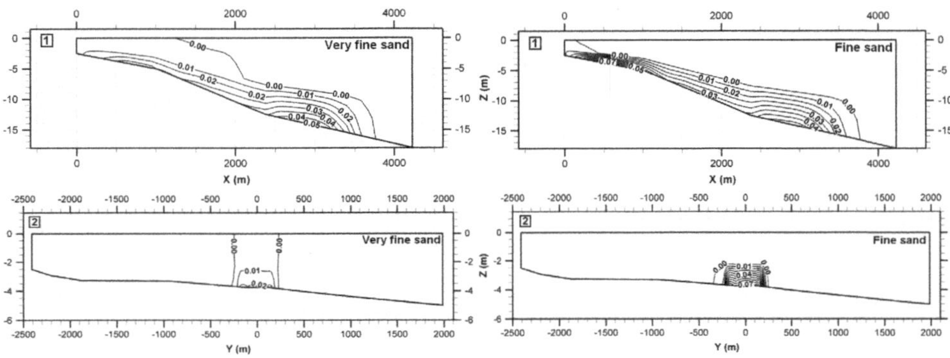

Fig. 16. Volume fractions of fine sand and very fine sand phases in vertical sections 1 and 2 (Flow time, t = 4 hr).

As it is observed, both of the suspended sediment phases have similar concentration distributions, with increasing values from the interface with the ambient fluid to the bottom of the receiving basin. However, the top and more dilute concentration layer of the very fine sand phase (volume fraction values 0.00 to 0.01), occupies more height within the flow field, than in the case of the fine sand phase. This fact is obviously due to the different settling velocities between the particles of the very fine sand and the fine sand. It is also evident that the part of the suspended sediment - fresh water mixture that spreads out along the free surface of the receiving basin, contains mainly fresh water and very fine sand particles. This can be seen in vertical section 1, where the plunge point of the very fine sand phase is traced in a considerably longer distance, from the inflow point (~1250 m), than in the case of the fine sand phase (~150 m).

Summarizing, from the results of the present subsection it is concluded that for the assumed initial conditions, the inflow of the suspended sediment - fresh water mixture from Evros River into the North Aegean Sea Basin, forms a turbidity current, which plunges to the bottom of the receiving basin. The effects of the bottom topography as well as the Coriolis force, in the root and dispersal of the studied turbidity current, are highly evident. Finally, different responses, of the different types of suspended sediment particles (fine sand and very fine sand) within the flow field, are also evident.

5. Conclusion

In the present chapter an "uncommon", CFD-based, three-dimensional, multiphase numerical approach for the simulation and study of the hydrodynamic and depositional characteristics of turbidity currents is presented. The main advantages of the proposed multiphase numerical approach, in relation to previous quasi single-phase approaches are the following:

- Separate velocity fields are calculated for each phase (ambient water, inflow/carrier water and various classes of suspended sediment), since the laws for the conservation of mass and momentum are accordingly modified, in order to be satisfied by each phase individually.
- The use of the RNG k-ε turbulence model significantly increases the applicability of the proposed numerical approach, as it can also account for turbidity current flows with low Reynolds numbers.
- The total number of flow phases that can be simulated is only limited by the available memory of the computational resources. Hence, it can also be used for the simulation of polydisperse turbidity currents that contain many classes of suspended sediment particles, which are more close to natural turbidity current flows.
- It can handle a wide range of particulate loading, and therefore is capable for the simulation of both dilute and dense turbidity current flows.
- It is based on the finite volume method, and therefore it can be applied in situations with complex geometries, like in the case of turbidity currents that are formed at natural, water basin beds (sea, lakes, reservoirs), where morphological anomalies are usually present.

The method is tested against published laboratory data that are available in the literature and the comparison of the numerical and experimental results indicate that its predictions are realistic and reliable.

The overall results of the laboratory scale application contribute considerably in the understanding of the dependence of the suspended sediment transport and deposition mechanism, from fundamental flow controlling parameters of natural, continuous, high-density turbidity currents that are usually formed during flood discharges at river outflows. It is found that the investigated controlling parameters affect with a different way and in a comparably different degree the dynamic and depositional characteristics of turbidity currents. In more detail, the increase of the channel slope causes an increase in the flow front advance velocity and a reduction in the deposit density. The increase of the initial suspended sediment concentration causes an increase both in the flow front advance velocity and in the deposit density. The increase of the suspended sediment grain diameter causes an increase both in the flow front advance velocity and in the deposit density, while the increase of the bed roughness causes a reduction in the flow front advance velocity and an increase in the deposit density. Finally, from the comparison of the relative percentage effect of each of the examined controlling parameters in the main hydrodynamic and depositional characteristics of the generated turbidity currents, it is found that the variation of the initial suspended sediment concentration as well as the suspended sediment grain diameter have the biggest effect, the variation of the bed roughness has the smallest effect, while the variation of the channel slope causes a moderate effect in the turbidity current flows, in relation to the rest controlling parameters.

From the field scale application it can be concluded that for the assumed initial conditions, the inflow of the suspended sediment - fresh water mixture from Evros River into the North Aegean Sea Basin, forms a turbidity current, which plunges to the bottom of the receiving basin. The effects of the bottom topography as well as the Coriolis force, in the root and dispersal of the studied turbidity current, are highly evident. More specifically, a big part of the turbidity current deviates to the left of the main inflow axis, due to the general slope gradient of the bottom relief in this direction. Another, smaller part of the turbidity current deviates to the right of the main inflow axis, due to the Coriolis force effect, encountering even negative slope gradients. The different responses, of the different types of suspended sediment particles (fine sand and very fine sand) within the flow field, are also characteristic. In more detail, in the concentration distributions, the upper, more dilute layer of the very fine sand concentration occupies more height within the flow field, than in the case of the fine sand case. This fact is obviously due to the different settling velocities, between the particles of the very fine sand and the fine sand.

Finally, the overall results presented in the present chapter indicate, the capabilities of the proposed numerical approach, as a possible and suitable tool for the further investigation of the hydrodynamic behavior of turbidity currents. It is shown that the proposed numerical approach can constitute a quite attractive alternative to laboratory experiments and field measurements since it allows the identification and the continuous monitoring of a wide range of flow parameters, with a relatively high accuracy.

6. References

ANSYS FLUENT Documentation, (2010). *User's Guide,* Version 13.0

Baas, J.H.; Van Kesteren, W. & Postma, G, (2004). Deposits of Depletive, Quasi-Steady High Density Turbidity Currents: a Flume Analogue of Bed Geometry, Structure and Texture. *Sedimentology,* Vol. 51, No. 5, (October 2004), pp. 1053-1089, ISSN 1365-3091

Blanchette, F.; Strauss, M.; Meiburg, E.; Kneller, B. & Glinsky, M.E. (2005). High-Resolution Numerical Simulations of Resuspending Gravity Currents: Conditions for Self-Sustainment. *Journal of Geophysical Research,* Vol.110, No.C12022, (December 2005), 15 pp., ISSN 2156–2202

Bombardelli, F.A.; Cantero, M.I.; Buscaglia, G.C. & Garcia, M.H. (2004). Comparative Study of Convergence of CFD Comercial Codes when Simulating Dense Underflows, In: *Mecánica Computacional,* G. Buscaglia, E. Dari & O. Zamonsky (Eds.), 1187 -1199, Vol. 23, Bariloche, Argentina

Britter, R.E. & Linden P.F. (1980). The Motion of the Front of a Gravity Current Travelling down an Incline. *Journal of Fluid Mechanics,* Vol.99, (April 2006), pp. 531-543, ISSN 1469-7645.

Cantero, M.; Garcia, M.; Buscaglia, G.; Bombardelli, F. & Dari, E. (2003). Multidimensional CFD Simulation of a Discontinuous Density Current, *Proceedings of the XXX IAHR International Congress,* pp. 405-412, ISBN 9602435968-9789602435960, Thessaloniki, Greece, August, 2003

Cantero, M.I.; Balachandar, S. & Garcia, M.H. (2008a). An Eulerian-Eulerian Model for Gravity Currents Driven by Inertial Paricles. *International Journal of Multiphase Flow,* Vol.34, No.5, (May 2008), pp. 484-501, ISSN 0301-9322

Cantero, M.I.; Balachandar, S.; Garcia, M.H. & Bock, D. (2008b). Turbulent Structures in Planar Gravity Currents and Their Influence on the Flow Dynamics. *Journal of Geophysical Research,* Vol.113, No. C08018, (August 2008), 22 pp., ISSN 2156–2202

Cebeci, T. & Bradshaw, P. (1977). Momentum transfer in boundary layers. *Hemisphere Publishing Corporation,* New York.

Farrell, G.J. & Stefan, H.G. (1986). Buoyancy induced plunging flows into reservoirs and coastal regions. Tech.Rep. No.241, At. Anthony Falls Hydr. Lab., Univ. of Minnesota, Minneapolis

Garcia, M. & Parker, G. (1989). Experiments on Hydraulic Jumps in Turbidity Currents near a Canyon-Fan Transition. *Science,* Vol.245, No.4916, (July 1989), ISSN 1095-9203

Garcia, M. (1994). Depositional Turbidity Currents Laden with Poorly Sorted Sediment. *Journal of Hydraulic Engineering,* Vol.120, No.11, (November 1994): pp. 1240-1263, ISSN 1943-7900

Georgoulas, A.; Tzanakis, T.; Angelidis, P.; Panagiotidis, T. & Kotsovinos, N. (2009) Numerical Simulation of Suspended Sediment Transport and Dispersal from Evros River into the North Aegean Sea, by the Mechanism of Turbidity Currents, *Proceedings of the 11th International Conference on Environmental Science and Technology,* Vol. A, pp. 343-350, Chania, Crete, Greece, September 3-5, 2009.

Georgoulas, A.; Panagiotidis, T.; Angelidis, P. & Kotsovinos, N. (2010). 3D Numerical Modelling of Turbidity Currents. *Environmental Fluid Mechanics,* Vol. 10, No.6, (August 2010), pp. 603-635, ISSN 1573-1510

Georgoulas, A. (2010). Study of the density currents at river outflows, due to suspened sediment particles. PhD Thesis, Democritus University of Thrace (November 2010), Department of Civil Engineering, Xanthi, Greece. Available online: http://thesis.ekt.gr/thesisBookReader/id/23438#page/8/mode/2up

Gergov, G. (1996). Suspended sediment load of Bulgarian rivers. *GeoJournal,* Vol.40, No.4, (December 1996), pp. 387-396, ISSN 1572-9893

Gladstone, C.; Phillips, J.C. & Sparks, R.S.J. (1998). Experiments on Bidisperse, Constant-Volume Gravity Currents: Propagation and Sediment Deposition. *Sedimentology,* Vol.45, No.5, (October 1998), pp. 833-843, ISSN 1365-3091

Hartel, C.; Meiburg, E. & Necker, F. (2000). Analysis and Direct Numerical Simulation of the Flow at a Gravity-Current Head: Part 1. Flow Topology and Front Speed for Slip and No-Slip Boundaries. *Journal of Fluid Mechanics,* Vol.418, (September 2000), pp. 189–212, ISSN 1469-7645

Heimsund, S.; Hansen, E.W.M. & Nemec, W. (2002). Computational 3D Fluid-Dynamics Model for Sediment Transport, Erosion and Deposition by Turbidity Currents, In: *Abstracts, International Association of Sedimentologists,* M. Knoper & B. Cairncross, (Ed.), 151-152, 16th International Sedimentological Congress, Rand Afrikaans Univ., Johannesburg

Huang, H.; Imran, J. & Pirmez, C. (2005). Numerical Model of Turbidity Currents with a Deforming Bottom Boundary. *Journal of Hydraulic Engineering,* Vol.131, No.4, (April 2005), pp. 283-293, ISSN 0733-9429

Imran, J.; Kassem, A. & Khan, S.M. (2004). Three-Dimensional Modelling of Density Current. I. Flow in Straight Confined and Unconfined Channels. *Journal of Hydraulic Research*, Vol.42, No.6, (August 2010), pp. 578–590, ISSN 1814-2079

Janbu, N. E.; Nemec, W.; Kirman, E. & Özaksoy, V. (2009) Facies Anatomy of a Sand-Rich Channelized Turbiditic System: The Eocene Kusuri Formation in the Sinop Basin, North-Central Turkey, In: *Sedimentary Processes, Environments and Basins: A Tribute to Peter Friend*, G. Nichols, E. Williams & C. Paola (Eds.), 457-517, Blackwell Publishing Ltd., ISBN 9781405179225, Oxford, UK

Kassem, A. & Imran, J. (2001). Simulation of Turbid Underflows Generated by the Plunging of a River. *Geology*, Vol.29, No.7, (July 2001), pp.655–658, ISSN 1943-2682

Kneller, B.C.; Bennett, S.J. & McCaffrey, W.D. (1997). Velocity and Turbulence Structure of Density Currents and Internal Solitary Waves: Potential Sediment Transport and the Formation of Wave Ripples in Deep Water. Sedimentary Geology, Vol.112, No.3-4, (September 1997), pp. 235-250, ISSN 0037-0738

Lavelli, A.; Boillat, J.L. & De Cesare, G. (2002). Numerical 3D Modelling of the Vertical Mass Exchange Induced by Turbidity Currents in Lake Lugano (Switzerland), *Proceedings of 5th International Conference on Hydro-Science and Engineering (ICHE-2002)*, Reference: LCH-CONF-2002-012 Note: [355]

Lovell, J.P.B. (1971). Control of Slope on Deposition from Small-Scale Turbidity Currents: Experimental Results and Possible Geological Significance. *Sedimentology*, Vol.17, No.1-2, (June 2006) pp. 81-88, ISSN 1365-3091

Mehdizadeh, A.; Firoozabadi, B. & Farhanieh, B. (2008). Numerical Simulation of Turbidity Current Using $\overline{v^2} - f$ Turbulence Model. *Journal of Applied Fluid Mechanics*, Vol.1, No. 2, pp. 45-55, ISSN 1735-3645

Mulder, T. & Alexander, J. (2001). The Physical Character of Subaqueous Sedimentary Density Flows and their Deposits. *Sedimentology*, Vol.48, No.2, (December 2001), pp. 269-299, ISSN 1365-3091

Necker, F.; Hartel, C.; Kleiser, L. & Meinburg, E. (2002). High-Resolution Simulations of Particle Driven Gravity Currents. *International Journal of Multiphase Flow*, Vol.28, No.2, (February 2002), pp. 279-300, ISSN: 0301-9322

Ooi, S.K.; Constantinescu, G. & Weber, L.J. (2007). 2D Large-Eddy Simulation of Lock-Exchange Gravity Current Flows at High Grashof Numbers. *Journal of Hydraulic Engineering*, Vol.133, No.9, (September 2007), pp. 1037-1047, ISSN 0733-9429

Posamentier, H.W. & Kolla, V. (2003). Seismic Geomorphology and Stratigraphy of Depositional Elements in Deep-Water Settings. *Journal of Sedimentary Research*, Vol.73, No.3, (May 2003), pp. 367-388, ISSN 1527-1404

Saller, A.; Werner, K.; Sugiaman, F.; Cebastiant, A.; May, R.; Glenn, D. & Barker, C. (2008). Characteristics of Pleistocene Deep-Water Fan Lobes and their Application to an Upper Miocene Reservoir Model, Offshore East Kalimantan, Indonesia. *AAPG Bulletin*, Vol.92, No.7, (July 2008), pp. 919-949, ISSN 0149-1423

Simpson, J.E. & Britter, R.E. (1979). The Dynamics of the Head of a Gravity Current Advancing over a Horizontal Surface. *Journal of Fluid Mechanics*, Vol. 94, No.3, (April 2006), pp. 477-495, ISSN 1469-7645

Singh, J. (2008). Simulation of Suspension Gravity Currents with Different Initial Aspect Ratio and Layout of Turbidity Fence. *Applied Mathematical Modelling*, Vol.32, No.11, (November 2008), pp. 2329–2346, ISSN 0307-904X

4

Numerical Simulation of the Unsteady Shock Interaction of Blunt Body Flows

Leonid Bazyma[1], Vasyl Rashkovan[2] and Vladimir Golovanevskiy[3]
[1]*National Aerospace University "Kharkov Aviation Institute"*
[2]*National Polytechnic Institute*
[3]*Western Australian School of Mines, Curtin University*
[1]*Ukraine*
[2]*Mexico*
[3]*Australia*

1. Introduction

Supersonic and hypersonic space vehicles are extremely sensitive to aerodynamic resistance. The combination of the two main rocket operation factors, low altitude and high velocity, produces considerable heat flows in the stagnation region of the nose. For this reason, passive heat-transfer analysis under such conditions is very important for understanding and solving rocket operation problems.

The possibility of use of energy supply as the method of the overall control of the airflow is defined in the experimental and theoretical research (Adegren et al., 2001, 2005; Bazyma & Rashkovan, 2005; Tret'yakov et al., 1994, 1996). As an example, reduction of head resistance of a supersonic aircraft can be achieved with the introduction of energy into the contrary incoming flow. On the other hand, supply of energy may be used to minimize negative consequences of the shock-wave interaction when the streamlining of the aerodynamic configurations of the compound form occurs. For example, oblique shock waves, which are distributed from the bow part of an airplane or a rocket, can interact with a bow shock wave of any part of the fuselage construction (tail unit, suspension, hood, diffuser, etc.). In certain cases the shock-wave interaction can result in significant negative and even catastrophic consequences for the aircraft.

The use of a laser to supply energy has been experimentally shown to be a good approach for both general and local flow control. Experimental research (Tret'yakov et al., 1994, 1996) shows that an extensive region of energy supply is realized in the supersonic flow when a powerful optical pulsating discharge is applied, with a thermal wake developing behind the area where the energy is supplied.

A cone or hemisphere in the thermal wake, located from 1.0 to 4.0 diameters distance from the focal plane of irradiation from a CO_2 laser, results in a reduction of aerodynamic drag of over a factor of two when a 100-kHz pulse frequency is applied (Tret'yakov et al., 1996). The thermal wake becomes continuous for 10–100-kHz radiation pulses (Tret'yakov et al., 1996).

Theoretical modeling results of the influence of a heat-release pulsating source on the supersonic flow around a hemisphere are presented by Guvernyuk & Samoilov (1997). The explicit total-variation-diminishing (TVD) method in Chakravarthy's) formulae (Chakravarthy & Osher, 1985; Chakravarthy, 1986) is applied in these calculations (Guvernyuk & Samoilov, 1997). In the case of M =3, γ=1.4 and a constant deposited energy per unit mass, the aerodynamic load upon the body exhibited a decrease. A pulse repetition rate corresponded to a minimum drag was determined and it was concluded that the use of the pulsating energy supply might be more effective than a constant energy source.

Other results (Georgievskii & Levin, 1988) show that pulse repetition rate, power supplied to the flow and the area into which this energy is supplied all greatly influence both the pressure distribution and the model surface and its flow regimes.

The energy supply parameters such as intensity and heat spot configuration influence the flow re-formation significantly as their combination determines the possibility of either airflow choking in the source (i.e. with the separated wave) or the choke-free flow. This considerably affects the spot properties behind the flow and consequently the stagnation pressure and configuration resistance.

This work presents the results of numerical simulation of the flow around a hemisphere at both the symmetric and asymmetric energy supply into the flow, when the energy supply is realized at 90^0 angle to the velocity vector of the incoming supersonic airflow.

The two types of the heat spot form considered were: the axis-symmetric spot (i.e. thin disk in the two-dimensional space) and the heat spot of the ellipsoidal form (in the three-dimensional space) with its main axis perpendicular to the symmetry axis. The results of the numerical simulation correspond well with the experimental data (Adegren et al., 2001).

2. Problem definition

Let us consider the energy supply from above the sphere and along the airflow at Mach number M_∞ = 3.45 and ratio of specific heats γ = 1.4 in the incoming supersonic gas flow, i.e. the same as in (Adegren et al., 2001). The energy supply scheme is illustrated in Figure 1. Assume that at the time t =0, a pulsating power supply source is initiated in front of the sphere.

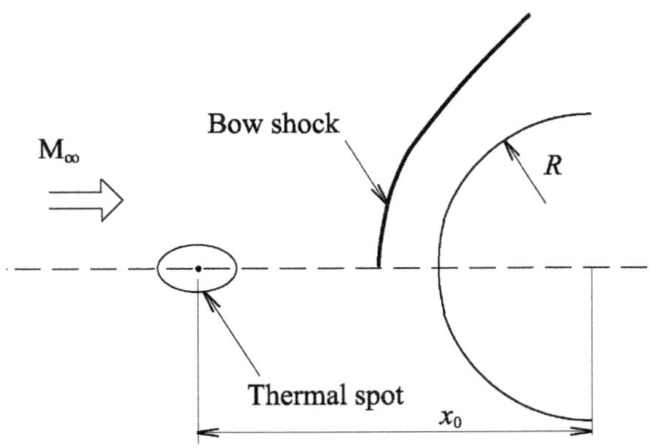

Fig. 1. Schematic of energy supply.

The equations of gas dynamics in the cylindrical coordinates, in contrast to (Bazyma & Rashkovan, 2005) are in the form that includes the azimuthal component

$$\frac{\partial \rho r}{\partial t}+\frac{\partial \rho u r}{\partial x}+\frac{\partial \rho \upsilon r}{\partial r}+\frac{\partial \rho \omega}{\partial \varphi}=0\,; \tag{1}$$

$$\frac{\partial \rho u r}{\partial t}+\frac{\partial (p+\rho u^2) r}{\partial x}+\frac{\partial \rho u \upsilon r}{\partial r}+\frac{\partial \rho u \omega}{\partial \varphi}=0\,; \tag{2}$$

$$\frac{\partial \rho \upsilon r}{\partial t}+\frac{\partial \rho u \upsilon r}{\partial x}+\frac{\partial (p+\rho \upsilon^2) r}{\partial r}+\frac{\partial \rho \upsilon \omega}{\partial \varphi}=p\,; \tag{3}$$

$$\frac{\partial \rho \omega r}{\partial t}+\frac{\partial \rho u \omega r}{\partial x}+\frac{\partial (p+\rho \upsilon^2) r}{\partial r}+\frac{\partial (p+\rho \omega^2)}{\partial \varphi}=0\,; \tag{4}$$

$$\frac{\partial \rho e r}{\partial t}+\frac{\partial \rho u(e+p/\rho) r}{\partial x}+\frac{\partial \rho \upsilon(e+p/\rho) r}{\partial r}+\frac{\partial \rho \omega(e+p/\rho)}{\partial \varphi}=\rho q r\,, \tag{5}$$

where p - pressure; ρ - density; u, υ, ω - components of the velocity vector on x, r and φ respectively; e – total energy of the mass unit of the gas; q – energy supplied to the mass of the gas by the external source; t – time. The system is completed with the perfect gas equation:

$$p=(\gamma-1)\rho e\,. \tag{6}$$

The energy supply is prescribed the same as in (Bazyma & Rashkovan, 2005; Guvernyuk & Samoilov, 1997):

$$q=W(x,r)\sum_{n=1}^{\infty}\frac{1}{f}\delta\left(t-\frac{n}{f}\right), \tag{7}$$

where δ – is the Dirak's impulse function; f – pulse repetition rate; W – average mass density of the energy supply. Here, in contrast to (Bazyma & Rashkovan, 2005; Guvernyuk & Samoilov, 1997), W was taken in the form that permits modeling different shapes of heat spot at the asymmetric energy supply:

$$W=W_0\left(\frac{p_\infty}{\rho_\infty}\right)^{3/2}\frac{1}{R}\exp\left(-\frac{k_1\left(r\cos\varphi\right)^2+k_2(x-x_0)^2+k_3\left(r\sin\varphi\right)^2}{L^2}\right), \tag{8}$$

where W_0, k_1, k_2, k_3 and L are constants defining deposited energy density and thermal spot shape.

3. Numerical method

Similar to (Bazyma & Rashkovan, 2005), the solution of the system of equations (1-4) was conducted using the Godunov's method (Godunov, 1976). The 110×60 and the 110×60×32 grids were used for the axis-symmetric and the three-dimensional problems respectivelly. In both cases the grid was designed with densening of the nodes near the body or in the areas of energy supply (i.e. the incoming flow disturbance areas).

The calculations were performed using the same finite difference scheme of the first-order approximation as that used by (Godunov, 1976). Computational grid used in calculations is shown in Figure 2.

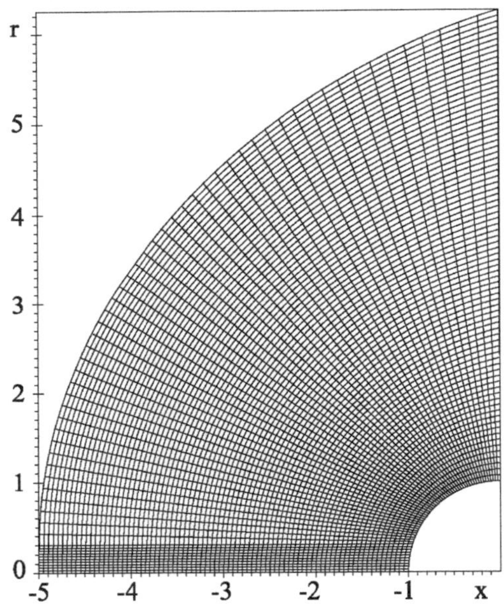

Fig. 2. Computational grid.

The necessary and sufficient condition for stability is that the permissible spacing in time τ must satisfy the inequality

$$\frac{\tau}{\tau_x}+\frac{\tau}{\tau_y}\leq 1 \tag{9}$$

resulting from a stability study of Godunov's difference scheme realized on the system of non-stationary acoustic equations on a uniform rectangular (or parallelogram) grid. Here, $\tau_x$and τ_y are the time spacing of the one-dimensional scheme. Physically τ_x and τ_y are mean time intervals, in which waves appearing at the break decomposition on the cell boundary reach the neighboring boundaries:

$$\tau_x = \frac{\Delta x}{\max(u+a, a-u)}, \tau_y = \frac{\Delta y}{\max(\upsilon+a, a-\upsilon)}, \tag{10}$$

where a is the velocity of the sound.

The stability condition so given is extended to the quasi-linear equations of gas dynamics. Calculations show that this condition (9) provides the necessary stability. Nevertheless, this condition is usually used with right-hand side less than one.

In our work, the time step was chosen from cell to cell according to the stability condition as follows:

$$\tau_{n-\frac{1}{2},m-\frac{1}{2}} = \left(\frac{\tau_x \tau_r}{\tau_x + \tau_r}\right)_{n-\frac{1}{2},m-\frac{1}{2}}, \bar{\tau} = \min_{n,m} \tau_{n-\frac{1}{2},m-\frac{1}{2}}, \tag{11}$$

In "k" space its value with respect to time is calculated using the spacing "k+1" as:

$$\tau^{k+1} = K\bar{\tau}^k, \tag{12}$$

where K is a safety factor similar in meaning to the Courant number.

In order to use dimensionless values, we make use of the following equalities:

$$\begin{gathered} r = \bar{r}R, \quad x = \bar{x}R, \quad t = \bar{t}R / a_\infty, \quad f = \bar{f}a_\infty / R, \quad a = \bar{a}a_\infty, \\ u = \bar{u}a_\infty, \quad \upsilon = \bar{\upsilon}a_\infty, \quad \omega = \bar{\omega}a_\infty \quad \rho = \bar{\rho}\rho_\infty, \quad p = \bar{p}\rho_\infty a_\infty^2, \quad W = \bar{W}a_\infty^3 / R, \end{gathered} \tag{13}$$

where a_∞ the velocity of the sound of the incident flow. In the following text, we have omitted bars above the dimensionless values $r, x, t, f, a, u, \upsilon, \rho, p, W$.

The boundary conditions are defined similar to those in (Guvernyuk & Samoilov, 1997). On the surface of the body and along the symmetry axis, solid-wall inviscid boundary conditions were applied. Along the external inflow boundary, undisturbed freestream conditions were utilized. On the downstream outflow boundary, extrapolation of the flow quantities from the adjacent internal boundary was performed.

The initial data in calculations without energy deposition corresponded to the dimensionless parameters of the incident stream:

$$p = p_\infty = 1/\gamma, \; \rho = \rho_\infty = 1, \; u = u_\infty = \mathrm{M}_\infty, \; \upsilon = 0, \; \omega = 0, \tag{14}$$

where γ is the ratio of specific heats.

Subsequent solutions of the flow field about a hemisphere with energy deposition were initialized using the flow field about a hemisphere solution without energy deposition. The solutions were advanced in time until the average flow conditions were stabilized.

Preliminary supersonic flow calculations around the hemisphere were conducted in (Bazyma & Rashkovan, 2005) to confirm the adequacy of our numerical scheme (see Fig. 3). Data reported in (Guvernyuk & Samoilov, 1997), where the explicit TVD method of

Chakravarthy's formulation (Chakravarthy & Osher, 1985; Chakravarthy, 1986) was used, showed good correspondence in all the observed flow regimes.

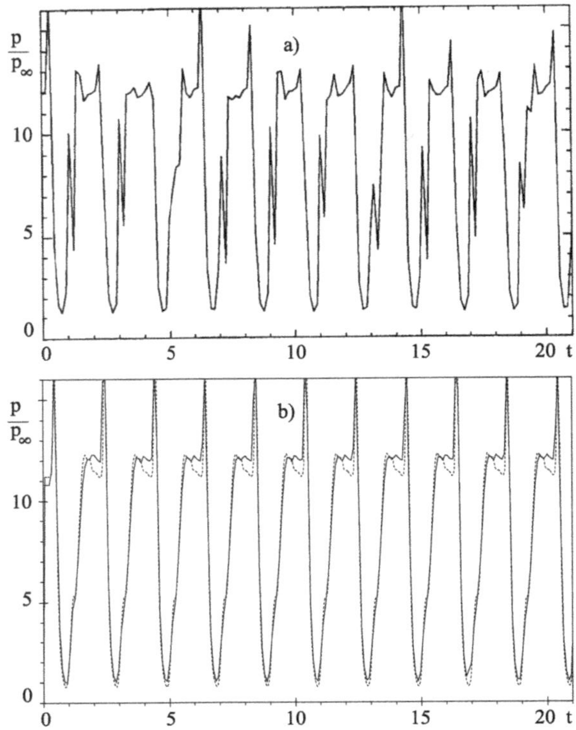

Fig. 3. Pressure at the hemisphere stagnation point versus dimensionless time at the pulse repetition rate f=0.5 (spherical heat spot; W_0 =20, x_0 = -3.5, L = 0.5): a) results of (Guvernyuk & Samoilov, 1997); b) results of (Bazyma & Rashkovan, 2005) (—— – 110×60 grid; - - - - – 219×119 grid).

Shown in Figure 3 is the dependence of the pressure at the stagnation point while flowing around the hemisphere (for M_∞ = 3, γ = 1.4) on the dimensionless time at pulse repetition rate f = 0.5. Results are shown both from (Guvernyuk & Samoilov, 1997) (a) and from the our previous work (b). As can be seen from Figure 3, the main pulsation parameters (i.e. period and amplitude) and their character obtained in our work correspond well with the data of (Guvernyuk & Samoilov, 1997). However, while the resolution of the numerical scheme used in (Guvernyuk & Samoilov, 1997) is somewhat higher the scheme used in our previous work (Bazyma & Rashkovan, 2005) allows simulation of the quasi-stationary pulsation process.

Shown in Figure 4a is the flow visualization near the hemisphere under the influence of the pulsating thermal source (spherical heat spot; W_0 =20, x_0 = -3.5, L = 0.5). These results were obtained in (Guvernyuk & Samoilov, 1997), and compared with the analogous data of the (Bazyma & Rashkovan, 2005) (Figure 4b). As can be seen in Figure 4, bow shock wave standoff distance, formed recirculation zones and the flow in general reported in (Guvernyuk & Samoilov, 1997) and derived in our work are in good agreement.

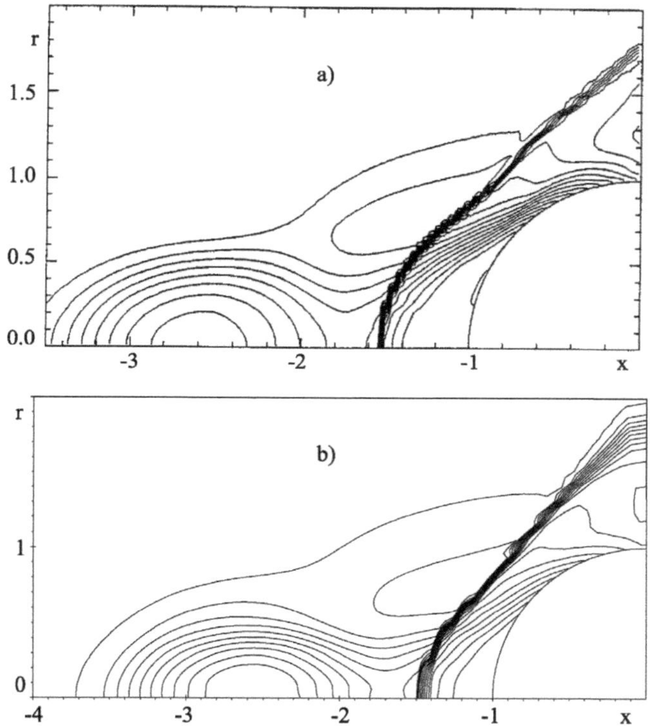

Fig. 4. Mach number isolines while flowing around the hemisphere with supersonic gas flow at pulse repetition rate *f*=2 (*t*=11.2): a) results of (Guvernyuk & Samoilov, 1997); b) results of (Bazyma & Rashkovan, 2005), 110×60 grid.

The grid resolution study was also conducted, with test calculations for hemisphere and cavity hemisphere carried out using the 219×119 grid. The 219×119 grid was obtained through twice the 110×60 grid spacing reduction. Minimum surface cell spacing values were 0.024 for the 110×60 grid (reduced to the sphere radius) and 0.012 for the 219×119 grid.

A comparison of the results derived with the use of the 110×60 (continuous line) and 219×119 grids can be seen in Figure 3b above. As the resolution of the 219×119 (dotted line) grid is higher than that of the 110×60 grid, the solution derived with the use of the 219×119 grid marked out some peculiarities of the pressure change on the compression stage. These peculiarities correspond to the solution in (Guvernyuk & Samoilov, 1997) as well; however they were not seen through the grid applied. It is worth noting that solution difference obtained with the 110×60 and 219×119 grids is rather small for the hemisphere with energy deposition.

The details of the numerical scheme, along with test examples, are given in (Godunov, 1976) and (Bazyma & Kholyavko, 1996; Bazyma & Rashkovan, 2006).

4. Results and discussion

When calculating the three-dimensional problem, the energy supply modeled was at 90^0 angle to the velocity vector of the incoming flow. The two types of the heat spot form considered were: the axis-symmetric spot (i.e. thin disk) and the heat spot of the ellipsoidal form with the main axis perpendicular to the symmetry axis.

The dimensionless parameters of the undisturbed contrary flow are assumed as the initial data in calculations without power supply.

The Table 1 shows the list of operational parameters for the facility used in the wind tunnel experiments (Adegren et al., 2001). This tunnel is a basic blowdown tunnel with an exhaust into atmospheric pressure.

Mach Number	3.45
Operating Stagnation Pressure	1.4 MPa
Typical Stagnation Temperature	290 K
Mass Flow Rate	9.8 Kg/s
Total Run Time	1.8 minutes
Test Area Cross Section	15 cm x 15 cm
Test Area Length	30 cm

Table 1. Operating Parameters for the Rutgers Mach 3.45 Supersonic Wind Tunnel.

4.1 General calculation procedure

The energy supplied to the mass unit of gas is prescribed in the form

$$q = \gamma^{-3/2} W_0 \exp\left(-\frac{k_1 (r\cos\varphi)^2 + k_2 (x - x_0)^2 + k_3 (r\sin\varphi)^2}{L^2} \right) t^* \delta(t - t^*) . \qquad (15)$$

Here x_0 = -3.0 (the energy is supplied at the distance of one diameter of the sphere from its surface, (Adegren et al., 2001)), $t^* = f^{-1}$ at the pulse frequency f=0.00068 (that corresponds to the frequency 10 Hz, (Adegren et al., 2001)). The form of the heat spot is defined by the parameters L, k_1, k_2, k_3. The value L =0.01 is fixed in all the calculations; the values k_1, k_2, k_3 are being variated, that permitted to obtain the heat spot dimensions characteristic for the experiment (Adegren et al., 2001) (the volume of the heat spot is evaluated approximately from 1 to 3 mm^3; The sphere radius in the experiment is 12.75 mm). Thus, for example, at $k_1 = k_2 = k_3$ one can obtain a spherical heat spot.

The parameter W_0 is being varied in the range 0.19 – 1.75 that in total with the selection of values k_1, k_2, k_3 provides the change of the energy density in the impulse that is provided in the experiment (Adegren et al., 2001) (13 mJ/pulse/1±0.5mm^3, 127 mJ/pulse/1.3±0.7mm^3, and 258 mJ/pulse/3±1mm^3).

The bow shock stand-off distance for the undisturbed model at Mach 3.45 was calculated and compared to the Lobb (Lobb, 1964) approximation to Van Dyke's (Van Dyke, 2003) shock stand-off model. The model predicted the stand-off distances within 3 percent of the calculated distances. The model for shock stand-off distance is given as

$$\Delta = 0.41D\frac{(\gamma-1)\ \ M_\infty^2+2}{(\gamma+1)\ \ M_\infty^2}, \tag{16}$$

where, Δ is the stand-off distance, D is the sphere diameter, and $M_\infty = 3.45$ is the freestream Mach number.

4.2 Symmetric energy supply

Naturally, obtaining the heat spot form similar to that used in the experiment (Adegren et al., 2001) for the two-dimensional case is impossible. However, with the heat spot size small compared to the size of the streamlined body (i.e. sphere) and low values of the energy density supplied to the incoming flow obtaining some similarity can be expected. For the spheroidal heat spot form (i.e. its axis of rotation coincides with the axis of rotation of the streamlined body), the character of the pressure change in the critical point of the body is sufficiently close (at the stage of compression and the first phase of expansion) to that obtained in the experiment (Adegren et al., 2001) for the value of energy supplied in the pulse in the order of 13mJ/1mm³ (see Figure 5).

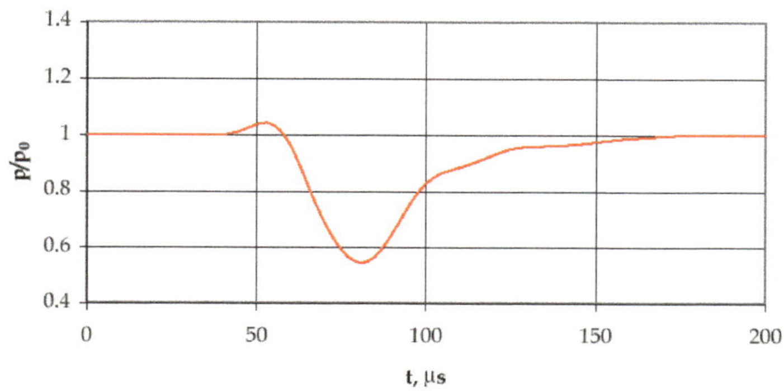

Fig. 5. Pressure variation at the critical point of the sphere versus time: $W_0 = 0.19$, $f = 0.00068$, ellipsoid heat point; $k_1 = k_3 = 0.016$, $k_2 = 0.39$.

Increasing the spot size and the energy supply density will not allow to obtain the conditions fully adequate to those of the physical experiment. At the same time, varying these two parameters (i.e. the size and configuration of the spot on the one hand and the energy supply density on the other hand) allow the character of the pressure change in the critical point of the body adequate to the experiment by the higher values of the energy supply density (127 mJ/pulse/1.3±0.7mm³, and 258 mJ/pulse/3±1mm³) to be obtained.

Character of pressure variation in the critical point of the sphere versus time for other values of k_1, k_2, and k_3 providing smaller volume of the heat spot (i.e. by the factor of 2 approximately) but twice the energy supply density (i.e. W_0= 0.38) is shown in Figure 6, curve 1. It is worth noting that the character of the pressure variation in both cases is similar. Curves 2 and 3 in Figure 6, with a considerable difference in pressure amplitude at the compression stage, are obtained for the form of the heat spot that is the flattened axis-symmetric disk with its radius comparable with the radius of the sphere.

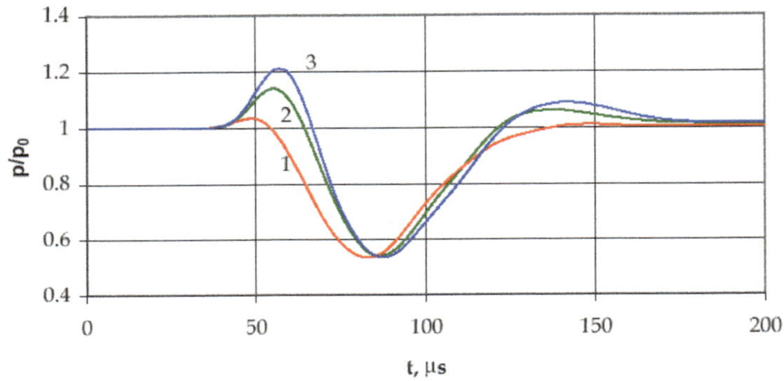

Fig. 6. Pressure variation at the critical point of the sphere versus time: 1 - W_0 = 0.38, f = 0.00068, ellipsoid heat point; $k_1 = k_3$ = 0.016, k_2 =0.56; 2 - W_0 = 0.38, f = 0.00068, disk heat point; $k_1 = k_3$ = 0.00045, k_2 =0.089;3 - W_0 = 0.38, f = 0.00068, disk heat point; $k_1 = k_3$ = 0.00023, k_2 =0.082.

As in the experiment (Adegren et al., 2001; refer figure 20) the time history of recorded pressure at centerline location of sphere surface for the three energy levels shows a common behavior comprised of an initial pressure rise, expansion, compression and transient decay. The expansion, compression and transient decay are similar to the ideal gas Euler simulations of (Georgievskii & Levin, 1993) for the interaction of a thermal spot with a sphere at Mach 3. The expansion lowers the surface pressure at the centerline by 40%. However, the initial compression phase observed in our research and in the experiment (Adegren et al., 2001) was not noted by (Georgievskii & Levin, 1993).

The interaction of the thermal spot, with the bow shock (Figure 7-9, t = 40-90 microseconds) causes a blooming of the bow shock (due to the lens effect of the thermal spot). This behavior is consistent with the simulations of Georgievski and Levin (Georgievskii & Levin, 1993).

Fig. 7. Time histories of pressure isolines:
$W_0 = 0.38$, $f = 0.00068$, disk heat point; $k_1 = k_3 = 0.00045$, $k_2 = 0.089$.

Fig. 8. Time histories of pressure isolines:
$W_0 = 0.38$, $f = 0.00068$, disk heat point; $k_1 = k_3 = 0.00045$, $k_2 = 0.089$.

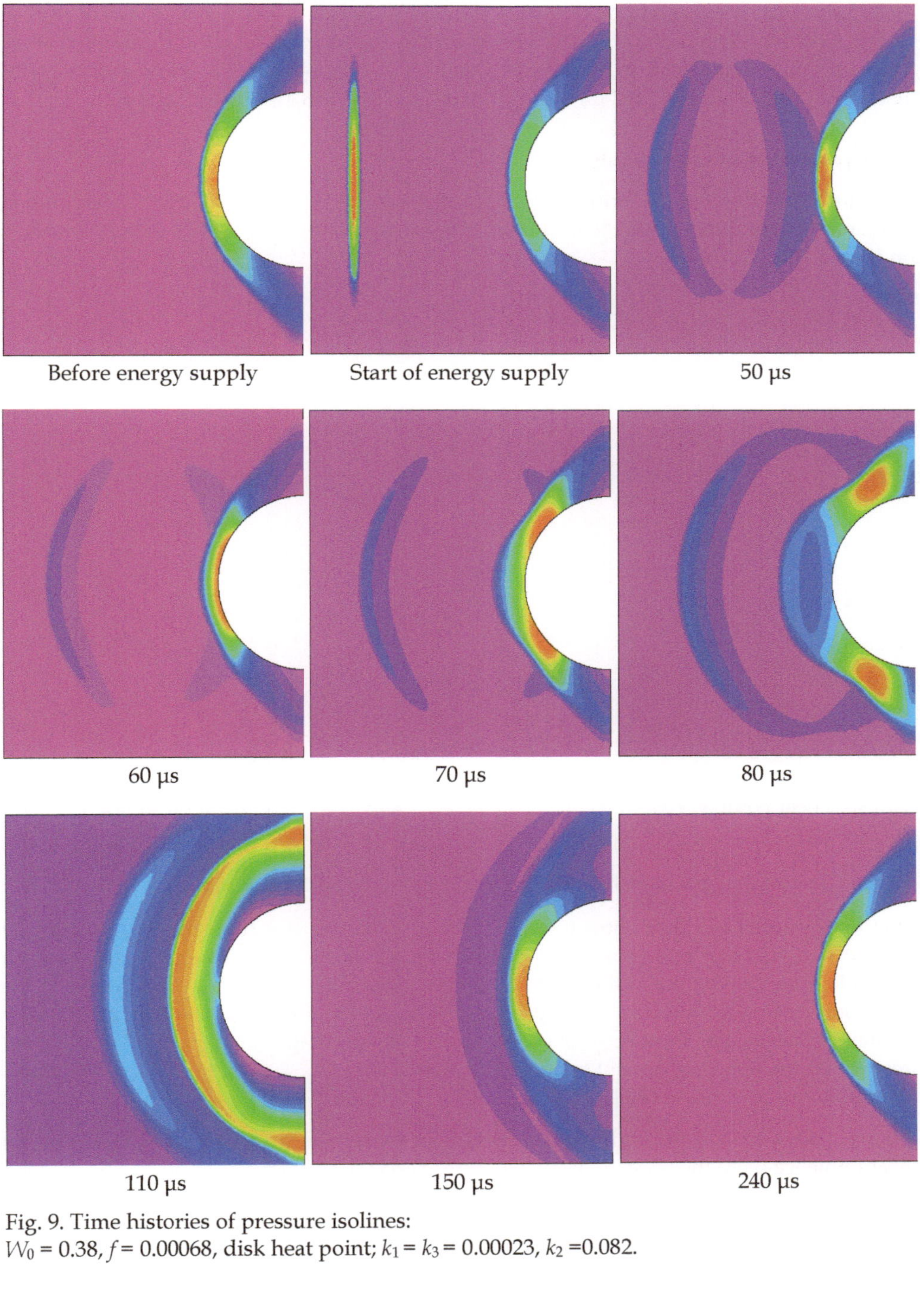

Fig. 9. Time histories of pressure isolines:
$W_0 = 0.38$, $f = 0.00068$, disk heat point; $k_1 = k_3 = 0.00023$, $k_2 = 0.082$.

Naturally, with these conditions satisfied the energy supply density was even less than in the experiment (Adegren et al., 2001). However, the character of the pressure variation at the stage of compression and expansion was similar to that obtained in the experiment for energy supply densities of 127 mJ/pulse/1.3±0.7mm³ and 258 mJ/pulse/3±1mm³.

4.3 Asymmetric energy supply

Pressure variation at the critical point of the sphere, relevant to the corresponding pressure obtained before the heat influence, versus time after the energy supply is shown in Figure 10.

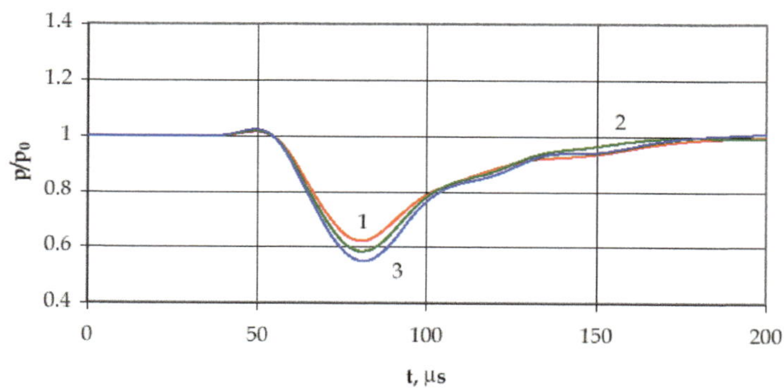

Fig. 10. Pressure variation at the critical point of the sphere versus time:
1 - $W_0 = 0.5, f = 0.00068$, ellipsoid heat point, $k_2 = k_3 = 0.67, k_1 = 0.007$; 2 - $W_0 = 1.5, f = 0.00068$, ellipsoid heat point, $k_2 = k_3 = 1, k_1 = 0.06$; 3 - $W_0 = 1.75, f = 0.00068$, ellipsoid heat point, $k_2 = k_3 = 1, k_1 = 0.17$.

Figure 11 shows the pressure history for the asymmetric energy deposition.

The Mach number in the contrary flow is M=3.45. The distance from the heat spot to the sphere was equal to one diameter of the sphere. The zone of the energy supply was an ellipsoid with the volume in the order of 1mm³ with its rotational axis perpendicular to the incoming flow velocity vector. The process of interaction of the heat track and the sphere contained two stages: short stage of compression and long stage of expansion. The obtained results are in good agreement with the experimental data (Adegren et al., 2001), both quantitatively and qualitatively.

Figure 12 shows the fields of the equal pressure on the sphere surface at 500 μs time after the heat influence (here the pressure is related to the contrary flow pressure). As can be seen, by this time the symmetrical vortex areas on the surface of the sphere still exist. It is worthwhile to point out that the flow rotation velocity is sufficiently high; with its maximum value reaching 0.3 of the contrary flow sound velocity (see Figure 13).

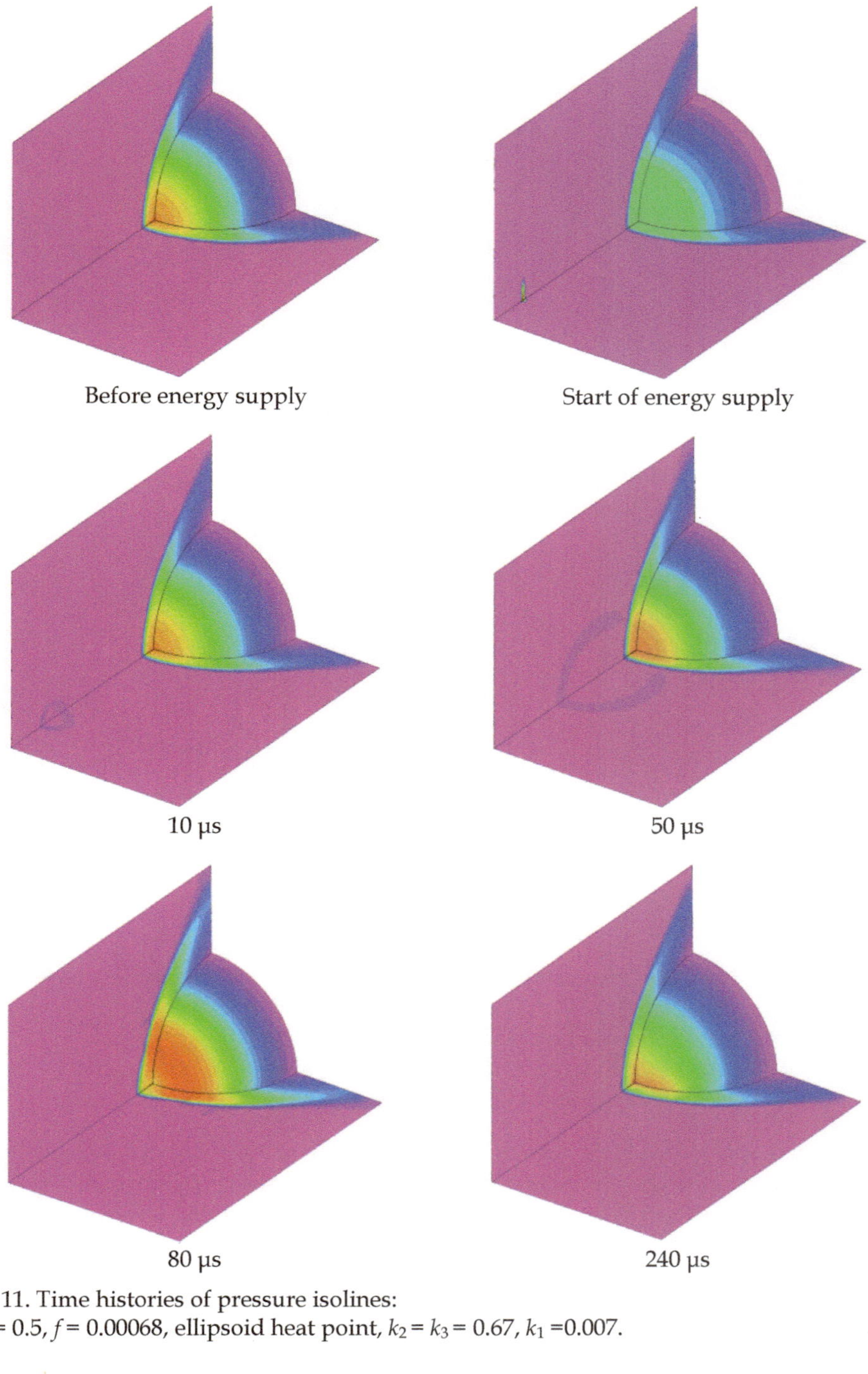

Fig. 11. Time histories of pressure isolines:
$W_0 = 0.5$, $f = 0.00068$, ellipsoid heat point, $k_2 = k_3 = 0.67$, $k_1 = 0.007$.

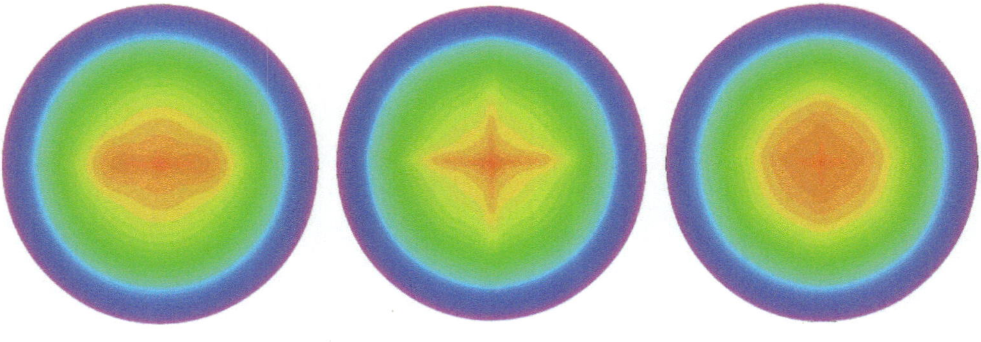

a b c

Fig. 12. Pressure on the sphere surface at 500 μs time after the energy supply:
a - $W_0 = 0.5$, $f = 0.00068$, ellipsoid heat point, $k_2 = k_3 = 0.67$, $k_1 = 0.007$;
b - $W_0 = 1.5$, $f = 0.00068$, ellipsoid heat point, $k_2 = k_3 = 1$, $k_1 = 0.06$;
c - $W_0 = 1.75$, $f = 0.00068$, ellipsoid heat point, $k_2 = k_3 = 1$, $k_1 = 0.17$.

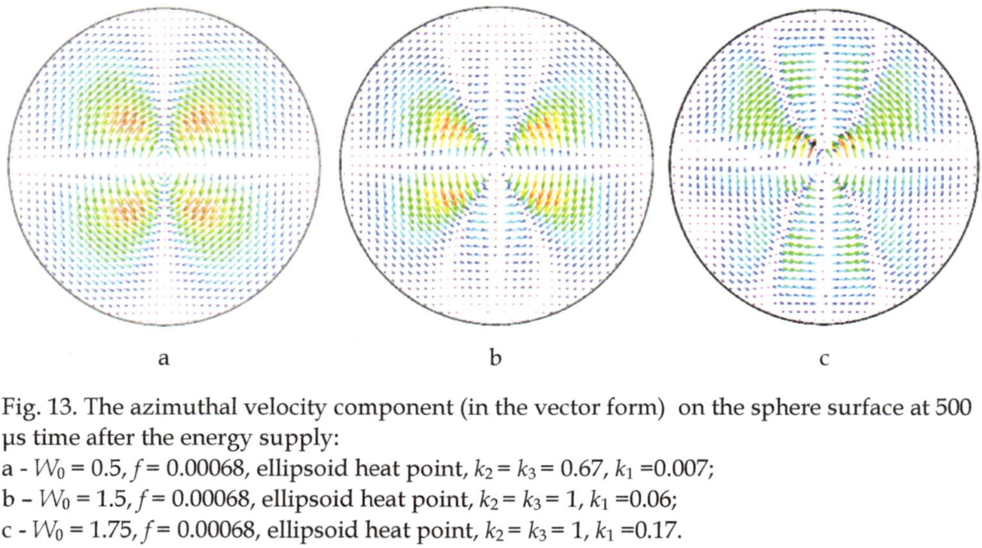

a b c

Fig. 13. The azimuthal velocity component (in the vector form) on the sphere surface at 500 μs time after the energy supply:
a - $W_0 = 0.5$, $f = 0.00068$, ellipsoid heat point, $k_2 = k_3 = 0.67$, $k_1 = 0.007$;
b - $W_0 = 1.5$, $f = 0.00068$, ellipsoid heat point, $k_2 = k_3 = 1$, $k_1 = 0.06$;
c - $W_0 = 1.75$, $f = 0.00068$, ellipsoid heat point, $k_2 = k_3 = 1$, $k_1 = 0.17$.

The energy supply parameters, i.e. the heat spot configuration and the energy supply intensity, influence the flow reconstruction significantly as their combination defines the possibility of either flow choking in the source (i.e. with the separated wave) or the choke-free flow. This considerably affects the track properties behind the flow and consequently the braking pressure and the configuration resistance. Energy supply into the incoming airflow allows the loads caused by the shock-wave influence on the construction elements of aircraft to be minimised.

5. Conclusion

The results of our research can be summarized as follows:

- An initial flow compression stage during the lens and blooming interaction process has been observed for both symmetric and asymmetric energy supply conditions;
- A 40% decrease in surface pressure during the 40 microsecond thermal spot interaction time was observed for the case of a sphere in Mach 3.45 flow with varying levels of energy deposition upstream of the bow shock for both symmetric and asymmetric energy supply conditions;
- For the case of symmetric energy supply, the process of heat track and sphere interaction has been found to consist of three stages i.e. an initial compression, expansion, compression and transient decay regardless of the energy input density magnitude;
- For asymmetric energy supply, the process of heat track and sphere interaction has been found to consist of two stages i.e. a short stage of compression followed by a long stage of expansion;
- Thermal spot shape and intensity of energy supply influence reorganization of flow, with their combination affecting properties of the track and stagnation pressure in both symmetric and asymmetric energy supply cases.

The application of energy deposition for local flow control requires low power in terms of the energy deposition into the flow. This low power requirement could potentially translate into small, low weight energy generation systems utilizing optical lasers or electric arc units etc. for effective and efficient flow control. This warrants consideration of the use of energy supply as effective means for solution of local issues of supersonic flow of various aircraft.

6. Acknowledgment

The authors would like to express their gratitude to Professor Alexander V. Gaydachuk of the National Aerospace University "Kharkov Aviation Institute" for his input and guidance during the manuscript preparation.

7. References

Adegren, R.; Elliot, G.; Knight D.; Zheltovodov, A. & Beutner, T. (2001). Energy deposition in supersonic flows, *AIAA Paper* N2001–0885, 2001

Adelgren, R.; Yan, H.; Elliott, G.; Knight D.; Beutner, T. & Zheltovodov, A. (2005). Control on Edney IV Interaction by Pulsed Laser Energy Deposition, *AIAA J.*, Vol. 43, No. 2, pp. 256-269, ISSN 0001-1452

Bazyma, L. & Rashkovan, V. (2005). Stabilization of Blunt Nose Cavity Flows Using Energy Deposition. *Journal of Spacecraft and Rockets*, Vol.42, No.5, (September-October 2005), pp. 790-794, ISSN 0022-4650

Bazyma, L. & Kholyavko V.I. (1996). A modification of Godunov's finite difference scheme on a mobile grid. *Computational Mathematics and Mathematical Physics*, Vol. 36, No.4, pp. 525-532, ISSN 0965-5425

Bazyma, L. & Rashkovan, V. (2006). Separation Flow Control by the Gas Injection Contrary Supersonic Stream, *AIAA J.*, Vol. 44, No. 12, pp. 2887-2895, ISSN 0001-1452

Chakravarthy, S. & Osher, S. (1985). New Class of High Accuracy TVD Schemes for Hyperbolic Conservation Laws, AIAA Paper 85-0363

Chakravarthy, S. (1986). The Versality and Reliability of Euler Solvers Based on High-Accuracy TVD Formulations," AIAA Paper 86-0243

Georgievskii, P. & Levin, B. (1988). Supersonic Flow Around Bodies in the Presence of External Heat Pulsed Sources, *Technical Physics Letters*, Vol. 14, No. 8, pp. 684–687 (ISSN 1063-7850)

Georgievski, P. & Levin, V. (1993). Unsteady Interaction of a Sphere with Atmospheric Temperature In homogeneity at Supersonic Speed, *Akademiya Nauk SSSR, Izvestiya, Mekhanika Zhidkosti i Gaza*, No. 4, June, pp. 174-183, ISSN 0568-5281

Godunov, S. Ed(s). (1976). *Numerical Solution of Multidimensional Problems in Gas Dynamics*, Nauka, Moscow (in Russian)

Guvernyuk, S. & Samoilov, A. (1997). Control of supersonic flow around bodies by means of a pulsed heat source. *Technical Physics Letters*, Vol.23, No.5, pp. 333-336, ISSN 1063-7850

Lobb, R.K. (1964). Experimental Measurement of Shock Detachment Distance on Spheres Fired in Air at Hypervelocities, In *The High Temperature Aspects of Hypersonic Flow*, Nelson, WC (ed.), pp. 519-527, Pergamon Press, New York, NY

Tret'yakov, P.; Grachev, G.; Ivanchenko, A.; Krainev, V.; Ponomarenko, A. & Tishenko, V. (1994). Optical breakdown stabilization in the supersonic argon flow. *Physics-Doclady*, Vol.39, No.6, pp. 415-416, ISSN 1063-7753

Tret'yakov, P.; Garanin, A.; Grachev, G.; Krainev, V.; Ponomarenko, A.; Tishenko, V. & Yakovlev, V. (1996). Control of supersonic flow around bodies by means of high-power recurrent optical breakdown. *Physics-Doclady*, Vol.41, No.11, pp. 566-567, ISSN 1063-7753

Van Dyke, M.D. (2003). The Supersonic Blunt-Body Problem - Review and Extension. *AIAA J.*, Vol. 41, No. 7, pp. 265-276, ISSN 0001-1452

5

Modelling Dynamics of Valley Glaciers

Surendra Adhikari and Shawn J. Marshall
Department of Geography, University of Calgary
Canada

1. Introduction

Ever since the Paleoproterozoic snowball Earth era, ca. 2.4 billion years ago (e.g. Hoffman & Schrag, 2000; Kirschvink, 1992), and beyond, the landscape of the planet Earth has been shaped up by the tremendous amount of scouring due to the repeated waxing and waning of ice masses. Over time, the dynamics of ice masses – a major part of Earth's cryosphere – has played a crucial role in global climate through complex interactions and feedbacks between the atmosphere, biosphere, and oceans. The cryosphere remains as one of the major dynamical components of the Earth system, participating in the geomorphologic and climatic evolution of the planet.

Presently, glaciers and ice sheets occupy ca. 10% of the Earth's land surface in the annual mean (Lemke et al., 2007). If it were to melt out completely, the mean sea level would rise by more than 64 m. The majority of this contribution comes from the large ice sheets of Antarctica, 56.6 m (Lythe et al., 2001), and Greenland, 7.3 m (Bamber et al., 2001). Glaciers and ice caps outside of Greenland and Antarctica contribute in a range between 0.15 m (Ohmura, 2004) and 0.37 m (Dyurgerov & Meier, 2005). In the ongoing warm epoch of climate since the little ice age, beginning in the late 19th century, glaciers and ice sheets have been retreating in most regions of the world (e.g. Cook et al., 2005; Krabill et al., 1999; Zemp et al., 2006). Such a response of the cryosphere creates a high-degree of disequilibrium, with positive feedbacks on the Earth's climate system, whereby the planet is likely to face ongoing and accelerated ice loss. Giving proper attention to the cryospheric component of climate system, most climate models forecast continued warming and glacier retreat at least until the end of 21st century (e.g. Christensen et al., 2007; Gillett et al., 2011).

On this premise, glaciological studies bear a tremendous importance; they are useful, for instance, (1) to understand the complex interaction between the ice and climate (e.g. Goelzer et al., 2011; Kaser, 2001), (2) to trace out the past climatic signals (e.g. Oerlemans, 2005; Thompson et al., 2003), (3) to assess the glacier-related hazards (e.g. Allen et al., 2009), and (4) to estimate glacial contributions to sea level rise (e.g. Leclercq et al., 2011; Meier, 1984; Raper & Braithwaite, 2006). To make future projections and to understand the intrinsic dynamical phenomena underlying glacier-climate interactions, such as the thermomechanical evolution of ice masses, numerical modelling, supplemented by field data, is the only option.

In this chapter, we discuss the physics and numerics of ice flow models with various degrees of complexity and we simulate the corresponding dynamics of a valley glacier. While valley glaciers make up only a tiny fraction ($< 1.0\%$) of the global cryosphere, proper understanding of glacier dynamics is essential for several reasons. First, valley glaciers are in close proximity to human settlement; any alteration in their dynamics affects society immediately. Second, valley glaciers and ice caps are of significant concern for watershed- and regional-scale water resources (e.g. Jansson et al., 2003; Viviroli et al., 2003); they, for instance, provide fresh water supply for municipal, agricultural, and industrial purposes. Third, the dynamical response of glaciers leaves footprints of past climate in their moraines (e.g. Beedle et al., 2009); they have hence become proven indicators of climate change. More importantly, the ice flow in valley glaciers and icefields comprises a high degree of complexity, primarily due to the irregular valley geometry. This demands a high-order treatment of glacier dynamics, thereby posing a challenge to numerical modellers. Finally, the fundamental physics of glaciers (i.e. mechanisms of ice flow) do not differ from those of larger ice sheets; experience in modelling valley-glacier dynamics can be directly extended to modelling of continental-scale ice sheets.

This chapter is hence designed to focus on the dynamics of valley glacier and its modelling. We (1) introduce ice rheology and briefly summarize the history of numerical modelling in glaciology, (2) describe the model physics and analyze the various approximations associated with the low-order (reduced) models, (3) provide an overview of numerical methods, concentrating on the finite element approach, and (4) present a numerical comparison of several models with various degrees of sophistication.

2. Ice rheology and glacier modelling

The rheological properties of glacier ice are practically independent of the isotropic pressure (e.g. Rigsby, 1958), and are therefore commonly described using deviatoric stresses rather than Cauchy stresses. The constitutive equation that relates deviatoric stresses to strain-rates in randomly oriented polycrystalline ice (under secondary creep) is given by the linearized inversion of Glen's flow law (Glen, 1955), i.e.

$$\tau_{ij} = 2\eta\dot{\epsilon}_{ij}, \tag{1}$$

where τ is the deviatoric stress tensor, $\dot{\epsilon}$ is the strain-rate tensor, and η is the effective viscosity. The viscosity of glacier ice is strain-rate dependent and is given by,

$$\eta = \frac{1}{2}A^{-\frac{1}{n}}\dot{\epsilon}_e^{\left(\frac{1-n}{n}\right)}, \tag{2}$$

where A is the flow law rate factor, n is the flow law exponent, and $\dot{\epsilon}_e$ is the effective strain-rate that can be understood from the second invariant of $\dot{\epsilon}$, i.e.

$$2\dot{\epsilon}_e^2 = \dot{\epsilon}_{ij}\dot{\epsilon}_{ji}. \tag{3}$$

By defining

$$\dot{\epsilon}_{ij} = \frac{1}{2}(u_{i,j} + u_{j,i}), \tag{4}$$

deviatoric stresses in Equation (1) can easily be expressed in terms of ice velocity, u, – the readily observable glaciological field variable.

Hypotheses and experimental foundations of this theory of ice rheology are given by Glen (1958) and are reviewed in detail by, e.g. Alley (1992), Budd & Jacka (1989), Cuffey & Paterson (2010), Hooke (1981), and Marshall (2005). Since the form of the constitutive relation (Eq. 1) is well-established and can be explained in terms of dislocation theory, these discussions revolve around the suitable parameterizations of A and n. The flow law rate factor, A, is thought to depend primarily on ice temperature, as well as on crystal size and orientation (anisotropy), water and impurity content, ice density and pressure, and perhaps on other several factors. Only the thermal dependence of A has been parameterized, following an Arrhenius relation (e.g. Hooke, 1981; Paterson & Budd, 1982), and coupled successfully with dynamical ice-flow models (e.g. Huybrechts & Oerlemans, 1988; Marshall & Clarke, 1997). Several attempts have also been made to account for anisotropic effects through the introduction of a "flow enhancement factor", both empirically (e.g. Wang & Warner, 1999) and through physically-based parameterizations (e.g. Gillet-Chaulet et al., 2005; Morland & Staroszczyk, 2003).

Similarly, the choice of flow law exponent, n, is also not obvious, as it varies in a range $1.5 - 4.2$ under different stress regimes (Weertman, 1973). For the realistic scenarios, i.e. $\tau \approx 50 - 200$ kPa, $n = 3$ is representative (e.g. Cuffey & Paterson, 2010). Assuming isothermal and isotropic ice masses for purposes here, we use $n = 3$ and $A = 10^{-16}$ $\mathrm{Pa}^{-3}\,\mathrm{a}^{-1}$ as in, e.g. Pattyn et al. (2008) and Sargent & Fastook (2010).

In the early 1950s, the power-law relation between τ and $\dot{\epsilon}$ (Eqs. 1–3) was formulated based on laboratory experiments (Glen, 1952) and field observations on the closure of boreholes (Nye, 1953). Subsequently, the Glen-Nye law, commonly known as the Glen's law after Glen (1955), emerged to describe the glacier ice as a quasi-viscous fluid with non-Newtonian flow behaviour. This not only discarded the then-prevailing theory of "extrusion flow" (see Waddington, 2010), but also opened the door for investigating the theoretical and mathematical foundations of modern glaciology. Nye's works (e.g. Nye, 1952; 1959) mark the beginning of such investigations, particularly focusing on the motion of glacier ice. Robin (1955) was the first to calculate ice temperature by considering glacial thermodynamics. Due to the lack of computational power, these early works were primarily based on the semi-analytical methods used in contemporary fluid mechanics. A nice summary of these early works that form the foundation of physical glaciology is given in Clarke (1987).

With the dawn of digital computing, model-based studies of glacier dynamics started in late 1960s (e.g. Campbell & Rasmussen, 1969). Soon after, several numerical models (e.g. Budd & Jenssen, 1975; Mahaffy, 1976; Oerlemans, 1982) were developed, including the ones with thermomechanical coupling (e.g. Jenssen, 1977). These pioneer models were based on the "shallow-ice" theory of glacier mechanics (e.g. Nye, 1959), which assumes that ice thickness is much less than the horizontal length scale over which a domain is discretized. This theory was later developed rigorously by Hutter (1983) and Morland (1984), which is now known formally as the shallow-ice approximation (SIA). SIA models have been used extensively for simulating large ice sheets (e.g. Calov & Hutter, 1996; Huybrechts & Oerlemans, 1988), as well as valley glaciers (e.g. Adhikari & Huybrechts, 2009; Oerlemans et al., 1998). In general, ice sheet models need to thermomechanically coupled, since the polar ice sheets span a range of temperature from the melting point to ca. –50°C, whereas models for temperate valley glaciers are commonly isothermal. This is reasonable outside of the polar regions, as most of the world's valley glaciers are temperate: at the pressure-melting point throughout. For SIA models with and without the coupled thermodynamics, benchmark numerical experiments are presented respectively by Payne et al. (1996) and Huybrechts et al. (1996).

SIA theory is strictly valid only where horizontal gradients in ice thickness and velocity are negligibly small and bedrock slopes are sufficiently gentle. These criteria are clearly violated in valley glaciers (e.g. Le Meur et al., 2004; Leysinger Vieli & Gudmundsson, 2004), fast-flowing ice streams (e.g. Whillans & Van der Veen, 1997), and at the ice divides and grounding zones of ice-sheet/ice-shelf systems (e.g. Baral et al., 2001). Several attempts have therefore been made to capture high-order dynamics in ice flow models; effects of longitudinal stress gradients (e.g. Adhikari & Marshall, 2011; Shoemaker & Morland, 1984; Souček & Martinec, 2008) and lateral drag (e.g. Adhikari & Marshall, in preperation; Nye, 1965) are particularly accounted via physically-based or numerical/empirical parameterizations. More complete representations of glacier dynamics are provided by high-order (e.g. Blatter, 1995; Pattyn, 2003) and Stokes (e.g. Jarosch, 2008; Jouvet et al., 2008; Zwinger et al., 2007) models. The development history of such models is nicely summarized by Blatter et al. (2010); corresponding benchmark experiments are presented by Pattyn et al. (2008).

With the material nonlinearity (see Eqs. 1–2), even SIA models are not analytically tractable; the coupled evolution of glacier temperatures, rheology, and high-order velocities therefore requires a numerical solution. A few attempts have been made to obtain analytical solutions (e.g. Bueler et al., 2007; Sargent & Fastook, 2010), however, at least as a tool for verification of numerical models in simple geometric and climatic settings.

3. Model physics, approximations, and boundary conditions

For full simulations of Earth's climate system, the dynamical models of glaciers and ice sheets are coupled with those of other climatic components, namely the atmosphere, the biosphere, and the ocean (see, for example, Fig. 1 in Huybrechts et al., 2011). The models of glaciers and ice sheets are usually accompanied by, for instance, (1) a mass balance model that describes the physics of mass exchange at the ice/atmosphere and ice/ocean interface, (2) a model of glacial isostatic processes whereby the underlying bed deforms due to the load of ice, and (3) a model of subglacial till deformation that yields the associated basal motion of ice. The intrinsic processes of ice flow can be described by a combination of gravitational creep deformation and decoupled basal sliding. To simulate creep deformation (i.e. effective viscosity) and to predict the regions where ice masses are warm-based (permitting basal sliding), a proper account of energy balance is essential. Along with the companion models listed above, a full three-dimensional ice flow model, equipped with thermomechanical coupling, is therefore required for the realistic simulations of glacier dynamics.

We outline a simple flowchart (Fig. 1) depicting the major components of a typical ice flow model. Given the boundary conditions and some description of mass budget, we accomplish ice flow modelling in a two-step simulation: (1) diagnostic simulation of a set of steady-state problems in order to obtain the quasi-stationary englacial velocity/stress and temperature fields at time t, and (2) prognostic simulation satisfying kinematic boundary conditions to update the glacier geometry at a subsequent time, $t + \Delta t$. Below, we describe the physics and associated low-order approximations of several ice flow models.

3.1 Diagnostic equations

Dynamical models for ice flow are based on the fundamental physics of conservation of mass, momentum, and energy. Glacier velocities are so small that we can remove the acceleration term from the momentum balance equation; the dynamical problem in glaciology therefore reduces to a Stokes problem. For isothermal glacier domains in $k(\geq 2)$-dimensional Euclidean

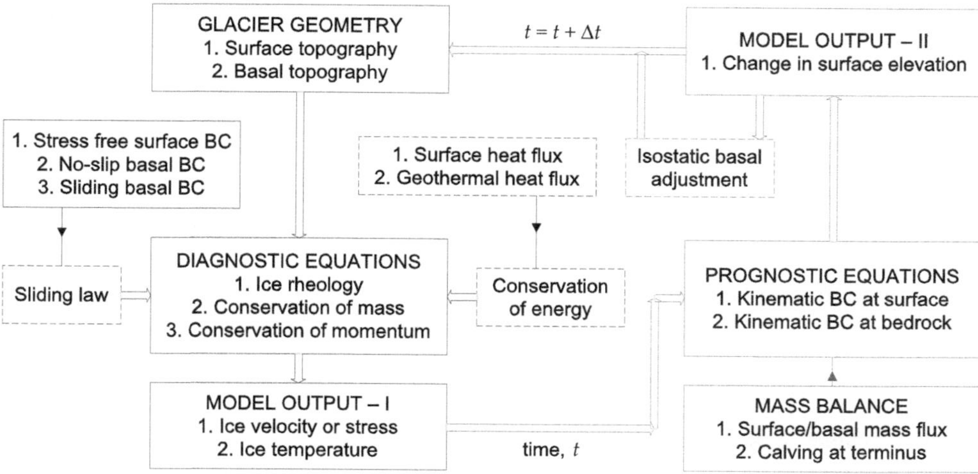

Fig. 1. Flowchart of an ice flow model. The processes associated with the diagnostic and prognostic simulations are listed in the LHS and RHS boxes, respectively; black and red colors are used for clarity. Dotted boxes enclose the processes that are not considered in this study. Mid-arrows are used to depict the corresponding boundary conditions.

space, $\Re^k$, the Stokes problem can be stated as,

$$u_{i,i} = 0, \tag{5}$$

$$\sigma_{ij,j} + \rho g_i = 0, \tag{6}$$

where u is the velocity vector, σ is the Cauchy stress tensor, ρ is ice density, and g is the gravity vector. We split σ into its deviatoric part, τ, and an isotropic pressure, p, i.e.

$$\sigma_{ij} = \tau_{ij} + p\delta_{ij}, \tag{7}$$

whereby the momentum balance equation (Eq. 6) can be expressed in terms of the velocity vector, as explained in Section 2. The isotropic pressure is dependent on the trace of Cauchy stress tensor, i.e. $p = \sigma_{ii}/k$, and is activated via the Kronecker delta, δ_{ij}, only when normal stresses are being considered ($\delta_{ij} = 1$ for $i = j$, and $\delta_{ij} = 0$ otherwise).

Intuitively, three-dimensional (3D) Stokes models, which solve a complete set of Stokes equations (Eqs. 5–6), describe the most sophisticated treatment of glacier dynamics. Virtually all models developed to date (e.g. SIA or high-order) can be considered as approximations of a Stokes model. Hindmarsh (2004) compares the numerical solutions of various approximations to the Stokes equations. Apart from the standard SIA model, he considers an 'L' family of models that include some of most common models, such as those of Blatter (1995), MacAyeal (1989) and Pattyn (2003). These L-models differ from each other in: (1) how they reduce the definitions of momentum balance (Eq. 6), strain-rate tensor (Eq. 4) and its second invariant (Eq. 3), and (2) how they obtain the approximate solutions of ice velocities from the previous iteration step to calculate effective viscosity (Eq. 2). Since such approximations are made primarily to optimize the solution accuracy and computational efficiency, it is not always obvious how to choose a particular model for a given glaciological scenario. Here, we present new brands of models, whose associated approximations clearly define the distinct physical mechanisms of glacier dynamics; the scope of each model therefore becomes apparent.

In a 3D domain of land-based glacier, key physical processes that act to balance a gravity-driven ice-flow consist of basal drag, τ_b, resistance associated with longitudinal stress gradients, τ_{lon}, and lateral drag, τ_{lat}. Mathematical details of these resistances can be found, for example, in Van der Veen (1999) and Whillans (1987). Based on the physical mechanisms associated with each resistance, we define three families of models. Ice flow in the first family of models is controlled collectively by $(\tau_b + \tau_{lon} + \tau_{lat})$, in the second one by $(\tau_b + \tau_{lon})$ only, and in the third one by τ_b alone. For a fairly wide glacier (such that $\tau_{lat} \approx 0$) resting on steep and undulating bedrock (such that τ_{lon} is significant), for example, the second family of models should be optimal to yield realistic simulations. Members of a given model family differ from each other mainly in that they deal with different spatial dimensions. Below, we describe each of them; relevant governing equations are given in Appendix A.

3.1.1 Full-system Stokes (FS) model

If a model solves the Stokes equations (Eqs. 5–6) in 3D space, $\Re^3$, we call it the full-system Stokes model (FS). This is the only member of the first model family. Here, the indices (i, j) refer to Cartesian coordinates (x, y, z); x is the horizontal coordinate along the principal flow direction, y is the second horizontal coordinate along the lateral direction, and z is the vertical coordinate opposite to gravity.

3.1.2 Plane-strain Stokes (PS) model

The 3D plane-strain Stokes model (PS3) does not strictly follow the plane-strain approximations as its name suggests. It rather excludes the lateral gradients of stress deviators, i.e. $\tau_{ij,y} = 0$, in the momentum balance equation (Eq. 6), and those of ice velocities, i.e. $u_{i,y} = 0$, in the strain-rate definition (Eq. 4). The flowline version of this model (PS2) solves the Stokes equations in a two-dimensional (2D) space, $\Re^2$, and hence follows the plane-strain approximations. In a flowline model, the indices (i, j) refer to Cartesian coordinates (x, z), where x and z are once again the horizontal and vertical coordinates, respectively.

3.1.3 Shear-deformational (SD) model

In 3D shear-deformational model (SD3), the vertical shear stresses, i.e. τ_{iz} with $i = (x, y)$, are the only non-zero stress components. As for the PS3 model, it also excludes the lateral gradients of stress deviators and ice velocities. In its flowline counterpart (SD2), τ_{xz} is the only non-zero stress component; no further assumption is needed.

The standard zeroth-order SIA models (Hutter, 1983) also belong to the SD family. The three-dimensional (SIA3) and flowline (SIA2) shallow-ice models can be derived respectively from the SD3 and SD2 models, by further assuming that the horizontal gradients in vertical shear stresses and ice velocities are negligible, i.e. $\tau_{iz,x} = 0$ and $u_{i,x} = 0$. The laminar-flow model (LF) is the simplest of SD models. There exist analytical solutions for ice velocities in isothermal, laminar flow (e.g. Cuffey & Paterson, 2010; Van der Veen, 1999); the horizontal velocity, u_x, at any point on the flowline plane (x, z), for example, is given by

$$u_x(x,z) = u_x(x,b) + \frac{2A}{n+1} \left[\rho g_z \alpha_s(x)\right]^n h(x)^{n+1} \left[1 - \left(\frac{s-z}{h(x)}\right)^{n+1}\right], \tag{8}$$

where $u_x(x,z)$ and $u_x(x,b)$ are respectively the velocities at any depth z and at the bedrock $z = b$. Similarly, g_z is the vertical component of the gravity vector, $h(x)$ is the ice thickness, s is the glacier surface, and $\alpha_s(x)$ is the surface slope.

3.2 Prognostic equations

In prognostic simulations of each model, the glacier surface, $z = s$, evolves satisfying the kinematic boundary condition,

$$s_{,t} + u_i(s)s_{,i} = u_z(s) + m(s,t), \tag{9}$$

where subscript t represents time, index i refers to the horizontal coordinates, i.e. $i = (x,y) \subset \Re^3$, and $i = x \subset \Re^2$, $u_z(s)$ is the vertical velocity at the glacier surface, and m is the mass balance function. To compute the unknown s at a new time $t + \Delta t$, we obtain glacier surface velocities from the diagnostic simulation of domain at an antecedent time t, and prescribe $m(s,t)$ as a vertical flux with units m ice eq. a^{-1}.

3.3 Boundary conditions

In addition to the kinematic boundary condition (Eq. 9), the upper ice surface satisfies the stress free criterion, i.e. $\sigma_{ij}(s) \approx 0$. This involves an assumption that the role of atmospheric pressure on the overall dynamics of glacier ice is negligibly small. The kinematic boundary condition (similar to Eq. 9) is also applied at the ice/bedrock interface. However, we impose a no-slip basal criterion, i.e. $u_i(b) = 0$. In doing so, we assume no mass exchange due to ice melting/refreezing, i.e. $m(b,t) = 0$, thus restricting the evolution of basal ice, i.e. $b_{,t} = 0$.

The lateral boundary condition (glacier margin) is typically free in glacier simulations, with ice free to advance or retreat within a domain. Domain extent is designed such that the ice mass does not reach the boundary; it then freely evolves within the domain, with a zone with ice thickness $h = 0$ around the periphery. If the combined snow accumulation rates and ice flux into an empty grid cell exceed local mass loss (ablation), the grid cell becomes glacierized and is included in the overall glacier continuum.

4. Numerical methods applied to glacier dynamics

The existence of numerical solutions to strongly nonlinear Stokes problem discussed in Section 3.1 is proven in Colinge & Rappaz (1999). Various approaches have been used to obtain solutions in 3D space; numerical schemes such as finite difference (e.g. Colinge & Blatter, 1998; Huybrechts et al., 1996; Marshall & Clarke, 1997; Pattyn, 2003), finite element (e.g. Gudmundsson, 1999; Hanson, 1995; Jarosch, 2008; Jouvet et al., 2008; Picasso et al., 2008; Zwinger et al., 2007), control volume (e.g. Price et al., 2007), and spectral (e.g. Hindmarsh, 2004) methods have all been employed. Finite difference (FD) and finite element (FE) methods are more common; we use the latter one. Along with a brief overview of the FD method, below we provide theoretical and numerical details of the FE method.

4.1 Finite difference method (FDM)

Most "classical" models of glacier and ice sheet dynamics are based on FDM, where the 3D domain is split into a series of regular grid cells on a Cartesian or spherical (i.e. Earth) grid. Staggered grids are used, solving ice thickness, stress and temperature in the cell centre and

3D velocity fields at the cell interfaces, with discretization of the governing equations through standard second-order FD approximations (e.g. Huybrechts & Oerlemans, 1988). Vertical resolution is usually fine compared to horizontal resolution, with 10-40 layers in the vertical, on an adaptive, stretched grid that is updated each time step as the glacier thins or thickens. Sometimes a nonlinear vertical grid transformation is also introduced in order to increase resolution near the bed, where velocity gradients are strongest.

For SIA models, the system of FD equations is particularly efficient to solve, as the governing equation at a point is dependent on only local conditions (ice thickness, surface slope). In this case the solution depends only on the nearest neighbours to a grid cell, presenting a banded matrix system that is amenable to sparse matrix techniques. With a more complete representation of the physics, i.e. the Stokes and high-order models, solutions are non-local and computational costs increase by at an order of magnitude or more (Blatter, 1995).

FD discretizations are also limited in their ability to describe complex geometries, as found in valley glaciers, ice shelves, and fjords: many of the most interesting glaciological situations. Much of the most interesting dynamics is at the glacier or ice sheet margin, where increased resolution is desirable. We therefore turn to FE method for glaciological simulations for the remainder of this chapter.

4.2 Finite element method (FEM)

The existence and convergence of FE solutions for the glaciological (Stokes) problem have been proven in Chow et al. (2004) and Glowinski & Rappaz (2003). Let $\Omega \in \Re^k | k \geq 2$ be a continuous glacier domain, enclosed by a boundary Γ. In FE schemes, we decompose Ω into a finite number of elemental domains Ω_e, thereby generating a finite number of boundary domains Γ_e. Satisfying the relevant boundary conditions, we first seek the approximate solutions of field variables within each sub-domain, Ω_e. We then assemble them over the continuous domain, Ω, to obtain the required solutions.

There are several methods, such as variational and weighted residuals, to formulate the FE counterparts of the governing equations. We use the standard Galerkin method – a weighted residual approach in which the weighted sum of system residuals arising from the FE approximations of a continuous domain is set to zero.

4.2.1 Diagnostic equations

Let $\tilde{u}$ and $\tilde{p}$ be the approximate solutions of field variables that vary within Ω_e, according to the respective interpolation functions ψ^u and ψ^p, such that

$$\tilde{u}(x,y,z) = \psi_i^u(x,y,z)u_i = [\psi^u(x,y,z)]\{u\}, \tag{10}$$

$$\tilde{p}(x,y,z) = \psi_i^p(x,y,z)p_i = [\psi^p(x,y,z)]\{p\}, \tag{11}$$

where index i refers to the elemental degrees of freedom associated with the velocity vector and pressure, respectively. Hence, $\{u\}$ and $\{p\}$ denote the nodal velocities and pressure.

By plugging in the approximations of field variables (Eqs. 10–11), we obtain the residuals of the diagnostic equations presented in Section 3.1. The weighted sum of these residuals, according to the standard Galerkin method, is then set to zero, i.e.

$$\int_{\Omega_e} \psi^p \left[\tilde{u}_{i,i}\right] \, d\Omega_e = 0, \tag{12}$$

$$\int_{\Omega_e} \psi^u \left[\sigma_{ij,j}(\tilde{u}_i, \tilde{p}) + \rho g_i\right] \, d\Omega_e = 0. \tag{13}$$

Here the weights are chosen to be the assumed interpolation functions; this is unique to the Galerkin method (e.g. Rao, 2005). See Appendix B for the details of construction of a linear system of these equations in 3D Cartesian coordinates.

As the governing equations comprise a one-degree high-order of derivative for the velocity vector than that for the isotropic pressure (see Appendix A), a typical Taylor-Hood element (Hood & Taylor, 1974) with quadratic interpolation function for velocities and linear one for pressure is recommended. For simplicity, however, we use the same order of interpolation function, so that $\psi^u = \psi^p \equiv \psi$. Any instability arising as a result is accommodated by using the stabilized finite elements (Franca & Frey, 1992).

Stabilization involves addition of mesh-dependent terms to the Galerkin formulation. These additional terms are the Euler-Lagrange equations evaluated elementwise, so that exact solutions satisfy both the Galerkin and these additional terms. The additional terms are,

$$\left\langle \sigma_{ij,j}, \delta_1 \sigma_{ij,j} \right\rangle + \left\langle u_{i,i}, \delta_2 u_{i,i} \right\rangle, \tag{14}$$

$$\left\langle \rho g_i, \delta_1 \sigma_{ij,j} \right\rangle. \tag{15}$$

Equation (14) is added to the elemental coefficient matrix and Equation (15) is added to the RHS force vector. In Equation (14), the first term inside the first inner product is the residual of momentum balance equation (Eq. 13), excluding the force term, and the first term inside the second inner product is the residual of continuity equation (Eq. 12). The second terms associated with stability parameters δ_1 and δ_2 are the stabilization contributions to the weight functions. Here, these contributions are assumed to be the same as the respective system residuals. The stability parameters are chosen following (Franca & Frey, 1992),

$$\delta_1 = \frac{m_k h_k^2}{4\eta}, \; \delta_2 = \frac{2\eta}{m_k}, \tag{16}$$

where m_k depends on the type of the element and h_k on its size. Details of diagnostic system stabilization (for the FS model) are given in Appendix C.

4.2.2 Prognostic equations

In prognostic simulations, we seek the approximate solution, $\tilde{s}$, of s along the ice surface. Over each relevant Γ_e, $\tilde{s}$ varies according to the chosen interpolation function ψ^s, such that

$$\tilde{s}(x, y, t) = \psi_i^s(x, y, t) s_i = \left[\psi^s(x, y, t)\right] \{s\}. \tag{17}$$

With this approximation of field variables, we obtain the residuals of the prognostic equation (Eq. 9), whose weighted sum is set to zero,

$$\int_{\Gamma_e} \psi^s \left[\tilde{s}_{,t} + u_i \tilde{s}_{,i} - (u_z + m)\right] \, d\Gamma_e = 0. \tag{18}$$

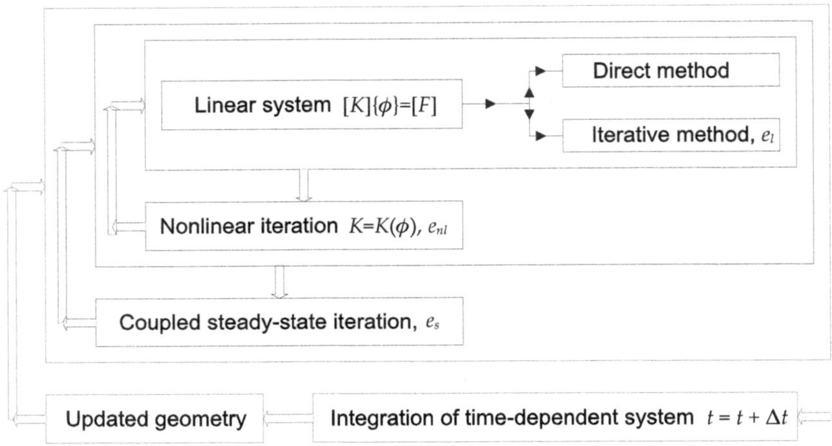

Fig. 2. Flowchart of the solution scheme. Here also, red color is used to distinguish the prognostic simulations from the diagnostic ones

This hyperbolic equation is stabilized by adding the element-wise terms (Donea & Huerta, 2003) to the mass and coefficient matrices, as well as to the force vector. Mathematical details of FE formulation and stabilization of the prognostic equation are given in Appendix D.

4.2.3 Elmer and model numerics

We use the open source FEM code Elmer (http://www.csc.fi/elmer), adapted for Glen's flow law for ice (Glen, 1955). Elmer gives approximate (numerical) solutions for both the FS and reduced models by solving the weak forms of the respective governing equations. The solutions from the FS and PS2 models are tested by Gagliardini & Zwinger (2008), against the ISMIP-HOM (Ice Sheet Model Intercomparison Project for Higher-Order Models; Pattyn et al., 2008) benchmark experiments. We solve additional subroutines to obtain FE solutions for the PS3 and SD family of models. We validate SIA2 model by comparing results with the corresponding analytical solutions (Eq. 8; see Adhikari & Marshall, 2011).

We sketch a flowchart of the solution scheme employed in Elmer (Fig. 2). The linear system of equations obtained from the Galerkin formulation (Eq. B16) is in the core of the solver. This can be solved by using either direct or iterative methods. The direct method yields an exact solution up to the machine precision; this, however, is not feasible for large problems. We therefore use an iterative method, the Krylov subspace method method (biconjugate gradient stabilized method, BiCGStab) with with an incomplete lower-upper factorization (ILU4) as the system pre-conditioner, and obtain the approximate solutions. Given the mesh density and element type, the accuracy of such solutions relies on the chosen convergence criterion, e_l; the smaller the value of e_l, the more accurate the solutions. However, too small a choice of e_l makes the job computationally inefficient.

We then solve the material nonlinearity associated with the constitutive relation. We apply a fixed point iteration scheme (the Picard linearization) to linearize the system by expressing η in terms of u_i from the previous iteration step. Here also, a suitable convergence criterion, e_{nl}, should be satisfied. For a given transient domain, we integrate the prognostic equation until the steady-state criterion, e_s, is reached; we use implicit scheme (first-order backward differentiation formula, BDF, scheme) for such time-dependent integrations. We advise maintaining $e_l < e_{nl} < e_s$ for good convergence. Other aspects of Elmer (e.g. effects of mesh density on solution accuracy and computational efficiency, and parallel simulations) are given by Gagliardini & Zwinger (2008).

For each experiment considered in this study, we generate the structured mesh by using ElmerGrid. ElmerGrid is basically a 2D mesh generator, but is also capable of extruding and manipulating the mesh in the third dimension. Since 3D experiments require a large amount of memory and computation time, we perform parallel runs in a high-performance computing cluster provided by the Western Canadian Research Grid (WestGrid).

5. Numerical comparison of physical approximations

We consider a $10 \times 2.5\,\mathrm{km}^2$ glacial valley. To mimic a typical real-world glacier scenario, we include meanders and bumps in the subglacial topography, $b(x,y)$, as defined below,

$$b(x,y) = 5000 - x \tan\alpha_b + a_x \sin\left(0.5\pi + \frac{4\pi x}{L}\right) - a_y \sin\left[(y+\theta)\frac{\pi}{W}\right], \tag{19}$$

where α_b is the mean bedrock slope in radians, L and W are the longitudinal and lateral extents of the valley, a_x and a_y are the amplitudes of the topographical variation in x and y directions, and θ is the sinusoidal offset of the flowline. Here, we use $\alpha_b = 12°$, $L = 10\,\mathrm{km}$, $W = 2.5\,\mathrm{km}$, $a_x = 200\,\mathrm{m}$, $a_y = (500 + 0.05x)\,\mathrm{m}$, and $\theta = 500\sin(2\pi x/L)\,\mathrm{m}$. The plan view of basal topography is shown in Figure 3a; the central flowline is also depicted.

Next, we define the climatic regime. For $z \leq 4.6\,\mathrm{km}$ and over a 500-m wide corridor around the central flowline (see Fig. 3a), the mass balance function, $m(s,t)$, is chosen as,

$$m(s,t) = \beta\left[s(x,y) - E\right] + \Delta m(t), \tag{20}$$

where β is the linear mass balance gradient, E is the equilibrium line altitude (ELA), and $\Delta m(t)$ is the time-dependent mass balance perturbation; $m(s,t) = 0$ elsewhere. For now, we choose $\beta = 0.01$ m ice eq. $\mathrm{a}^{-1}\,\mathrm{m}^{-1}$, $E = 3.7\,\mathrm{km}$, and $\Delta m(t) = 0\,\mathrm{m}$ ice eq. a^{-1}. Since $s(x,y)$ evolves in prognostic simulations (Eq. 9), the parameterization of $m(s,t)$ (Eq. 20) ensures that our models capture the height/mass-balance feedback inclusively.

Under these geometric and climatic settings, we grow glaciers to steady state using several models. In order to illustrate the importance of each physical mechanism of ice flow, we consider three different models (one from each model family), namely the FS, PS3 and SIA3 models. We denote them respectively by FS, PS and SD, unless otherwise specified. Each model domain consists of 50 k bilinear quadrilateral elements, with average horizontal dimensions $50 \times 50\,\mathrm{m}^2$. The vertical dimension of element varies according to the ice thickness; we use five vertical layers. Below, we carry out numerical comparison of these 3D models in terms of steady state geometry, surface velocity, basal shear stress, and response timescales. Considering pragmatic flowline models (PS2 and SIA2), we also present a brief tutorial on modelling valley glacier dynamics, which involves (1) sensitivity tests for a glacier, (2) reconstruction of past climate or glacier extent, and (3) projection of a glacier's future.

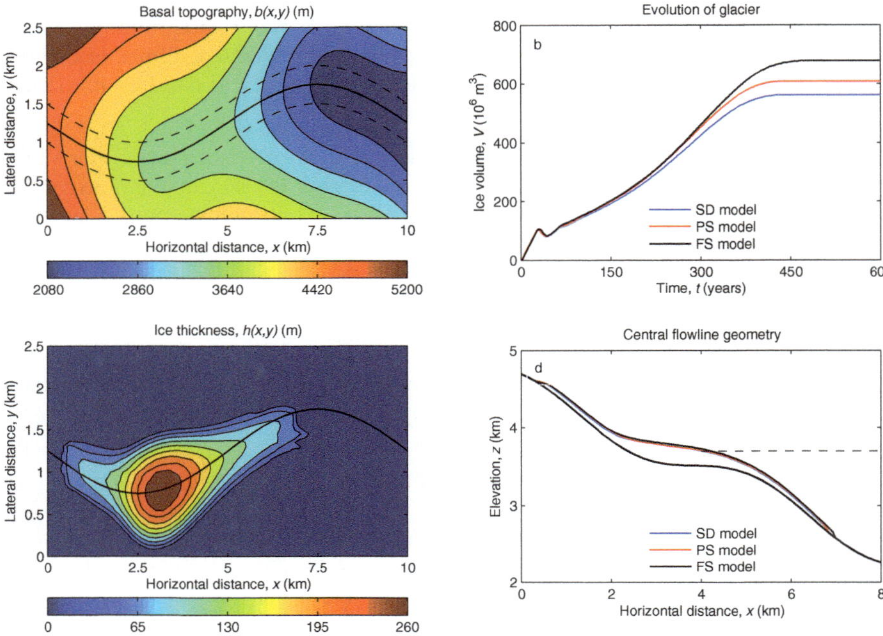

Fig. 3. (a) Basal topography, showing the central flowline. The mass balance function (Eq. 20) is applied only over a passage enclosed by dotted lines. (b) Evolution of ice volume. (c) Steady state ice thickness obtained from the FS model. (d) Longitudinal profiles of steady state geometry along the central flowline; the corresponding ELA is shown with a dotted line.

5.1 Geometry and field variables

The evolution of a glacier from zero ice volume to steady state is shown for each model case (Fig. 3b). By accounting for the high-order physical mechanisms, FS and PS models hold more ice mass than does the SD model. To assess the importance of high-order dynamics, i.e. the role of τ_{lat} and/or τ_{lon}, we compute errors between the models. We denote the error, for example, by $e_{PS.FS}$ to explain a difference between the PS and FS models with respect to the latter one. The errors $e_{PS.FS}$ and $e_{SD.PS}$ therefore illustrate the sole role of τ_{lat} and τ_{lon}, respectively; while $e_{SD.FS}$ explains their collective effects. The steady state ice volume obtained from each model and the associated errors are listed in Table 1. For the chosen geometric setting, the role of τ_{lat} ($e_{PS.FS} = -10.6\%$) is relatively more pronounced than that of τ_{lon} ($e_{SD.PS} = -7.6\%$).

The plan view of the steady state ice thickness obtained from the FS model is shown in Figure 3c. The maximum ice thickness is observed along the central flowline, and specifically around the basal depression at $x \approx 3$ km. This is true for each model case, as shown in Figure 3d. Although the longitudinal profiles of surface elevation appear to superimpose on each other, there is a considerable difference in both the mean (Table 1) and maximum values of ice thickness. The SD model generates a glacier that is 6.4% and 15.9% thinner than the PS

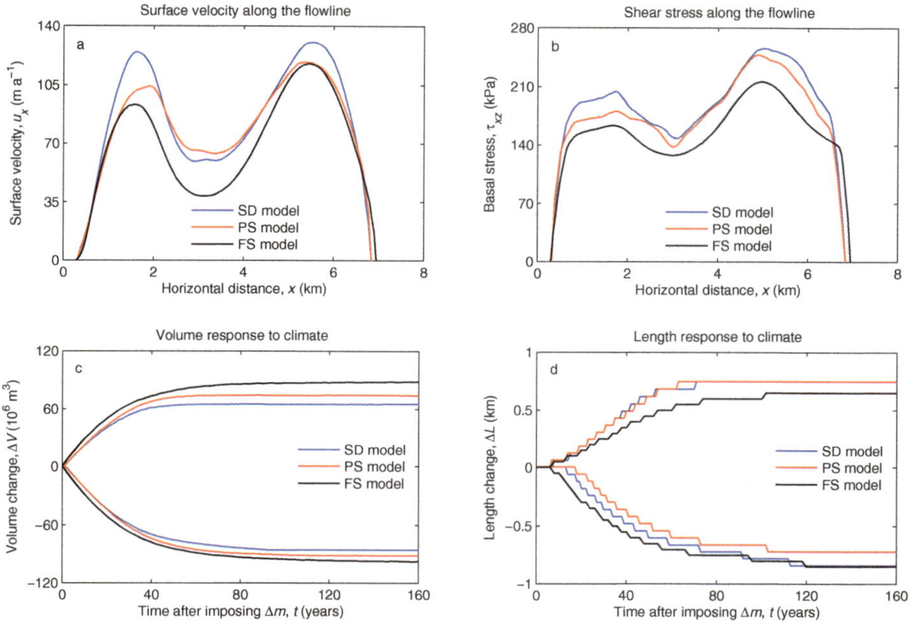

Fig. 4. (a) Surface velocity and (b) basal shear stress along the central flowline. Evolutions of (c) ice volume and (d) glacier length in response to step changes in climate, imposed on the steady-state glaciers whose geometries are shown in Fig. 3.

and FS models, respectively. This indicates the significance of resistances associated with the high-order dynamics; τ_{lat} once again appears to be more crucial than τ_{lon}.

The steady state horizontal velocity at the upper ice surface, $u_x(x, y, s)$, along the central flowline is plotted in Figure 4a. In each model case, we find smaller velocities around the basal depression at $x \approx 3$ km; while ice flows faster at places with steeper basal slopes (see Fig. 3d). Due to resistive effects of high-order dynamics, the FS and PS models yield relatively smaller ice velocities. The mean surface velocity along the central flowline in the SD model is higher by 5.9% and 23.6% than in the PS and FS models, respectively (Table 1).

We also calculate the deviatoric stresses from the englacial velocity field, using Equations (1–4). The vertical shear stress at the ice/bedrock interface, $\tau_{xz}(x, y, b)$, in a steady state longitudinal profile along the central flowline is shown in Figure 4b. As expected, basal shear stress is smaller in the PS and FS models, where other stress components are also active (i.e. τ_{xz} is not the only non-zero component) to control the glacier ice flow. In the SD model, $\tau_{xz}(x, y, b)$ characterizes τ_b (e.g. Adhikari & Marshall, 2011), which is the sole resistance to the gravitational driving stress, τ_d; it follows that $\tau_{xz}(x, y, b) \approx \tau_d$. Therefore, we calculate the errors as the difference between the SD and FS/PS models with respect to the former one. This gives a rough idea about the fractional contributions of high-order resistances, i.e. other than τ_b, to balance the gravity-driven ice flow; we find $\tau_{lon} \approx e_{SD.PS} = 6.1\%$ and

$(\tau_{lon} + \tau_{lat}) \approx e_{SD.FS} = 17.6\%$ (Table 1). These figures represent lower-limit estimates, as $\tau_d = \rho g h \alpha_s$ (e.g. Van der Veen, 1999) should be larger for the high-order models, which hold thicker ice masses.

5.2 Response of the glacier to climate change

Before simulating the past and future dynamics of a glacier, it is useful to conduct a simple sensitivity test by imposing a step change in climate, i.e. mass balance, on the steady state geometry. This yields the characteristic timescales of a glacier, specifically the response times, which explain the length of time over which the glacier carries in its memory the mass balance history. On the corresponding steady state geometry of each model (Section 5.1), we impose $\Delta m(t) = \pm 1\,\mathrm{m}$ ice eq. a^{-1} in turn and we let the glacier respond until it attains a new steady state. The volume and length response of a glacier in each model case are plotted respectively in Figure 4c and 4d. Based on the e-folding concept (e.g. Jóhannesson et al., 1989), we calculate both the volume, t_v, and length response time, t_l, as the time required for a glacier to adjust $(1 - e^{-1}) \approx 63\%$ of total change in volume, ΔV, and length, ΔL, respectively.

Response times, t_v and t_l, are listed in Table 1; values are given for 2D flowline models as well. Based on these values, we note a few important points. First, a glacier takes less time, by ca. 17 (3D) and ca. 8 years (2D), to adjust its ice volume vs. its length. The relatively shorter t_v is primarily due to the instantaneous response of ice thickness to the climatic perturbation. Secondly, all models with a given spatial dimension yield nearly the same response times, i.e. $t_v \approx 25$ and $t_l \approx 42$ years (3D), and $t_v \approx 12$ and $t_l \approx 21$ years (2D). Leysinger Vieli & Gudmundsson (2004) find the same for 2D models, and they suggest that simpler models are sufficient for the purpose of estimating response times. This however does not imply that 2D models yield representative timescales for 3D cases. For the chosen geometry, 3D models appear to take twice as long to respond as the flowline models. This is mainly because the flowline models only capture the maximum velocity, along the central flowline, and hence adjust its geometry more quickly; whereas 3D models have an integrated ice flux across the glacier, which gives a slower average velocity, and consequently take longer time to adjust its geometry. Therefore, the flowline models, which lack the proper account of effects of varying glacier width, do not yield realistic estimates of response times for valley glaciers.

5.3 Projecting the glacier's future using flowline models

One of the key reasons for using dynamical ice flow models is to simulate the future states of a glacier under several possible climatic scenarios. To be able to obtain realistic predictions of ice volume, ice flow models should be constrained properly, ensuring that they truly represent the dynamics of the glacier at hand. Depending upon the availability of field data, this is usually accomplished through simulations of the ice surface velocities and/or historical front variations (e.g. Adhikari & Huybrechts, 2009). As we have considered synthetic glaciers, we take advantage to assume that the FS dynamics represents a real-world scenario and we constrain the reduced models accordingly. The majority of valley glacier simulations are based on flowline dynamics (e.g. Oerlemans et al., 1998); we choose PS2 and SIA2 models to reconstruct the past and project the future of a glacier.

By mimicking a real-world climatic history with inter-decadal variability, we impose a pre-defined $\Delta m(t)$ upon the FS model (Fig. 5a), and record the corresponding changes in the terminus position (Fig. 5b). For each of the PS2 and SIA2 models, we tune $\Delta m(t)$ so that

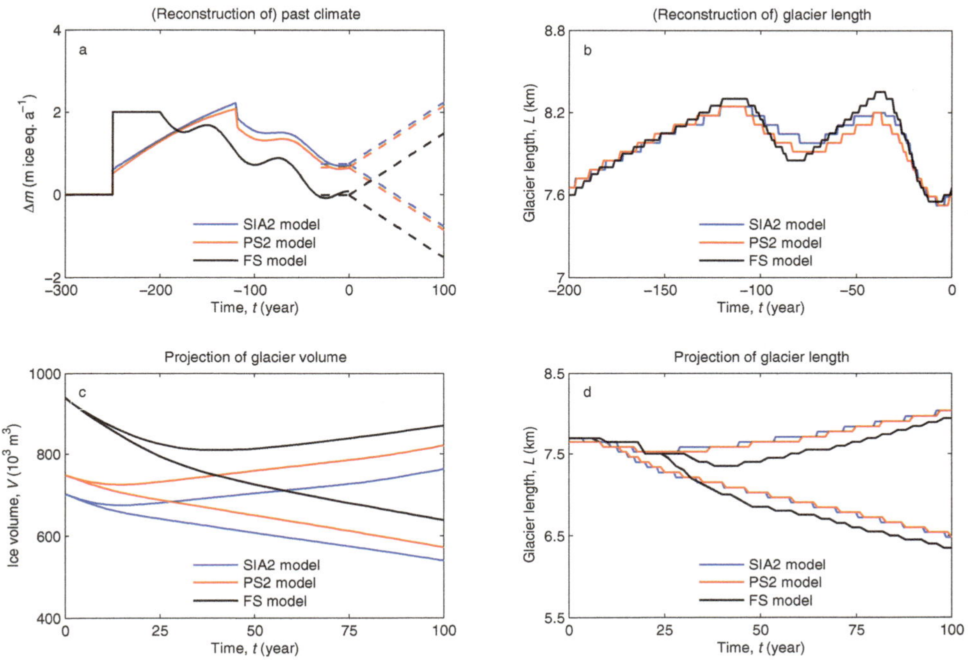

Fig. 5. Reconstructions of the past (a) climate and (b) glacier extent. A pre-defined $\Delta m(t)$ is imposed on the FS model to induce the evolution of glacier length. Using each of the PS2 and SIA2 models, the past $\Delta m(t)$ is then reconstructed so that the FS glacier extent is properly simulated. Future projections of (c) flowline ice volume and (d) glacier extent. Negative and positive times respectively indicate the past and future.

the model simulates such variations in the glacier length properly. The reconstructed glacier lengths and associated past climate are shown in Figure 5b and 5a, respectively. Due to the lack of high-order resistances, the PS2 and SIA2 models require a larger $\Delta m(t)$ in order to generate the (flowline) mass flux large enough to maintain the FS glacier lengths. With the corresponding climate history, the PS2 and SIA2 models are now ready to project the glacier's future.

With respect to the present climate, we consider two distinct scenarios for the future. The present climate is obtained by averaging the data of past 30 years. In addition to this, the future scenarios are defined as the linear changes in climate with $\Delta m(t) = \pm 1.5\,\mathrm{m}$ ice eq. a^{-1} by the end of $t = 100$ years (Fig. 5a). Under such climatic scenarios, we simulate each model to project the future dynamics of glacier. The evolutions of ice volume along the central flowline in a longitudinal band of unit width and glacier length are shown in Figures 5c and 5d, respectively. It is clear that the present time ($t = 0$ years) glacier lengths are similar in each model case; the corresponding ice volumes differ considerably, however. Short of tuning the ice dynamics (i.e. altering the ice viscosity or basal sliding rates), it is not possible to constrain the reduced models to obtain both the FS length and volume at the same time. As discussed before (Section 5.1), the reduced models hold relatively smaller ice thickness and hence underestimate the ice volume, even if they yield the same glacier length. We could constrain the models so that they simulate the evolution of ice volume properly; in such cases the associated glacier lengths, however, would have been overestimated. This is not recommended, becasue ice volumes are unknown for most real-world glaciers.

Based on the reactions of ice volume and glacier length (Figs. 5c–5d), we note the following: (1) initially both the ice volume and glacier length decrease in each model case, even in the positive mass balance scenario, as a consequence of the generally decreasing trend of the past mass balance; (2) the times after which the glaciers start gaining ice volume and length (for the positive $\Delta m(t)$ scenario) are roughly equivalent to the corresponding values of t_v and t_l, respectively; (3) in the latter half of simulations, all models follow the similar trend for both the volume and length evolutions; (4) interestingly, the PS2 and SIA2 models slightly overestimate the glacier lengths, when they forecast lower ice volumes. The projected ice volumes and lengths after $t = 100$ years are summarized in Table 1. Due to its quick response to climate change and lack of resistances associated with high-order dynamics, the SIA2 model predicts reduced ice volumes, respectively by 6.4% and 14.0% on an average, than the PS2 and FS models. The reduced models, however, predict longer glaciers by 2.1%; this is probably because the FS model needs a longer time ($t_l = 42$ years, compared to flowline models, i.e. $t_l \approx 21$ years) to fully adjust the glacier length.

6. Summary and outlook

Based on the distinct physical mechanisms of land-based glacier dynamics, we present a simplified classification of models, namely the SD, PS and FS models. SD models are simplest and are based on the shallow-ice approximation, where local driving stress is solely balanced by the basal drag. The FS models, which solve a complete set of Stokes equations, are the most comprehensive. In addition to the basal drag, they also capture the effects of longitudinal stress gradients and lateral drag. The PS models, the intermediate complexity models, only deal with the basal drag and the longitudinal stress gradients. The main advantage of having these classes of models is that the scope of each model is clear. Since the dominant physical mechanisms in a given glaciological condition can easily be identified, one can choose the optimal model accordingly to yield realistic simulations. In the interior of the large ice sheets where the shallow-ice theory holds, for example, the computationally expedient SD models are optimal. For proper simulations of a narrow and steep valley glacier, however, FS models are essential, although they are numerically intensive.

We consider three different models, one from each model family, and compare them numerically at several stages of valley glacier modelling. Results are summarized in Table 1. We find that: (1) the high-order resistances are crucial to control the dynamics of gravity-driven ice flow, (2) absence of such resistances makes the reduced models yield larger velocities in diagnostic simulations and reduced ice thickness in prognostic simulations, (3) simpler flowline models, without accounting for the effects of varying glacier width, are not sufficient to yield realistic estimates of response times; they predict response times that are 50% too rapid to 3D models, and (4) constraining a model for the particular glacier application is not straightforward, as it is difficult to properly simulate the ice volume, glacier length and surface velocities altogether.

	FS model	PS model	SD model	$e_{PS.FS}$	$e_{SD.FS}$	$e_{SD.PS}$
	Steady state geometry and field variables					
Volume, V (10^6 m^3)	679.5	607.5	561.3	−10.6	−17.4	−7.6
Thickness, h (m)	133.0	119.4	111.8	−10.2	−15.9	−6.4
Velocity, u_x (m a^{-1})	69.2	80.8	85.5	16.7	23.6	5.9
Stress, τ_{xz} (kPa)	154.6	176.3	187.7	-	17.7	6.1
	Response timescales					
Timescale, t_v (a)	26.0	25.0	23.5	−3.8	−9.6	−6.0
$	-	12.5	12.0	−51.9	−53.8	-
Timescale, t_l (a)	42.0	42.0	41.0	0.0	−2.4	−2.4
$	-	21.5	20.5	−48.8	−51.2	-
	Future projections (after $t = 100$ years)					
Volume, V (10^3 m^3)*	871.8	821.1	762.9	−5.8	−12.5	−7.1
#	637.7	571.6	539.0	−10.4	−15.5	−5.7
Length, L (km)*	7.9	8.0	8.0	1.8	1.8	0.0
#	6.4	6.5	6.5	2.1	2.1	0.0

Table 1. Numerical comparison of several models, one from each model family. Errors between the models, e.g. $e_{PS.FS}$, are given in percentage. Response timescales are also calculated for flowline models ($). The future projection results represent for the central flowline; values are listed for both the positive (*) and negative (#) mass balance (Fig. 5a)

There are more than 200 k small glaciers and ice caps on Earth; it is not feasible to use the numerically intensive FS model to simulate every valley glacier. Therefore, we mostly encounter simple flowline models, e.g. the SIA2 and PS2 models, being used in the valley glacier applications. Without adding the numerical complexity, the dynamical reach of such models can be extended through the introduction of parameterized correction factors. The effects of longitudinal stress gradients can be accounted for by embedding L-factors (Adhikari & Marshall, 2011); the lateral drag associated with the valley walls (Nye, 1965) and stick/slip basal interface (Adhikari & Marshall, in preperation) can also be captured via analogous correction factors. This offers a pragmatic middle ground for simulating glacier response to climate change.

Undoubtedly, the biggest challenge in glacier modelling is the lack of sufficient field data. Geometric and climatic data (e.g. basal and surface topographies, glacier length records, and mass balance fields), as well as observations of ice velocities, are not available in most cases. They are essential to justify the cost of using complex, 3D, FS models. Furthermore, the lack of proper theories and associated data to describe basal processes, e.g. basal sliding, is also a subject of concern that we have not discussed in the text; in many cases, poor characterization of basal flow is the limiting factor in modelling glacier dynamics.

7. Acknowledgements

We acknowledge support from the Western Canadian Cryospheric Network (WC^{2}N), funded by the Canadian Foundation for Climate and Atmospheric Sciences (CFCAS), and the Natural Sciences and Engineering Research Council (NSERC) of Canada.

8. Appendix

Appendix A: Diagnostic equations

We present the governing equations, with field variables u and p, for each model introduced in Section 3.1. The continuity equation (Eq. 5) holds in all cases; it takes the following respective form for 3D and 2D (flowline) models,

$$\frac{\partial u_x}{\partial x} + \frac{\partial u_y}{\partial y} + \frac{\partial u_z}{\partial z} = 0, \tag{A1}$$

$$\frac{\partial u_x}{\partial x} + \frac{\partial u_z}{\partial z} = 0. \tag{A2}$$

Hereafter, the velocity components, $\{u_x \ u_y \ u_z\}^T$, are simply denoted by $\{u \ v \ w\}^T$.

The differences between the models arise from the approximations made in the momentum balance equation itself (Eq. 6) and in the definition of strain-rate tensor (Eq. 4). Assuming that g_z (hereafter denoted by g) is the only non-zero component of gravity vector, we obtain the governing equations associated with the momentum balance via Equations (7) and (1–4). The algebraic process is straightforward; we simply quote the results for each model.

A1: The FS model

This full-system 3D model has no such approximations at all. Along with Equation (A1), followings are the governing equations for the FS model,

$$2\frac{\partial}{\partial x}\left(\eta\frac{\partial u}{\partial x}\right) + \frac{\partial}{\partial y}\left(\eta\frac{\partial u}{\partial y}\right) + \frac{\partial}{\partial z}\left(\eta\frac{\partial u}{\partial z}\right) + \frac{\partial}{\partial y}\left(\eta\frac{\partial v}{\partial x}\right) + \frac{\partial}{\partial z}\left(\eta\frac{\partial w}{\partial x}\right) + \frac{\partial p}{\partial x} = 0, \tag{A3}$$

$$\frac{\partial}{\partial x}\left(\eta\frac{\partial u}{\partial y}\right) + \frac{\partial}{\partial x}\left(\eta\frac{\partial v}{\partial x}\right) + 2\frac{\partial}{\partial y}\left(\eta\frac{\partial v}{\partial y}\right) + \frac{\partial}{\partial z}\left(\eta\frac{\partial v}{\partial z}\right) + \frac{\partial}{\partial z}\left(\eta\frac{\partial w}{\partial y}\right) + \frac{\partial p}{\partial y} = 0, \tag{A4}$$

$$\frac{\partial}{\partial x}\left(\eta\frac{\partial u}{\partial z}\right) + \frac{\partial}{\partial y}\left(\eta\frac{\partial v}{\partial z}\right) + \frac{\partial}{\partial x}\left(\eta\frac{\partial w}{\partial x}\right) + \frac{\partial}{\partial y}\left(\eta\frac{\partial w}{\partial y}\right) + 2\frac{\partial}{\partial z}\left(\eta\frac{\partial w}{\partial z}\right) + \frac{\partial p}{\partial z} + \rho g = 0, \tag{A5}$$

where the nonlinear effective viscosity, η, is given by Equation (2) with

$$2\dot{\epsilon}_e^2 = \frac{\partial^2 u}{\partial x^2} + \frac{\partial^2 v}{\partial y^2} + \frac{\partial^2 w}{\partial z^2} + \frac{1}{2}\left[\left(\frac{\partial u}{\partial y} + \frac{\partial v}{\partial x}\right)^2 + \left(\frac{\partial v}{\partial z} + \frac{\partial w}{\partial y}\right)^2 + \left(\frac{\partial w}{\partial x} + \frac{\partial u}{\partial z}\right)^2\right]. \tag{A6}$$

A2: The PS model

In this family of models, we make two approximations to ensure that the effects of lateral drag are completely absent. We neglect (1) the lateral variation of stress deviators, i.e. $\tau_{ij,y} = 0$, and (2) reduce the definition of $\dot{\epsilon}$ by excluding the lateral variation of ice velocities, i.e. $u_{i,y} = 0$. Along with Equation (A1), the **PS3 model** has the following governing equations,

$$2\frac{\partial}{\partial x}\left(\eta\frac{\partial u}{\partial x}\right) + \frac{\partial}{\partial z}\left(\eta\frac{\partial u}{\partial z}\right) + \frac{\partial}{\partial z}\left(\eta\frac{\partial w}{\partial x}\right) + \frac{\partial p}{\partial x} = 0, \tag{A7}$$

$$\frac{\partial}{\partial x}\left(\eta\frac{\partial v}{\partial x}\right) + \frac{\partial}{\partial z}\left(\eta\frac{\partial v}{\partial z}\right) + \frac{\partial p}{\partial y} = 0, \tag{A8}$$

$$\frac{\partial}{\partial x}\left(\eta\frac{\partial u}{\partial z}\right)+\frac{\partial}{\partial x}\left(\eta\frac{\partial w}{\partial x}\right)+2\frac{\partial}{\partial z}\left(\eta\frac{\partial w}{\partial z}\right)+\frac{\partial p}{\partial z}+\rho g=0, \tag{A9}$$

where the nonlinear effective viscosity, η, is given by Equation (2) with

$$2\dot{\epsilon}_e^2=\frac{\partial^2 u}{\partial x^2}+\frac{\partial^2 w}{\partial z^2}+\frac{1}{2}\left[\frac{\partial^2 v}{\partial x^2}+\frac{\partial^2 v}{\partial z^2}+\left(\frac{\partial w}{\partial x}+\frac{\partial u}{\partial z}\right)^2\right]. \tag{A10}$$

The **PS2 model** is a 2D Stokes model, strictly following the plane-strain approximations (after neglecting the non-zero τ_{yy}, which is required to maintain $\dot{\epsilon}_{yy}=0$). Equations (A2), (A7) and (A9) form the governing equations, where η is given by Equation (2) with

$$2\dot{\epsilon}_e^2=\frac{\partial^2 u}{\partial x^2}+\frac{\partial^2 w}{\partial z^2}+\frac{1}{2}\left(\frac{\partial w}{\partial x}+\frac{\partial u}{\partial z}\right)^2. \tag{A11}$$

A3: The SD model

This family of models does not account for the effects of longitudinal stress gradients, as well as those of lateral drag. Consequently, the vertical shear stresses are the only non-zero stress components. In addition, the reduced definition of $\dot{\epsilon}$ approximated in the PS models is also applied here. Along with Equation (A1), the governing equations for the **SD3 model** are,

$$\frac{\partial}{\partial z}\left(\eta\frac{\partial u}{\partial z}\right)+\frac{\partial}{\partial z}\left(\eta\frac{\partial w}{\partial x}\right)+\frac{\partial p}{\partial x}=0, \tag{A12}$$

$$\frac{\partial}{\partial z}\left(\eta\frac{\partial v}{\partial z}\right)+\frac{\partial p}{\partial y}=0, \tag{A13}$$

$$\frac{\partial}{\partial x}\left(\eta\frac{\partial u}{\partial z}\right)+\frac{\partial}{\partial x}\left(\eta\frac{\partial w}{\partial x}\right)+\frac{\partial p}{\partial z}+\rho g=0, \tag{A14}$$

where the nonlinear effective viscosity, η, is given by Equation (2) with

$$2\dot{\epsilon}_e^2=\frac{1}{2}\left[\frac{\partial^2 v}{\partial z^2}+\left(\frac{\partial w}{\partial x}+\frac{\partial u}{\partial z}\right)^2\right]. \tag{A15}$$

Equations (A2), (A12) and (A14) are the governing equations for the **SD2 model**, where η is given by Equation (2) with

$$2\dot{\epsilon}_e^2=\frac{1}{2}\left(\frac{\partial w}{\partial x}+\frac{\partial u}{\partial z}\right)^2. \tag{A16}$$

Further approximations are needed to obtain the standard SIA models; (1) the horizontal gradients in stresses are negligible, i.e. $\tau_{ij,x}=0$, and (2) the definition of $\dot{\epsilon}$ is further reduced by excluding the horizontal gradients of velocities, i.e. $u_{i,x}=0$. Hence, in addition to Equation (A1), the **SIA3 model** has the following governing equations,

$$\frac{\partial}{\partial z}\left(\eta\frac{\partial u}{\partial z}\right)+\frac{\partial p}{\partial x}=0, \tag{A17}$$

$$\frac{\partial}{\partial z}\left(\eta\frac{\partial v}{\partial z}\right)+\frac{\partial p}{\partial y}=0, \tag{A18}$$

$$\frac{\partial p}{\partial z} + \rho g = 0, \tag{A19}$$

where the nonlinear effective viscosity, η, is given by Equation (2) with

$$2\dot{\epsilon}_e^2 = \frac{1}{2}\left(\frac{\partial^2 v}{\partial z^2} + \frac{\partial^2 u}{\partial z^2}\right). \tag{A20}$$

The set of Equations (A2), (A17) and (A19) form the governing equations for the **SIA2 model**, where η is given by Equation (2) with

$$2\dot{\epsilon}_e^2 = \frac{1}{2}\frac{\partial^2 u}{\partial z^2}. \tag{A21}$$

With a simple algebraic manipulation of the governing equations for SIA2 model, one can obtain the analytical solution for $u(x,z)$ in a laminar flow (Eq. 8). The corresponding solution for $w(x,z)$ follows directly from the incompressibility criterion (Eq. A2).

Appendix B: FE formulation of diagnostic equations

Here, we construct a linear system of Equations (12) and (13) for the FS model; those for the reduced models can be obtained in a similar manner. The LHSs of governing equations (Eqs. A3–A5 and A1), after imposing the approximations of field variables, represent the system residuals, i.e. the terms inside the large parentheses in Equations (12–13). So, expanding Equation (13) in horizontal x axis, for example, gives

$$\int_{\Omega_e} \psi_i \left[2\frac{\partial\left(\eta\frac{\partial\tilde{u}}{\partial x}\right)}{\partial x} + \frac{\partial\left(\eta\frac{\partial\tilde{u}}{\partial y}\right)}{\partial y} + \frac{\partial\left(\eta\frac{\partial\tilde{u}}{\partial z}\right)}{\partial z} + \frac{\partial\left(\eta\frac{\partial\tilde{v}}{\partial x}\right)}{\partial y} + \frac{\partial\left(\eta\frac{\partial\tilde{w}}{\partial x}\right)}{\partial z} + \frac{\partial\tilde{p}}{\partial x} \right] \mathrm{d}\Omega_e = 0. \tag{B1}$$

Similar equations can be obtained for the second horizontal and vertical directions, and also for the incompressibility criterion (Eq. 12).

Applying Green-Gauss theorem to integrate Equation (B1) by parts (Rao, 2005), we obtain,

$$\begin{aligned}
&-\int_{\Omega_e} 2\frac{\partial\psi_i}{\partial x}\left(\eta\frac{\partial\tilde{u}}{\partial x}\right) \mathrm{d}x\mathrm{d}y\mathrm{d}z + \int_{\Gamma_e} 2\psi_i\left(\eta\frac{\partial\tilde{u}}{\partial x}\right) n_x \,\mathrm{d}S - \int_{\Omega_e} \frac{\partial\psi_i}{\partial y}\left(\eta\frac{\partial\tilde{u}}{\partial y}\right) \mathrm{d}x\mathrm{d}y\mathrm{d}z + \\
&+\int_{\Gamma_e} \psi_i\left(\eta\frac{\partial\tilde{u}}{\partial y}\right) n_y \,\mathrm{d}S - \int_{\Omega_e} \frac{\partial\psi_i}{\partial z}\left(\eta\frac{\partial\tilde{u}}{\partial z}\right) \mathrm{d}x\mathrm{d}y\mathrm{d}z + \int_{\Gamma_e} \psi_i\left(\eta\frac{\partial\tilde{u}}{\partial z}\right) n_z \,\mathrm{d}S + \\
&-\int_{\Omega_e} \frac{\partial\psi_i}{\partial y}\left(\eta\frac{\partial\tilde{v}}{\partial x}\right) \mathrm{d}x\mathrm{d}y\mathrm{d}z + \int_{\Gamma_e} \psi_i\left(\eta\frac{\partial\tilde{v}}{\partial x}\right) n_y \,\mathrm{d}S - \int_{\Omega_e} \frac{\partial\psi_i}{\partial z}\left(\eta\frac{\partial\tilde{w}}{\partial x}\right) \mathrm{d}x\mathrm{d}y\mathrm{d}z + \\
&+\int_{\Gamma_e} \psi_i\left(\eta\frac{\partial\tilde{w}}{\partial x}\right) n_z \,\mathrm{d}S - \int_{\Omega_e} \frac{\partial\psi_i}{\partial x}\tilde{p}\,\mathrm{d}x\mathrm{d}y\mathrm{d}z + \int_{\Gamma_e} \psi_i\tilde{p}n_x \,\mathrm{d}S = 0,
\end{aligned} \tag{B2}$$

where $\mathrm{d}S \in \Gamma_e$ is a boundary surface, and $\{n_x \;\; n_y \;\; n_z\}^T$ are the x, y and z components of the unit normal at the boundary surface.

By substituting approximations of field variables (Eqs. 10–11) into Equation (B2), we obtain

$$\int_{\Omega_e} \eta \left(2\frac{\partial \psi_i}{\partial x}\frac{\partial [\psi]}{\partial x} + \frac{\partial \psi_i}{\partial y}\frac{\partial [\psi]}{\partial y} + \frac{\partial \psi_i}{\partial z}\frac{\partial [\psi]}{\partial z} \right) \{u\}\, \mathrm{d}x\mathrm{d}y\mathrm{d}z + \int_{\Omega_e} \eta \frac{\partial \psi_i}{\partial y}\frac{\partial [\psi]}{\partial x} \{v\}\, \mathrm{d}x\mathrm{d}y\mathrm{d}z$$
$$+ \int_{\Omega_e} \eta \frac{\partial \psi_i}{\partial z}\frac{\partial [\psi]}{\partial x} \{w\}\, \mathrm{d}x\mathrm{d}y\mathrm{d}z + \int_{\Omega_e} \frac{\partial \psi_i}{\partial x} [\psi]\{p\}\, \mathrm{d}x\mathrm{d}y\mathrm{d}z = \tag{B3}$$
$$\int_{\Gamma_e} \psi_i \left[\left(2\eta \frac{\partial \tilde{u}}{\partial x} + \tilde{p} \right) n_x + \eta \left(\frac{\partial \tilde{u}}{\partial y} + \frac{\partial \tilde{v}}{\partial x} \right) n_y + \eta \left(\frac{\partial \tilde{u}}{\partial z} + \frac{\partial \tilde{w}}{\partial x} \right) n_z \right] \mathrm{d}S.$$

Here, the RHS surface integral forms a Neumann or natural boundary condition. With Equations (4), (1), and (7), the terms inside large parentheses take the following forms,

$$\left(2\eta \frac{\partial \tilde{u}}{\partial x} + \tilde{p} \right) n_x + \eta \left(\frac{\partial \tilde{u}}{\partial y} + \frac{\partial \tilde{v}}{\partial x} \right) n_y + \eta \left(\frac{\partial \tilde{u}}{\partial z} + \frac{\partial \tilde{w}}{\partial x} \right) n_z = \sigma_{xx} n_x + \sigma_{xy} n_y + \sigma_{xz} n_z. \tag{B4}$$

The RHS term above indicates the horizontal x component of the Cauchy stress tensor on the boundary surface. Denoting this by σ_x, we rewrite Equation (B3) as,

$$\int_{\Omega_e} \eta \left(2\frac{\partial \psi_i}{\partial x}\frac{\partial [\psi]}{\partial x} + \frac{\partial \psi_i}{\partial y}\frac{\partial [\psi]}{\partial y} + \frac{\partial \psi_i}{\partial z}\frac{\partial [\psi]}{\partial z} \right) \{u\}\, \mathrm{d}x\mathrm{d}y\mathrm{d}z + \int_{\Omega_e} \eta \frac{\partial \psi_i}{\partial y}\frac{\partial [\psi]}{\partial x} \{v\}\, \mathrm{d}x\mathrm{d}y\mathrm{d}z$$
$$+ \int_{\Omega_e} \eta \frac{\partial \psi_i}{\partial z}\frac{\partial [\psi]}{\partial x} \{w\}\, \mathrm{d}x\mathrm{d}y\mathrm{d}z + \int_{\Omega_e} \frac{\partial \psi_i}{\partial x} [\psi]\{p\}\, \mathrm{d}x\mathrm{d}y\mathrm{d}z = \int_{\Gamma_e} \psi_i \sigma_x\, \mathrm{d}S. \tag{B5}$$

Similar equations can be written, respectively for horizontal y and vertical z directions,

$$\int_{\Omega_e} \eta \frac{\partial \psi_i}{\partial x}\frac{\partial [\psi]}{\partial y} \{u\}\, \mathrm{d}x\mathrm{d}y\mathrm{d}z + \int_{\Omega_e} \eta \left(\frac{\partial \psi_i}{\partial x}\frac{\partial [\psi]}{\partial x} + 2\frac{\partial \psi_i}{\partial y}\frac{\partial [\psi]}{\partial y} + \frac{\partial \psi_i}{\partial z}\frac{\partial [\psi]}{\partial z} \right) \{v\}\, \mathrm{d}x\mathrm{d}y\mathrm{d}z$$
$$+ \int_{\Omega_e} \eta \frac{\partial \psi_i}{\partial z}\frac{\partial [\psi]}{\partial y} \{w\}\, \mathrm{d}x\mathrm{d}y\mathrm{d}z + \int_{\Omega_e} \frac{\partial \psi_i}{\partial y} [\psi]\{p\}\, \mathrm{d}x\mathrm{d}y\mathrm{d}z = \int_{\Gamma_e} \psi_i \sigma_y\, \mathrm{d}S, \tag{B6}$$

$$\int_{\Omega_e} \eta \frac{\partial \psi_i}{\partial x}\frac{\partial [\psi]}{\partial z} \{u\}\, \mathrm{d}x\mathrm{d}y\mathrm{d}z + \int_{\Omega_e} \eta \frac{\partial \psi_i}{\partial y}\frac{\partial [\psi]}{\partial z} \{v\}\, \mathrm{d}x\mathrm{d}y\mathrm{d}z$$
$$\int_{\Omega_e} \eta \left(\frac{\partial \psi_i}{\partial x}\frac{\partial [\psi]}{\partial x} + \frac{\partial \psi_i}{\partial y}\frac{\partial [\psi]}{\partial y} + 2\frac{\partial \psi_i}{\partial z}\frac{\partial [\psi]}{\partial z} \right) \{w\}\, \mathrm{d}x\mathrm{d}y\mathrm{d}z + \int_{\Omega_e} \frac{\partial \psi_i}{\partial z} [\psi]\{p\}\, \mathrm{d}x\mathrm{d}y\mathrm{d}z \tag{B7}$$
$$= \int_{\Omega_e} \rho g \psi_i\, \mathrm{d}x\mathrm{d}y\mathrm{d}z + \int_{\Gamma_e} \psi_i \sigma_z\, \mathrm{d}S,$$

where $\sigma_y \left(= \sigma_{xy} n_x + \sigma_{yy} n_y + \sigma_{yz} n_z\right)$ and $\sigma_z \left(= \sigma_{xz} n_x + \sigma_{yz} n_y + \sigma_{zz} n_z\right)$ are y and z components of σ on the boundary surface, respectively. For the stress free boundary condition, the surface integrals in Equations (B5–B7) become zeros.

The incompressibility criterion (Eq. 12) can simply be expanded as,

$$\int_{\Omega_e} \psi_i \frac{\partial [\psi]}{\partial x} \{u\}\, \mathrm{d}x\mathrm{d}y\mathrm{d}z + \int_{\Omega_e} \psi_i \frac{\partial [\psi]}{\partial y} \{v\}\, \mathrm{d}x\mathrm{d}y\mathrm{d}z + \int_{\Omega_e} \psi_i \frac{\partial [\psi]}{\partial z} \{w\}\, \mathrm{d}x\mathrm{d}y\mathrm{d}z = 0. \tag{B8}$$

Now, we write the linear equations (Eqs. B5–B8) in the following elemental matrix form,

$$[K]\{\phi\} = \{F\}, \tag{B9}$$

where $[K]$ is the coefficient matrix, $\{\phi\}$ is comprised of unknown field variables, and $\{F\}$ is the RHS force vector. They are given by,

$$\begin{bmatrix} [K_*^{11}] & [K^{12}] & [K^{13}] & [K^{10}] \\ [K^{21}] & [K_*^{22}] & [K^{23}] & [K^{20}] \\ [K^{31}] & [K^{32}] & [K_*^{33}] & [K^{30}] \\ [K^{10}]^T & [K^{20}]^T & [K^{30}]^T & [0] \end{bmatrix} \begin{Bmatrix} \{u\} \\ \{v\} \\ \{w\} \\ \{p\} \end{Bmatrix} = \begin{Bmatrix} \{F^1\} \\ \{F^2\} \\ \{F^3\} \\ \{0\} \end{Bmatrix}, \tag{B10}$$

Components of the coefficient matrix and the force vector are as follow,

$$[K_*^{\alpha\alpha}] = [K^{\alpha\alpha}] + \left[K^{11}\right] + \left[K^{22}\right] + \left[K^{33}\right]; \ \alpha = 1,2,3, \tag{B11}$$

$$K_{ij}^{\alpha\beta} = \int_{\Omega_e} \eta \frac{\partial\psi_i}{\partial\chi_\beta} \frac{\partial\psi_j}{\partial\chi_\alpha} \,\mathrm{d}x\mathrm{d}y\mathrm{d}z; \ \alpha,\beta = 1,2,3, \tag{B12}$$

$$K_{ij}^{\alpha 0} = \int_{\Omega_e} \eta \frac{\partial\psi_i}{\partial\chi_\alpha} \psi_j \,\mathrm{d}x\mathrm{d}y\mathrm{d}z; \ \alpha = 1,2,3, \tag{B13}$$

$$F^\alpha = \int_{\Gamma_e} \psi_i \sigma_\alpha \,\mathrm{d}S; \ \alpha = 1,2, \tag{B14}$$

$$F^3 = \int_{\Omega_e} \rho g \psi_i \,\mathrm{d}x\mathrm{d}y\mathrm{d}z + \int_{\Gamma_e} \psi_i \sigma_z \,\mathrm{d}S, \tag{B15}$$

where χ denotes the Cartesian coordinates with subscripts $(1,2,3)$ for (x,y,z). Similarly, (σ_1,σ_2) in Equation (B14) denote (σ_x,σ_y), respectively.

Now, we assemble the elemental equation (Eq. B9) to get the global equation,

$$[\mathbf{K}]\{\mathbf{\Phi}\} = \{\mathbf{F}\}, \tag{B16}$$

where $[\mathbf{K}] = \sum_{e=1}^{E}[K]$, $\{\mathbf{\Phi}\} = \sum_{e=1}^{E}\{\phi\}$, $\{\mathbf{F}\} = \sum_{e=1}^{E}\{F\}$, and E is the total number of elements e within the global domain Ω.

Appendix C: Diagnostic system stabilization

We use superscript of *stb* to refer to the stabilization contributions to the elemental matrices and vectors. With stabilization terms, Equation (B9) takes the following form

$$\left[[K] + \left[K^{stb}\right]\right]\{\phi\} = \left\{\{F\} + \left\{F^{stb}\right\}\right\}. \tag{C1}$$

For the FS model, the residual matrix, $[R_\sigma]$, associated with the momentum balance equation (see Eq. 14) can be obtained from Equations (A3–A5),

$$[R_\sigma] = \begin{bmatrix} [R_*^{11}] & [R^{12}] & [R^{13}] & [R^{10}] \\ [R^{21}] & [R_*^{22}] & [R^{23}] & [R^{20}] \\ [R^{31}] & [R^{32}] & [R_*^{33}] & [R^{30}] \\ [0] & [0] & [0] & [0] \end{bmatrix}, \tag{C2}$$

where

$$[R_*^{\alpha\alpha}] = [R^{\alpha\alpha}] + \left[R^{11}\right] + \left[R^{22}\right] + \left[R^{33}\right]; \ \alpha = 1,2,3, \tag{C3}$$

$$R_i^{\alpha\beta} = \frac{\partial}{\partial \chi_\beta}\left(\eta \frac{\partial \psi_i}{\partial \chi_\alpha}\right); \ \alpha, \beta = 1,2,3, \tag{C4}$$

$$R_i^{\alpha 0} = \frac{\partial \psi_i}{\partial \chi_\alpha}; \ \alpha = 1,2,3, \tag{C5}$$

Similarly, the corresponding residual matrix, $[R_u]$, associated with the continuity equation is obtained from Equation (A1),

$$[R_u] = \left[\begin{array}{cccc} \frac{\partial \psi_i}{\partial x} & \frac{\partial \psi_i}{\partial y} & \frac{\partial \psi_i}{\partial z} & [0] \end{array}\right]. \tag{C6}$$

Referring to Equations (14–15), the stabilization contributions to the elemental matrix and the RHS force vector take the following respective forms,

$$\left[K^{stb}\right] = \delta_1 \left[R_\sigma\right]^T \left[R_\sigma\right] + \delta_2 \left[R_u\right]^T \left[R_u\right], \tag{C7}$$

$$\left[F^{stb}\right] = \delta_1 \left[F\right]^T \left[R_\sigma\right], \tag{C8}$$

where $[F]$ is the force vector, given in the RHS of Equation (B10).

Appendix D: FE formulation and stabilization of prognostic equations

Equation (18), in a three-dimensional space, can be expanded as,

$$\int_{\Gamma_e} \psi_i^s \left[\frac{\partial \tilde{s}}{\partial t} + u\frac{\partial \tilde{s}}{\partial x} + v\frac{\partial \tilde{s}}{\partial y} - (w+m)\right] \mathrm{d}\Gamma_e = 0. \tag{D1}$$

With approximation of field variable (Eq. 17), Equation (D1) becomes,

$$\int_{\Gamma_e} \psi_i \left[\psi\right] \frac{\partial \{s\}}{\partial t} \,\mathrm{d}S + \int_{\Gamma_e} \psi_i \left(u\frac{\partial [\psi]}{\partial x} + v\frac{\partial [\psi]}{\partial y}\right) \{s\} \,\mathrm{d}S = \int_{\Gamma_e} \psi_i \left(w+m\right) \mathrm{d}S, \tag{D2}$$

which can be expressed in the following elemental matrix form,

$$\left[[M] + \left[M^{stb}\right]\right] \frac{\partial \{\phi\}}{\partial t} + \left[[K] + \left[K^{stb}\right]\right] \{\phi\} = \left\{\{F\} + \left\{F^{stb}\right\}\right\}, \tag{D3}$$

where

$$[M] = \left[\int_{\Gamma_e} \psi_i \psi_j \,\mathrm{d}S\right], \ \left[M^{stb}\right] = \langle \psi_i, \delta_3 R_s \rangle, \tag{D4}$$

$$[K] = \left[\int_{\Gamma_e} \psi_i \left(u\frac{\partial \psi_j}{\partial x} + v\frac{\partial \psi_j}{\partial y}\right) \mathrm{d}S\right], \ \left[K^{stb}\right] = \langle R_s, \delta_3 R_s \rangle, \tag{D5}$$

$$\{F\} = \left\{\int_{\Gamma_e} \psi_i \left(w+m\right) \mathrm{d}S\right\}, \ \left\{F^{stb}\right\} = \langle (w+m), \delta_3 R_s \rangle, \tag{D6}$$

$$[R_s] = \left[u\frac{\partial \psi_i}{\partial x} + v\frac{\partial \psi_i}{\partial y}\right], \tag{D7}$$

Here, δ_3 is the stability parameter that is a function of element size, h_k, and the norm of velocity vector, such that $\delta_3 = \frac{h_k}{2\|u\|}$. The mass matrix, $[M]$, deals with the evolution of glacier surface. Matrices and vectors with superscript *stb* are once again the stabilization contributions. The elemental equation (Eq. D3) can be assembled into the global matrix, as for Equation (B9).

9. References

Adhikari, S. & Huybrechts, P. (2009). Numerical modelling of historical front variations and the 21st-century evolution of glacier AX010, Nepal Himalaya, *Ann. Glaciol.* 50(52): 27–34.

Adhikari, S. & Marshall, S. (2011). Improvements to shear-deformational models of glacier dynamics through a longitudinal stress factor, *J. Glaciol.* 57(206): 1003–1016.

Adhikari, S. & Marshall, S. (in preperation). Parameterization of slip-induced lateral drag in flowline models of glacier dynamics.

Allen, S., Schneider, D. & Owens, I. (2009). First approaches towards modelling glacial hazards in the Mount Cook region of New Zealand's Southern Alps, *Nat. Hazard Earth Sys.* 9(2): 481–499.

Alley, R. (1992). Flow-law hypotheses for ice-sheet modeling, *J. Glaciol.* 38(129): 245–256.

Bamber, J., Layberry, R. & Gogineni, S. (2001). A new ice thickness and bed data set for the greenland ice sheet, 1. Measurement, data reduction, and errors, *J. Geophys. Res.* 106: 33733–33780.

Baral, D., Hutter, K. & Greve, R. (2001). Asymptotic theories of large-scale motion, temperature, and moisture distribution in land-based polythermal ice sheets: a critical review and new developments, *Appl. Mech. Rev.* 54: 215–256.

Beedle, M., Menounos, B., Luckman, B. & Wheate, R. (2009). Annual push moraines as climate proxy, *Geophys. Res. Lett.* 36: L20501.

Blatter, H. (1995). Velocity and stress fields in grounded glaciers: a simple algorithm for including deviatoric stress gradients, *J. Glaciol.* 41(138): 333–344.

Blatter, H., Greve, R. & Abe-Ouchi, A. (2010). A short history of the thermomechanical theory and modelling of glaciers and ice sheets, *J. Glaciol.* 56(200): 1087–1094.

Budd, W. & Jacka, T. (1989). A review of ice rheology for ice sheet modelling, *Cold Re. Sci. Technol.* 16: 107–144.

Budd, W. & Jenssen, D. (1975). Numerical modelling of glacier systems, *IASH Publ.* 104: 257–291.

Bueler, E., Brown, J. & Lingle, C. (2007). Exact solutions to the thermomechanically coupled shallow-ice approximation: effective tools for varification, *J. Glaciol.* 53(182): 499–516.

Calov, R. & Hutter, K. (1996). Thermo mechanical response of the Greenland ice sheet to various climate scenarios, *Clim. Dynam.* 12(4): 243–260.

Campbell, W. & Rasmussen, L. (1969). Three-dimensional surges and recoveries in a numerical glacier model, *J. Glaciol.* 6(4): 979–986.

Chow, S., Carey, G. & Anderson, M. (2004). Finite element approximations of a glaciological problem, *ESAIM-Math. Model Num.* 38(5): 741–756.

Christensen, J., Hewitson, B., Busuioc, A., Chen, A., Gao, X., Held, I., Jones, R., Kolli, R., Kwon, W.-T., Laprise, R., Magaña Rueda, V., Mearns, L., Menéndez, C., Räisänen, J., Rinke, A., Sarr, A. & Whetton, P. (2007). Regional climate projections, *in* S. Solomon & 7 others (eds), *Climate Change 2007: The Physical Science Basis. Contribution of WG I to the Fourth Assessment Report of the IPCC*, Cambridge Univ. Press, Cambridge, pp. 847–940.

Clarke, G. (1987). A short history of scientific investigations on glaciers, *J. Glaciol.* Special Issue: 4–24.

Colinge, J. & Blatter, H. (1998). Stress and velocity fields in glaciers: Part I. Finite-difference schemes for higher-order glacier models, *J. Glaciol.* 44: 448–456.

Colinge, J. & Rappaz, J. (1999). A strongly nonlinear problem arising in glaciology, *ESAIM-Math. Model Num.* 33(2): 395–406.

Cook, A. J., Fox, A. J., Vaughan, D. G. & Ferrigno, J. G. (2005). Retreating glacier fronts on the antarctic peninsula over the past half-century, *Science* 308(5721): 541–544.

Cuffey, K. & Paterson, W. (2010). *The Physics of Glaciers*, fourth edn, Butterworth Heinemann, Elsevier.

Donea, J. & Huerta, A. (2003). *Finite Element Methods for Flow Problems*, John Wiley and Sons.

Dyurgerov, M. & Meier, M. (2005). *Glaciers and the Changing Earth System: A 2004 snapshot*, Occasional paper 58, Institute of Arctic and Alpine Research, U. Colorado, Boulder.

Franca, L. & Frey, S. (1992). Stabilized finite element methods: II the incompressibility Navier-Stokes equations, *Comput. Method Appl. M.* 99: 209–233.

Gagliardini, O. & Zwinger, T. (2008). The ISMIP-HOM benchmark experiments performed using the Finite-Element code Elmer, *The Cryosphere* 2(1): 67–76.

Gillet-Chaulet, F., Gagliardini, O., Meyssonnier, J., Montagnat, M. & Castelnau, O. (2005). A user-friendly anisotropic fow law for ice-sheet modelling, *J. Glaciol.* 51(172): 3–14.

Gillett, N., Arora, V., Zickfeld, K., Marshall, S. & Merryfield, W. (2011). Ongoing climate change following a complete cessation of carbon dioxide emissions, *Nature Geosci.* 4: 83–87.

Glen, J. (1952). Experiments on the deformation of ice, *J. Glaciol.* 2(12): 111–114.

Glen, J. (1955). The creep of polycrystalline ice, *Proc. R. Soc. London, Ser. A* 228: 519–538.

Glen, J. (1958). The flow law of ice: a discussion of the assumptions made in glacier theory, their experimental foundations and consequences, *Int. Assoc. Hydrol. Sci. Pub.* 47: 171–183.

Glowinski, R. & Rappaz, J. (2003). Approximation of a nonlinear elliptic problem arising in a non-Newtonian fluid flow model in glaciology, *ESAIM-Math. Model Num.* 37(1): 175–186.

Goelzer, H., Huybrechts, P., Loutre, M., Goosse, H., Fichefet, T. & Mouchet, A. (2011). Impact of Greenland and Antarctic ice sheet interactions on climate sensitivity, *Clim. Dynam.* 37: 1005–1018.

Gudmundsson, G. (1999). A three-dimensional numerical model of the confluence area of Unteraargletscher, Bernese Alps, Switzerland, *J. Glaciol.* 45(150): 219–230.

Hanson, B. (1995). A fully three-dimensional finite-element model applied to velocities on Storglaciaren, Sweden, *J. Glaciol.* 41 (137): 91–102.

Hindmarsh, R. (2004). A numerical comparison of approximations to the stokes-equations used in the ice sheet and glacier modeling, *J. Geophys. Res.* 109: F01012.

Hoffman, P. & Schrag, D. (2000). Snowball earth, *Sci. Am.* 282: 62–75.

Hood, P. & Taylor, C. (1974). Navier-Stokes equations using mixed interpolation, *in* J. Oden, O. Zienkiewicz, R. Gallagher & C. Taylor (eds), *Finite Element Methods in Flow Problems*, UAH Press, pp. 121–132.

Hooke, R. (1981). Flow law for polycrystalline ice in glaciers: comparison of theoretical predictions, laboratory data and field measurements, *Rev. Geophys. Space Phys.* 19: 664–672.

Hutter, K. (1983). *Theoretical Glaciology: Material Science of Ice and the Mechanics of Glaciers and Ice Sheets*, Reidel, Dordrecht, Netherlands.

Huybrechts, P., Goelzer, H., Janssens, I., Driesschaert, E., Fichefet, T., Goosse, H. & Loutre, M. (2011). Response of the Greenland and Antarctic ice sheets to multi-millennial greenhouse warming in the Earth system model of intermediate complexity LOVECLIM, *Surv. Geophy.* pp. 1–20. Published online, doi: 10.1007/s10712-011-9131-5.

Huybrechts, P. & Oerlemans, J. (1988). Evolution of the East Antarctic ice sheet: a numerical study of thermo-mechanical response patterns with changing climate, *Ann. Glaciol.* 11: 52–59.

Huybrechts, P., Payne, T. & the EISMINT intercomparison group (1996). The EISMINT benchmarks for testing ice sheets models, *Ann. Glaciol.* 23: 1–12.

Jansson, P., Hock, R. & Schneider, T. (2003). The concept of glacier storage: a review, *J. Hydrol.* 282(1–4): 116–129.

Jarosch, A. (2008). Icetools: a full stokes finite element model for glaciers, *Comput. Geosci.* 34: 1005–1014.

Jenssen, D. (1977). A three-dimensional polar ice-sheet model, *J. Glaciol.* 18(80): 373–389.

Jóhannesson, T., Raymond, C. & Waddington, E. (1989). Time-scale for adjustment of glaciers to changes in mass balance, *J. Glaciol.* 35(121): 355–369.

Jouvet, G., Picasso, M., Rappaz, J. & Blatter, H. (2008). A new algorithm to simulate the dynamics of a glacier: theory and applications, *J. Glaciol.* 54(188): 801–811.

Kaser, G. (2001). Glacier-climate interaction at low latitudes, *J. Glaciol.* 47(157): 195–204.

Kirschvink, J. (1992). Late Proterozoic low-latitude global glaciation: the snowball earth, *in* J. W. Schopf & C. Klein (eds), *The Proterozoic Biosphere*, Cambridge Univ. Press, Cambridge, pp. 51–52.

Krabill, W., Frederick, E., Manizade, S., Martin, C., Sonntag, J., Swift, R., Thomas, R., Wright, W. & Yungel, J. (1999). Rapid thinning of parts of the southern greenland ice sheet, *Science* 283(5407): 1522–1524.

Le Meur, E., Gagliardini, O., Zwinger, T. & Ruokolainen, J. (2004). Glacier flow modeling: a comparison of the shallow ice approximation and the full-stokes solution, *CR Physique* 5: 709–722.

Leclercq, P., Oerlemans, J. & Cogley, J. (2011). Estimating the glacier contribution to sea-level rise for the period 1800-2005, *Surv. Geophy.* pp. 1–17. Published online, doi: 10.1007/s10712-011-9121-7.

Lemke, P., Ren, J., Alley, R., Allison, I., Carrasco, J., Flato, G., Fujii, Y., Kaser, G., Mote, P., Thomas, R. & Zhang, T. (2007). Observations: Changes in snow, ice and frozen ground, *in* S. Solomon & 7 others (eds), *Climate Change 2007: The Physical Science Basis. Contribution of WG I to the Fourth Assessment Report of the IPCC*, Cambridge Univ. Press, Cambridge, pp. 337–383.

Leysinger Vieli, G. & Gudmundsson, G. (2004). On estimating length fluctuations of glaciers caused by changes in climate forcing, *J. Geophys. Res.* 109: F0 1007.

Lythe, M., Vaughan, D. & the BEDMAP Group (2001). BEDMAP: a new ice thickness and subglacial topographic model of Antarctica, *J. Geophys. Res.* 106(B6): 11335–11351.

MacAyeal, D. (1989). Large-scale ice flow over a viscous basal sediment: Theory and application to Ice Stream B, Antarctica, *J. Geophys. Res.* 94(B4): 4071–4087.

Mahaffy, M. (1976). A three-dimensional numerical model of ice sheets: tests on the Barnes Ice Cap, Northwest Territories, *J. Geophys. Res.* 81(6): 1059–1066.

Marshall, S. (2005). Recent advances in understanding ice sheet dynamics, *Earth Planet Sc. Lett.* 240: 191–204.

Marshall, S. & Clarke, G. (1997). A continuum mixture model of ice stream thermomechanics in the Laurentide Ice Sheet 1. Theory, *J. Geophys. Res.* 102(B9): 20599–20613.

Meier, M. (1984). Contribution of small glaciers to global sea level, *Science* 226(4681): 1418–1421.

Morland, L. (1984). Thermo mechanical balances of ice sheet flows, *Geophys. Astrophys. Fluid Dyn.* 29(1–4): 237–266.

Morland, L. & Staroszczyk, R. (2003). Strain-rate formulation of ice fabric evolution, *Ann. Glaciol.* 37: 35–39.

Nye, J. (1952). The mechanics of glacier flow, *J. Glaciol.* 2(12): 82–93.

Nye, J. (1953). The flow law of ice from measurements in glacier tunnels, laboratory experiments and the Jungfraufirn borehole experiment, *Proc. R. Soc. London, Ser. A* 219(1139): 477–489.

Nye, J. (1959). The motion of ice sheets and glaciers, *J. Glaciol.* 3(26): 493–507.

Nye, J. (1965). The flow of a glacier in a channel of rectangular, elliptical or parabolic cross-section, *J. Glaciol.* 5(41): 661–690.

Oerlemans, J. (1982). A model of the Antarctic ice sheet, *Nature* 297(5867): 550–553.

Oerlemans, J. (2005). Extracting a climate signal from 169 glacier record, *Science* 308(5722): 675–677.

Oerlemans, J., Anderson, B., Hubbard, A., Huybrechts, P., Jóhannesson, T., Knap, W., Schmeits, M., Stroeven, A., de Wal, R. V., Wallinga, J. & Zuo, Z. (1998). Modelling the response of glaciers to climate warming, *Clim. Dynam.* 14: 267–274.

Ohmura, A. (2004). Cryosphere during the twentieth century, *in* R. Sparks & C. Hawkesworth (eds), *The state of the planet: Frontiers and challenges in geophysics*, Geophysical monograph 150, pp. 239–257.

Paterson, W. & Budd, W. (1982). Flow parameters for ice-sheet modeling, *Cold Reg. Sci. Technol.* 6: 175–177.

Pattyn, F. (2003). A new three-dimensional higher-order thermomechanical ice sheet model: basic sensitivity, ice stream development, and ice flow across subglacial lakes, *J. Geophys. Res.* 108: B82382.

Pattyn, F., Perichon, L. & 19 others (2008). Benchmark experiments for higher-order and full-stokes ice sheet models (ISMIP-HOM), *The Cryosphere* 2: 95–108.

Payne, A., Huybrechts, P. & 9 others (1996). Results from the EISMINT model intercomparison: the effects of thermomechanical coupling, *J. Glaciol.* 46(153): 227–238.

Picasso, M., Rappaz, J. & Reist, A. (2008). Numerical simulation of the motion of a three-dimensional glacier, *Ann. Math. Blaise Pascal* 15: 1–28.

Price, S., Waddington, E. & Conway, H. (2007). A full-stress, thermomechanical flow band model using the finite volume method, *J. Geophys. Res.* 112(F3): F03020.

Rao, S. (2005). *The Finite Element Method in Engineering*, Elsevier Butterworth–Heinemann.

Raper, S. & Braithwaite, R. (2006). Low sea level rise projections from mountain glaciers and icecaps under global warming, *Nature* 439: 311–313.

Rigsby, G. (1958). Effect of hydrostatic pressure on the velocity of shear deformation of single ice crystals, *J. Glaciol.* 3: 273–278.

Robin, G. (1955). Ice movement and temperature distribution in glaciers and ice sheets, *J. Glaciol.* 2(18): 523–532.

Sargent, A. & Fastook, J. (2010). Manufactured analytical solutions for isothermal full-stokes ice sheet models, *The Cryosphere* 4: 285–311.

Shoemaker, E. & Morland, L. (1984). A glacier flow model incorporating longitudinal deviatoric stress, *J. Glaciol.* 30(106): 334–340.

Souček, O. & Martinec, Z. (2008). Iterative improvement of the shallow-ice approximation, *J. Glaciol.* 54(188): 812–822.

Thompson, L., Mosley-Thompson, E., Davis, M., Lin, P.-N., Henderson, K. & Mashiotta, T. (2003). Tropical glacier and ice core evidence of climate change on annual to millennial time scales, *Climatic Change* 59: 137–155.

Van der Veen, C. (1999). *Fundamentals of Glacier Dynamics*, AA Balkema, Rotterdam.

Viviroli, D., Weingartner, R. & Messerli, B. (2003). Assessing the hydrological significance of the world's mountains, *Mt. Res. Dev.* 23: 32–40.

Waddington, E. (2010). Life, death and afterlife of the extrusion flow theory, *J. Glaciol.* 56(200): 973–996.

Wang, W. & Warner, R. (1999). Modelling of anisotropic ice flow in Law Dome, East Antarctica, *Ann. Glaciol.* 29: 184–190.

Weertman, J. (1973). Creep of ice, *in* E. Whalley, S. Jones & L. Gold (eds), *Physics and Chemistry of Ice*, Royal Society of Canada, Ottawa, pp. 320–337.

Whillans, I. (1987). Force budget of ice sheets, *in* C. Van der Veen & J. Oerlemans (eds), *Dynamics of the West Antarctic Ice Sheet*, Reidel Publ. Co., Dordrecht, pp. 17–36.

Whillans, I. & Van der Veen, C. (1997). The role of lateral drag in the dynamics of Ice Stream B, Antarctica, *J. Glaciol.* 43(144): 231–237.

Zemp, M., Haeberli, W., Hoelzle, M. & Paul, F. (2006). Alpine glaciers to disappear within decades?, *Geophys. Res. Lett* 33: L13504.

Zwinger, T., Greve, R., Gagliardini, O., Shiraiwa, T. & Lyly, M. (2007). A full stokes-flow thermo-mechanical model for firn and ice applied to the Gorshkov crater glacier, Kamchatka, *Ann. Glaciol.* 45: 29–37.

6

Numerical Modelling of Heavy Metals Transport Processes in Riverine Basins

Seyed Mahmood Kashefipour[1] and Ali Roshanfekr[2]
[1]*Department of Hydraulic Structures, Shahid Chamran University, Ahwaz, Khuzestan,*
[2]*Department of Civil and Resource Engineering, Dalhousie University, Halifax, NS*
[1]*Iran*
[2]*Canada*

1. Introduction

There has been a growing concern in the international community and an increased awareness of riverine pollution problems, particularly with regard to water pollution (Falconer & Lin, 2003). Human and aquatic life is often threatened by the transport of pollutants through riverine systems to coastal waters and it is therefore not surprising to find that, from a water quality point of view, rivers have been studied very extensively and for longer than any other water bodies (Thomann & Mueller, 1987). This is probably due to the fact that people live close to, or interact with, rivers and streams. Many rivers and estuaries have suffered environmental damage due to discharges from manufacturing processes and wastewater from centres of pollution over several decades. In recent years these environmental concerns have made the development of computer models that predict the dispersion of pollutants in natural water systems more urgent. The main attraction of such models, in contrast with physical models, is their low cost and the fact that they easily adapt to new situations. Thus the widespread popularity of mathematical modelling techniques for the hydrodynamic and pollutant transport in rivers justifies any attempt to develop new models based on novel and rigorous approaches (Nassehi & Bikangaga, 1993).

This chapter describes numerical modelling of heavy metals in a riverine basin. It would be necessary to recognize and introduce the heavy metals behaviours and different processes during their transportation along the rivers; for example their sources, chemical and physical reactions, and also introducing the environmental conditions affecting the rate of concentration variability of these substances. The one-dimensional (1D) partial differential governing equations (PDE) of hydrodynamic and water quality will be fully described with the corresponding numerical solution methods. As a part of water quality PDE equations the 1D Advection-Dispersion Equation (ADE) will be described and it will be shown that how the dissolved heavy metals, *i.e.* lead and cadmium, may be numerically modelled through the source term of this equation. Details of the development of a modelling approach for predicting dissolved heavy metal fluxes and the application of the model to the Karoon River, located in the south west of Iran are also provided in this chapter. The model was calibrated and verified against field-measured time series data for discharges, water levels and dissolved lead and cadmium heavy metal concentrations.

It is found that pH and EC play an essential role in adsorption and desorption of heavy metals by the particles in solution (Roshanfekr *et al.*, 2008a & 2008b). Based on the effect of different substances such as pH and salinity on dissolved heavy metal concentrations in rivers, the necessity of heavy metal modelling with more accuracy in predicting the concentration is inevitable.

This chapter provides a methodology predicting a varying reaction coefficient for dissolved lead and cadmium heavy metals using pH and EC (as a function of salinity) which affects the reaction coefficient in the ADE for improved accuracy. Also the procedure for dissolved heavy metal modelling and finding the best relationship between pH and EC with the reaction coefficient is described. Finally, the best relationships for dissolved lead and cadmium reaction coefficients were introduced and the results were successfully compared with the corresponding measured values.

2. Heavy metals and their transport processes in a riverine basin

Modelling dissolved heavy metal transport in rivers requires a good understanding of the phenomenon. Heavy metals generally exist in two phases in river waters, *i.e.* in the dissolved phase in the water column and in the particulate phase adsorbed on the sediments. The behaviour of heavy metals in the aquatic environment is strongly influenced by adsorption on organic and inorganic particles. The dissolved fraction of heavy metals may be transported via the process of advection-dispersion (Wu *et al.*, 2005). These pollutants are non-conservative in nature and their concentrations depend on salinity and pH, which may vary with time and along a river (Pafilippaki *et al.*, 2008). As a result, the dissolved metal may come out of solution or even re-dissolve, depending on conditions along the time or channel (Nassehi & Bikangaga, 1993).

Figure 1 illustrates the dissolved heavy metal transport process in a riverine basin. This process is very complicated, since presence and mobility of the heavy metal is highly depended on the environmental conditions, *e.g.* bed and suspended sediments. A quick review in the literature for the couple of recent years shows that the main attention was focused mostly on the measurements of heavy metals in alluvial rivers with or without sediments. For example, Rauf *et al.* (2009), Akan *et al.* (2010) and Kumar *et al.* (2011) investigated the effect of sediments on the transport of heavy metals; the seasonal variations of the heavy metals in rivers were investigated by Papafilippaki *et al.* (2008) and Sanayei *et al.* (2009). The effect of heavy metals on the river self purification process was studied by Mala & Maly (2009).

It should be noted that much attention should be given to the study of heavy metal transport dynamics. In fact, the main factors closely related to the heavy metal pollution transport-transformation in natural bodies are: water flow, sediment motion, pH value, water salinity, water temperature, sediment size distribution, sediment concentration, mineral composition for sediments, degree of mineralization of water and time. Thus theoretically the mathematical model of heavy metal transport dynamics should include equations describing all factors mentioned above (Haung *et al.*, 2007; Haung, 2010).

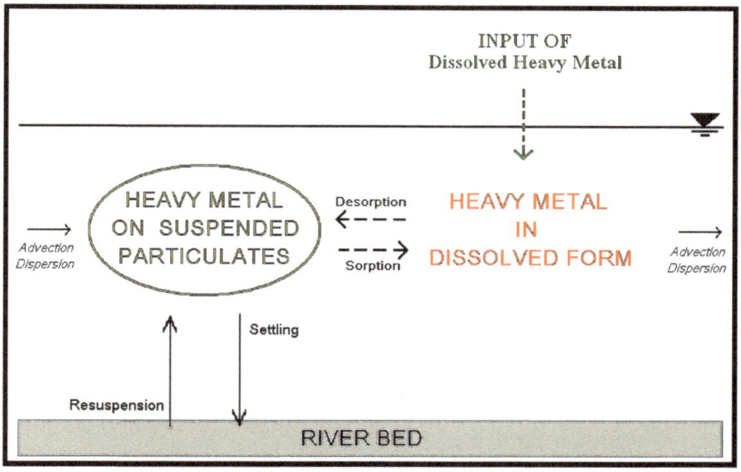

Fig. 1. Schematic illustration of dissolved heavy metal process in riverine waters.

One way to numerically model heavy metals in riverine basins is to assume a reaction coefficient in the source term of the ADE (Advection-Dispersion Equation), which will be described later, showing the presence of the desired substance in the solution. In many studies (such as: Nassehi & Bikangaga, 1993; Shrestha & Orlob, 1996; Wu *et al.,* 2001 & 2005; etc.) the researchers assumed a constant reaction coefficient with time, whereas in the field this coefficient may vary according to the rate of pH, salinity, temperature or even other chemical substances and other hydraulic characteristics of the river. Roshanfekr *et al.* (2008a & 2008b) found that pH and EC play an essential role in adsorption and desorption of heavy metals by the particles in solution.

3. Theoretical background

Numerical models provide a valuable tool for predicting the fate and transport of dissolved heavy metals in river environments and are increasingly used for such hydro-environmental management studies of river waters. However, computer-based tools used for predicting such heavy metal concentrations are still used infrequently, even though they can support decision-making by the regulatory authorities, marine environment agencies and industry (Ng *et al.,* 1996).

The use of computers has provided the opportunity to better understand and assess our water resources through comprehensive numerical model simulations and testing of various schemes or options. The numerical model allows the user to assess the hydraulic conditions in the river basin and thus, establish a better understanding of human impacts upon a natural or modified river system.

Any numerical model used to predict the flow and dissolved heavy metal transport processes in rivers depends primarily on solving the governing hydro-environmental equations. In most riverine systems, the basin is regarded as a 1D system, with longitudinal flow dominating throughout the system. Any type of hydro-environmental model commonly used by environmental engineers and water managers to predict the dissolved heavy metals concentrations in rivers generally involves solving the hydrodynamic and water quality equations as described below.

3.1 Equations for hydrodynamic modelling

In order to dynamically model the heavy metals in riverine basins, the governing hydrodynamic partial deferential equations must be numerically solved. The governing equations and their numerical solutions for modelling the river hydrodynamics (*i.e.* velocity and water elevation at any point and time) are therefore presented in this section.

For unsteady flow and hydrodynamic modelling the velocity and the water-level at any point of the basin and any time is of interest. The velocity component in rivers is usually assumed as a one-dimensional vector. The one-dimensional governing hydrodynamic equations describing flow and water elevations in rivers are based on the well-known St. Venant equations, applicable to 1D unsteady open-channel flows. Various forms of the St. Venant equations have been formulated in the field for unsteady open-channel flows since the 1950s, when numerical model simulations were first developed. The most widely used form in practice is generally written as (Cunge *et al.*, 1980; Wu, 2008):

$$T\frac{\partial \xi}{\partial t}+\frac{\partial Q}{\partial x}=\frac{Q_L}{\delta x} \tag{1}$$

$$\underbrace{\frac{\partial Q}{\partial t}}_{1}+\underbrace{\beta\frac{\partial}{\partial x}\left(\frac{Q^2}{A}\right)}_{2}+\underbrace{gA\frac{\partial \xi}{\partial x}}_{3}+\underbrace{g\frac{Q|Q|}{C_Z^2AR}}_{4}=0 \tag{2}$$

The individual terms in the momentum equation can be defined as: (1) local acceleration, (2) advective acceleration, (3) pressure gradient and (4) bed resistance, where:

T =top width of the channel;

ξ =water elevation above datum;

Q =discharge;

β =momentum correction factor due to non uniform velocity over the cross-section;

A =wetted cross-section area;

$R = A / P$ =hydraulic radius;

P =wetted parameter of the cross-section;

Q_L =lateral inflow or outflow (positive for inflow and negative for outflow);

C_Z =Chezy coefficient;

δx =longitudinal distance between two consecutive nodes;

g =acceleration due to gravity;

x,t =river flow direction and time respectively.

Equations (1) and (2) are solved numerically to provide the varying values of discharge and water elevations.

3.2 Equations for water quality modelling

Water quality modelling involves the prediction of water pollution using mathematical simulation techniques. The following sections describe the governing equations and their numerical solution for heavy metals modelling in riverine basins. The model was then applied to Karoon River for lead and cadmium modelling.

The transport of heavy metals in the dissolved phase can be described by the following one-dimensional advection-dispersion equation (ADE) (Kashefipour, 2002):

$$\underbrace{\frac{\partial SA}{\partial t}}_{1}+\underbrace{\frac{\partial SQ}{\partial x}}_{2}-\underbrace{\frac{\partial}{\partial x}\left[AD_x\frac{\partial S}{\partial x}\right]}_{3}=\underbrace{S_0^d}_{4}+\underbrace{S_t^d}_{5} \tag{3}$$

The individual terms in the advection-dispersion equation refer to: (1) local effects, (2) transport by advection, (3) longitudinal dispersion and turbulent diffusion, (4) sources or sinks of dissolved heavy metals and (5) transformation term defining absorbed and desorbed particulate fluxes to or from sediments (source term), where:

S =cross-sectional averaged dissolved heavy metal concentration;
D_x =longitudinal dispersion coefficient;
S_0^d =source or sink of dissolved heavy metal;
S_t^d =transformation term defining absorbed and desorbed particulate fluxes to or from sediments (source term).

Sources or sinks of dissolved heavy metals can be defined as:

$$S_0^d=\frac{Q_L S_L}{\Delta x} \tag{4}$$

where:

Q_L =lateral inflow or outflow discharge;
S_L =lateral inflow or outflow dissolved heavy metal concentration;
Δx =distance between two consecutive cross-sections which can be either constant or variable.

The longitudinal dispersion coefficient (D_x) in natural rivers is dependent upon many hydrodynamic parameters including: depth, width, velocity and shear velocity (Fischer *et al.* 1979). There are many empirical and/or semi-empirical equations describing this very important dynamic coefficient. Kashefipour & Falconer (2002) presented two empirical equations based on applying the dimensional analysis procedure to more than 80 data sets in 30 natural rivers, to estimate longitudinal dispersion coefficient in natural channels and showed that these equations performed relatively better than the other existing equations. In this chapter the Kashefipour & Falconer (2002) relationship has been used to estimate the longitudinal dispersion coefficient. This relationship is written as:

$$D_x=\left[7.428+1.775\left(\frac{T}{H}\right)^{0.620}\left(\frac{U_*}{U}\right)^{0.572}\right]HU\left(\frac{U}{U_*}\right) \tag{5}$$

where:

H =averaged depth over the cross-section;
U = cross-sectional average velocity;
U_* =local shear velocity.

In addition Tavakolizadeh (2006) used this dispersion coefficient for water quality modelling in Karoon River and achieved acceptable results for different water quality parameters.

3.3 Heavy metals modelling

Heavy metals can exist in both the dissolved and adsorbed particulate phases in rivers. The distribution between these two phases may be expressed by a partition coefficient. In recent years much effort has been focused on correlating the partitioning rate of heavy metals in particulate and dissolved phases to several environmental factors and water properties. It would be possible to numerically model each type of heavy metals in the water column, separately (Wu *et al.*, 2001 & 2005). However, it is sometimes important to the environmental managers to have a good understanding of the ratio of these two types of heavy metal presence in water body. A few researchers assume a reaction coefficient in the transformation term, *i.e.* S_t^d , in the form of Equation (6) to model either dissolved or particulate heavy metal in the water column.

Since in outfalls a proportion of a pollutant that is added to the water column generally decays, and settles according to the chemical and hydraulic characteristics of the flow, it can be concluded that the pollutant may also be added from or to the sediments. Therefore, for water bodies close to outfalls the conditions are not generally consistent with equilibrium conditions. For equilibrium conditions it can be assumed that the parameter S_t^d in Equation (3) is equal zero. On the other hand, the transformation of heavy metal from dissolved phase to the particulate phase and vice versa is assumed to be equal. A review of the literature has shown that a number of researchers include this type of assumption in their models, such as Wu *et al.* (2005). However, another group of researches, for example Nassehi & Bikangaga (1993), assumed a decay term having a form of Equation (6) with a constant coefficient.

The fate and decay of toxic subtances can result from physical, chemical, and/or biological reaction. Transformation processes are those in which toxic subtances are essentialy irreversibly destroyed, changed, or removed from the water system. These transformation processes are often described by kinematic equations. Most decay processes are expressed as first-order reactions. Therefore, in this chapter the first-order chemical reaction was used as the transformation parameter in Equation (3) for dissolved heavy metal modelling and is written as follows (Zhen-Gang, 2008):

$$S_t^d = -\kappa SA \tag{6}$$

κ is a reaction coefficient rate, which may have a positive or negative value as the dissolved heavy metals disappears or accumulates in a given river section.

Since the exchange of the heavy metal substance between particulate and dissolved phases is a chemical process and is highly dependent on the environmental conditions, it seems that assuming zero value or a constant value for the reaction coefficient κ may not provide an accurate simulation. Therefore, the following sections describe the procedure for calculating varied reaction coefficients for dissolved lead and cadmium modelling using pH and EC changes in the water column. The key point is that the chemical characteristics of the flow, such as pH and EC, can affect the dissolved heavy metals from sorption and desorption, to or from the sediments, and these characteristics can have an important effect on the dissolved heavy metal concentrations. For more accurate heavy metal modelling, varied reaction coefficients has been suggested linking pH and EC to the kinetic processes.

Based on the different characteristics of each heavy metal (such as lead, cadmium and etc.) the varied reaction coefficient should be computed and the corresponding relation of the reaction coefficient should be used separately for each metal.

Reaction coefficient, also known as the decay coefficient, is the ratio for the number of atoms that decay in a given period of time compared with the total number of atoms of the same kind present at the beginning of that period (Zhen-Gang, 2008). There are different environmental parameters, such as: temperature, pH, salinity and etc., which generally affect the reaction coefficient in heavy metals. Therefore, κ can be defined as:

$$\kappa = f(pH, Salinity, Temperature, ...) \tag{7}$$

The κ value may be related with to temperature as given by the following equation (see Orlob, 1983):

$$\kappa = \kappa_{20} \times O^{(TEMP-20)} \tag{8}$$

where:

κ_{20} = reaction coefficient at 20°C;
$TEMP$ = temperature of water;
O = temperature coefficient which it can vary from 1.047 to 1.135 (Orlob, 1983).

Theoretically the best mathematical model for heavy metal reaction coefficient should consist of all factors affecting the heavy metals concentration. Roshanfekr *et al.* (2008a & 2008b) found that pH and EC play a more essential role in adsorption and desorption of heavy metals by the particles in solution. Therefore assuming a variable reaction coefficient seems more reasonable.

The first step to calculate a variable reaction coefficient is to select the parameters that are most likely affecting the dissolved heavy metal concentration. Then the efforts can be focused on finding suitable functions to represent the reaction coefficient rate for dissolved heavy metals (*e.g.* lead and cadmium) in rivers. In calibrating the model against measured dissolved lead and cadmium data, five approaches for each dissolved metal can be used:

1. No rate of reaction for dissolved heavy metal (used by some researchers and models for equilibrium conditions).
2. A constant reaction coefficient for the rate of reaction during the whole simulation time (the general practice in dissolved heavy metals modelling used by many researchers).
3. A time varying reaction coefficient for the rate of the reaction using pH as a variable.

4. A time varying reaction coefficient for the rate of the reaction using EC as a variable.
5. A time varying reaction coefficient for the rate of the reaction using both pH and EC variables.

For each one of these five cases a number of simulation calibration runs were carried out and the initial reaction coefficient was subsequently adjusted by comparing the predicted dissolved lead and cadmium concentrations with the corresponding measured values at sites and for the times of measured values. Final values of the reaction coefficients for each indicator were adopted when the best fit occurred between the series of data. The adjusted rate of reaction coefficients were then correlated with pH, EC and both to find the best relationships for κ as a function of pH and/or EC. These equations (*i.e.* Equations (20) to (25)) were added to the model as a part of the numerical solution of the ADE (see Equation (3)). The model was then validated using the corresponding measured data for different time series at the survey site.

4. Numerical modelling

Following sections describe the numerical methods used to solve the hydrodynamic and water quality partial differential equations for heavy metals modelling.

4.1 Numerical methods for hydrodynamic equations solutions

There are many implicit and explicit numerical methods used for solving 1D hydrodynamic equations (*i.e.* Equations (1) and (2)), in which the stability, accuracy and consistency of the numerical solution are important. Almost all implicit methods are unconditionally stable, however the accuracy of model predictions is highly depended on the Courant number (*i.e.* $C_r = (U\Delta t)/\Delta x$). Different methods for numerical solution of the above equations may be found in Abbott and Basco (1997).

In this study, the numerical model FASTER (Flow and Solute Transport in Estuaries and Rivers) (Kashefipour *et al.*, 1999) was used. This model was first developed by Kashefipour (2002) and has since been extended and improved to predict the dissolved heavy metals concentrations for different reaction coefficients. The hydrodynamic module of FASTER model numerically solves the Saint Venant equations using Crank Nicolson with an implicit staggered scheme (Wu, 2008). This model uses the influenced line technique, enabling the model to remain implicit and thereby unconditionally stable and accurate over the whole domain, especially in river confluences. This model can be applied for complex channel networks with complex geometry and has been successfully applied to many research projects in Cardiff University, UK (Kashefipour *et al.*, 2002). In the numerical method used for this model, the hydrodynamic equations were formulated on a staggered grid to provide advantages in treating the typical hydrodynamic boundary conditions that are commonly used in such models. The implicit finite difference solution of the governing hydrodynamic equations is second order accurate in space and time and is unconditionally stable. However, where reasonable precision is required the Courant number, expressed in the form of $C_r = (\Delta t U)/\Delta x$, should be less than five. The scheme remains stable for higher Courant numbers, but the accuracy may reduce particularly at wave peaks.

A summary of numerical solution method is described here. The difference form of the continuity equation using the Crank-Nicholson central scheme around the node i (Figure 2) can be written as:

$$\frac{1}{\Delta t}T^{n}_{Wi}\left(\xi^{n+1}_{i}-\xi^{n}_{i}\right)\left(x_{i+1/2}-x_{i-1/2}\right)+\theta\left(Q^{n+1}_{i+1/2}-Q^{n+1}_{i-1/2}\right)+\left(1-\theta\right)\left(Q^{n}_{i+1/2}-Q^{n}_{i-1/2}\right)=Q^{n+1}_{Li} \quad (9)$$

where:

n and n+1=refer to time t and $t+\Delta t$, respectively;

θ =a weighting coefficient between 0 and 1 to split the spatial derivatives between the upper and lower time levels ($0\leq\theta\leq1$).

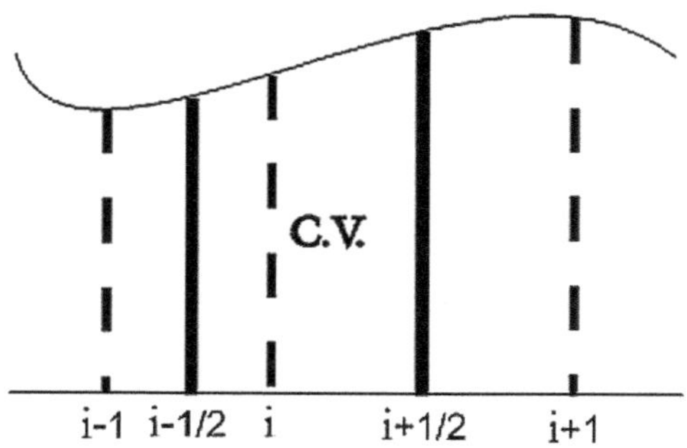

Fig. 2. Domain of the discretization of continuity and momentum equations.

The non-conservative form of the momentum equation may be discretised using the finite difference central scheme around the node (i+1/2) as shown in Figure 2, yields:

$$\begin{aligned}&\frac{1}{\Delta t}\left(Q^{n+1}_{i+1/2}-Q^{n}_{i+1/2}\right)+\frac{2\beta}{\left(x_{i+1}-x_{i}\right)A^{n}_{i+1/2}}\left[\theta Q^{n+1}_{i+1/2}+\left(1-\theta\right)Q^{n}_{i+1/2}\right]Q^{n+1}_{Li+1/2}\\&-\beta\frac{Q^{n}_{i+1/2}T^{n}_{Wi+1/2}}{\Delta t A^{n}_{i+1/2}}\left(\xi^{n+1}_{i+1}+\xi^{n+1}_{i}-\xi^{n}_{i+1}-\xi^{n}_{i}\right)-\frac{\beta Q^{n+1}_{i+1/2}Q^{n}_{i+1/2}}{\left(A^{n}_{i+1/2}\right)^{2}\left(x_{i+1}-x_{i}\right)}\left(A^{n}_{i+1}-A^{n}_{i}\right)\\&+\frac{gA^{n}_{i+1/2}}{\left(x_{i+1}-x_{i-1}\right)}\left[\theta\left(\xi^{n+1}_{i+1}-\xi^{n+1}_{i}\right)+\left(1-\theta\right)\left(\xi^{n}_{i+1}-\xi^{n}_{i}\right)\right]+\frac{gQ^{n+1}_{i+1/2}\left|Q^{n}_{i+1/2}\right|}{\left(C^{n}_{Zi+1/2}\right)^{2}A^{n}_{i+1/2}R^{n}_{i+1/2}}=0\end{aligned} \quad (10)$$

By rearranging Equations (9) and (10) the following algebraic linear equations may be written, respectively:

$$a_{i}Q^{n+1}_{i-1/2}+b_{i}\xi^{n+1}_{i}+c_{i}Q^{n+1}_{i+1/2}=d_{i} \quad (11)$$

$$a_{i+1/2}\xi^{n+1}_{i}+b_{i+1/2}Q^{n+1}_{i+1/2}+c_{i+1/2}\xi^{n+1}_{i+1}=d_{i+1/2} \quad (12)$$

The staggered varying grid size with the numerical scheme is alternatively applied to the continuity and momentum equations to produce a set of linear algebraic equations (*i.e.*

Equations (11) and (12)) for each three consecutive ξ and Q points. Applying Equations (11) and (12) simultaneously at all grid points in the discrete solution domain, from time $n\Delta t$ to (n+1) Δt, yields a matrix system of linear algebraic equations based on ξ^{n+1} and Q^{n+1}. A general equation, which contains all of the linear algebraic equations and may be solved by the Thomas algorithm or Gauss elimination procedure for the numerical solution of the governing partial differential equations, may be defined using the following equation:

$$[B][X]=[D] \tag{13}$$

where:

$[B]$ = matrix of the coefficients;

$[X]$ = matrix of the variables;

$[D]$ = matrix of the constants of the linear equations.

Due to the staggered method, the matrix $[B]$ is usually given as a tri-diagonal matrix and then Equation (13) can be generally solved using the well known Thomas algorithm. More information regarding the numerical solution and application of the influence line technique in FASTER model to keep the whole numerical solution in implicit form, specially in river confluences and junctions, can be found in Kashefipour (2002).

4.2 Numerical methods for ADE solution

In order to solve the ADE, an implicit algorithm has been developed and used in the FASTER model. This finite volume based solution procedure calculates the advection of a concentrate of solute, or suspended sediments at each face of any control volume, by means of a modified form of the highly accurate ULTIMATE QUICKEST[1] scheme (Lin & Falconer, 1997). As before, a space staggered grid system is used to solve the finite volume form of the ADE, in which the variable S is located at the center of the control volume (Falconer *et al.*, 2005).

Double integration of the one-dimensional ADE, *i.e.* Equation (3), with respect to time and volume over the control volume, as shown in Figure 2 gives:

$$\underbrace{\int_t^{t+\Delta t}\int_V \frac{\partial SA}{\partial t}dVdt}_{1}+\underbrace{\int_t^{t+\Delta t}\int_V \frac{\partial SQ}{\partial x}dVdt}_{2}-\underbrace{\int_t^{t+\Delta t}\int_V \frac{\partial}{\partial x}AD_l\frac{\partial S}{\partial x}dVdt}_{3}= \\ \underbrace{\int_t^{t+\Delta t}\int_V \frac{Q_L S_L}{\Delta x}dVdt}_{4}+\underbrace{\int_t^{t+\Delta t}\int_V S_t^d dVdt}_{5} \tag{14}$$

where:

V= volume.

---

1 Universal Limiter Transient Interpolation Modelling for Advection Term Equation - Quadratic Upstream Interpolation for Convective Kinematics with Estimated Streaming Terms (Leonard, 1979 & 1991).

In Equation (14) the term (1) describes the local change of the solute concentration within the control volume from time (t) to time (t+Δt). Terms (2) to (5) refer to changes in the solute concentration due to: advection, diffusion, lateral inputs and transformation, respectively. The discrete forms of the terms in Equation (14) using finite volume method can be written as follows:

$$\int_t^{t+\Delta t}\int_V \frac{\partial SA}{\partial t} dVdt = \left[(SA)_i^{n+1} - (SA)_i^n\right]\Delta V \tag{15}$$

$$\begin{aligned}\int_t^{t+\Delta t}\int_V \frac{\partial SQ}{\partial x} dVdt &= \left[\psi\left((SQA)_{i+1/2}^{n+1} - (SQA)_{i-1/2}^{n+1}\right) + (1-\psi)\left((SQA)_{i+1/2}^{n} - (SQA)_{i-1/2}^{n}\right)\right]\Delta t = \\ &\frac{\Delta t}{2}\left[\begin{aligned}&\psi\left((S_i+S_{i+1})^{n+1}(QA)_{i+1/2}^{n+1} - (S_i+S_{i-1})^{n+1}(QA)_{i-1/2}^{n+1}\right) + \\ &(1-\psi)\left((S_i+S_{i+1})^{n}(QA)_{i+1/2}^{n} - (S_i+S_{i-1})^{n}(QA)_{i-1/2}^{n}\right)\end{aligned}\right]\end{aligned} \tag{16}$$

$$\begin{aligned}&\int_t^{t+\Delta t}\int_V \frac{\partial}{\partial x} AD_l \frac{\partial S}{\partial x} dVdt = \\ &\Delta t\left\{\begin{aligned}&\psi\left[\left(A^2D_l\right)_{i+1/2}^{n+1}\frac{(S_{i+1}-S_i)^{n+1}}{x_{i+1}-x_i} - \left(A^2D_l\right)_{i-1/2}^{n+1}\frac{(S_i-S_{i-1})^{n+1}}{x_i-x_{i-1}}\right] \\ &+(1-\psi)\left[\left(A^2D_l\right)_{i+1/2}^{n}\frac{(S_{i+1}-S_i)^{n}}{x_{i+1}-x_i} - \left(A^2D_l\right)_{i-1/2}^{n}\frac{(S_i-S_{i-1})^{n}}{x_i-x_{i-1}}\right]\end{aligned}\right\}\end{aligned} \tag{17}$$

$$\int_t^{t+\Delta t}\int_V \frac{Q_L S_L}{\Delta x} dVdt = \overline{Q_L S_L A}\Delta t \tag{18}$$

$$\int_t^{t+\Delta t}\int_V -S_t^d dVdt = -\overline{S_t^d}\Delta V\Delta t = -\frac{1}{4}\left\{\begin{aligned}&(\kappa AS)_{i+1/2}^{n+1} + (\kappa AS)_{i-1/2}^{n+1} \\ &(\kappa AS)_{i+1/2}^{n} + (\kappa AS)_{i-1/2}^{n}\end{aligned}\right\}\Delta V\Delta t \tag{19}$$

More information regarding the FASTER model may be found in Kashefipour (2002) and Yang *et al.* (2002).

In the current chapter the dissolved heavy metal reaction coefficient comprises two parameters, including pH and EC. The coefficient was formulated using a linear regression relationship. These varied reaction coefficients were then added to the model for predicting dissolved lead and cadmium. The procedure of development and the equations added to the model can be followed in the modelling application and dissolved heavy metal results sections respectively.

5. Case study

The Karoon River is the largest and the only navigable river in south west of Iran (see Figure 3(a)). In this study the Mollasani-Farsiat reach of the Karoon River, a distance of 110Km was selected due to the high amount of heavy metal concentrations along this reach (see Figure

3(b)). The Karoon River basin has a network of gauging stations and there are several effluent inputs to the river between gauging stations at Mollasani and Farsiat, including industrial units such as: piping, steel, paint making, agriculture, paper mill, fish cultivation and power plant industries draining from wastewater works into the river (see Figure 3(c)) (Diagomanolin *et al.*, 2004). Hydrodynamic and water quality data were acquired via Khuzestan Water and Power Authority (KWPA). A set of six field-measured data were available from March 2004, including discharge and water levels measurements at the Mollasani, Ahwaz and Farsiat gauging stations and pH, EC, dissolved lead and cadmium concentrations at the Mollasani and Shekare gauging stations (see Figure 3(c)).

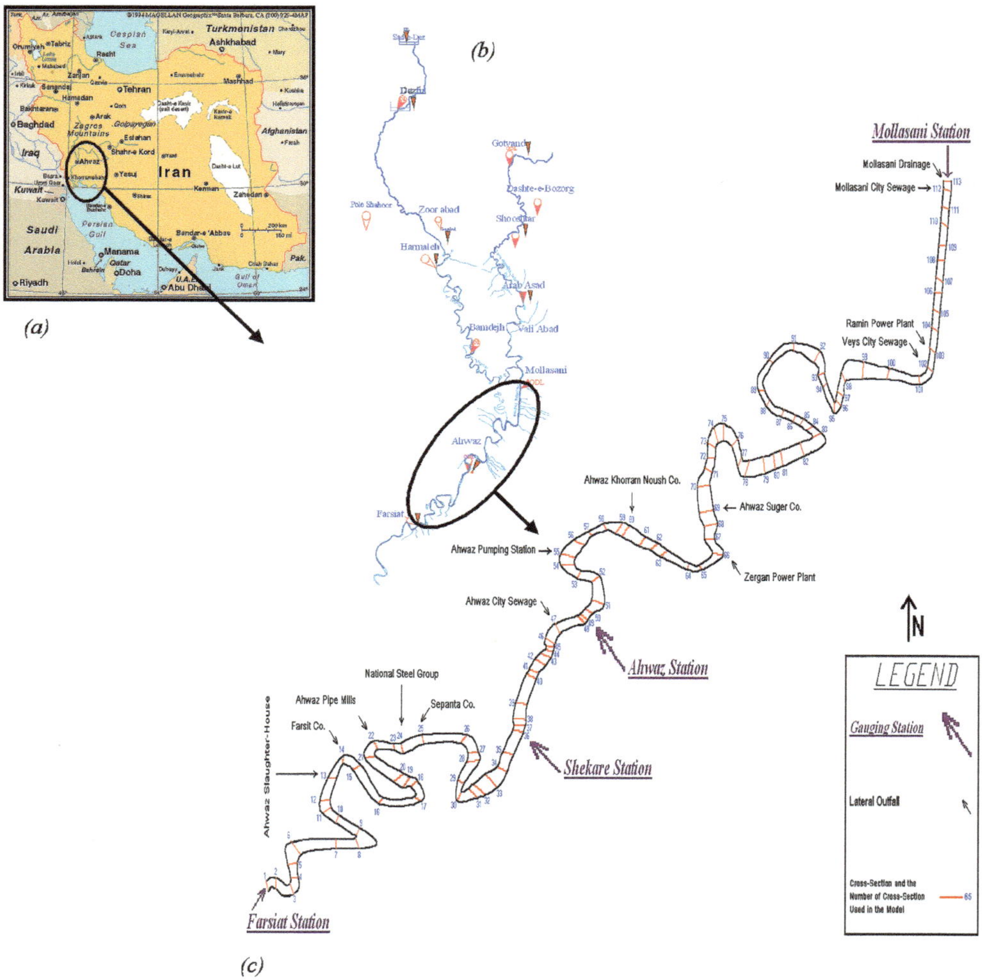

Fig. 3. (a) Location of Karoon River, (b) Karoon river network and gauging stations and (c) Outfalls, gauging stations and cross-sections used in the model between Mollasani and Farsiat reach.

The 1D grid, covering the region from Mollasani to Farsiat, was represented using 113 segments, with extensive bathymetric data at each cross-section being collected during the most recent bathymetric survey conducted by Khuzestan Water and Power Authorities in 2000.

The time series water elevations recorded at the Farsiat hydrometric station were chosen as the downstream boundary and the measured discharges and heavy metal concentrations at the Mollasani station were used as the upstream boundary conditions for flow and water quality modules of the main model, respectively. Also concentrations of dissolved lead and cadmium were measured from more than fifteen outfalls and industrial locations along the Mollasani and Farsiat reach. Cross-sections No.1, 36, 49 and 113 corresponded to the cross-sections at the gauging stations of Farsiat, Shekare, Ahwaz and Mollasani, respectively.

5.1 Application of hydrodynamic modelling

The hydrodynamic module of the FASTER model was calibrated against the data provided for the year 2004, starting from the month of March. The main hydrodynamic parameter used for calibration was the manning roughness coefficient. The river was separated into 4 parts, with the manning coefficient varying from 0.026 to 0.050. Good agreement was obtained between the predicted water levels and corresponding field data at the Ahwaz gauging station as the hydrometric survey site, with a difference in results being less than 3% (see Figure 4(a)) and also the model discharges agreed well with the field data obtained at the Ahwaz gauging station with the difference being less than 16% (see Figure 4(b)). The hydrodynamic module was then validated using another series of measured data (see Figures 5(a) and (b)). As can be seen from these figures the predicted data also gave relatively good correlation with the corresponding measured values. A summary of the statistical analysis of the model results is illustrated in Table 1.

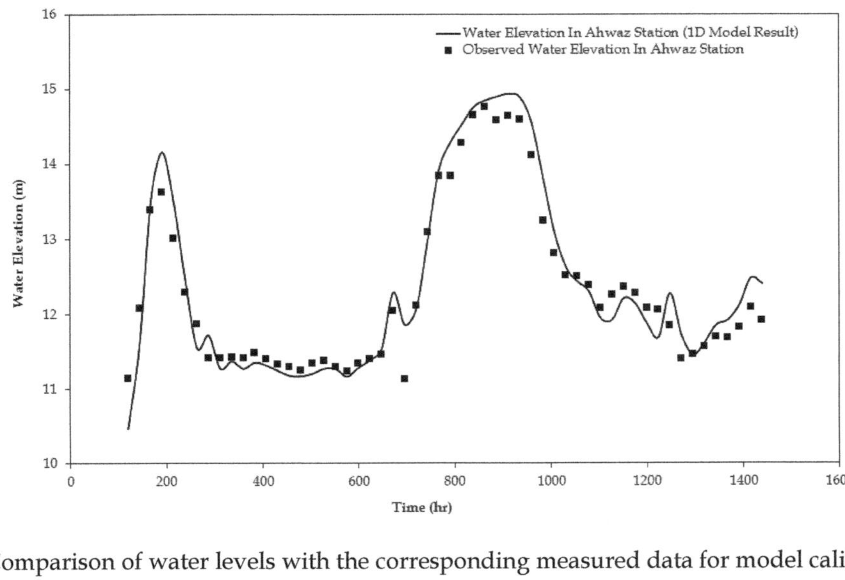

(a) Comparison of water levels with the corresponding measured data for model calibration

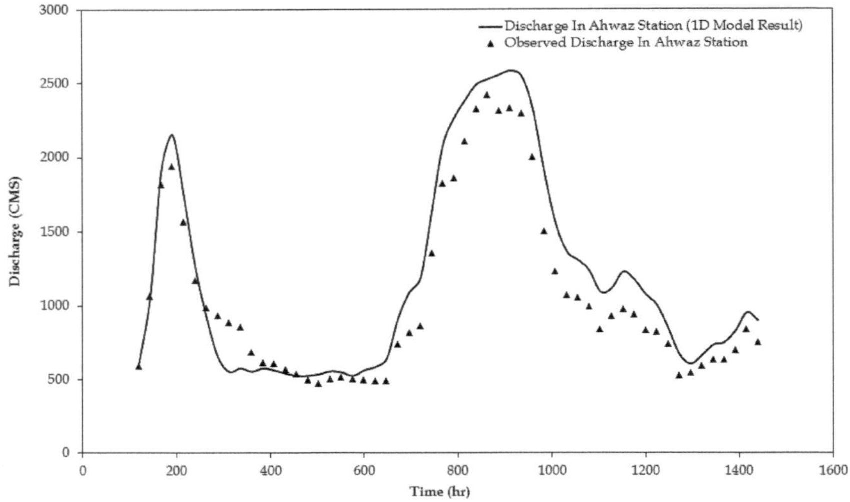

(b) Comparison of discharges with the corresponding measured data for model calibration

Fig. 4. Results of hydrodynamic model calibration.

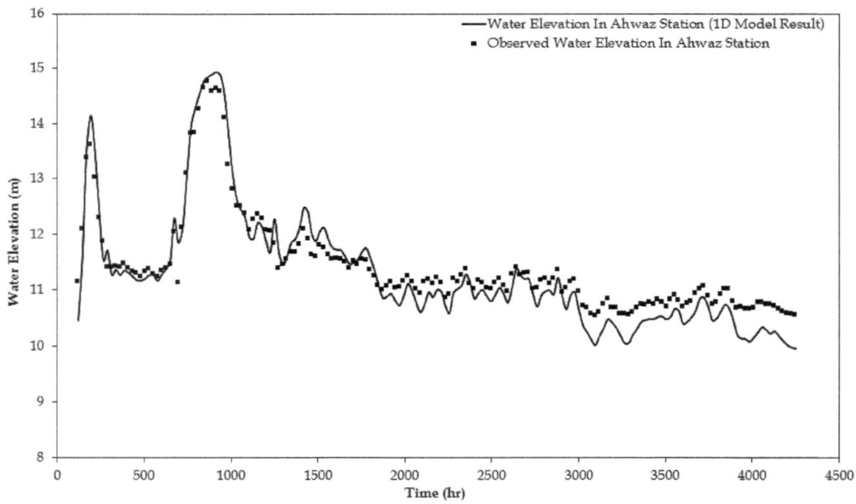

(a) Comparison of water levels with the corresponding measured data for model verification

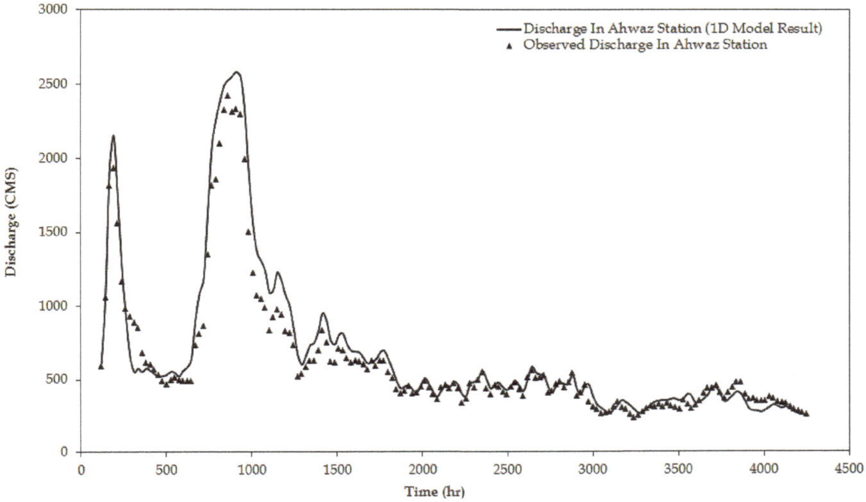

(b) Comparison of discharges with the corresponding measured data for model verification

Fig. 5. Results of hydrodynamic model verification.

	CALIBRATION			VERIFICATION		
	RMSE [(a)]	R^2 [(b)]	*%Error* [(c)]	*RMSE*	R^2	*%Error*
Water Elevation	0.350	0.935	2.17	1.013	0.869	2.98
Discharge	1.580	0.960	15.20	1.870	0.930	13.21

(a) Root Mean Square Error $RMSE = \left[\sum_{i=1}^{n} \frac{\left(X_{ip} - X_{im}\right)^2}{n} \right]^{0.5}$

(b) Coefficient of Determination (r-Square) $R^2 = \frac{\left(\sum_{i=1}^{n} X_{ip} X_{im}\right)^2}{\sum_{i=1}^{n} X_{ip}^2 \sum_{i=1}^{n} X_{im}^2}$

(c) Average Absolute Error $\%Error = \frac{\sum_{i=1}^{n} \left|\left(X_{ip} - X_{im}\right)\right|}{\sum_{i=1}^{n} X_{im}} \times 100$

where:

X_{ip} =Predicted Data; X_{im} = Measured Data and n =Number of Data (Azmathullah *et al.*, 2005).

Table 1. A summary of the hydrodynamic model results.

5.2 Application of dissolved heavy metals modelling

As discussed above, the rate of reaction plays an important role in predicting the concentration distribution of the dissolved heavy metals for river, estuarine and coastal waters. In the current section effort was made to find suitable functions to represent the reaction coefficient rate for dissolved lead and cadmium metal modelling in rivers. These functions were established using a comparison of the predicted heavy metal concentrations with the corresponding measured values at the Shekare gauging station (see Figure 3). In calibrating the model against measured dissolved lead and cadmium levels, five approaches for each dissolved metal were used. The model with the adjusted rate of reaction was then validated using the corresponding measured data for different time series at the survey site (Kashefipour *et al.*, 2006). In the following sections the equations and the results of the modelling for dissolved lead and cadmium, with the derived equations for the reaction coefficients are illustrated.

5.2.1 Results of dissolved lead modelling

For the first run a conservative dissolved lead was assumed, leading to a zero value for the rate of reaction coefficient. The fit between the predicted and measured data showed 25.2% and 33.3% errors for calibration and verification of the model respectively. As can be seen from Figures 6(a) and (b) the predicted dissolved lead in this case did not agree well with the corresponding measured data at the survey site (*i.e.* Shekare gauging station).

In the second run for predicting the dissolved lead concentration, the dissolved metal concentration was assumed to be non-conservative with the reaction coefficient in Equation (6) being constant. The best fit between the predicted and measured dissolved lead concentrations occurred for a reaction coefficient of 0.12 day^{-1}. This assumption led to a prediction error of 3.4% and 17.1% for calibration and verification of the model, respectively (see Figures 6(a) and (b)). However, some research results suggest that the reaction coefficients for different pH and salinity conditions were not constant. A more detailed investigation is being planned to determine the rate of reaction coefficient for different pH and EC (*i.e.* as a function of salinity).

According to the above findings, it seems that using a variable reaction coefficient, which can be adjusted automatically within a numerical model, depending on the pH, EC or pH and EC values may give better calibration results. A number of simulations were carried out to find a formulation for describing the relationship between the reaction coefficient and the pH value. Using the measured dissolved lead concentrations, it was found that the most suitable relationship between the reaction coefficient for dissolved lead and pH of the river was of the following form:

$$\kappa = -0.1646 \times pH + 1.4934 \qquad (R^2 = 0.643) \tag{20}$$

where:

pH =the mean pH of the river at the site for each time.

The predicted results, for which the reaction coefficient was calculated using Equation (20) in the model, were compared with the corresponding measured values for calibration and verification in Figures 6(a) and (b), respectively. The comparison showed that the error of simulation had reduced to 1.9% and 15% for calibration and verification of the model, respectively.

Based on the fact that the reaction coefficient relates to the EC value, a number of simulations were also carried out to find a suitable formulation for describing the reaction coefficient with the EC value. Using the measured dissolved lead concentration, it was found that the most suitable relationship between the reaction coefficient for dissolved lead and the EC of the river was of the following form:

$$\kappa = -0.00023 \times EC + 0.581 \qquad (R^2 = 0.924) \tag{21}$$

where:

EC = (Electrical Conductivity), the mean EC of the river at the site for each time (micro mhos/cm).

The predicted lead concentration, for which the reaction coefficients were calculated using Equation (21) in the model, were compared with the corresponding measured values for calibration and verification in Figures 6(a) and (b), respectively. This showed that the error of simulation had also declined to 0.8% and 10.8% for calibration and verification of the model, respectively.

For the last run of the dissolved lead, a number of simulations were carried out to find a formulation for describing the relationship between the reaction coefficient and both the pH and EC variables. Using the measured dissolved lead concentrations, it was found that the most suitable relationship between the reaction coefficient for dissolved lead and pH and EC as variables for the river was of the following form:

$$\kappa = 0.160 \times pH - 0.000402 \times EC - 0.401 \qquad (R^2 = 1.000) \tag{22}$$

The predicted results, for which the reaction coefficient were calculated using Equation (22) in the model, were then compared with the corresponding measured values for calibration and verification in Figures 6(a) and (b) with the errors of 0.4% and 8.3% respectively. These results showed another improvement in the predicted dissolved lead concentrations.

A summary of the statistical analysis for the different model results is shown in Table 2. As it is clear from this table the predicted dissolved lead concentrations improved, giving lower errors when varying reaction coefficients were applied as a part of ADE.

	CALIBRATION		VERIFICATION	
	RMSE	*%Error*	*RMSE*	*%Error*
$\kappa = 0$	2.9544	25.24	4.2072	33.30
$\kappa = Const.$	0.4064	3.35	2.6855	17.13
$\kappa = -0.1646 \times pH + 1.4934$	0.2633	1.89	2.3984	14.97
$\kappa = -0.00023 \times EC + 0.581$	0.1396	0.84	1.7807	10.77
$\kappa = 0.160 \times pH - 0.000402 \times EC - 0.401$	0.0671	0.35	1.5329	8.29

Table 2. A summary of the dissolved lead model results.

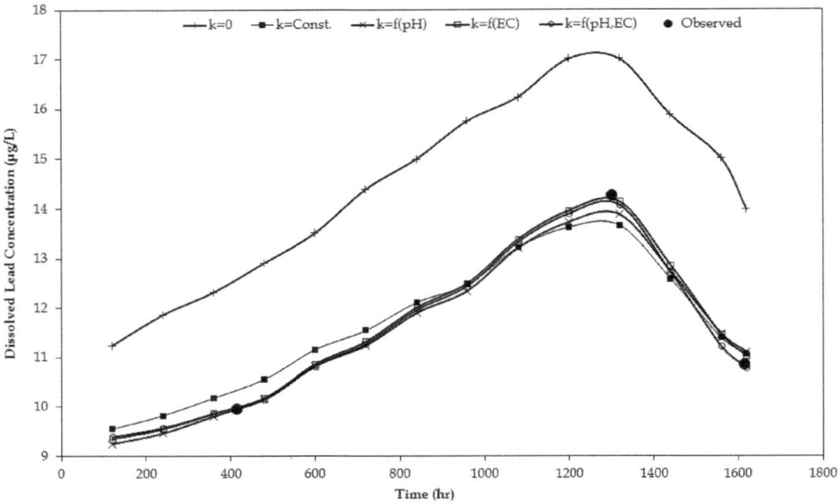

(a) Calibration

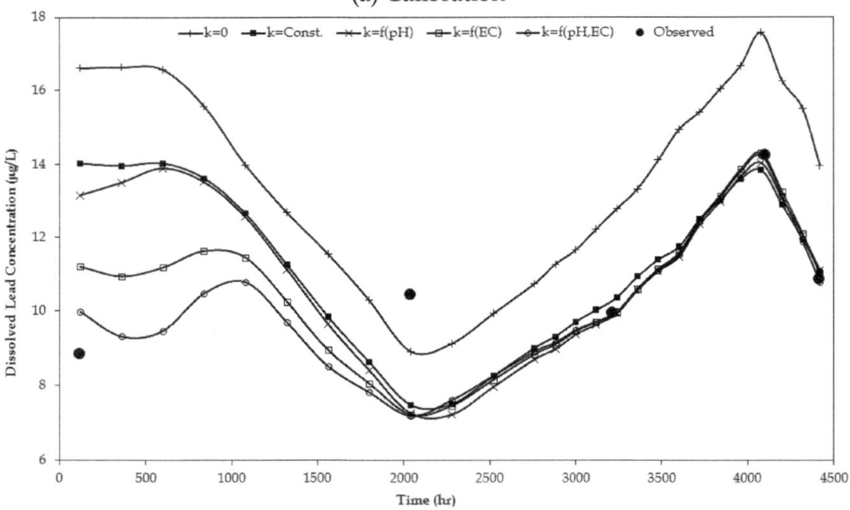

(b) Verification

Fig. 6. Comparison of predicted dissolved Lead with the corresponding measured values.

5.2.2 Results of dissolved cadmium modelling

The same procedure was carried out for dissolved cadmium modelling. For the first run cadmium was assumed to be conservative for which the predicted data did not show reasonable agreement with the measured data and the error was estimated to be 71.1% and 76.4% for calibration and verification of the model respectively (see Figures 7(a) and (b)).

For the second run cadmium was assumed to be non-conservative, with a constant reaction coefficient and the best fit between the predicted and measured data occurred when a reaction coefficient of 0.38 day^{-1} was used in the model. This assumption significantly reduce the error to 7.7% and 8.5% for calibration and verification of the model, respectively (see Figures 7(a) and (b)).

In the third run the reaction coefficient was assumed to be varying with pH. The most suitable relationship between the reaction coefficient for dissolved cadmium and pH in the river was found to be of following form:

$$\kappa = -0.2462 \times pH + 2.3738 \qquad (R^2 = 0.703) \tag{23}$$

The predicted dissolved cadmium for which the reaction coefficients were calculated using Equation (23) in the model, were compared with the corresponding measured values for calibration and verification in Figures 7(a) and (b), respectively. This comparison showed that the error of simulation had declined to 1.8% and 4.2% for calibration and verification of the model, respectively.

The fourth model run was carried out using the EC as a variable in computing the reaction coefficient. For the measured data of dissolved cadmium, the most suitable function for relating the reaction coefficient with EC in the river was found to be:

$$\kappa = -0.000201 \times EC + 0.7286 \qquad (R^2 = 0.350) \tag{24}$$

This function was added to the model for predicting the results of dissolved cadmium with EC. The results showed that the error of simulation was 2.5% and 2.6% for calibration and verification, respectively (see Figures 7(a) and (b)).

For the last run for dissolved cadmium a number of simulations were carried out to find a formulation for time varying reaction coefficients for the rate of reaction using both the pH and EC as variables. With using the measured data of dissolved cadmium, the most suitable function for relating the reaction coefficient with pH and EC in river was found to be:

$$\kappa = -0.1231 \times pH - 0.0001 \times EC + 1.5512 \qquad (R^2 = 0.560) \tag{25}$$

The predicted dissolved cadmium for which the reaction coefficients were calculated using Equation (25) in the model, were compared with the corresponding measured values for calibration and verification in Figures 7(a) and (b), respectively. This showed that the error of simulation had declined to 2.2% and 2.3% for calibration and verification of the model, respectively. A summary of the statistical analysis for the different model results is shown in Table 3.

	CALIBRATION		VERIFICATION	
	RMSE	*%Error*	*RMSE*	*%Error*
$\kappa = 0$	0.1086	71.11	0.1483	76.42
$\kappa = Const.$	0.0126	7.67	0.0171	8.47
$\kappa = -0.2462 \times pH + 2.3738$	0.0028	1.78	0.0114	4.16
$\kappa = -0.000201 \times EC + 0.7286$	0.0041	2.54	0.0050	2.56
$\kappa = -0.1231 \times pH - 0.0001 \times EC + 1.5512$	0.0035	2.18	0.0046	2.29

Table 3. A summary of the dissolved cadmium model results.

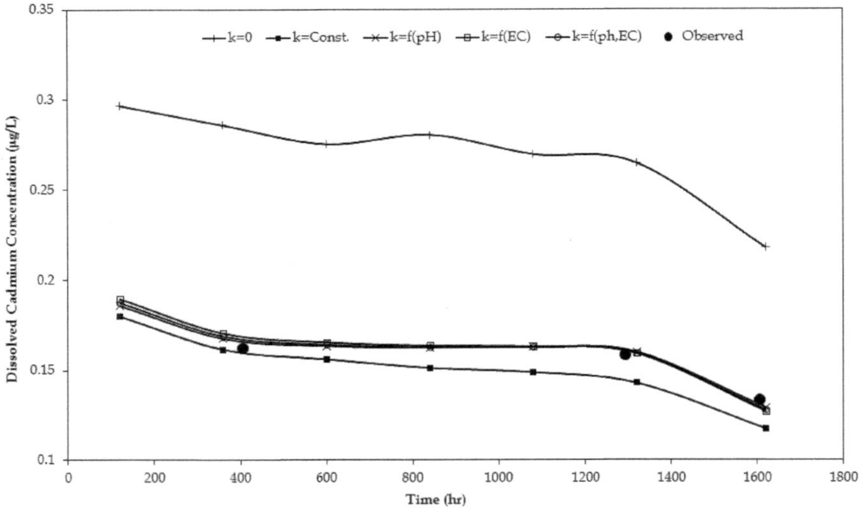

(a) Calibration

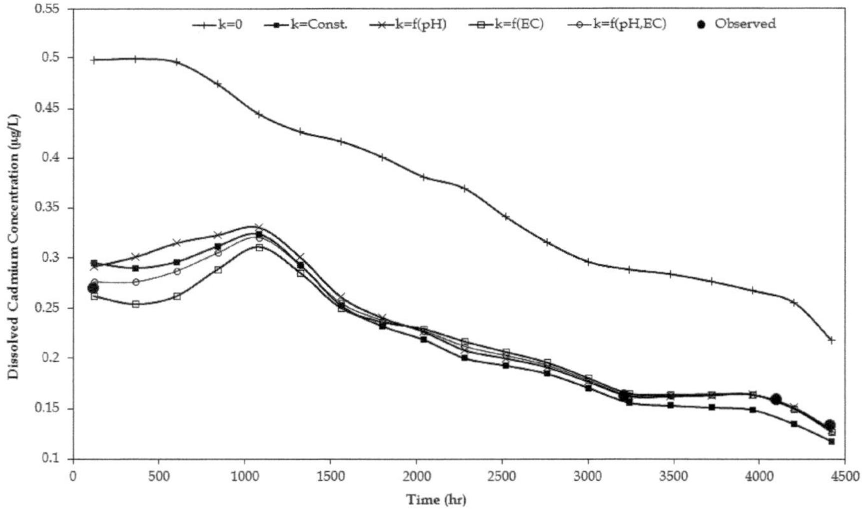

(b) Verification

Fig. 7. Comparison of predicted dissolved Cadmium with the corresponding measured values.

6. Discussion

Salinity has been found by many investigators to be more influential on the reaction coefficient than any other environmental or water properties in riverine and estuarine waters. The results published by Turner *et al.* (2002) showed that the trace metal distribution coefficient in estuarine waters is primarily a function of salinity. Nassehi & Bikangaga (1993) calculated the value of the reaction coefficient for dissolved zinc in different elements of a river. Wu *et al.* (2005) used salinity for modelling the partitioning coefficient of heavy metals in the Mersey estuary and concluded that the modelling results agreed well with the measured data.

It should be noted that the proposed method in this chapter is valid for rivers with large variations in salinity and pH. Therefore, this method could be used for rivers either close to the coastal waters, and thus affected by tides, or such rivers that have many agricultural inputs from saline soils draining into them. The chosen reach of the Karoon River in this research was an example of the second type of river. The average minimum and maximum EC for three years data collection (2002 - 2004) at the Ahwaz hydrometric station (see Figure 3) were 707 and 2254 $\mu\Omega^{-1}/cm$, respectively. The pH values also ranged from a minimum of 7.3 to a maximum of 8.5 at this station.

In deriving Equations (20) to (22) for lead and the similar ones for cadmium (Table 3), it was assumed that the environmental factors and water properties remained constant during the whole simulation period. Since the model was calibrated using measured dissolved lead and cadmium at the site this assumption was thought to be valid. However, there are some limitations in using these equations. Firstly, simultaneous measurements of dissolved lead and cadmium were only made at one site and for six months. More field-measured data are needed to validate and improve the formulae, which relate the pH and EC values to the reaction coefficient for dissolved lead and cadmium. Secondly, a one-dimensional model was used. Although one-dimensional models have been successfully used in riverine hydrodynamic and water quality studies, it seems that applying a two- or three-dimensional model may improve the derived equations. However, using two- or three-dimensional models needs extensive field-measured data. The importance of the models is to estimate the desirable variables as accurately as possible. Measuring some special environmental variables, such as heavy metals, in the field is sensitive and ideally needs extensive laboratory studies with sophisticated instruments and with large investments. Measuring pH and EC in riverine systems is relatively straightforward and can be done with even portable instruments. The main idea from this research work is therefore to introduce a procedure that relates the pH and EC values to reaction coefficients of heavy metal substances, such as lead and cadmium, for model predictions. Hence, for heavy metals modelling studies, measurements of pH and EC would be a suitable tool for relatively accurate estimation of these substances.

The results show an average improvement of 25% and 71.5% in error estimations of lead and cadmium, respectively, when using pH and EC as two variables affecting the dynamic processes of these heavy metals.

7. Summary and conclusions

Details are given of the hydro-environmental model to predict the dissolved heavy metals concentration along rivers using a varied reaction coefficient approach to the source term of the Advection-Dispersion Equation (ADE). The main purpose of this chapter was to describe the dissolved heavy metals modelling procedure and assess the impact of pH and EC on the reaction coefficient used in dissolved lead and cadmium modelling. The hydrodynamic module was first calibrated and validated using the field-measured data taken at a site located along the Karoon River, the largest river in the south west of Iran. In order to find the best equation between pH and EC with the reaction coefficient used in the ADE too, many model runs were carried out and the water quality module was subsequently calibrated by adjusting the reaction coefficient. For each measured lead or cadmium value at any time the most appropriate reaction coefficient was specified and from there for the considered heavy metals a few equation between pH and EC with the reaction coefficient were proposed and added to the water quality module of the model. The main findings from the model simulations can be summarized as follows:

1. Five different procedures were used for estimating the rate of reaction coefficient for dissolved lead and cadmium, including: a zero reaction coefficient, a constant reaction coefficient, a varying reaction coefficient with pH, a varying reaction coefficient with EC and a varying reaction coefficient with both pH and EC.
2. Improvements were achieved in the predicted dissolved lead and cadmium concentration distributions when varying reaction coefficients were used.
3. The best fit between the predicted and measured values for simulation with a constant reaction coefficient was obtained when the coefficient was set to 0.12 and 0.38 day^{-1} for dissolved lead and cadmium, respectively.
4. According to Equations (20) to (22) for lead and the similar ones in Table 3 for cadmium and the measured pH and EC values, the ranges of reaction coefficients were calculated to be: (0.11-0.18, 0.10-0.29, 0.10-0.43) and (0.31-0.40, 0.31-0.48, 0.31-0.44) for lead and cadmium for the three suggested procedures, respectively. The error estimation was decreased from an average of 30% to 4% for lead and 74% to 2.2% for cadmium when pH and EC were used as two variables affecting the reaction coefficient.

8. References

Abbott, M.B. & Basco, D.R. (1997). *Computational Fluid Dynamics: An introduction for engineers,* Longman Singapore Publishers (Pte) Ltd., 425pp.

Akan, J.C.; Abdulrahman, F.I.; Sodipo, O.A.; Ochanya, A.E. & Askira, Y.K. (2010). Heavy metals in sediments from river Ngada, Maiduguri, Metropolis, Borno State, Nigeria. *Journal of Environmental Chemistry and Ecotoxicology,* Vol.2, No.9, pp.131-140

Azmathullah, H.Md.; Deo, M.C. & Deolalikar, P.B. (2005). Neural networks for estimating the scour downstream of a ski-jump bucket. *Journal of Hydraulic Engineering,* Vol.131, No.10, pp.898-908

Cunge, J.A.; Holly, Jr.F.M.; & Verwey, A. (1980). *Practical aspects of computational river hydraulics,* Pitman Publishing Limited, Boston

Diagomanolin, V.; Farhang, M.; Ghazi-Khansari, M. & Jafarzadeh, N. (2004). Heavy Metals (Ni, Cr, Cu) in the Karoon waterway river, Iran. *Toxicology Letters,* Vol.151, pp.63-68

Falconer, R.A. & Lin, B. (2003). Hydro-environmental modelling of riverine basins using dynamic rate and partitioning coefficients. *International Journal of River Basin Management,* Vol.1, No.1, pp.81-89

Falconer, R.A.; Lin, B. & Kashefipour, S.M. (2005). *Modelling water quality processes in estuaries.* In: *Computational Fluid Dynamics, Application in Environmental Hydraulics* Edited By: P.D. Bates, S.N. Lane & R.I. Ferguson, Chapter 12, John Wiley & Sons, Ltd., pp.305-328

Fischer, H.B.; List, E.J.; Koh, R.C.J.; Imberger, J. & Brooks, N.H. (1979). *Mixing in inland and coastal waters.* Academic Press, Inc., San Diego, 483pp.

Haung, S.L. (2010). Equations and their physical interpretation in numerical modeling of heavy metals in fluvial rivers. *Science China, Technological Sciences,* Vol.53, No.2, pp.548-557.

Haung, S.L.; Wan, Z.H. & Smith, P. (2007). Numerical modelling of heavy metal pollution transport-transformation in fluvial rivers: A Review. *International Journal of Sediment research,* Vol.22, No.1, pp.16-26.

Kashefipour, S.M., (2002). *Modelling flow, water quality and sediment transport processes in riverine basins.* Ph.D. Thesis, Department of Civil Engineering, Cardiff University, UK., 295pp.

Kashefipour, S.M. & Falconer, R.A. (2002). Longitudinal dispersion coefficient in natural channels. *Water Research,* Vol.36, pp.1596-1608

Kashefipour, S.M.; Falconer, R.A. & Lin, B. (1999). *FASTER model reference manual.* Environmental Water Management Research Centre Report, Cardiff University.

Kashefipour, S.M.; Lin, B. & Falconer, R.A. (2006). Modelling the fate of faecal indicators in a coastal basin. *Water Research,* Vol.40, pp.1413-1425

Kashefipour, S.M.; Lin, B.; Harris, E.L. & Falconer, R.A. (2002). Hydro-Environmental modeling for bathing water compliance of an estuarin basin. *Water Research,* Vol.36, No.7, pp.1854-1868

Kumar, B.; Kumar, S.; Mishra, M.; Prakash, D.; Singh, S.K.; Sharma, C.S. & Mukerjee, D.P. (2011). An assessment of heavy metals in sediments from two tributaries of lower stretch of Hugli estuary in West Bengal. *Archives of Applied Science Research,* Vol.3, No.4, pp.139-146

Leonard, B.P. (1979). A stable and accurate convective modeling procedure based on quadratic upstream interpolation. *Computer Methods in Applied Mechanics and Engineering,* Vol.19, pp.59-98

Leonard, B.P. (1991). The ULTIMATE conservative difference scheme applied to unsteady one-dimensional advection. *Computer Methods in Applied Mechanics and Engineering,* Vol. 88, pp.17-74.

Lin, B. & Falconer, R.A. (1997). Tidal flow and transport modelling using the ULTIMATE QUICKEST scheme. *Journal of Hydraulic Engineering,* Vol.123, No.4, pp.303-314

Mala, J. & Maly, J. (2009). Effect of heavy metals on self purification processes in rivers. *Applied Ecology and Environmental Research,* Vol.7, No.4, pp.333-340

Nassehi, V. & Bikangaga, J.H. (1993). A mathematical model for hydrodynamics and pollutant transport in long and narrow tidal rivers. *Appli. Math Modelling Journal,* Vol.17, pp.415-422

Ng, B.; Turner, A.; Tyler, A.O.; Falconer, R.A. & Millward, G.E. (1996). Modelling contaminant geo-chemistry in estuaries. *Water Research,* Vol.30, pp.63-74

Orlob, G.T. (1983). *Mathematical modelling of water quality: Streams, Lakes and Reservoirs,* John Wiley & Sons, Great Britain

Papafilippaki, A.K.; Kotti, M.E. & Stavroulakis, G.G. (2008). Seasonal variations in dissolved heavy metals in the Keritis river, Chania, Greece. *Global Nest Journal,* Vol.10, No.3, pp.320-325

Rauf, A.; Javed, M.; Ubaidyllah, M. & Abdullah, S. (2009). Assessment of heavy metals in sediments of river Ravi, Pakistan. *International Journal of Agriculture & Biology*, Vol.11, No.2, pp.197-200

Roshanfekr, A.; Kashefipour, S.M. & Jafarzadeh, N. (2008a). A new approach for modeling dissolved lead using an integrated 1D and 2D model, *Journal of Applied Sciences*, Vol.8, No.12, pp.2242-2249

Roshanfekr, A.; Kashefipour, S.M. & Jafarzadeh, N. (2008b). Numerical modelling of heavy metals for riverine systems using a new approach to the source term in the ADE, *IWA Journal of Hydroinformatics*, Vol.10, No.3, pp.245-255

Sanayei, Y.; Ismail, N. & Talebi, S.M. (2009). Determination of heavy metals in Zayandeh Rood river, Isfahan, Iran. *World Applied Sciences Journal*, Vo.6, No.9, pp.1209-1214

Shrestha, P.L. & Orlob, G.T. (1996). Multiphase distribution of cohesive sediments and heavy metals in estuarine systems. *Journal of Environmental Engineering*, Vol.122, No.8, pp.730-740

Tavakolizadeh, A.A. (2006). *Modelling hydrodynamics and water quality in riverine systems.* MSc. Thesis, Department of Hydraulic Structures, Shahid Chamran University, Iran. 136 pp. (In Farsi)

Thomann, A. & Mueller, J.A. (1987). *Principles of surface water quality modelling control.* Harper Collins, New York

Turner, A.; Martino, M. & Le roux, S.M. (2002). Trace metal distribution coefficients in the Mersey estuary, UK: Evidence got salting out of metal complexes. *Environmental Science and Technology Journal*, Vol.36, No.21, pp.4578-4584

Wu, W. (2008). *Computational River Dynamics*, Taylor & Francis Group, London, 494pp.

Wu, Y.; Falconer, R.A. & Lin, B. (2001). Hydro-environmental modelling of heavy metal fluxes in an estuary. *Proceedings of XXIX IAHR Congress, Theme b: Environmental Hydraulics*, Beijing, China, pp.732-739

Wu, Y.; Falconer, R.A. & Lin, B. (2005). Modelling trace metal concentration distributions in estuarine waters. *Journal of Estuarine, Costal and Shelf Science*, Vol. 64, pp. 699-709

Yang, L.; Lin, B.; Kashefipour, S.M. & Falconer, R.A. (2002). Integration of a 1-D river model with object-oriented methodology. *Environmental Modelling and Software*, Vol.17, pp.693-701

Zhen-Gang, J. (2008). *Hydrodynamics and Water Quality: Modeling Rivers, Lakes, and Estuaries.* John Wiley & Sons, Inc., Hoboken, NJ, 676pp.

7

Numerical Modelling of Dynamic Nitrogen at Atmospheric Pressure in a Negative DC Corona Discharge

A.K. Ferouani, B. Liani and M. Lemerini
Laboratory of Theoretical Physics, University Abou Bakr Belkaid, Tlemcen Algéria

1. Introduction

Plasmas are generated by supplying energy to a neutral gas causing the formation of charge carriers. Electrons and ions are produced in the gas phase when electrons or photons with sufficient energy collide with the neutral atoms and molecules in the feed gas (electron impact ionization or photoionization). The most widely used method for plasma generation utilizes the electrical breakdown of a neutral gas in the presence of an external electric field. Charge carriers accelerated in the electric field couple their energy into the plasma via collisions with other particles. Electrons retain most of their energy in elastic collisions with atoms and molecules because of their small mass and transfer their energy primarily in inelastic collisions. Discharges are classified as DC discharges, AC discharges, or pulsed discharges on the basis of the temporal behaviour of the sustaining electric field. The spatial and temporal characteristics of plasma depend to a large degree on the particular application for which the plasma will be used [1].

Today, plasmas are increasingly used in industry [1–3]. There are two types of plasma, the so-called thermal plasmas and cold plasmas said. The corona is a process that could lead to the creation of the latter. The use of techniques involving the corona tends to grow in importance. Indeed, they are out and already widely used in the areas of destruction of pollutants and waste gas, the surface treatment (cleaning and surface erosion, deposition of films, modifying the surface chemistry). They are also used in other applications such as ozone generation and elimination of static electricity. Also, to minimize development costs, recent research attempting to model the phenomena involved

In this work, we study the thermodynamics of the neutral gas subjected to energy injection as the result of electric discharge in the considered medium. This approach to the problem allows considering the discharge only on its energetic aspect. The discharge plays the role of an injection in the gas. To define the profile of this energy injection, we propose a mathematical function that represents the spatial dependence of the discharge density. The spatio-temporal evolution of the neutral gas particles is studied on the basis of hydrodynamic set of equations, i.e. equations of transport for mass, momentum and energy [4]. The hydrodynamic set of equations is solved by the F.C.T method (Flux Corrected Transport).

2. Description of corona discharge

Under the action of an electric field, the gas molecules undergo electron collisions, according to the complex mechanisms associated with shock [5]. The reactivity of the gas depends mainly on the shape of the energy delivered to the electrode system and generating the corona called "reactor ". Geometries are often very divergent and energy sources can be of multiple origins [6].

A corona discharge occurs when a current, power is created between two electrodes brought to a high potential and separated by an inert gas, usually air ionization plasma is created and the electric charges propagate through ions with neutral gas molecules. When the electric field at a point of a gas is sufficiently large, the gas ionizes around this point and becomes conductive. In particular, if one has been charged peaks, the electric field will be greater than elsewhere, this is usually as a corona discharge will occur, the phenomenon will tend to stabilize itself as the region becomes ionized conductive tip will apparently tend to disappear. The charged particles dissipate while under the influence of the electric force and neutralize an object in contact with opposite charge.

Corona discharges therefore generally occur between an electrode of small radius of curvature (for example: fault of the conductor forming a point) as the electric field surrounding area is large enough to allow the formation of a plasma. Corona discharge can be positive or negative depending on the polarity of the electrode with a small radius of curvature. If positive, it is called positive corona, otherwise negative crown [7]. Because of the difference in mass between electrons (negative) and ions (positive), the physics of these two types of corona is radically different.

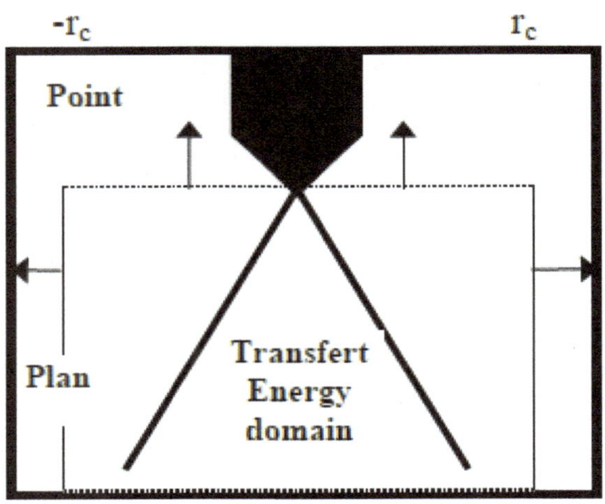

Fig. 1. domain study.

Landfills crown are guy characterized by asymmetry of the electrodes, at least one of the two electrodes with high curvature. Reduces the electric field produced in the electrode gap, when applying a high voltage is strongly inhomogeneous. The name of corona discharge is the luminous halo-shaped crown that appears around the electrode with high curvature at the initiation of the discharge. Deferent types of geometry are used in the experiments: tip-up, wire up, wire-wire and wire-cylinder. The high voltage applied to the electrode with high curvature can be positive or negative [8-9].

One of the main difficulties encountered with landfill type crown is the transition to the arc. This phenomenon is characterized by a strong rise in the current flowing in the discharge and a significant increase in the gas temperature. The plasma is then generated close to thermodynamic equilibrium [10].

In a point-to-plane configuration at atmospheric pressure, with the sharp electrode being supplied with a negative discharge DC [8], the corona discharge inception is principally due to the acceleration of background electrons (resulting from cosmic radiation) in the high electric field created by the small curvature radius of the point. The resulting space charge field, added to the 'geometrical' initial one, allows the electrons situated a little farther away to be accelerated [11].

The corona discharge is initiated when the electric field near the wire is sufficient to ionize the gaseous species. The minimum electric field is a function of the wire radius, the surface roughness of the wire, Nitrogene temperature, and pressure. The free electrons produced in the initial ionization process are accelerated away from the wire in the imposed electric field. More frequent inelastic collisions of electrons and neutral gas molecules occur [8].

Numerous models of corona discharge have been proposed. In [5] a wire-to-cylinder corona discharge is modelled by means of electronic injectors with azimuth symmetry, assimilating the coaxial discharge to a succession of elementary point-to-cylinder electrical discharges (Fig. 2).

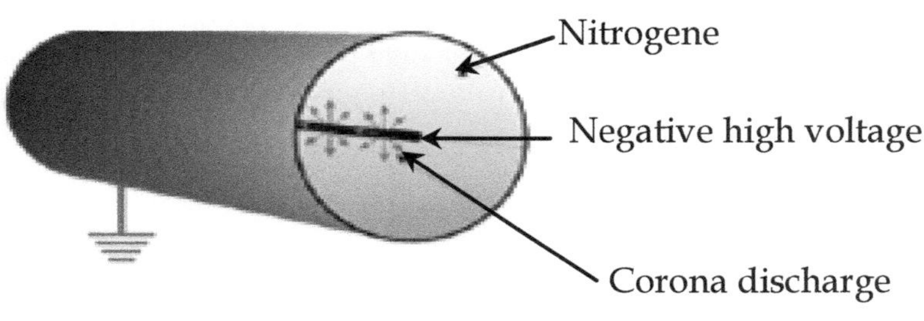

Fig. 2. Corona discharge in wire cylinder electrode geometry.

During the inception and development of the plasma in a point to plane gas discharge [10], a spatio-temporel evolution of the temperature of the neutral gas occurs as a result of plasma-neutral molecules energy interaction [11]. The temperature gradient causes a phenomenon of diffusion and convection as a result of the accompanied strong heterogeneity in the neutral gas density and pressure [12]. The fundamental role of neutral heating in the inception of gas breakdown has been shown by theoretical studies [13-14], as well as by experimental studies [15]. The behaviour of a point to plane discharge has been optically and electrically analysed for a centimeter gaps in Nitrogene at atmospheric pressure [16].

3. Introduction to kinetic theory

There are many phenomena in ionized gases for which we need to consider the velocity distribution function of the particles, or at least of some particles such as free electrons, and to use a treatment called kinetic theory. In fluid theory, the velocity distribution of each species is assumed to be Maxwellian everywhere and is therefore uniquely specified by the species temperature T. Because inelastic collisions, especially between electrons and neutral particles, play a major role in low-temperature plasmas, significant deviations from thermal equilibrium are usually present in such media, which justifies the need for using the kinetic theory. By definition, the velocity distribution function f(r, v, t) of a given species represents the number of particles of that species per unit volume of the six dimensional phase space at position (r, v) and time t. This means that the number of particles per unit volume in configuration space with velocity components between v_x and $v_x + dv_x$, v_y and $v_y + dv_y$, and v_z and $v_z + dv_z$ at time t is:

$$f(x,y,z,v_x,v_y,v_z).dv_x dv_y dv_z \tag{1}$$

When we consider velocity distributions, we therefore have seven independent scalar variables (r, v, t). The density in configuration space n = n(r, t), which is a function of only four scalar variables, is obtained by integration of f(r, v, t) over velocity space, that is

$$n(r,t) = \int_{-\infty}^{\infty} dv_x \int_{-\infty}^{\infty} dv_y \int_{-\infty}^{\infty} f(r,v,t)\, dv_z \tag{2}$$

The distribution function of simple speed allows us to calculate, for each position r and time t, the average value of certain physical properties, resulting in the so-called macroscopic or hydrodynamic quantities. Let A (r, v, t) a molecular property of any kind. The most general definition of its average value, denoted by A (r, t), is given by the expression:

$$A(\vec{r},t) = \frac{1}{n(\vec{r},t)} \int_{-\infty}^{\infty} A(\vec{r},\vec{v},t)\, f(\vec{r},\vec{v},t) dv^3 \tag{3}$$

3.1 The Boltzmann equation

The Boltzmann equation is derived rigorously from the Liouville theorem,. However, we can get this equation quickly, but in a formal way, by first assuming the absence of collisions between particles and, in a second time, taking into account the effect of collisions. The

Boltzmann equation is useful to describe the evolution of a gas of charged particles in an external electromagnetic field, and it obviously implies that the particles are small enough density does not change the outfield. It is written as follows [17]:

$$\frac{df}{dt}=\frac{\partial f}{\partial t}+\vec{v}.\vec{\nabla} f+\frac{\vec{F}}{m}.\vec{\nabla}_v f=\left(\frac{\partial f}{\partial t}\right)_{coll} \tag{4}$$

Here F is the force acting on the particles, and $(\partial f/\partial t)_{coll}$ is the time rate of change of f due to collisions. Considering, for example, free electrons, this collision term must account for elastic and inelastic electron–neutral collisions, and, at relatively high degrees of ionization, for electron–electron and electron–ion collisions. The symbol ∇ stands, as usual, for the gradient in configuration space (x,y,z) while the symbol ∂/∂v or ∇v stands for the gradient in velocity space.

Here, ∂f/∂t is the rate of change due to the explicit dependence on time. The next three terms are just v.∇f while the last three terms, taking into account Newton's third law m(dv/dt) = F are recognized as (F/m).(∂f/∂v). The total derivative df/dt can be interpreted as the rate of change as seen in a frame moving with the particles in the six dimensional (r, v) space. The Boltzmann equation simply says that df/dt is zero unless there are collisions. Collisions have the effect of removing a particle from one element of velocity space and replacing it in another, or even creating a new particle in the case of ionization. One provides for this by the collision term $(\partial f/\partial t)_{coll}$.

3.2 The conservation equations

The conservation equations of density number, momentum and energy for a single species may be obtained by the method of moments. In this method, f (r, v, t) is multiplied with a function g(v) of the velocity and integrated over the entire velocity space. For the case that g(v) = 1, we obtain the continuity equation, if g(v) = m × v we obtain the momentum conservation equation, and if $g(v)= mv^2/2$, we obtain the energy conservation equation. The fluid equations are simply moments of the Boltzmann equation. The lowest moment is obtained just by integrating this equation over velocity space [17-19]:

$$\int_{-\infty}^{\infty}\frac{\partial f}{\partial t}\,d^3v+\int_{-\infty}^{\infty}\vec{v}\vec{\nabla} f\,d^3v+\int_{-\infty}^{\infty}\frac{\vec{F}}{m}\vec{\nabla}_v f\,d^3v=\int_{-\infty}^{\infty}\left(\frac{\partial f}{\partial t}\right)_{coll}d^3v \tag{5}$$

where dv stands for a three-dimensional volume element in velocity space. By transforming the third term on the left-hand side by Green's theorem and after straightforward calculations one obtains the continuity equation:

$$\frac{\partial n}{\partial t}+\vec{\nabla}(nu)=S \tag{6}$$

where u is the average (fluid) velocity:

$$u(\vec{r},t)=\frac{1}{n(\vec{r},t)}\int_{-\infty}^{\infty}\vec{v}\,f(\vec{r},\vec{v},t)dv^3 \tag{7}$$

S represents the net creation rate of particles per unit volume as a result of collisions (for example, in the case of electrons, this term takes into account new electrons created by ionization and electron losses due to recombination with ions or attachment).

The next moment of the Boltzmann equation is obtained by multiplying it by mv and integrating over dv. We have:

$$m\int_{-\infty}^{\infty}\vec{v}\frac{\partial f}{\partial t}\,d^3v + m\int_{-\infty}^{\infty}\vec{v}\,(v\vec{\nabla})f\,d^3v + m\int_{-\infty}^{\infty}\vec{v}(\vec{F}.\vec{\nabla}_v)f\,d^3v = \int_{-\infty}^{\infty}\left(\frac{\partial f}{\partial t}\right)_{coll}d^3v \tag{8}$$

After calculation we obtain [4]:

$$\frac{\partial nm\vec{u}}{\partial t} + \vec{\nabla}.(n\vec{u}\vec{u}) + \vec{\nabla}\Pi + \vec{\nabla}.p - n\vec{F} = \int_{-\infty}^{\infty} m\vec{v}\left(\frac{\partial f}{\partial t}\right)_{coll}d^3v \tag{9}$$

where Π denotes the viscosity, p denotes the pressure, and F the specific external forces exerted on the species. The first term of (9) represents the accumulation of the specific momentum, which is generally nonzero in a transient system. The second term denotes the momentum transport caused by the flow. The third term represents the viscous forces. The fourth term is the pressure gradient. Formany flowing systems, including the plasmas treated in this work, this is the driving force that causes the various plasma species to flow. The fifth term represents the external forces, thus the combined action of the electric force, the Lorentz force and gravity. Tight-hand side term represents the momentum gained and lost trough collisions with other species. This may include the transfer of momentum from other species, or the creation of species with nonzero momentum.

We have deduced the form of the equation of conservation of energy as a function of thermal energy, using the Fourier law for thermal conductivity and the ideal gas law [1,4]:

$$\frac{3}{2}\frac{\partial nk_BT}{\partial t} + \frac{3}{2}\vec{\nabla}(nk_BT\vec{u}) + p\vec{\nabla}.\vec{u} + (\vec{\nabla}.\vec{u})\Pi - \vec{\nabla}(\lambda\vec{\nabla}T) = \int_{-\infty}^{+\infty} E^T\left(\frac{\partial f}{\partial t}\right)_{coll}d^3v \tag{10}$$

with k_B Boltzmann's Constant, λ the thermal conductivity and E^T the thermal energy. By assuming the existence of a temperature T for the species, we implicitly assume Maxwell-Boltzmann equilibrium. However, (10) can readily be rewritten in terms of average particle energies if deviations from Maxwell-Boltzmann equilibrium are relevant.

The first term on the left-hand side of (10) denotes the accumulation of thermal energy, and generally is nonzero in the transient systems treated in this work. The second term represents the convective transport of energy by means of the systematic velocity of the species. The third term represents the expansion work. The fourth term is the production of thermal energy by viscous dissipation, which is in fact the transfer of directed kinetic energy to random thermal energy in the species. The fifth term represents the diffusive heat transport (thermal conduction).

The term on the right side represents the transfer of thermal energy by collision

4. Mathematical model

The discharge column is considered to be cylindrically symmetric and longitudinally uniform. It is quasi-neutral, weakly-ionized, and collision-dominated. The column is characterized by a current density distribution j(r,t), witch is a function of radius r as well as time t, and a longitudinal voltage gradient E, which is assumed independent of both position in the column and time. The current density j goes to zero at a fixed radius R_C. The ionized region, which is initially diffuse, is contained in an infinite background of perfect gas of particle density (at t=0) uniform at N_0 and T_0, respectively.

The rate at which thermal energy is added to the gas per unit volume is given by j(r,t) E(r,t). That is, all the input power is assumed to be transferred from the electron to the background gas. As the temperature increases, the gas expands and its density decreases near the axis. Where the gas density decreases the electrical conductivity and current density increase, thus enhancing the subsequent rate of heating and expansion [13].

The gas dynamics are described by the conservation equations for a viscous compressible fluid and the equation of state for a perfect gas. The equations are written in cylindrical coordinates, written rotational symmetry and axial uniformity, with gas flow in the radial direction only, and with zero body forces. The fluid equations are, for the conservation of masse (continuity equation).

$$\frac{\partial N}{\partial t}+\frac{1}{r}\frac{\partial (N v_r r)}{\partial r}=0 \qquad (11)$$

where N is particle density, r is function of radius as will as time t and v_r is the radial velocity.

For momentum (equation of motion):

$$MN\left[\frac{\partial v_r}{\partial t}+v_r\frac{\partial v_r}{\partial r}\right]=-\frac{\partial p}{\partial r}+\frac{\partial}{\partial r}\left[\mu\left(2\frac{\partial v_r}{\partial r}-\frac{3}{2}\frac{1}{r}\frac{\partial v_r r}{\partial r}\right)\right]+\frac{2\mu}{r}\left(\frac{\partial v_r}{\partial r}-\frac{v_r}{r}\right) \qquad (12)$$

where M is the masse of a gas molecule and μ is the coefficient of viscosity.

and energy, for a perfect gas,

$$\begin{aligned} &MN\left[C_v\frac{\partial T}{\partial t}+v_r C_v\frac{\partial T}{\partial r}+\frac{p}{M}\frac{\partial (1/N)}{\partial t}+\frac{p}{M}v_r\frac{\partial (1/N)}{\partial t}\right] \\ &=jE+\frac{1}{r}\frac{\partial}{\partial r}\left(\lambda\frac{\partial T}{\partial r}\right)+\frac{4\mu}{3}\left[\left(\frac{v_r}{r}\right)^2+\left(\frac{\partial v_r}{\partial r}\right)^2-\frac{v_r}{r}\frac{\partial v_r}{\partial r}\right] \end{aligned} \qquad (13)$$

where p is the gas pressure, C_v is the specific heat at constant volume, T temperature, λ is the coefficient of thermal conductivity.

The equation of state is:

$$p = N k_B T \qquad (14)$$

where k_B is Boltzmann constant

Energy transfer by radiation has been neglected. Eliminating the time derivative of N from the energy equation by using the continuity equation, replacing the pressure with Nk_BT, and rearranging terms, we can rewrite the conservation equations:

$$\frac{\partial N}{\partial t}=-\frac{1}{r}\frac{\partial(N v_r r)}{\partial r} \tag{15}$$

$$\frac{\partial v_r}{\partial t}=-v_r\frac{\partial v_r}{\partial r}-\frac{k}{MN}\frac{\partial(NT)}{\partial r}+\frac{1}{MN}\frac{\partial}{\partial r}\left[\mu\left(2\frac{\partial v_r}{\partial r}-\frac{2}{3}\frac{1}{r}\frac{\partial v_r r}{\partial r}\right)\right]+\frac{2\mu}{MNr}\left(\frac{\partial v_r}{\partial r}-\frac{v_r}{r}\right) \tag{16}$$

and

$$\begin{aligned}\frac{\partial T}{\partial t}&=-v_r\frac{\partial T}{\partial t}-\frac{kT}{MC_vN}\frac{1}{r}\frac{\partial(Nv_r r)}{\partial r}+\frac{kTv_r}{MC_vN}\frac{\partial N}{\partial r}\\&+\frac{1}{MNC_v}\left\{jE+\frac{1}{r}\frac{\partial}{\partial r}\left(\lambda\frac{\partial T}{\partial r}\right)+\frac{4\mu}{3}\left[\left(\frac{v_r}{r}\right)^2+\left(\frac{\partial v_r}{\partial r}\right)^2-\frac{v_r}{r}\frac{\partial v_r}{\partial r}\right]\right\}\end{aligned} \tag{17}$$

If viscosity and thermal conduction are neglected, the speed of sound in the gas v_s can be written:

$$v_s=\left(\frac{\gamma kT}{M}\right)^{1/2} \tag{18}$$

Where γ is the ratio C_p/C_v of the specific heat at constant pressure to that at constant volume [13].

5. Numerical analysis

The discharge studied in our work requires that the method used to solve the equations of transport is efficient and has the ability to follow the strong density gradients while keeping a reasonable computation time. To this end, we opted for the scheme of Flux Corrected Transport Low phase error has already been used successfully in several areas such as solving the Boltzmann equation in weakly ionized gases. The diagram FCT (Flux Corrected Transport) is certainly one of the best choices to make while it is quite complex. Among its advantages are: the absence of spurious oscillations, numerical diffusion minimum; It can also calculate the evolution of profiles with very sharp spatial variations [20].

Our work of the simulation of the discharge in space is two-dimensional with cylindrical symmetry. The hydrodynamic set of equations is solved by the F.C.T method (Flux Corrected Transport) using the procedure of time splitting for the two space variables. An FCT algorithm consists conceptually of two major stages, a transport or convective stage (Stage I) followed by an antidiffusive or corrective stage (Stage II) [20-22]. All transport equations of the charged or the neutral particles defined previously obey the same generic form:

$$\frac{\partial}{\partial t}\varphi(r,z,t)+\frac{1}{r}\frac{\partial r\,\varphi(r,z,t)v_r}{\partial r}+\frac{\partial\varphi(r,z,t)v_z}{\partial z}=S(r,z,t) \tag{19}$$

where r, z are space variables, t is temporal variable, φ(r,z,t) is the transported size (density, momentum or energy) and S(r,z,t) indicates the source term of the corresponding transport equation.

The transport equations which are narrowly coupled are discretized by the method of volumes finished and are corrected by the method of the finished volume and corrected by the method of corrections of flow developed by Boris and Book [20].

The transport equations were discretized on the mesh nodes using numerical schemes to avoid the problems of digital broadcasting, which is especially important. To simplify the presentation of the method we consider a time step Δt constant, we divide the two-dimensional space into cells infinitely small. The application of this method involves three steps:

- The transport step: we calculate the value of the quantity transported and distributed in each node of the cell.
- The diffusion step to ensure that the solution is positive
- The next step is to reverse the spread where it is not necessary. Such an anti-diffusion step is necessary to find the accuracy of the transport step

The study domain is defined by figure 2. The limit velocity of the molecules on the surface is assumed equal to zero. As it is necessary to take into account the local heating effects, the temperature of the surface is assumed equal to the averaged temperature of the surrounding gas, and the temperature of the electrode body is assumed invariable and equal to the ambient temperature.

$$\frac{\partial N}{\partial r}(0,0,t)=\frac{\partial T}{\partial r}(0,0,t)=\frac{\partial v}{\partial r}(0,0,t)=0 \text{ and } T=293\,K,\quad v_r=0,\quad U=12\ kV \tag{20}$$

6. Results and discussion

In Figs. 3-6, the spatio-temporal evolution of temperature, density, pressure and speed of neutrals are shown, respectively, for the case of a negative point discharge, cold wall and constant injection of energy.

In Figure 3, we observe a growing neutral heating in the function of time. This transfer of heat is important for the discharge was near the center. Indeed ~ 10 mm (from the point), the temperature passes from the value 350 K at t = 1 μs to 650 K at t_6 = 50μs, whereas it remains almost constant near the edge and varies slowly near the cathode.

On the other hand the temperature increases rapidly with time, there is also a shift of the maximum temperature in the direction of the anode

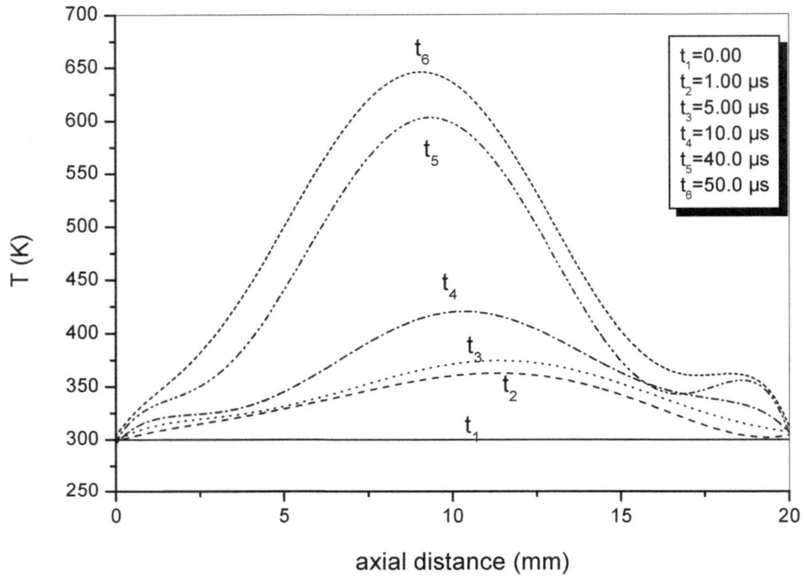

Fig. 3. The axial evolution of neutral temperature for several laps (negative point discharge, cold wall, constant injection of energy).

Figure 4 shows the evolution of the density of neutral in the function of time and space. We notice on all the curves a neutral depopulation in Inter electrode space. This decline results from the thermal footprint caused by the passage of the streamer discharge, it is more important at 10 mm (from the point), where there is a rate of 40% at t_6 = 50 µs , while there is a decline up to 5 % in t_2 =1µs. In the middle of the discharge are the ionization phenomena responsible for the decrease in the density of neutral or figures 5 and 6, which represents the evolution of the pressure and the module of the neutral speed, we notice, because of the inertia of molecules of gas, a phase shift between the maximum module speed and total maximum pressure. This gap is especially well marked on the axis, and at the beginning of the discharge. For other parts of the field, this phase shift is less accentuated because the disturbance created by the discharge is less important for intensity. As the time elapses, the evolution of pressure and speed module becomes constant.

From the moment 20 µs, we see a trend toward stationarity for all sizes (temperature, density, pressure and speed), because the heating in a comprehensive manner (contribution of all terms), decreases in intensity over time and the dissipation of energy becomes important. The result of all these processes, that all occurs as if a heating effect (known as heat wave) begins at the tip to spread towards the plan.

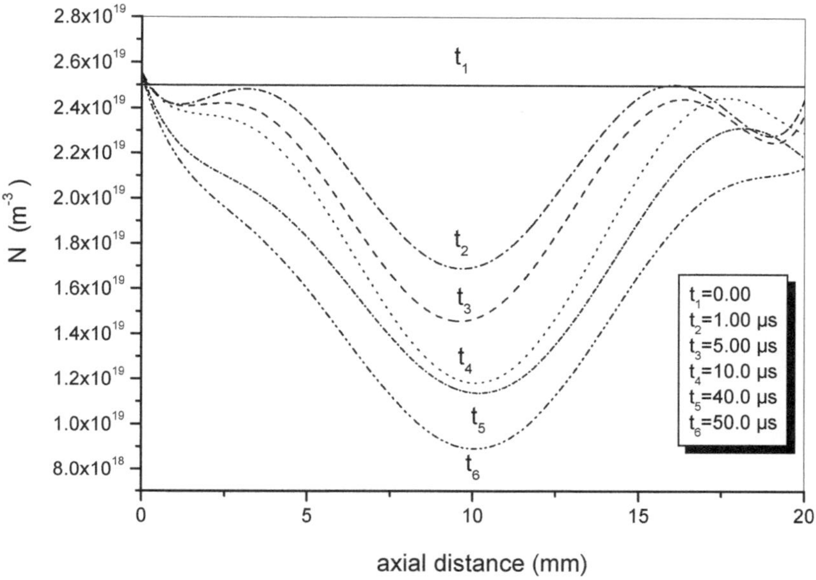

Fig. 4. The axial evolution of neutral density for several laps (negative point discharge, cold wall, constant injection of energy).

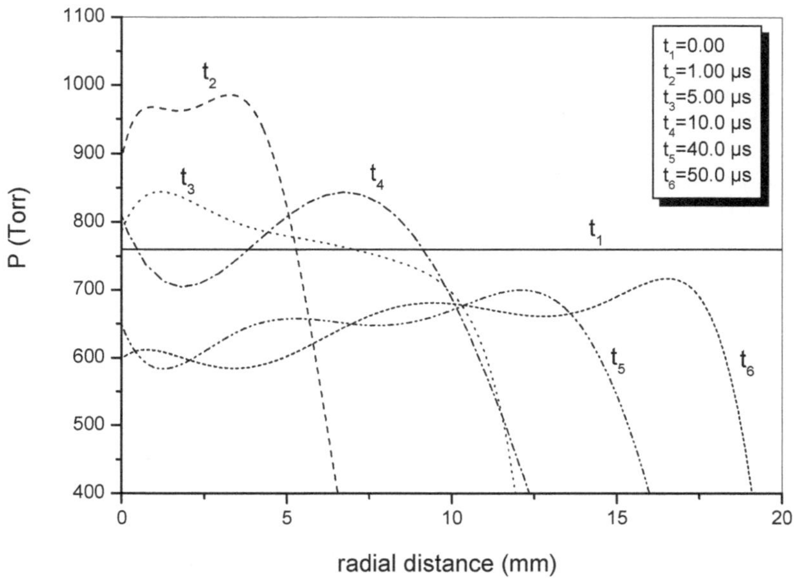

Fig. 5. The radial evolution of neutral pressure for several laps (negative point discharge, cold wall, constant injection of energy).

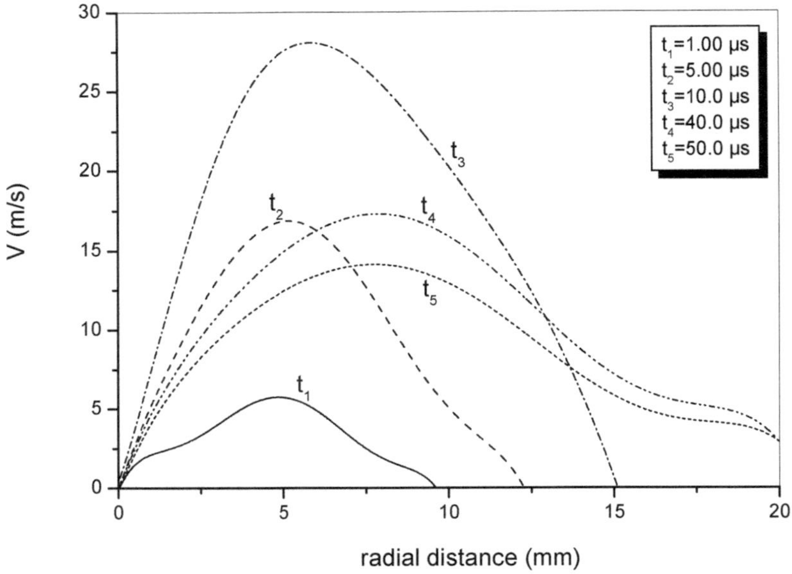

Fig. 6. The radial evolution of total speed module of neutral for several laps (negative peak discharge, cold wall constant injection of energy).

7. Conclusion

In this paper, we have presented numerical calculations for neutral thermal effects produced by the negative dc corona discharge DC at atmospheric pressure, is conducted.

The objective of this study is to develop an efficient numerical model for solving transport equations. This allowed us to study the evolution of temperature and density of neutral particles as a function of axial distance, in the case of a negative corona discharge at atmospheric pressure for a better understanding of the evolution and heat transfer in situations of large variations in density and electric field. We completed this approach by a numerical parametric study on the behavior of radial profiles of pressure and velocity neutral particles.

The results obtained reveal the existence of the phenomena of interaction between charged particles and neutral particles, which are causing instability reaction electric shocks. These instabilities can come from two sources:

- Electrostatic origin since the space charge occurring in the gas, change the local electric field.
- Thermal, since the energy transfer between gas ions and the neutral gas causes local variations in temperature and density of the neutrals.

The results show that the stabilization of the neutral gas is mainly on the function of the energy injection distribution, and depopulation is more important than the plane advance. So, as soon as a current goes through the neutral gas, obviously a Joule heating effect increases the temperature locally. These results also show that:

- Temperature increases with time, the middle of the discharge is warmer.
- The neutral density varies inversely as the temperature
- The appearance of a phase difference between the maximum speed of neutral and maximum total pressure module, due to the inertia of the gas molecules.

Due to its qualities of stability, accuracy and speed compared to digital technologies that preceded it, we can say that using the FCT method has opened new perspectives for modeling non-equilibrium discharges in general and in particular corona.

8. References

[1] Michel Moisan and Jacques Pelletier, *Physique des plasmas collisionnels: Applications aux décharges hautes fréquence*, EDP Sciences, 2006, ISBN 2-86883-822-7

[2] S.I. Medjahdi, A.K. Ferouani , M. Lemerini and S. Belhour, *International Review of Physique* ,2, (October 2008), 291-295 ISSN 1971-680X

[3] Alexandre Labergue 1992, *Etude de décharges électriques dans l'air pour le développement d'actionneurs plasmas, Application au contrôle de décollements d'écoulements* [PhD] (2005) University of Poitiers, France

[4] B. Held, Physique des plasmas froids, Paris: Masson, 1994, ISBN 978-2-225-84580-2

[5] C. Soria, F. Pontiga and A. Castellanos , *Plasma Sources Sci Technol.*, 13, (November 2004), PP. 95-107, ISSN 1361-6595

[6] J. Zhang, K. Adamiak and G.S.P Castle, *Journal of Electrostatics*, 65, (September 2007), PP. 174–181, ISSN: 0304- 3886

[7] J Dupuy and A Gibert *J. Phys. D: Appl. Phys.*, 15, (April 1982),PP. 655-664. ISSN 1361-6463

[8] A.K.Ferouani, M. Lemerini and S. Belhour, *Plasma Science and Technology*, 12, (April 2010), PP. 208-211, ISSN 1009-0630

[9] Junhong Chen and Jane H. Davidson, Model of the Negative DC Corona Plasma: Comparison to the Positive DC Corona Plasma *Plasma Chemistry and Plasma Processing*, 23, (March 2003), 83-101

[10] M. Lemerini,B. Bouhafs,B. Benyoucef and A Belaidi, *Rev. Energ. Ren.* 2, (September 1999),123-130

[11] J.C Mateo-Velez, P. Degond, F. Rogier, A. Seraudie and F. Thivet, *J. Phys. D: Appl. Phys.*, 41, (January 2008), PP. 1-11, ISSN 1361-6463

[12] K. Yanallah,S. Hadj-Ziane, A. Belasri and Y. Meslem , *Journal of Molecular Structure*. 777, (August 2006), PP.125-129. ISSN: 0022-2860

[13] G.L Rogoff , *J. Phys. Fluids.*,15, (1972), PP. 1931-1940, ISSP 1089-7666

[14] R. Morrow , *J. Phys. D: Appl. Phys.*, 30, (June1997), PP. 3099-3114, ISSN 1361-6463

[15] O. Ducasse, O. Eichwald, N. Merbahi , D. Dubois, and M. Yousfi , *J. Appl. Phys*,101, (2007), PP. 1046-1049, ISSN 1361-6463

[16] A. Luque, U. Ebert, and W. Hundsdorfer, physical review letters,(December 2007),101, PP. 1-4, ISSN. 0031- 9007

[17] C.M. Ferreira and J Loureiro, Electron kinetics in atomic and molecular plasmas, *Plasma Sources Sci. Technol.* 9 (2000) 528–540.

[18] J.L. Delcroix , A. Bers Physique des plasmas, volume 1, EDP Sciences, 1994. ISBN, 2271051266

[19] A.K.Ferouani, M.Lemerini, F.Boudahri and S.Belhour, *Conference Proceedings American Institute of Physics* (September 2008),1047, 232-235. ISSN 0094-243X

[20] J.P Boris and D. L. Book , journal of computational physics, 11,(November 1972), PP. 38-69 ISSN 0021-9991

[21] T. S .Sergey and j. S. Shang, *Journal of Computational Physics* 199, (September 2004),PP. 437–464 ISSN:0021-9991

[22] D. Kuzmin R. Löhner S. Turek, Flux-Corrected Transport : Principles, Algorithms, and Applications Springer-Verlag Berlin Heidelberg 2005, ISBN-10: 3540237305

Part 2

Maxwell's Equations

8

Numerical Study of Diffusion of Interacting Particles in a Magnetic Fluid Layer

Olga Lavrova[1], Viktor Polevikov[1] and Lutz Tobiska[2]
[1]*Belarusian State University*
[2]*Otto-von-Guericke University*
[1]*Belarus*
[2]*Germany*

1. Introduction

Magnetic fluids are stable colloidal suspensions of ferromagnetic nano-particles (of size 10-20 nm) in a nonmagnetic liquid carrier. An initially uniform particle distribution in the carrier becomes spatially inhomogeneous in nonuniform magnetic fields. Motion of particles in magnetic fluids under the action of magnetic fields is of particular interest for contemporary mathematical and numerical modelling in ferrohydrodynamics.

The most theoretical models for the diffusion process in magnetic fluids assume no interaction between particles (Bashtovoi et al., 2007; 2008; Lavrova et al., 2010; Polevikov & Tobiska, 2008; 2011), which is valid for dilute fluids only. This assumption allows to construct an explicit dependence between equilibrium particle concentration and the magnetic field distribution, simplifying significantly the modelling. In case of concentrated magnetic fluids another theoretical model should be considered. Recently, a dynamic mass transfer equation for describing diffusion of interacting ferromagnetic particles in magnetic fluids was derived (Pshenichnikov et al., 2011). In this paper it is mentioned that

> "...In the case of high particle concentrations, the magnetic and diffusion problems are strictly interrelated, and the concentration profile depends markedly on steric, magnetodipole, and hydrodynamic interparticle interactions, whose counting is a problem of great concern..."

The present study is devoted to the classical problem of ferrohydrostatics on stability (known as the normal field instability or the Rosensweig instability) of a horizontal semi-infinite layer of a magnetic fluid under the influence of gravity and a uniform magnetic field normal to the plane free surface of the layer (Rosensweig, 1998). A periodic peak-shaped structure is formed on the fluid surface when the applied magnetic field exceeds a critical value. This phenomenon was observed first experimentally (Cowley & Rosensweig, 1967).

A number of papers are devoted to numerical investigations of the Rosensweig instability. The numerical results concern equilibrium states of the ferrofluid layer in (Aristidopoulou et al., 1996; Bashtovoi et al., 2002; Boudouvis et al., 1987; Gollwitzer et al., 2007; 2009; Lange et al., 2007; Lavrova et al., 2003; 2008; 2010) and analyze dynamical properties of the ferrofluid behavior in (Knieling et al., 2007; Matthies & Tobiska, 2005). A non-uniform equilibrium distribution of particles in the ferrofluid layer is computed in (Lavrova et al., 2010) for the first time.

In order to reach the equilibrium between concentration and the magnetic field, quite a long time is needed. The concentration remains almost constant for much shorter time scales. That is why, the validity of the results, mentioned in the previous paragraph, will not be abolished by the preset contribution. The aging process of the Rosensweig instability was experimentally studied over the long time (up to 50 days) in (Sudo et al., 2006). The effect of evaporation and non-evaporation of magnetic fluid on the pattern aging were examined. It was found that the interfacial spike pattern of magnetic fluids changes with time dramatically. Namely, the cell pattern gradually bifurcates at the constant magnetic field, and the number of spikes increases with time.

The particles are in Brownian motion inside the ferrofluid layer, when no magnetic field is applied, and the particle concentration is constant over the fluid volume. This is correct under an assumption that the gravity force has a negligible influence to the diffusion of particle. When the applied field is switched on but the field intensity is too weak to perturb the plane surface, then the magnetic field inside the layer remains constant and the particle concentration is constant, as a consequence. The situation changes when the applied field intensity is strong enough to perturb the free surface. A nonuniformity of the magnetic field inside the fluid causes a redistribution of the particles. This is due to interactions between field and particles and interparticle interactions. An interaction between particles and the magnetic field of the fluid was taken into account for the modelling of the Rosensweig instability in our previous research in (Lavrova et al., 2010). This interaction plays a dominant role in dilute magnetic fluids. The main objective of the contribution is the extension to the case of interacting particles, which should be taken into account for concentrated magnetic fluids.

Mathematical model of the coupled problem consists of the magnetostatic subproblem, the concentration subproblem and the free-surface subproblem. The concentration subproblem is based on a recently developed mass transfer equation for describing diffusion of interacting ferromagnetic particles in magnetic fluids (Pshenichnikov et al., 2011). Three subproblems are strictly interrelated to each other and should be solved simultaneously to resolve equilibrium states of the system. An iterative decoupling strategy is applied for solving the coupled system of equations. The finite-element method is used for discretization of the magnetostatic subproblem in a fixed domain. The Newton method is applied to find an element-wise distribution of the concentration at the finite-element mesh. The finite-difference approach is used for the free-surface subproblem. Numerical results of three models (model 1 - nonuniform particle distribution without particle interaction, model 2 - nonuniform particle distribution with particle interaction, model 3 - uniform particle distribution) are compared. The effect of particle interaction shows a considerable influence on behavior of the ferrofluid layer in a uniform applied magnetic field.

2. Mathematical model

We consider a semi-infinite ferrofluid layer with a horizontal plane free surface bounded from above by a nonmagnetic gas (air). The unperturbed free surface is defined by equation $z = 0$, whereas the fluid corresponds to a region $z < 0$. The system is regarded under the action of gravity $\boldsymbol{g} = (0, 0, -g)$ and a uniform magnetic field $\boldsymbol{H}_0 = (0, 0, H_0)$ normal to the plane free surface of the layer. We consider a single peak in the surface pattern with a cell Ω_{cell} and a free surface Γ. The problem will be formulated in a cylinder $\Omega_{cell} \times (-\infty, +\infty)$, see Fig. 1 for the case of a hexagonal cell.

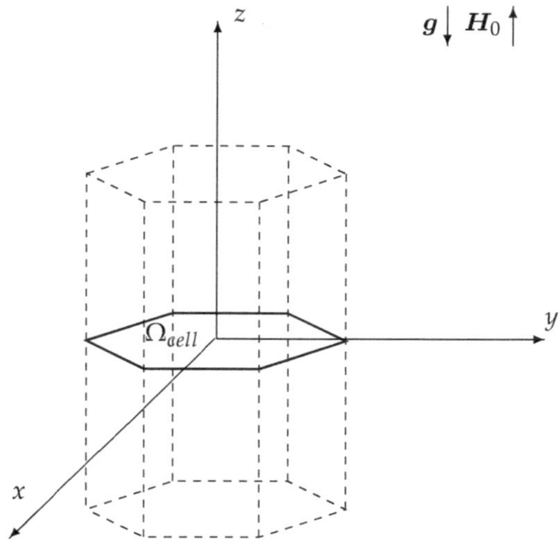

Fig. 1. The computational domain.

The mathematical model for a non-uniform equilibrium distribution of ferromagnetic particles in a magnetic fluid with a free surface leads to a coupled problem formulation consisting of three subproblems. The first subproblem describes the magnetic field structure inside the fluid and in the surrounding air by the Maxwell's equations. The second subproblem concerns the diffusion of particles in the bulk of the fluid as a steady-state concentration problem. Finally, the third subproblem is given by the generalized Young-Laplace equation for equilibrium free-surface shapes of the fluid-air interface.

The Maxwell's equations inside the magnetic fluid and in the air are

$$\nabla \times \mathbf{H} = \mathbf{0}, \quad \nabla \cdot \mathbf{B} = 0 \quad \text{in } \Omega_{cell} \times (-\infty, +\infty), \tag{1}$$

see e.g. (Rosensweig, 1998). Here $\mathbf{H}$ denotes the magnetic field, $\mathbf{B} = \mu_0(\mathbf{H} + \mathbf{M})$ the magnetic induction, $\mu_0 = 4\pi \times 10^{-7}\text{H/m}$ the permeability of vacuum. The magnetization vector $\mathbf{M}$ is parallel to the field vector and follows a magnetization law $M = M(H, C)$ dependent on the field intensity H and a particle concentration C. An equilibrium magnetization of magnetic fluids with account for interparticle interaction is derived in (Ivanov & Kuznetsova, 2001) in the framework of the modified model of the effective field

$$M(H, C) = M_s \frac{C}{C_0} L(\gamma H_e), \quad H_e = H + \frac{M_s}{3} \frac{C}{C_0} L(\gamma H), \quad \gamma = \frac{3\chi_L}{M_s}. \tag{2}$$

Here M_s is the saturation magnetization, C the volumetric concentration and $C_0 = \frac{1}{|\Omega_f|} \int\limits_{\Omega_f} C dx$ for the fluid domain Ω_f with the volume $|\Omega_f|$. $L(\xi) = \coth(\xi) - 1/\xi$ is the Langevin function, γ the Langevin parameter, χ_L the initial susceptibility of the Langevin magnetization, H_e the effective field. The magnetization of air equals to zero. The magnetic field satisfies continuity conditions at the interface Γ between ferrofluid and air, see (Rosensweig, 1998)

$$[\mathbf{H} \cdot \tau_k] = 0, k = 1,2 \qquad [\mathbf{B} \cdot \mathbf{n}] = 0 \quad \text{on } \Gamma. \tag{3}$$

Here $[\cdot]$ denotes a jump over the interface, τ_k and $\mathbf{n}$ are tangential and normal vectors to the interface. A symmetry condition is specified at the side of a cylinder domain

$$\mathbf{H} \cdot \mathbf{n} = 0 \quad \text{on } \partial\Omega_{cell} \times (-\infty, +\infty). \tag{4}$$

The uniform applied field $\boldsymbol{H}_0$ is perturbed only in a neighborhood of the interface Γ. That is why we introduce asymptotic boundaries $z = \pm\delta$, far enough from the interface, and specify there a uniform magnetic field

$$\boldsymbol{H} = \boldsymbol{H}_0 \quad \text{for } z = \delta, \quad \boldsymbol{H} = \boldsymbol{H}_0^1 \quad \text{for } z = -\delta, \tag{5}$$

where $\delta > 0$. The intensity of the applied field H_0 presents a control parameter of the model, whereas the field $\boldsymbol{H}_0^1 = (0, 0, H_0^1)$ will be computed from the second transmission condition (3), satisfied at the flat interface $z = 0$. The Maxwell's equations (1) together with conditions (3)-(5) present the first subproblem of the mathematical model.

The second subproblem describes a magnetophoresis process, i.e. the diffusion of ferromagnetic particles in the magnetic fluid under the action of a nonuniform magnetic field. A dynamic mass transfer equation for describing diffusion of interacting ferromagnetic particles in magnetic fluids was derived in (Pshenichnikov et al., 2011)

$$\frac{\partial C}{\partial t} = -\nabla \cdot \left[D_0 K(C) \left\{ - \left(1 + \frac{2C(4-C)}{(1-C)^4} - C \frac{\partial^2 (C^2 G(\lambda, C))}{\partial C^2} \right) \nabla C + C L(\gamma H_e) \nabla (\gamma H_e) \right\} \right]$$

in $\Omega_f, t > 0$. The equation is presented with an assumption that the gravity force has a negligible influence to the diffusion of particles. The constant D_0 denotes Einstein's value of the diffusion coefficient for dilute solutions,

$$K(C) = (1 - 6.55C)$$

is the relative mobility of particles in the magnetic fluid. A function

$$G(\lambda, C) = \frac{4}{3} \lambda^2 \frac{(1 + 0.04\lambda^2)}{(1 + 0.308\lambda^2 C)} \frac{(1 + 1.28972C + 0.72543C^2)}{(1 + 0.83333\lambda C)}$$

specifies the contribution of a magnetodipole interaction to the free energy of the dipolar hard sphere (Pshenichnikov et al., 2011). Here λ is the dipolar coupling constant or the aggregation parameter, estimating the intensity of the magnetodipole interaction in comparison with thermal energy. The modified effective field model, which is used to describe the equilibrium magnetization (2), is applicable for $\lambda \leq 2$ (Pshenichnikov & Lebedev, 2004). We take $\lambda = 1$ in our model and get

$$G(1,C) = \frac{1.38667(1+1.28972C+0.72543C^2)}{(1+0.308C)(1+0.83333C)}.$$

The static distribution of particles in the cavity is obtained by equating the full particle flux to zero. According to (Pshenichnikov et al., 2011), it gives

$$\ln C + \frac{3-C}{(1-C)^3} - \frac{\partial(C^2 G(\lambda, C))}{\partial C} = \ln\left(\frac{\sinh(\gamma H_e)}{\gamma H_e}\right) + c_c. \tag{6}$$

To fix the constant c_c the condition of conservation for the concentration over the fluid domain

$$\int_{\Omega_f} C d\Omega = C_0 |\Omega_f| \tag{7}$$

will be used. Let us substitute the function $G(1,C)$ to equation (6), then

$$\ln C + R(C) = \ln\left(\frac{\sinh(\gamma H_e)}{\gamma H_e}\right) + c_c, \quad H_e = H + \frac{\chi_L}{\gamma}\frac{C}{C_0}L(\gamma H), \tag{8}$$

where $R(C)$ is a rational function of the concentration

$$R(C) = \frac{7.838(1.517+C)(4.844+C)(2.258-2.862C+C^2)(0.509-0.222C+C^2)(0.688+1.282C+C^2)}{(1-C)^3(1.2+C)^2(3.247+C)^2}.$$

The second subproblem of the mathematical model is presented by equation (8) and condition (7) to fix the constant c_c.

Remark 2.1. *Equation (8) can be reformulated as*

$$Ce^{R(C)} = const\frac{\sinh(\gamma H_e)}{\gamma H_e}, \quad H_e = H + \frac{\chi_L}{\gamma}\frac{C}{C_0}L(\gamma H).$$

Such a representation is similar in form to the solution for the concentration in dilute approximation (Polevikov & Tobiska, 2008)

$$C = const\frac{\sinh(\gamma H)}{\gamma H}, \quad const = C_0|\Omega_f| / \int\limits_{\Omega_f} \frac{\sinh(\gamma H)}{\gamma H} d\Omega. \tag{9}$$

Thus, for concentrated fluids the explicit dependence of C on H (9) is replaced by the implicit dependence (8).

The third subproblem of the model defines a shape of the interface Γ between the magnetic fluid and the air. Equilibrium shapes of a free magnetic-fluid surface are described by the generalized Young-Laplace equation, which presents the force balance at the fluid-air interface

$$\sigma\mathcal{K} + p_0 = p + \frac{\mu_0}{2}\left(M\frac{H_n}{H}\right)^2 \quad \text{on } \Gamma. \tag{10}$$

Here σ is the surface tension coefficient, $\mathcal{K}$ the sum of principal curvatures, p_0 the pressure in the air and p is the fluid pressure. The equation of hydrostatics for magnetic fluids is

$$\nabla p = -\rho g e_z + \mu_0 M \nabla H - \nabla \left[-\mu_0 \int_0^H C \left(\frac{\partial M}{\partial C} \right)_H dH + \mu_0 \int_0^H M dH \right],$$

see e.g. in (Rosensweig, 1998), where ρ is the fluid density and $e_z = (0, 0, 1)$. This equation allows us to express in explicit form the fluid pressure as

$$p = -\rho g z + \mu_0 \int_0^H C \left(\frac{\partial M}{\partial C} \right)_H dH + c_f, \tag{11}$$

where c_f is an integration constant. The equation (10) for the fluid pressure (11) is

$$\sigma \mathcal{K} = -\rho g z + f(C, H) + c_f \quad \text{on } \Gamma, \tag{12}$$

where

$$f(C, H) = \mu_0 \int_0^H C \left(\frac{\partial M}{\partial C} \right)_H dH + \frac{\mu_0}{2} \left(M \frac{H_n}{H} \right)^2 .$$

The constant c_f will be determined by integrating equation (12) over the surface Γ, see Section 3.3 for details.

A solution of the magnetostatic subproblem (1)-(5) depends on the particle concentration and on the shape of the fluid-air interface. The concentration distribution is a solution of the nonlinear equation (8) and depends on the magnetic field configuration inside the ferrofluid. The fluid-air interface satisfies equation (12), which depends on the magnetic field and the concentration at the free surface of the ferrofluid. Three subproblems are strictly interrelated to each other and should be solved simultaneously to resolve equilibrium states of the system.

3. Numerical methods and tools

Experiments show that the surface pattern of the Rosensweig instability is presented by a hexagonal or square array of spikes, see e.g. (Gollwitzer et al., 2006). For sake of simplicity, we assume that the cell has a circular shape of radius a inscribed to a hexagonal cell of the pattern, see Fig. 2. It allows us to study axisymmetric solutions of the presented model in a two-dimensional geometry of cylindrical coordinates. Axisymmetric solutions on a circular cell were compared with three-dimensional ones on a hexagonal cell in (Lavrova et al., 2003). It was found that the profile shapes of both models nearly completely coincides. The different geometry of the cell results in a 10 % smaller amplitude of an axisymmetric peak in comparison with a three-dimensional solution.

3.1 Magnetostatic subproblem

Let us consider the magnetostatic subproblem (1)-(5) in a domain with the fixed interface and assume that the spacial distribution of the particle concentration is given. We express the magnetic field in terms of a scalar magnetic potential ϕ as $\mathbf{H} = \nabla \phi$ and reformulate the magnetostatic subproblem (1)-(5). The first Maxwell's equation (1) is exactly satisfied, whereas the second one takes form of an elliptic partial differential equation with jumping coefficients,

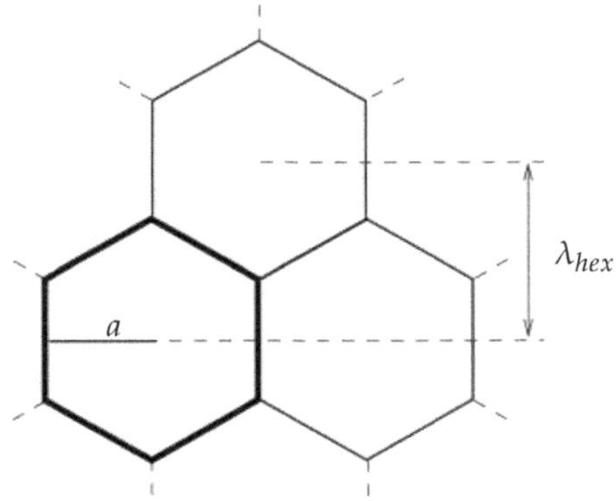

Fig. 2. Top view of the hexagonal surface pattern with wavelength λ_{hex}.

due to different magnetization in the fluid and the air

$$-\nabla \cdot (\mu(r,z,C,|\nabla\phi|)\nabla\phi) = 0, \quad \text{in } \Omega_{ax}, \tag{13}$$

$$\mu(r,z,C,|\nabla\phi|) = \begin{cases} 1 + \frac{3\chi_L}{\gamma}\frac{C}{C_0}\frac{L\left(\gamma|\nabla\phi| + \chi_L \frac{C}{C_0} L(\gamma|\nabla\phi|)\right)}{|\nabla\phi|} & \text{in } \Omega_1, \\ 1 & \text{in } \Omega_2. \end{cases}$$

Here $\Omega_{ax} := \Omega_1 \cup \Omega_2$ is a meridional cross-section of the 3D-domain Ω, where Ω_1 corresponds to the fluid and Ω_2 to the air. Differential operators are though in cylindrical coordinates with an assumption of an axial symmetry for the potential. The magnetostatic problem is closed by a set of conditions

$$[\phi] = 0, \quad \left[\mu \frac{\partial \phi}{\partial n}\right] = 0 \quad \text{for } (r,z) \in \Gamma, \tag{14}$$

$$\phi = \phi_D \quad \text{for } (r,z) \in \Gamma_D, \tag{15}$$

$$\frac{\partial \phi}{\partial n} = 0 \quad \text{for } (r,z) \in \Gamma_N. \tag{16}$$

The boundary Γ_D consists of the top and bottom boundaries of the rectangular domain $\Omega_{ax} = (0,a) \times (-\delta,\delta)$ and $\Gamma_N = \partial\Omega_{ax} \setminus \Gamma_D$. We have from condition (5) that

$$\phi_D(r,z) = \begin{cases} H_0 z & \text{for } z = \delta, \\ H_0^1 z & \text{for } z = -\delta. \end{cases}$$

Remark 3.1. *The constant H_0^1 is defined from the second transmission condition (14), satisfied for the undisturbed interface $z = 0$. The potential is given in this case as*

$$\phi = H_0 z \quad \textit{for } z > 0 \quad \textit{and} \quad \phi = H_0^1 z \quad \textit{for } z < 0.$$

We get

$$\left[\mu \frac{\partial \phi}{\partial z}\right] = 0 \quad \Rightarrow \quad \mu(C_0, H_0^1) H_0^1 = H_0 \quad \Rightarrow$$

$$H_0^1 + \frac{3\chi_L}{\gamma} L(\gamma H_0^1 + \chi_L L(\gamma H_0^1)) = H_0. \tag{17}$$

The nonlinear equation (17) is solved by the Newton method. Starting from the value H_0*, the method converges for 3-5 iterations for the relative error* 10^{-10}.

Due to the problem reformulation in cylindrical coordinates and the assumption of axial symmetry, a corresponding variational problem is formulated in weighted Sobolev spaces:

Find $\phi \in V(\Omega)$ such that

$$\int_\Omega \mu(r,z,C,|\nabla\phi|)\nabla\phi \cdot \nabla v r dr dz = 0 \quad \text{for any } v \in V_D(\Omega), \tag{18}$$

and $\phi - \phi_0 \in V_D(\Omega)$ for any $\phi_0 \in V(\Omega)$ such that $\phi_0|_{\Gamma_D} = \phi_D$.

$$V(\Omega) = \{v| \int_\Omega v^2 r dr dz < \infty, \int_\Omega |\nabla v|^2 r dr dz < \infty\}, \quad V_D(\Omega) = \{v| v \in V(\Omega), v|_{\Gamma_D} = 0\}.$$

A structured triangular mesh is used for discretization. A schematic representation of the mesh is shown in Fig. 3. An algorithm of bilinear interpolation is applied for the mesh construction. This algorithm defines every interior grid point (filled circular marker in Fig. 3) by interpolation of eight boundary points (empty circular markers)

$$\xi(s,t) = (1-s)\xi_W + s\xi_E + (1-t)(\xi_S - (1-s)\xi_{SW} - s\xi_{SE}) + t(\xi_N - (1-s)\xi_{NW} - s\xi_{NE}),$$

where parameters $s, t \in (0,1)$ and indices represent the quarter directions of boundary points (N - north, S - south, W - west, E - east), relative to the inner point x. To produce an interface-fitted mesh, the algorithm is applied in Ω_1 and Ω_2 separately, based on the pointwise representation of the interface, as a solution of the free-surface subproblem, and a uniform distribution of grid points at the boundary sides of Ω_1 and Ω_2, see Fig. 3. A finite element mesh is reconstructed every time, when the interface is changed. By construction all meshes have the same topology. It allows to define an initial approximation of the potential at the new mesh as a solution of the magnetostatic problem at the old mesh without any interpolation.

The variational problem (18) is discretized by continuous piecewise linear functions on triangles for the given concentration C. Nonlinearities in the discrete equations are treated by a fixed-point iteration

$$\int_{\Omega_{ax}} \mu(r,z,C,|\nabla\phi_h^k|)\nabla\phi_h^{k+1} \cdot \nabla v_h r dr dz = 0. \tag{19}$$

One Gauss-Seidel step is applied to the resulting system of linear equations in each iteration. The iterative process (19) continues until the relative a-posteriori error will be smaller than ϵ_p (generally 10^{-5})

$$\frac{\max\limits_{1 \le i \le N_p} \left|\phi_i^{k+1} - \phi_i^k\right|}{\frac{1}{N_p}\sum_{i=1}^{N_p} |\phi_i^{k+1}|} < \epsilon_p.$$

Here N_p denotes the number of unknowns and $\phi_i = \phi_h(\xi_i)$ is the nodal value of the potential at the grid point ξ_i. The iterative process needs 5-10 iterations to converge independent of the mesh size.

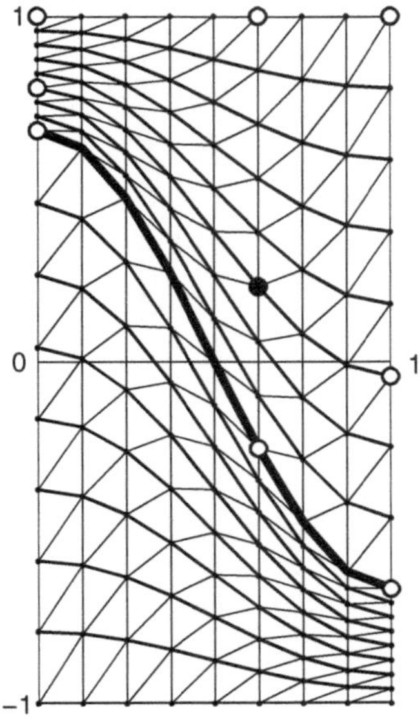

Fig. 3. Schematic representation of the finite element mesh. Thick solid line denotes the fluid-air interface. Bilinear interpolation is used for the mesh construction: inner node (filled circular marker) is defined by eight boundary nodes (empty circular markers).

3.2 Concentration subproblem

Let us consider algebraic equation (8)

$$\Phi(C,H) := \ln C + R(C) - \ln\left(\frac{\sinh\left(\gamma H + \chi_L \frac{C}{C_0} L(\gamma H)\right)}{\gamma H + \chi_L \frac{C}{C_0} L(\gamma H)}\right) = c_c. \tag{20}$$

and assume that a spacial configuration of the magnetic field H is given. The magnetic field is computed from a solution of the magnetostatic problem (13)-(16) as $H_h = |\nabla \phi_h|$. The function H_h is a piecewise-constant approximation of the field H over the finite-element mesh and is given by the values $H_1, \ldots H_M$, where H_i corresponds to the cell T_i and M is a number of cells in the fluid domain Ω_1. For the given field values $H_1, \ldots H_M$ we have to find $C_1, \ldots C_M$ and the constant c_c satisfying (20)

$$F_i(C_i, c_c) := \Phi(C_i, H_i) - c_c = 0, \quad i = 1, \ldots, M \tag{21}$$

and the integral condition (7)

$$F_{M+1}(C_1,\ldots,C_M) := \sum_{i=1}^{M} \omega_i C_i - C_0 \left(\sum_{i=1}^{M} \omega_i\right) = 0, \quad \omega_i = 2\pi \int_{T_i} r dr dz. \tag{22}$$

Here ω_i is the area of a circular element, obtained by rotating of the cell T_i. A system of M nonlinear equations (21) and one linear equation (22)

$$\mathbf{F}(\mathbf{x}) = \mathbf{0},$$

where $\mathbf{F} = (F_1,\ldots,F_{M+1})^T, \mathbf{x} = (C_1,\ldots,C_M,c_c)^T$, will be solved by the Newton method

$$\mathbf{x}^{k+1} = \mathbf{x}^k - (\mathbf{F}'(\mathbf{x}^k))^{-1}\mathbf{F}(\mathbf{x}^k) \tag{23}$$

Here $\mathbf{x}^k = (C_1^k,\ldots,C_M^k,c_c^k)^T$ and

$$\mathbf{F}'(\mathbf{x}) = \begin{pmatrix} \frac{\partial\Phi}{\partial C}(C_1,H_1) & 0 & 0 & \cdots & 0 & -1 \\ 0 & \frac{\partial\Phi}{\partial C}(C_2,H_2) & 0 & \cdots & 0 & -1 \\ & & & \cdots & & \\ 0 & 0 & 0 & \cdots & \frac{\partial\Phi}{\partial C}(C_M,H_M) & -1 \\ \omega_1 & \omega_2 & \omega_3 & \cdots & \omega_M & 0 \end{pmatrix}.$$

Eliminating $C_1^{k+1},\ldots,C_M^{k+1}$ from the system (23) we get

$$c_c^{k+1} = \sum_{i=1}^{M} \frac{\Phi(C_i^k,H_i)}{\frac{\partial\Phi}{\partial C_i}(C_i^k,H_i)} \Big/ \sum_{i=1}^{M} \frac{\partial\Phi}{\partial C}(C_i^k,H_i).$$

Finally we compute C_i^{k+1} from the i-th equation of the system (23) for the given c_c^{k+1}

$$C_i^{k+1} = C_i^k - \frac{\Phi(C_i^k,H_i) - c_c^{k+1}}{\frac{\partial\Phi}{\partial C}(C_i^k,H_i)}, \quad i = 1,\ldots,M.$$

The computations for $C_1^{k+1},\ldots,C_M^{k+1}$ can be realized in parallel.

The Newton method requires $3-5$ iterations to converge for the relative a-posteriori error $|\mathbf{x}^{k+1} - \mathbf{x}^k|$ to be smaller than ϵ_c (generally 10^{-9}). We set $C_i^0 = C_0$ at the initial Newton step. An initial value c_c^0 is defined from the condition that far from the interface the concentration and the magnetic field intensity are known and equal C_0 and H_0^1, respectively. From equation (20) we get

$$c_c^0 = \ln C_0 + R(C_0) - \ln\left(\frac{\sinh(\gamma H_0^1 + \chi_L L(\gamma H_0^1))}{\gamma H_0^1 + \chi_L L(\gamma H_0^1)}\right). \tag{24}$$

3.3 Free-surface subproblem

Let us write equation (12) for space variables dimensionless over a

$$\sigma\frac{\mathcal{K}}{a} = -a\rho g z + \mu_0 \int_0^H C\left(\frac{\partial M}{\partial C}\right)_H dH + \frac{\mu_0}{2}\left(M(H,C)\frac{H_n}{H}\right)^2 + c_f \quad \text{on } \Gamma. \tag{25}$$

We assume that the magnetic field H and the concentration C are given at the interface Γ. The magnetic field is determined from the fluid side as $H = |\nabla\phi_1|$ and $H_n = \nabla\phi_1 \cdot \mathbf{n}$ for the normal vector $\mathbf{n}$, external to the fluid domain Ω_1. We introduce dimensionless parameters

$$\lambda = a\sqrt{\rho g/\sigma}, \quad \text{Si} = \mu_0 M_s^2/(2\sqrt{\rho g \sigma}).$$

Equation (25) takes a new form

$$\mathcal{K} = -\lambda^2 z + f(C,H) + c_f \quad \text{on } \Gamma, \tag{26}$$

where

$$f(C,H) = \lambda \text{Si}\frac{2\gamma}{3\chi_L}\int_0^H C\left(\frac{\partial}{\partial C}\left[CL(\gamma H_e)\right]\right)_H dH + \lambda\text{Si}\left(CL(\gamma H_e)\frac{H_n}{H}\right)^2 \tag{27}$$

and the effective field $H_e = H + \frac{\chi_L}{\gamma}\frac{C}{C_0}L(\gamma H)$. The unknown constant c_f is different in equations (25) and (26) and will be fixed later. The integrand in (27) is further transformed

$$\begin{aligned} f_I(C,H) &:= C\left(\frac{\partial}{\partial C}\left[CL(\gamma H_e)\right]\right)_H \\ &= C\left(L(\gamma H_e) + \frac{C}{C_0}\chi_L L(\gamma H) L'(\gamma H_e)\right) \\ &= C\left(L(\gamma H_e) + \frac{C}{C_0}\chi_L L(\gamma H)\left[\frac{1}{(\gamma H_e)^2} - \frac{1}{(\sinh(\gamma H_e))^2}\right]\right). \end{aligned}$$

We approximate the integral term in (27) by a composite trapezoidal rule on a uniform grid

$$\int_0^H f_I(C,H)dH \approx h\left(\frac{1}{2}\left(f_I(C_0,0) + f_I(C_n,H)\right) + \sum_{i=1}^{n-1} f_I(C_i,H_i)\right). \tag{28}$$

Here $h = H/n$, $(n+1)$ is the number of the grid points, $H_i = ih$ and C_i denotes the concentration, corresponding to the field H_i. $f_I(C_0,0) = 0$, because $L(0) = 0$ and $L'(0) = 1/3$. The final form of the function $f(C,H)$, used in computations, is

$$f(C,H) \approx \lambda\text{Si}\frac{2\gamma}{3\chi_L}h\left(\frac{1}{2}f_I(C_n,H) + \sum_{i=1}^{n-1} f_I(C_i,H_i)\right) + \lambda\text{Si}\left(CL(\gamma H_e)\frac{H_n}{H}\right)^2. \tag{29}$$

To find concentration C_i, corresponding to the integration points H_i, we apply the Newton method to equation (20) for the given field $H = H_i$ and the given c_c. The value of constant c_c is defined in the process of solving the concentration subproblem. For details of computations see Section 3.2. Test computations of the integral approximation (28) for $n = 10$ and $n = 20$ show changes in the fifth significant digit. We use $n = 20$ for our computations.

The free boundary Γ is represented by an arc-length parametrization

$$\Gamma = \{(r,z) \mid r = r(s), z = z(s), s = [0,\ell]\},$$

where the parameter s is measured from the top of the peak ($s = 0$) to the peak foot ($s = \ell$). Following the approach in (Polevikov, 2004), we reformulate equation (26) as a system of second-order ordinary differential equations

$$\begin{aligned} \bar{r}'' &= -\bar{z}'F, \quad \bar{z}'' = \bar{r}'F \qquad 0 < \bar{s} < 1; \\ F &= -\frac{\bar{z}'}{\bar{r}} + \lambda^2\ell^2\bar{z} - \ell f(H) + c_f. \end{aligned} \tag{30}$$

Here $\bar{r}(\bar{s}) = r(s)/\ell$, $\bar{z}(\bar{s}) = z(s)/\ell$ and $\bar{s} = s/\ell$ are scaled versions of the space variables, introduced to have a formulation at the fixed domain $[0,1]$ instead of the changing and a-priori unknown domain $[0,\ell]$. The boundary condition $r(\ell) = 1$, transformed to $\bar{r}(1) = 1/\bar{\ell}$ in new variables, allows us to express the unknown parameter ℓ of the problem as

$$\ell = \frac{1}{\bar{r}(1)}.$$

The constant c_f is determined by integrating equation (30) over $\bar{s}$

$$\int_0^1 (\bar{r}\bar{z}')'\, d\bar{s} = \int_0^1 \left(\lambda^2\ell^2\bar{z} - \ell f(H) + c_f\right) \bar{r}\bar{r}' d\bar{s}.$$

The left-hand side equals zero, because $\bar{r}(0) = 0$ and $\bar{z}'(1) = 0$. The right-hand side gives that

$$c_f = 2\ell^3 \int_0^1 f(H)\bar{r}\bar{r}' d\bar{s},$$

using the volume conservation condition

$$\int_0^1 \bar{z}\bar{r}\bar{r}' d\bar{s} = 0. \tag{31}$$

Equations (30) are augmented by boundary conditions

$$\begin{aligned} &\bar{r}(0) = 0, \\ &\bar{r}'(1) = 1, \\ &\bar{z}'(0) = 0, \\ &\bar{z}(1) = \ell^2 \textstyle\int_0^1 \bar{r}^2\bar{z}' d\bar{s}. \end{aligned} \tag{32}$$

The nonlocal boundary condition is due to the integration by parts of the volume conservation condition (31).

An iterative finite-difference scheme of the second order approximation for the parametric Young-Laplace equations was developed in (Polevikov, 2004). We apply the developed approach to equations (30) with boundary conditions (32)

$$\frac{1}{\tau_f}\left(\bar{r}^{k+1}_{ss,i} - \bar{r}^{k}_{ss,i}\right) + \bar{r}^{k}_{ss,i} + \bar{z}^{k}_{\overset{\circ}{s},i} F^k_i = 0, \quad i = 1, ..., N-1$$

$$\bar{r}^{k+1}_0 = 0, \quad \frac{\bar{r}^{k+1}_N - \bar{r}^{k+1}_{N-1}}{h} = 1;$$

$$\frac{1}{\tau_f}\left(\bar{z}^{k+1}_{ss,i} - \bar{z}^{k}_{ss,i}\right) + \bar{z}^{k}_{ss,i} - \bar{r}^{k}_{\overset{\circ}{s},i} F^k_i = 0, \quad i = 1, ..., N-1$$

$$\frac{\bar{z}^{k+1}_1 - \bar{z}^{k+1}_0}{h} = \frac{h}{2}F^k_0, \; \bar{z}^{k+1}_N = \left(\frac{1}{\bar{r}^k_N}\right)^2 \sum_{i=1}^{N}\left[\left(\bar{z}^k_i - \bar{z}^k_{i-1}\right)\left(\frac{\bar{r}^k_{i-1} + \bar{r}^k_i}{2}\right)^2\right];$$

$$F^k_i = -\frac{\bar{z}^k_{\overset{\circ}{s},i}}{\bar{r}^k_i} + \lambda^2\left(\frac{1}{\bar{r}^k_N}\right)^2 \bar{z}^k_i - \frac{1}{\bar{r}^k_N} f(H^k_i) + \left(\frac{1}{\bar{r}^k_N}\right)^3 \sum_{i=1}^{N}\left[\left(\left(\bar{r}^k_i\right)^2 - \left(\bar{r}^k_{i-1}\right)^2\right) f(H^k_i)\right].$$

Here $\{\bar{r}_i\}_{i=0}^N$ and $\{\bar{z}_i\}_{i=0}^N$ are grid-functions uniformly distributed over the free surface with a step size $h = 1/N$. The difference quotients correspond to the central derivatives ($\bar{r}_{\overset{\circ}{s}}$, $\bar{z}_{\overset{\circ}{s}}$) and the second derivatives ($\bar{r}_{ss}$, $\bar{z}_{ss}$). Nonlinearities of equations (30) are resolved by iterations, resulting in a three-diagonal system for the unknown grid functions at the $(k+1)$-th iteration. A relaxation technique with a parameter τ_f is applied to improve numerical stability. We took $\tau_f = 0.01$ in computations.

3.4 Decoupling strategy

The model under study consists of the magnetostatic subproblem, the concentration subproblem and the free-surface subproblem. The magnetostatic subproblem is described by a nonlinear elliptic partial differential equation with jumping coefficients (13) for the magnetostatic potential, augmented by transmission and Dirichlet-Neumann boundary conditions (14)-(16). The concentration subproblem is presented by nonlinear algebraic equation (20) for the concentration, augmented by integral condition (7). The free-surface subproblem is described by a system of two nonlinear ordinary differential equations (30) for the parametric representation of the free surface, augmented by integral and boundary conditions (31)-(32).

We apply an iterative decoupling strategy for solving the coupled system of equations. It consists of three steps at every iteration. The first step is a numerical solution of the magnetostatic problem in a fixed domain and for a given distribution of the concentration. The finite element method is applied for the discretization, see Section 3.1 for details. The second step is a numerical solution of system of nonlinear equations (21)-(22) for the element-wise concentration at the finite-element mesh for the given magnetic field distribution, as a solution at the first step. The Newton method is applied for a solution of the system, see Section 3.2 for details. The third step is a numerical solution of the free-surface subproblem for the given magnetic field from the first step and the given concentration from the second step of the iterative procedure. The finite-difference method is applied for the discretization, see Section 3.3 for details. A relaxation technique is applied to the free surface representation at every iteration

$$r_i^{n+1} = r_i^n + \tau(r_i^{n'} - r_i^n), \quad z_i^{n+1} = z_i^n + \tau(z_i^{n'} - z_i^n), \quad i = \overline{0, M}.$$

It allows to suppress a rapid change of free surface shapes during iterations. We take initially $\tau = 0.1$ and decrease this value to $\tau = 0.01$ in the case of strong shape changes.

An initial surface configuration at the first iteration of the presented iterative algorithm is defined as a small perturbation of the flat surface with an amplitude of around 1 % of the cell radius. An initial concentration equals C_0. The iterations are stopped when the change in the surface shape is smaller than a prescribed threshold ϵ (generally 10^{-7})

$$\max_{0 \le i \le M} \left(\left| r_i^{n+1} - r_i^n \right|, \left| z_i^{n+1} - z_i^n \right| \right) < \epsilon.$$

The iterative process is controlled by the threshold ϵ, whereas three subproblems are controlled by own thresholds ϵ_p, ϵ_c and ϵ_f.

All algorithms, discussed in Section 3, and the coupling of three subproblems were implemented in Fortran with the help of the software tools, earlier developed for the Rosensweig instability computations. Numerical results of the previous computations are published in (Bashtovoi et al., 2002; Lavrova et al., 2008; 2010).

4. Results of computations

Numerical calculations were performed for the magnetic fluid EMG 901 (Ferrotec) with the following characteristic properties: the initial susceptibility $\chi = 2.2$, the density $\rho = 1406$ kg/m^3, the surface tension coefficient $\sigma = 0.025$ kg/s^2, the saturation magnetization $M_s = 48$ kA/m, the volumetric concentration, corresponding to the uniform particle distribution, $C_0 = 0.1$. The initial susceptibility of the Langevin magnetization χ_L is related to the initial susceptibility of the effective-field magnetization (2) as

$$\chi = \chi_L(1 + \chi_L/3),$$

see e.g. (Pshenichnikov et al., 1996). For the considered magnetic fluid we have that $\chi = 2.2$ and $\chi_L \approx 1.47489$. The control parameter of the model is the applied field intensity H_0.

Computations are performed at the computational domain $\Omega_{ax} = (0, a) \times (-\delta, \delta)$ with $\delta = 5a$. Computations with different δ have shown that the error caused by replacing the unbounded domain by a bounded one is less than 1%.

A linear stability analysis for the Rosensweig instability was carried out under the assumption of a uniform particles distribution in a layer of infinite thickness (Rosensweig, 1998). The stability theory predicts a critical value of the magnetic field intensity H_c, corresponding to the onset of instability, as a solution of the nonlinear equation

$$M(H_c)^2 = \frac{2\sqrt{\rho g \sigma}}{\mu_0} \left(1 + \left(1 + \frac{M(H_c)}{H_c} \right)^{-1/2} \left(1 + \frac{\partial M}{\partial H}(H_c) \right)^{-1/2} \right).$$

The intensity H_c corresponds to the fluid domain. The critical field intensity at the air domain H^* is found from the transmission condition $[\mathbf{B} \cdot \mathbf{n}] = 0$, satisfied at the unperturbed interface $z = 0$

$$\left(1 + \frac{M(H_c, C_0)}{H_c}\right) H_c = H^*.$$

We get $H^* \approx 9.11$ kA/m for the considered ferrofluid. The stability theory predicts a critical value of the pattern wavelength

$$\lambda_c = 2\pi / \sqrt{\rho g / \sigma}.$$

We assume that the hexagonal pattern wavelength λ_{hex}, see Fig. 3, equals λ_c. Then for the radius a of the circular cell we have

$$a = \lambda_{hex} / \sqrt{3} = \lambda_c / \sqrt{3} = 2\pi / \sqrt{\rho g / \sigma} / \sqrt{3},$$

whereas

$$\lambda = a\sqrt{\rho g / \sigma} = 2\pi / \sqrt{3}.$$

We assume that the parameter λ is fixed for any applied field intensity.

Two meshes have been used for computations to analyze an influence of the discretization refinement to the computational predictions. Table 1 shows the critical field, the maximum value of the particle concentration over the fluid domain and z-coordinate of the equilibrium free surface at the peak axis and the peak foot for the applied field $H_0 = 9.2$ kA/m at different meshes. The found difference in values allows us to conclude that computations at the mesh with 160×1600 nodes are accurate enough. This mesh has been used to get results in Fig. 4-Fig. 6.

Mesh	H_2^*, kA/m	$\max(C)/C_0$	$z(0)/a$	$z(\ell)/a$
80×800	8.71	1.111383	1.186572	−0.258553
160×1600	8.65	1.115006	1.222114	-0.261704
% Difference	0.7	0.3	3	1.2

Table 1. Values of some control parameters and their difference at different meshes.

Results of two models, which account for a nonuniform particle distribution inside the ferrofluid layer will be compared with the results of the model with a uniform particle distribution. The first model assumes no interaction between particles and it was numerically studied in (Lavrova et al., 2010). The second model accounts for interaction between particles and is the subject of this contribution. The model with a uniform particle distribution is called model 3 in what follows.

Computations for the first and the third models in (Lavrova et al., 2010) show that the onset of the instability occurs at $H_1^* = 9.12 \pm 0.01$ kA/m and $H_1^* = H_3^*$. This value nearly coincides with the result of the linear stability analysis $H^* \approx 9.11$ kA/m, which assumes a uniform particle distribution. It means, that the concentration effect does not influence to the onset of the instability in the frame of the model without particle interactions. Computations for the second model, which takes into account particle interactions, show that the onset of the instability occurs in a weaker field $H_2^* = 8.65 \pm 0.01$ kA/m. A concentration effect in this case influences to the critical field. A possible reason for this effect is that the interparticle interaction can intensify considerably the fluid mangetization and a small initial surface perturbation is developed to a surface pattern in a weaker field if to compare with the models without particle interactions.

Fig. 4 shows equilibrium free-surface shapes for three models. A more elongated peak region is formed for solutions with a nonuniform particle distribution. Namely, the peak is 20 % higher for the model without particle interactions and 60 % higher for the model with particle interactions in comparison with the uniform case. A 20 % difference of the first model is due to the concentration effect, which intensifies a spacial nonuniformity of the fluid magnetization in the peak region. A 60 % difference of the second model is influenced also by the fact that the onset of the instability for the second model occurs at the weaker field $H_2^* < H_3^*$. A further increase in the field strength results in the increasing peak amplitude, which can lead to a sizable difference in the peak height if we compare results of the second and the third model at the overcritical field $H_0 = 9.2$ kA/m. Fig. 4 contains isolines of the equilibrium concentration for the second model. The main inhomogeneity of the particle distribution occurs at the peak region. The concentration takes the greatest value at the top of the peak and the smallest value at the peak foot.

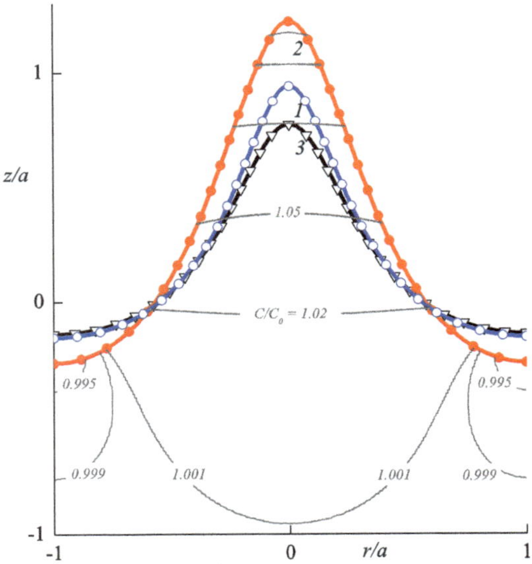

Fig. 4. Overcritical equilibrium free-surface shapes at the applied field $H_0 = 9.2$ kA/m. 1 - nonuniform particle distribution without particle interaction, 2 - nonuniform particle distribution with particle interaction, 3 - uniform particle distribution. Isolines of the concentration corresponds to $C/C_0 = \{0.995, 0.999, 1.001, 1.02, 1.05, 1.08, 1.1, 1.11\}$.

Fig. 5 shows the equilibrium distribution of the particle concentration over the peak axis for three models. The concentration increases monotonically in z-direction, moving along the peak axis, for the models with the nonuniform particle distribution and the concentration is constant for the third model. The concentration takes a value at the peak top which is about 25 % greater than in the fluid bulk for the first model and about 11 % greater for the second model. Taking into account the particle interaction, we get a smaller concentration at the peak but a more elongated shape. Fig. 5 shows that the concentration equals the volumetric value C_0 for $z/a < -1$ and the particle diffusion mechanism is present only near the free surface.

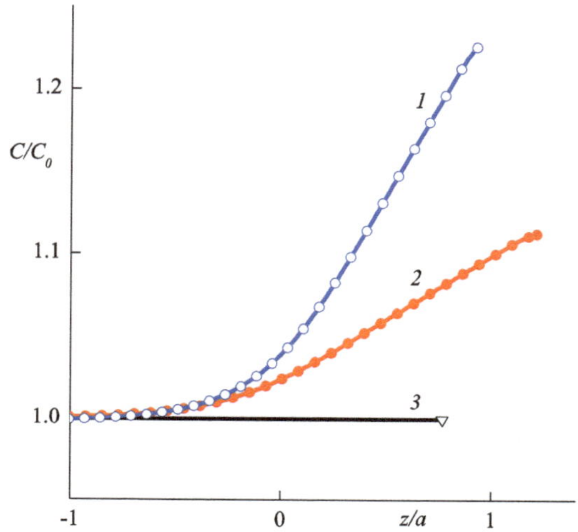

Fig. 5. Equilibrium distribution of the particle concentration over the peak axis at the applied field $H_0 = 9.2$ kA/m. 1 - nonuniform particle distribution without particle interaction, 2 - nonuniform particle distribution with particle interaction, 3 - uniform particle distribution.

The distribution of a z-component of the magnetic field vector inside of the ferrofluid and in the air is presented in Fig. 6 for three models. The magnetic field is uniform inside of

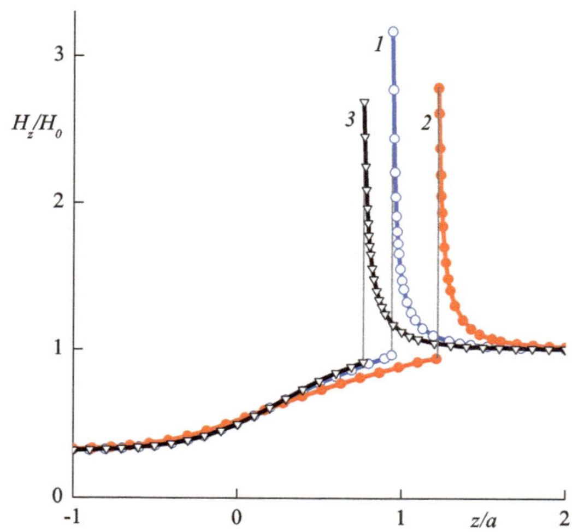

Fig. 6. Equilibrium distribution of the field intensity over the peak axis at the applied field $H_0 = 9.2$ kA/m. 1 - nonuniform particle distribution without particle interaction, 2 - nonuniform particle distribution with particle interaction, 3 - uniform particle distribution.

the ferrofluid far from the interface, $H_z = H_0^1$, and the particle diffusion mechanism is absent there. The field distribution for three models coincides for $z/a < -1$. The intensity of the field increases monotonically in z-direction, moving along the peak axis inside the ferrofluid, up to the value close to the applied field intensity H_0 at the interface. Crossing the interface, the field intensity jumps to the value $3.17H_0$ for the first model, $2.79H_0$ for the second model and $2.7H_0$ for the third model. The intensity of the field decreases monotonically in z-direction, moving along the peak axis inside the air, up to the value of the applied field intensity H_0 far from the interface. All considered models show that the field non-uniformity occurs at the region $-a < z < 2a$ and the field is nearly uniform outside of this region. It means that a restriction of the unbounded domain by a bounded one with $-5a < z < 5a$ will introduce an insignificant error in computations.

5. Conclusions

The effect of particle interaction in ferrofluid has been taken into account in numerical simulations of the Rosensweig instability for the first time. The mathematical model is based on a recently developed mass transfer equation for describing diffusion of interacting ferromagnetic particles in ferrofluids in (Pshenichnikov et al., 2011). The mathematical model for the Rosensweig instability in homogeneous ferrofluids is augmented by the concentration subproblem in the form of the nonlinear algebraic equation of the particle concentration and the magnetic field intensity inside the ferrofluid. We suggest an approach for numerical solution of the concentration subproblem and for the coupling of the concentration subproblem to the magnetic field and free surface computations.

Based on comparison between three models (model 1 - nonuniform particle distribution without particle interaction, model 2 - nonuniform particle distribution with particle interaction, model 3 - uniform particle distribution), it has been shown that the effect of particle interaction cannot be neglected or replaced by simpler models which try to capture the diffusion of particles. It was found that the onset of the instability occurs in a weaker field and the peak region is more elongated, if we take into account the particle interactions.

The effect of particle interaction on the pattern wavelength can not be numerically analyzed in the frame of the presented mathematical model. The pattern wavelength is fixed by the critical value obtained in the frame of the linear stability analysis. The experimental study of the aging variation of the Rosensweig instability in (Sudo et al., 2006) shows, however, that the pattern wavelength changes with time dramatically. Namely, the cell pattern gradually bifurcates at the constant magnetic field, and the number of spikes increases with time. That is why we plan for the future to consider a soliton-like surface configurations in the Rosensweig instability (Richter & Barashenkov, 2005). The pattern wavelength is a variable of this problem and the effect of particle interaction to the wavelength selection can be numerically analyzed.

6. Acknowledgements

The authors would like to thank Prof. A. Ivanov for stimulating discussions and Prof. A. Pshenichnikov for the helpful comments on the formulation of the mathematical model.

7. References

Aristidopoulou, A.A.; Papaioannou, A.T. & Boudouvis, A.G. (1996). Computational analysis of free surface magnetohydrostatic equilibrium: force versus energy formulation. *Magnetohydrodynamics*, Vol. 32, No. 4, 374–389.

Bashtovoi, V.G.; Lavrova, O.A.; Polevikov, V.K. & Tobiska, L. (2002). Computer modeling of the instability of a horizontal magnetic-fluid layer in a uniform magnetic field. *Journal of Magnetism and Magnetic Materials*, Vol. 252, 299–301.

Bashtovoi, V.G.; Polevikov, V.K.; Suprun, A.E.; Stroots, A.V. & Beresnev, S.A. (2007). Influence of Brownian diffusion on statics of magnetic fluids. *Magnetohydrodynamics*, Vol. 43, No. 1, 17–25.

Bashtovoi, V.G.; Polevikov, V.K.; Suprun, A.E.; Stroots, A.V. & Beresnev, S.A. (2008). The effect of magnetophoresis and Brownian diffusion on the levitation of bodies in a magnetic fluid. *Magnetohydrodynamics*, Vol. 44, No. 2, 121–126.

Boudouvis, A.G.; Puchala, J.L.; Scriven, L.E. & Rosensweig, R.E. (1987). Normal field instability and patterns in pools of ferrofluid. *Journal of Magnetism and Magnetic Materials*, Vol. 65, 307–310.

Cowley, M & Rosensweig, R. (1967). The interfacial instability of a ferromagnetic fluid. *Journal of Fluid Mechanics*, Vol. 30, No. 4, 671–688.

Gollwitzer, C.; Richter, R. & Rehberg, I. (2006). Via hexagons to squares in ferrofluids: experiments on hysteretic surface transformations under variation of the normal magnetic field. *Journal of Physics: Condensed Matter*, Vol. 18, S2643–2656.

Gollwitzer, C.; Matthies, G.; Richter, R.; Rehberg, I. & Tobiska, L. (2007). The surface topography of a magnetic fluid – a quantitative comparison between experiment and numerical simulation. *Journal of Fluid Mechanics*, Vol. 571, 455–474.

Gollwitzer, C.; Spyropoulos, A.N.; Papathanasiou, A.G.; Boudouvis, A.G. & Richter, R. (2009). The normal field instability under side-wall effects: comparison of experiments and computations. *New Journal of Physics*, Vol. 11, 053016.

Ivanov, A.O. & Kuznetsova, O.B. (2001). Magnetic properties of dense ferrofluids: An influence of interparticle correlations. *Physical Review E*, Vol. 64, 041405.

Knieling, H.; Richter, R.; Rehberg, I.; Matthies, G. & Lange, A. (2007). Growth of surface undulations at the Rosensweig instability. *Physical Review E*, Vol. 76, No. 6, 066301.

Lange, A.; Richter, R. & Tobiska, L. (2007). Linear and nonlinear approach to the Rosensweig instability. *GAMM-Mitteilungen*, Vol. 30, No. 1, 171–184.

Lavrova, O.; Matthies, G.; Mitkova, T.; Polevikov, V. & Tobiska, L. (2003). Finite element methods for coupled problems in ferrohydrodynamics, In: *Lecture Notes in Computational Science and Engineering*, Bänsch, E. (Ed.), Vol. 35, 160-183, Springer-Verlag.

Lavrova, O.; Matthies, G. & Tobiska, L. (2008). Numerical study of soliton-like surface configurations on a magnetic fluid layer in the Rosensweig instability. *Communications in Nonlinear Science and Numerical Simulation*, Vol. 13, 1302–1310.

Lavrova, O.; Polevikov, V. & Tobiska, L. (2010). Numerical Study of the Rosensweig Instability in a Magnetic Fluid Subject to Diffusion of Magnetic Particles. *Mathematical Modelling and Analysis*, Vol. 15, No. 2, 223–233.

Matthies, G. & Tobiska, L. (2005). Numerical simulation of normal-flied instability in the static and dynamic case. *Journal of Magnetism and Magnetic Materials*, Vol. 289, 346–349.

Polevikov, V. (2004). Methods for numerical modeling of two-dimensional capillary surfaces. *Computational Methods in Applied Mathematics*, Vol. 4, No. 1, 66–93.

Polevikov, V. & Tobiska, L. (2008). On the solution of the steady-state diffusion problem for ferromagnetic particles in a magnetic fluid. *Mathematical Modelling and Analysis*, Vol. 13, No. 2, 233–240.

Polevikov, V. & Tobiska, L. (2011). Influence of diffusion of magnetic particles on stability of a static magnetic fluid seal under the action of external pressure drop. *Communications in Nonlinear Science and Numerical Simulation*, Vol. 16, 4021–4027.

Pshenichnikov, A.F.; Mekhonoshin, V.V. & Lebedev, A.V. (1996). Magneto-granulometric analysis of concentrated ferrocolloids. *Journal of Magnetism and Magnetic Materials*, Vol. 161, 94–102.

Pshenichnikov, A.F. & Lebedev, A.V. (2004). Low-temperature susceptibility of concentrated magnetic fluids. *Journal of Chemical Physics*, Vol. 121, No. 11, 5455–5467.

Pshenichnikov, A.F.; Elfimova, E.A. & Ivanov, A.O. (2011). Magnetophoresis, sedimentation, and diffusion of particles in concentrated magnetic fluids. *Journal of Chemical Physics*, Vol. 134, 184508.

Richter, R. & Barashenkov, I. (2005). Two-dimensional solitons on the surface of magnetic fluids. *Physical Review Letters*, Vol. 94, 184503–184506.

Rosensweig, R.E. (1998). *Ferrohydrodynamics*, Dover Pubns.

Sudo, S.; Yano, T. & Nakagawa, A. (2006). The aging variation of interfacial instability of magnetic fluids. *Journal of Advanced Science*, Vol. 18, No. 1-2, 119–122.

9

Numerical Modelling and Design of an Eddy Current Sensor

Philip May[1] and Erping Zhou[2]
[1]*Elcometer Ltd. and*
[2]*University of Bolton*
UK

1. Introduction

Eddy current testing involves exciting a coil with a fixed frequency or pulse and bringing it into close proximity with a conductive material. The electrical impedance of the coil changes due to the influence of electrical 'eddy currents' in the material. Using an eddy current technique, the sizing of surface and sub-surface defects, measurements of thickness of metallic plates and of conductive and non-conductive coatings on metal substrates, assessment of corrosion, ductility, heat treatment and measurements of electrical conductivity and magnetic permeability are all possible and quantifiable. The eddy current method has become one of the most successful non-destructive techniques for testing conductive coatings on conductive substrates.

The data acquired from eddy current sensors however is affected by a large number of variables, which include sample conductivity; permeability; geometry and temperature as well as sensor lift off. The multivariable properties of sample coatings add an even greater level of complexity. Many of these problems have been overcome in the laboratory using precision wound air-cored coils, multiple excitation frequencies and theoretical inversion models. High levels of agreement between theoretical models and measurement however are only possible with accurately constructed coils, which are difficult to manufacture in practice. Coils are also prone to poor sensitivity, poor resolution, and a poor dynamic range as well as self-resonance at high frequencies, which make them unsuitable for online process control. Many of the problems associated with air-cored coils however can be overcome when the coils use ferrite cores or cup cores.

Inversion models often make use of simplifying assumptions, which include symmetrically wound coils, constant current distributions in coil regions and ideal test materials. Ideal coils simply do not exist outside the laboratory; ideal test materials do not exist outside the laboratory either. An example of non-ideal test materials is hot-dip galvanising, where molten zinc reacts with steel to form distinct eutectic alloy layers (Langhill, 1999). Another example is case hardening in steel. Steel also has a magnetic permeability that is frequency dependent and subject to localised variation (Bowler, 2006).

Other than non-ideal coils and test materials, a practical limit exists to the information that can be extracted through eddy current testing (Norton & Bowler, 1993). Eddy currents can only really sense the presence of layer boundaries owing to the integrating character of eddy current signals. Glorieux and co-workers give an example of this, observing that sharp material profile features appear smooth under reconstruction (Glorieux et al, 1999). Another limitation, which affects coatings on steel, is the permeability-conductivity ratio and coating conductivity-thickness product (Becker et al, 1988). One of these quantities must be known prior to inspection.

This chapter focuses on the development and testing of a new highly accurate and highly sensitive ferrite-cored sensor and a novel magnetic moment model of the sensor, which requires only the discretisation of the sensor core-air boundary interface. The chapter starts by developing a set of partial differential equations (PDE) to model the vector potential fields present in the regions bounding the sensor. Sensor regions were considered to be source-less with imaginary surface currents imposed at region interfaces. Green's functions were determined for all bounded regions. Basis functions were then used to represent the sensor cores surface current distribution, which were then formed into a set of $2N$ linearly independent equations by applying the relevant boundary conditions. A matrix method was finally developed to solve these equations using a moment method.

The matrix method was further developed in this chapter in order to calculate sensor coil impedance and induced voltage. An efficient material profile function $m(\alpha)$ for modelling the interaction between the sensor and test material was also developed and verified. A novel form of parameterisation was adopted for $m(\alpha)$. The accuracy and convergence of the vector potentials generated by the source coil and core-air boundary surface currents was reviewed and a new free-space Green's function introduced.

2. Sensor theoretical model

This section introduces a new ferrite-cored eddy current sensor and develops integral equations to characterise the source vector potential and core vector potential fields. Closed form solutions of the core equations are applied to the core-air boundary interface, generating $2N$ linearly independent equations with $2N$ unknown coefficients. The unknown coefficients are evaluated using the method of weighted residuals.

2.1 Basic sensor design

When a ferrite core is used in an eddy current sensor, the coil inductance, sensitivity and resolution increase significantly (Blitz, 1991, Moulder et al, 1992). A ferrite core is therefore incorporated into the sensor design used for this chapter, which is shown in figure 1. Coaxial to the ferrite core below are three coils, a central source coil, which carries a current I amps and two sense or pick-up coils. The sensor is a reflection sensor (or transformer style sensor) with pick-up coils in a differential configuration. Each coil is assumed to have n_c coil turns per unit area with a total of N_c turns. The sensor is located in free space and positioned above and orthogonal to a medium, which is comprised of M planar layers. Each layer is considered to be linear, isotropic and homogeneous, where the ith layer has conductivity σ_i and permeability μ_i.

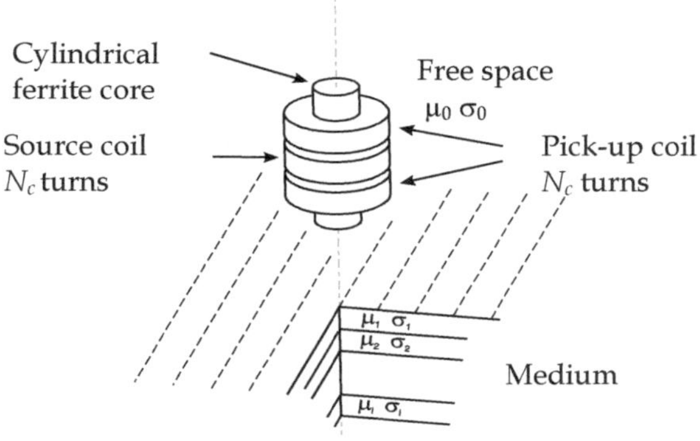

Fig. 1. Ferrite-Cored Eddy Current Sensor.

Certain assumptions are made about the sensor, which are listed below:

- The self-resonant frequency of each coil is greater than the maximum operating frequency of the sensor as a measuring system; corrections for coil self-capacitance or coil-ferrite capacitance is not considered necessary (Harrison et al, 1996).
- The source coil is considered to be a region of constant current density.
- Pick-up coils are matched and act into loads of infinite impedance. Pick-up coils generate no magnetic flux.
- The sensor core is soft magnetic ferrite. The core is assumed to be linear, isotropic and homogeneous; core conductivity is assumed to be negligible.

In order to begin an analysis of the sensor of Fig. 1, the ferrite core and pick-up coils were removed and the source coil replaced with a delta function coil. The free space region bounding the sensor was also divided into two regions, one above the plane of the delta function coil (region 1) and one below and extending to the surface of the medium (region 2). See figure 2.

Using Maxwell's equations and the homogeneous wave equation, the PDE defining the source vector potential field $\boldsymbol{A_S}$ in any of the regions of figure 2 is of the form:

$$\nabla^2 \boldsymbol{A_S} = \mu \boldsymbol{J_t} \tag{1}$$

μ represents medium permeability and $\boldsymbol{J_t}$ is the total electric current density. If $\boldsymbol{J_t}$ is comprised of an impressed current density $\boldsymbol{J_s}$ and an effective electric conduction current density $\boldsymbol{J_{ce}}$, then

$$\nabla^2 \boldsymbol{A_S} = \mu(\boldsymbol{J_s} + \boldsymbol{J_{ce}}) \tag{2}$$

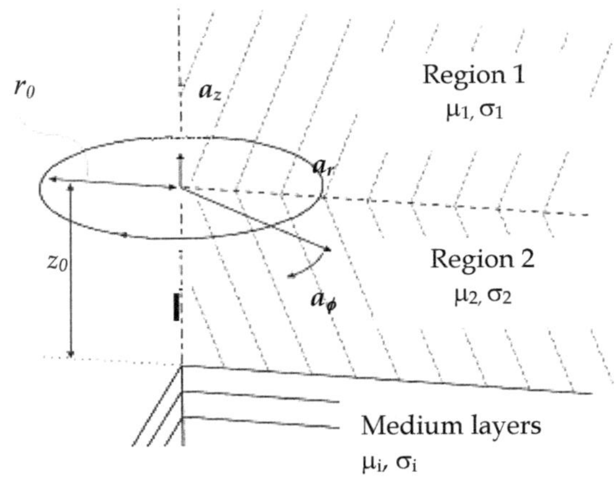

Fig. 2. Delta Function Source Coil Located above a Layered Medium.

If a delta function coil $\boldsymbol{I_0} = \delta(r - r_0)\delta(z - z_0)\boldsymbol{a}_\varphi$ is located at (r_0, z_0), where

$$\delta(r - r_0)\delta(z - z_0) = \begin{cases} \infty, & if \quad (r = r_0), \quad (z = z_0) \\ 0 & otherwise \end{cases} , \tag{3}$$

and current density $\boldsymbol{J_{ce}} = j\omega\mu\sigma\boldsymbol{A_S}$, then:

$$\nabla^2 \boldsymbol{A_S} - j\omega\mu\sigma\boldsymbol{A_S} + \mu\boldsymbol{I_0} = 0 \ . \tag{4}$$

Using the vector identity $\nabla^2\boldsymbol{B} = \nabla(\nabla \cdot \boldsymbol{B}) - \nabla \times \nabla \times \boldsymbol{B}$ gives

$$\nabla(\nabla \cdot \boldsymbol{A_S}) - \nabla \times \nabla \times \boldsymbol{A_S} - j\omega\mu\sigma\boldsymbol{A_S} + \mu\boldsymbol{I_0} = 0 \tag{5}$$

Since the coil excitation is azimuthal and since both the media and sensor core have axial symmetry, then vector potential $\boldsymbol{A_S}$ will also be azimuthal, hence let the source field be

$$\boldsymbol{A_S} = A_{S\varphi}(r,z)\boldsymbol{a}_\varphi \ . \tag{6}$$

Substituting the Coulomb gauge $\nabla \cdot \boldsymbol{A_S} = 0$ into PDE (5) gives

$$-\nabla \times \nabla \times \boldsymbol{A_S} - j\omega\mu\sigma\boldsymbol{A_S} + \mu\boldsymbol{I_0} = 0 \tag{7}$$

and using:

$$\nabla \times \nabla \times \boldsymbol{A_S} = \begin{vmatrix} \boldsymbol{a_r}/r & \boldsymbol{a_\varphi} & \boldsymbol{a_z}/r \\ \partial/\partial r & \partial/\partial\phi & \partial/\partial z \\ -\partial A_{S\varphi}/\partial z & 0 & (1/r)\partial(rA_{S\varphi})/\partial r \end{vmatrix} \tag{8}$$

gives the PDE for the delta function coil:

$$\frac{\partial^2 A_{s\varphi}}{\partial r^2} + \frac{1}{r}\frac{\partial A_{s\varphi}}{\partial r} - \frac{A_{s\varphi}}{r^2} + \frac{\partial A_{s\varphi}}{\partial z^2} - j\omega\mu\sigma A_{s\varphi} + \mu\delta(r - r_0)\delta(z - z_0) = 0\,. \tag{9}$$

Equation (9) is widely recognised as the PDE first used by Dodd and Deeds (Dodd & Deeds, 1968). If $G_s(r, z; r_0, z_0)$ is the Green's function for equation (9), then:

$$A_{s\varphi}(r,z) = \mu G_s(r,z;r_0,z_0)\,. \tag{10}$$

If the coil has a rectangular cross section and a source distribution $\boldsymbol{J_s} = J_{s\varphi}(r_0, z_0)\boldsymbol{a_\varphi}$, then:

$$A_{s\varphi}(r,z) = \mu\iint G_s(r,z;r_0,z_0)J_{s\varphi}(r_0,z_0)dr_0dz_0 \tag{11}$$

The Green's function for equation (11) first proposed by Cheng and co-workers (Cheng et al, 1971) and forming the basis of nearly all subsequent eddy current research, is given below for the generalised n^{th} media layer of figure 2:

$$G_s^{(n)}(r,z;r',z') = \int_0^\infty [B_n(\alpha)e^{-\alpha_n z} + C_n(\alpha)e^{\alpha_n z}]J_1(\alpha r)d\alpha \tag{12}$$

$B_n(\alpha)$ and $C_n(\alpha)$ are media dependent functions and $J_1(\alpha r)$ is a Bessel function of the first order and first kind; α is defined for each region as follows:

$$\alpha_n^2 = \alpha^2 + j\omega\mu_n\sigma_n\,. \tag{13}$$

2.2 The influence of the sensor core

The Cheng method imposes surface currents on media layer boundaries according to the surface equivalence theorem, where coefficients $B_n(\alpha)$ and $C_n(\alpha)$ are determined by enforcing boundary conditions at each layer interface. The sensor core can be treated in exactly the same way. Figure 3 shows surface current $\boldsymbol{J}$ impressed on closed surface S at the sensor core-air interface.

Figure 3 shows the sensor core partitioned into two separate regions, an external region (vector potential $\boldsymbol{A_E}$) and an internal source-less region (vector potential $\boldsymbol{A_I}$). Considering the internal region first, let a surface current $\boldsymbol{J_I}$ be impressed on the closed surface S. Let $\boldsymbol{J_I}$ be azimuthal, and at some arbitrary point ρ' on S, let the limiting value of $\boldsymbol{J_I}$ be a delta function source:

$$\boldsymbol{J}_I(\rho) = \int d^3\rho' \delta(\rho - \rho') \boldsymbol{J}_I(\rho') . \tag{14}$$

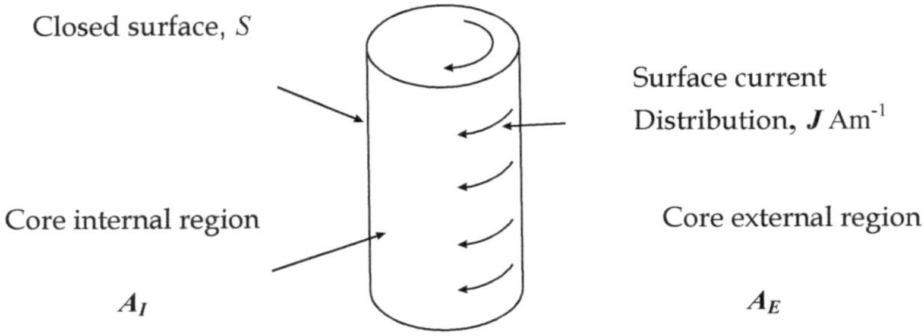

Fig. 3. Surface Current Distribution $\boldsymbol{J}$ on Core Surface S.

Since $\boldsymbol{J}_I$ is azimuthal it follows that vector potential $\boldsymbol{A}_I$ is likewise azimuthal, which leads

To the following PDE for the component $A_{I\varphi}$:

$$\frac{\partial^2 A_{I\varphi}}{\partial r^2} + \frac{1}{r}\frac{\partial A_{I\varphi}}{\partial r} - \frac{A_{I\varphi}}{r^2} + \frac{\partial A_{I\varphi}}{\partial z^2} + \mu_I \delta(r - r')\delta(z - z') = 0 \tag{15}$$

with the solution for $\boldsymbol{J}_I = J_{I\varphi}(r', z')\boldsymbol{a}_\varphi$

$$A_{I\phi}(r,z) = \mu_I \int_C G_I(r,z;r',z') J_{I\varphi}(r',z') ds' . \tag{16}$$

Integration is along a contour C on closed surface S. G_I $(r, z; r', z')$ and G_s $(r, z; r_0, z_0)$ are clearly identical and differ only in media dependent functions B_n (α) and C_n (α). A similar set of equations does not directly follow for the external core field $\boldsymbol{A}_E$ due to the presence of source current $\boldsymbol{J}_s$. The field in this region must be regarded as the vector sum of the source field $\boldsymbol{A}_s$ and a source-less scattered field $\boldsymbol{A}_R$ (Yildir et al, 1992):

$$\boldsymbol{A}_E = \boldsymbol{A}_R + \boldsymbol{A}_s \tag{17}$$

Concentrating on the source-less scattered field $\boldsymbol{A}_R$ and impressing a scattering current $\boldsymbol{J}_R$ on S, leads to the following:

$$A_{R\varphi}(r,z) = \mu \int_C G_R(r,z;r',z') J_{R\varphi}(r',z') ds' \tag{18}$$

G_R (r, z, r', z') and G_S $(r, z; r_0, z_0)$ are the same function as both are determined for the same $M + 2$ media layers. A solution to equation (18) proceeds by expanding surface current $J_{R\varphi}$ as follows (Balanis, 1989):

$$J_{R\varphi}(r',z') = \sum_{i=1}^{N} u_i a_i(r',z') \tag{19}$$

where a_i represents a basis function and u_i the basis function coefficient. Substituting equation (19) into (18) gives the following:

$$A_{R\varphi}(r,z) = \mu \sum_{i=1}^{N} u_i \int_C G_R(r,z;r',z') a_i(r',z') ds' \tag{20}$$

Basis functions were now chosen to accurately represent the anticipated unknown function $J_{R\varphi}$. A piecewise constant sub-domain function was chosen to do this, which is of the form shown below:

$$a_i(r',z') = \begin{cases} 1 & z' \in (z'_{i-1}+\Delta_z, z'_{i+1}-\Delta_z), \quad z'_{i+1} > z'_{i-1} \\ 1 & r' \in (r'_{i-1}+\Delta_r, r'_{i+1}-\Delta_r), \quad r'_{i+1} \neq r'_{i-1} \\ 0 & otherwise \end{cases} \tag{21}$$

$$\Delta_r = |r'_{i+1} - r'_{i-1}|/4 \text{ and } \Delta_z = |z'_{i+1} - z'_{i-1}|/4 .$$

Sub-domains were divided into N sub-intervals and evenly distributed along the sensor core-air interface. Observation points (r, z) were located at the centre of sub-domains (see figure 4) for greatest computational accuracy (Balanis, 1989).

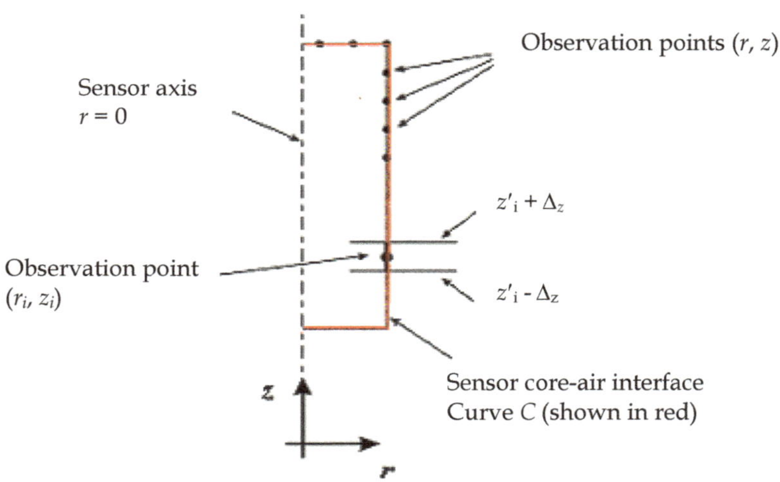

Fig. 4. Discretization of Current J_R on the Sensor Core Interface.

2.3 The total field and internal core field

When combined the source and scattered field gives the total vector potential $A_{E\varphi}$. For the nth media layer outside the sensor core:

$$A_{E\varphi}(r,z) = \mu_n \sum_{i=1}^{N} u_i \int_C G_R^{(n)}(r,z;r',z') a_i(r',z') ds' + \mu_n \iint G_s^{(n)}(r,z;r_0,z_0) J_{s\varphi}(r_0,z_0) dr_0 dz_0 \tag{22}$$

where $J_{s\varphi}$ is the source current distribution:

$$J_{s\varphi}(r_0,z_0) = IN_c / (l_2 - l_1)(r_2 - r_1) \tag{23}$$

(l_2, l_1) and (r_2, r_1) are the length and radial dimensions of the source coil, which is assumed to be rectangular in cross section. Since $A_{I\varphi}$ is solved in exactly the same way as scattered field $A_{R\varphi}$ expand $J_{I\varphi}$ into a similar N term series, letting b_i represent the series basis function and v_i the expansion coefficient. Given this b_i is defined as follows:

$$b_i(r',z') = \begin{cases} 1 & z' \in (z'_{i-1}+\Delta_z, z'_{i+1}-\Delta_z), \quad z'_{i+1} > z'_{i-1} \\ 1 & r' \in (r'_{i-1}+\Delta_r, r'_{i+1}-\Delta_r), \quad r'_{i+1} \neq r'_{i-1} \\ 0 & otherwise \end{cases} . \tag{24}$$

Substitution of equation (24) into equation (16) gives:

$$A_{I\varphi}(r,z) = \mu_I \sum_{i=1}^{N} v_i \int_C G_I(r,z;r',z') b_i(r',z') ds' . \tag{25}$$

2.4 Sensor core boundary conditions

Unknown expansion coefficients u_i and v_i are determined by applying the sensor core boundary conditions. Since the core is rod shaped, two surfaces exist where boundary conditions must be met. These surfaces are shown below in figure 5 with appropriate unit normal vectors $\boldsymbol{n}$:

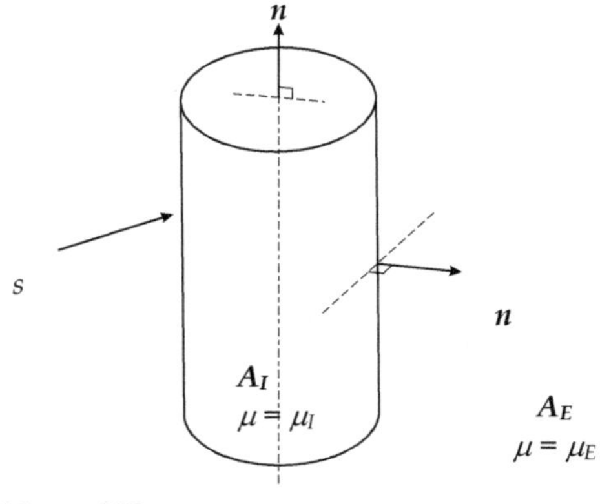

Fig. 5. Core Boundary Unit Normal Vectors $\boldsymbol{n}$.

If A_I is the vector potential inside the core and A_E outside, then the boundary conditions for the core-air interface can be shown to be:

$$(1/\mu_I)\nabla\times \boldsymbol{A_I} \times \boldsymbol{n} = (1/\mu_E)\nabla\times \boldsymbol{A_E} \times \boldsymbol{n} \tag{26}$$

and

$$\boldsymbol{A_I} = \boldsymbol{A_E} \tag{27}$$

Substituting $\boldsymbol{A_E} = \boldsymbol{A_R} + \boldsymbol{A_S}$ and assuming $\mu_E = 1$ for free space gives:

$$\nabla\times \boldsymbol{A_S} \times \boldsymbol{n} = ((1/\mu_I)\nabla\times \boldsymbol{A_I} - \nabla\times \boldsymbol{A_R})\times \boldsymbol{n} \tag{28}$$

and

$$\boldsymbol{A_S} = \boldsymbol{A_I} - \boldsymbol{A_R}\,. \tag{29}$$

Evaluating equation (28) gives the following for the core upper and lower flat faces:

$$\partial A_{s\varphi}(r,z)/\partial z = \partial(A_{I\varphi}(r,z)/\mu_I - A_{R\varphi}(r,z))/\partial z \tag{30}$$

and for the core's central cylindrical face:

$$(1/r+\partial/\partial r)A_{s\varphi}(r,z) = (1/r+\partial/\partial r)(A_{I\varphi}(r,z)/\mu_I - A_{R\varphi}(r,z)) \tag{31}$$

2.5 Evaluation of expansion coefficients using the method of weighted residuals

The two sensor core boundary equations define the relationship between unknown basis function coefficients u_i and v_i. The method of weighted residuals was used to solve these equations, which proceeds by grouping u_i and v_i together into a single $2N\times1$ column matrix $\boldsymbol{K}$ with the following elements:

$$k_p = \begin{cases} u_p & p = 1,\ldots,N \\ v_p & p = N+1,\ldots 2N \end{cases}. \tag{32}$$

Given $\boldsymbol{K}$ above, define a new $1\times 2N$ row matrix $\boldsymbol{\Psi}$ such that:

$$`\,\boldsymbol{\Psi K} = \left(k_1\psi_1 + \ldots + k_{2N}\psi_{2N}\right). \tag{33}$$

The following can be seen to apply for element ψ_p:

$$\psi_p = \begin{cases} \nabla\times(\mu_I^{-1}\int_c G_I(\boldsymbol{\rho_q};\boldsymbol{\rho}')b_p(\boldsymbol{\rho}')ds' - \int_c G_R^{(n)}(\boldsymbol{\rho_q};\boldsymbol{\rho}')a_p(\boldsymbol{\rho}')ds')\boldsymbol{a_\varphi}\times\boldsymbol{n} & p = 1,\ldots,N \\ \int_c G_I(\boldsymbol{\rho_q};\boldsymbol{\rho}')b_p(\boldsymbol{\rho}')ds' - \int_c G_R^{(n)}(\boldsymbol{\rho_q};\boldsymbol{\rho}')a_p(\boldsymbol{\rho}')ds' & p = N+1,\ldots,2N \end{cases} \tag{34}$$

where ρ_q is a field point at (r_q, z_q), ρ' is a source point at (r', z') in sub-domain p and $\boldsymbol{n}$ is a unit normal vector at field point (r_q, z_q) on the core boundary surface. Since 2N unknowns in $\boldsymbol{K}$ require the formation of 2N linearly independent equations, define a 2N×1 column matrix $\boldsymbol{F}$ for the source field at N points $\{(r_q, z_q); q = 1,..., N\}$ on C, with elements:

$$f_q = \begin{cases} \nabla \times (\iint G_s^{(n)}(r_q, z_q; r_0, z_0) J_{s\phi}(r_0, z_0) dr_0 dz_0) \boldsymbol{a}_{\varphi} \times \boldsymbol{n} & q = 1,..., N \\ \iint G_s^{(n)}(r_q, z_q; r_0, z_0) J_{s\varphi}(r_0, z_0) dr_0 dz_0) & q = N+1,...,2N \end{cases} . \tag{35}$$

In order to form 2N linearly independent equations, introduce a further 1×2N row matrix $\boldsymbol{W}$ and take the inner product $\langle \boldsymbol{W}^T, \boldsymbol{\Psi} \rangle$. Integration is along the entire length of the core-air interface (curve C) to minimise any residual error, giving:

$$\left[\int_c \boldsymbol{W}^T \boldsymbol{\Psi} ds \right] \cdot \boldsymbol{K} = \int_c \boldsymbol{W}^T \boldsymbol{F} ds \tag{36}$$

Weight vector $\boldsymbol{W}$ is selected according to one of the following methods (Sadiku, 1992):

- Point collocation.
- Sub-domain collocation.
- Least square.
- Galerkin.

Point collocation was selected because it provided acceptable accuracy for computational effort (Balanis, 1989). Point collocation uses the following weight vector $\boldsymbol{W}$:

$$\boldsymbol{W} = [\delta(c - c_1),..., \delta(c - c_{2N})], \tag{37}$$

Collocation points on C are chosen to coincide with basis function observation points, where the following applies for weight element w_q:

$$\delta(c - c_q) = \begin{cases} \infty & if \quad (c = c_q) \\ 0 & otherwise \end{cases} \tag{38}$$

Recognising that $\int_{x_i^-}^{x_i^+} \delta(x - x_i) dx = 1$ and inserting this into row q of equation (36), gives:

$$f_q \int_{C_q^-}^{C_q^+} \delta(c - c_q) ds = \psi_1 u_1 \int_{C_q^-}^{C_q^+} \delta(c - c_q) ds + \cdots + \psi_{2N} v_{2N} \int_{C_q^-}^{C_q^+} \delta(c - c_q) ds . \tag{39}$$

Evaluating equation (39) for all N collocation points leads to the matrix equation for u_i and v_i:

$$\begin{pmatrix} f_1(r_1,z_1) \\ .. \\ .. \\ .. \\ f_1(r_N,z_N) \\ f_1(r_{N+1},z_{N+1}) \\ .. \\ .. \\ .. \\ f_{2N}(r_{2N},z_{2N}) \end{pmatrix} = \begin{bmatrix} \psi_1(r_1,z_1) & .. & .. & .. & .. & \psi_{2N}(r_1,z_1) \\ .. & & & & & .. \\ .. & & .. & .. & & .. \\ .. & & & & & .. \\ \psi_1(r_N,z_N) & .. & .. & .. & .. & \psi_{2N}(r_N,z_N) \\ \psi_1(r_{N+1},z_{N+1}) & .. & .. & .. & .. & \psi_{2N}(r_{N+1},z_{N+1}) \\ .. & & & & & .. \\ .. & & .. & .. & & .. \\ .. & & & & & .. \\ \psi_1(r_{2N},z_{2N}) & .. & .. & .. & .. & \psi_{2N}(r_{2N},z_{2N}) \end{bmatrix} \begin{pmatrix} u_1 \\ .. \\ .. \\ .. \\ u_N \\ v_{N+1} \\ .. \\ .. \\ .. \\ v_{2N} \end{pmatrix} \quad (40)$$

3. Numerical model implementation

In previous section, a set of partial differential equations were developed to model the vector potential fields present in the regions bounding a ferrite-cored eddy current sensor; sensor regions were considered to be source-less with imaginary surface currents imposed at region interfaces. Green's functions were determined for all bounded regions. A novel set of Basis functions were introduced to reproduce the surface currents present on the sensor core-air interface, which were then formed into a system of $2N$ linearly independent equations. A matrix method was finally developed to solve these equations using the method of weighted residuals.

The matrix method was further developed in this section in order to calculate sensor coil impedance and induced voltage. An efficient material profile function $m(\alpha)$ for modelling the interaction between the sensor and test material was also developed and verified. A novel form of parameterisation is adopted for $m(\alpha)$. The accuracy and convergence of vector potential fields generated is reviewed and a new free-space Green's function introduced.

3.1 The material profile function $m(\alpha)$ for stratified layers

Section 2.1 introduced the medium as being comprised of M planar layers. Each layer had medium properties that were considered to be isotropic, homogeneous and linear. Any change between the electrical and magnetic properties of the layers was a step change. This approach enabled Cheng and co-workers to successfully model non-homogeneous materials using piecewise constant approximations (Cheng et al, 1971). Such non-linear material profiles might be produced, as an example, by coating a substrate, by case hardening, heat treatment, ion bombardment, or by chemical processing. A recently developed alternative method used hyperbolic tangential profiles to represent near surface changes in conductivity (Uzal et al, 1993). The method proposed by Cheng was adopted here due to it being more flexible. Applying boundary equations (26) and (27) to Green's function (12) for the nth media layer below the coil gives:

$$B_{n-1} = \frac{1}{2} e^{\alpha_{-1} z_n} [(1+\beta_{n-1,n}) e^{-\alpha_n z_n} B_n + (1-\beta_{n-1,n}) e^{\alpha_n z_n} C_n] \quad (41)$$

and

$$C_{n-1}=\frac{1}{2}e^{\alpha_{n-1}z_n}[(1-\beta_{n-1,n})e^{-\alpha_n z_n}B_n+(1+\beta_{n-1,n})e^{\alpha_n z_n}C_n] \tag{42}$$

where:

$$\beta_{n-1,n}=\beta_n/\beta_{n-1} \tag{43}$$

and

$$\beta_n=\alpha_n/\mu_n\ . \tag{44}$$

Taking the coil region (regions 1 and 2) to have the properties α_0, β_0 and μ_0, and giving consideration to the presence of a source give:

$$-B_2e^{-\alpha_0 z_0}+C_2e^{\alpha_0 z_0}=-B_1e^{-\alpha_0 z_0}+C_1e^{\alpha_0 z_0}+\frac{\alpha}{\beta_0}r_0J_1(\alpha r_0)\ . \tag{45}$$

and

$$-B_2e^{-\alpha_0 z_0}+C_2e^{\alpha_0 z_0}=-B_1e^{-\alpha_0 z_0}+C_1e^{\alpha_0 z_0} \tag{46}$$

Cheng and co workers showed that unknown coefficients B_n and C_n were dramatically simplified using a matrix method (Cheng et al, 1971). This method defined the following matrices, a 2x1 coefficient column matrix:

$$\boldsymbol{A}_n=\begin{bmatrix}B_n\\ C_n\end{bmatrix} \tag{47}$$

and a 2x2 transformation matrix $\boldsymbol{T}_{n-1,n}$, with the elements:

$$(\boldsymbol{T}_{n-1,n})_{11}=\frac{1}{2}(1+\beta_{n-1,n})e^{(\alpha_{n-1}-\alpha_n)z_n} \tag{48}$$

$$(\boldsymbol{T}_{n-1,n})_{12}=\frac{1}{2}(1-\beta_{n-1,n})e^{(\alpha_{n-1}+\alpha_n)z_n} \tag{49}$$

$$(\boldsymbol{T}_{n-1,n})_{21}=\frac{1}{2}(1-\beta_{n-1,n})e^{-(\alpha_{n-1}+\alpha_n)z_n} \tag{50}$$

$$(\boldsymbol{T}_{n-1,n})_{22}=\frac{1}{2}(1+\beta_{n-1,n})e^{-(\alpha_{n-1}-\alpha_n)z_n} \tag{51}$$

Successive multiplication of $\boldsymbol{T}_{n-1,n}$ gives the following:

$$\boldsymbol{A}_2=\begin{bmatrix}B_2\\ C_2\end{bmatrix}=\boldsymbol{T}_{2,3}\cdot\boldsymbol{T}_{3,4}\ldots\boldsymbol{T}_{M-2,M-1}\cdot\boldsymbol{T}_{M-1,M}\cdot\boldsymbol{A}_M \tag{52}$$

where

$$A_M = \begin{bmatrix} 0 \\ C_M \end{bmatrix}. \tag{53}$$

If $V(n,M)$ is a 2x2 matrix, then define:

$$V(n,M) = T_{2,3} \cdot T_{3,4} ... T_{M-2,M-1} \cdot T_{M-1,M} \tag{54}$$

Evaluating the above gives the Green's functions for the regions bounding the coil:

Region 1

$$G_S^{(1)}(r,z,r_0,z_0) = \int_0^\infty \alpha r_0 J_1(\alpha r_0) J_1(\alpha r) \left(\frac{v_{12}(2,M)}{v_{22}(2,M)} \cdot e^{-\alpha_0 z_0} + e^{\alpha_0 z_0} \right) \cdot e^{-\alpha_0 z} d\alpha \tag{55}$$

Region 2

$$G_S^{(2)}(r,z,r_0,z_0) = \int_0^\infty \alpha r_0 J_1(\alpha r_0) J_1(\alpha r) \left(\frac{v_{12}(2,M)}{v_{22}(2,M)} \cdot e^{-\alpha_0 z} + e^{\alpha_0 z} \right) \cdot e^{-\alpha_0 z_0} d\alpha \tag{56}$$

A new coil region (region 12) can also be defined for coils with finite length and radial dimensions by adding vector potential $A^{(1)}$ and $A^{(2)}$ and applying relevant boundary conditions on coil length Z (Dodd & Deeds, 1968).

From the above, the following relationship is evident:

$$m = \frac{v_{12}(2,M)}{v_{22}(2,M)} \tag{57}$$

Material profile function m is dependent only on the media properties and is independent of coil geometry and coil lift-off; m not only applies to the source field $\boldsymbol{A_S}$, but also to the scattered field $\boldsymbol{A_R}$, which makes this function a universal profile function. If the material under test is comprised of two layers for simplicity, a conductive coating of thickness Z_c (region 3) deposited on a magnetic substrate (region 4), then the material profile function m can be shown to be (Dodd & Deeds, 1968):

$$m(\alpha) = \frac{(\alpha + \alpha_3)(\alpha_3 - \alpha_4 / \mu_4) + (\alpha - \alpha_3)(\alpha_3 + \alpha_4 / \mu_4) \cdot e^{2 \cdot Z_c \cdot \alpha_3}}{(\alpha - \alpha_3)(\alpha_3 - \alpha_4 / \mu_4) + (\alpha + \alpha_3)(\alpha_3 + \alpha_4 / \mu_4) \cdot e^{2 \cdot Z_c \cdot \alpha_3}} \tag{58}$$

where

$$\alpha_3 = \sqrt{\alpha^2 + j\omega\mu_3\sigma_3} \text{ and } \alpha_4 = \sqrt{\alpha^2 + j\omega\mu_4\sigma_4}.$$

3.2 Implementation of the material profile function

The material profile function m is actually a function of many variables. Most of these variables however can be regarded as constant for a given test, making the material profile function a function of spatial frequency α only. For a large number of medium layers, at least 40 to accurately represent continuously varying profiles (Uzal et al, 1993), the evaluation of V_{12} and V_{22} begins to become computationally prohibitive. Not only does the amount of matrix algebra required to calculate $m(\alpha)$ dramatically increase, but this calculation must be repeated for every element of matrix equation (40). A more efficient approach replaces $m(\alpha)$ in its matrix form with a spline curve. The oscillatory nature of high degree polynomial approximations, such as least squares regression, discounts their use. In order to assess the suitability of cubic spline interpolation it is necessary to determine the general form and amount of variation expected for $m(\alpha)$. Given this, consider a two layer medium defined by equation (58), where angular frequency ω ranges from $\omega = 2\pi{\cdot}100$ rads/sec to $\omega = 2\pi{\cdot}3{\cdot}10^4$ rads/sec and coating thickness Z_c (region 3) ranges from $Z_c = 0$ µm to 300 µm. A worse case of copper plating on steel is assumed. The following two graphs show real and imaginary components of $m(\alpha)$ for these conditions.

Examination of figure 6a and 6b clearly shows that $m(\alpha)$ has considerable variation below $\alpha = 10^4$, but that above this it is relatively smooth. Assuming that $m(\alpha)$ is defined on the interval $\alpha \in \{a, b\}$ and that a clamped boundary is used, let cubic polynomial S_j occur on subinterval $[\alpha_j, \alpha_{j+1}]$. Given this it can be shown that the maximum error occurs when: (Burden & Faires, 1989):

$$\max_{a \le \alpha \le b} \left| m(\alpha) - S(\alpha) \right| \le \frac{5 \cdot M_\alpha}{384} \max_{0 \le j \le n-1} (\alpha_{j+1} - \alpha_j)^4 \tag{59}$$

where

$$M_\alpha \ge \max_{a \le \alpha \le b} \left| m^4(\alpha) \right| \text{ and } a = \alpha_0 \langle \alpha_1 \langle \ldots \langle \alpha_n = b \, .$$

From above, interpolation error can be linked with the maximum subinterval step size $\max[\alpha_j, \alpha_{j+1}]$ and the maximum 4th derivative of $m(\alpha)$. Since the maximum derivative error is always below $\alpha = 2{\cdot}10^3$ and since $m(\alpha)$ is almost linear above $\alpha = 10^4$, it seems reasonable to reduce subinterval step size for low α and increase it above $\alpha = 10^4$. This adjustment enables the interpolating cubic polynomials S_j to more accurately reproduce data in regions of maximum variation, whilst minimising the total number of subinterval domains. Empirical study showed that the optimum choice for α_j is:

$$\alpha_j = 0.035 \cdot ((j+1)^3 + j^3) \tag{60}$$

where

$$j \in \{0, 1, \ldots, 111\}.$$

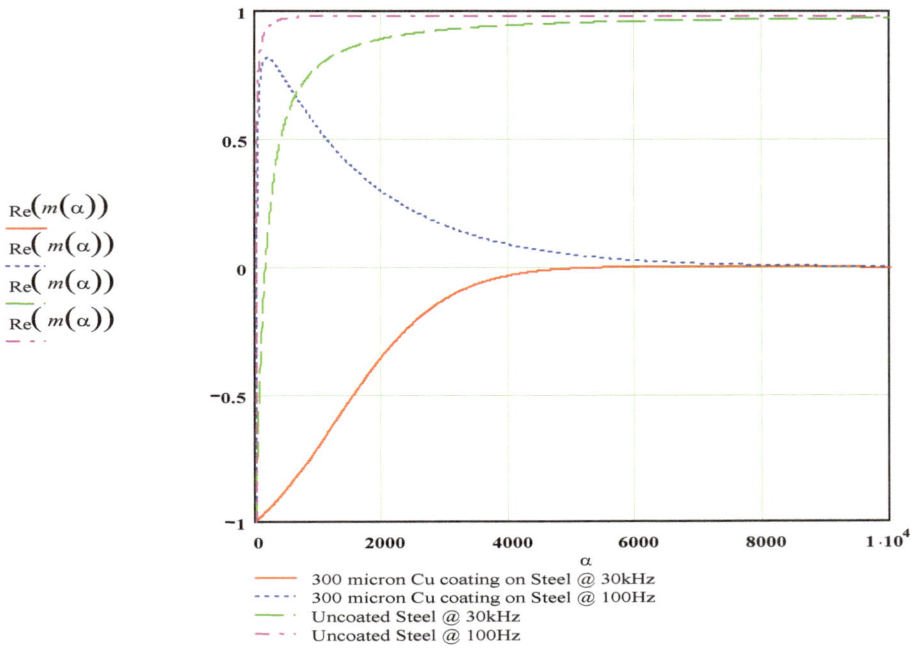

Fig. 6a. Real Component of Material Profile Function $m(\alpha)$.

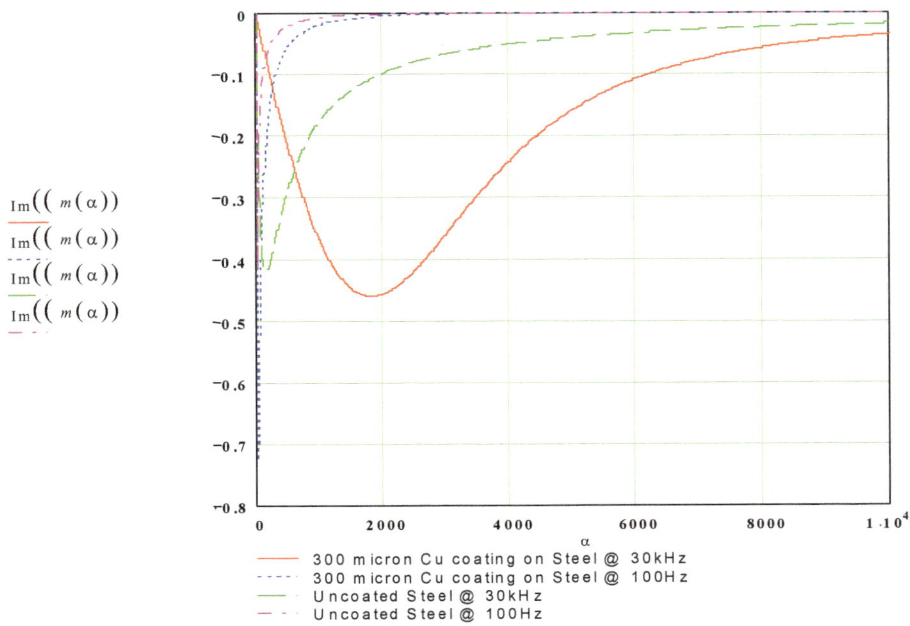

Fig. 6b. Imaginary Component of Material Profile Function $m(\alpha)$.

3.3 Material profile function testing and evaluation

A typical spline curve is given in figure 7 for a two layer material: substrate (region 4: μ_4 = 100, σ_4 = 10 MS/m) and coating (region 3: μ_3 = 1, σ_3 = 58 MS/m, coating thickness Z_c= 300 μm) for an excitation frequency $\omega = 2\pi \cdot 30 \cdot 10^3$ rads/sec.

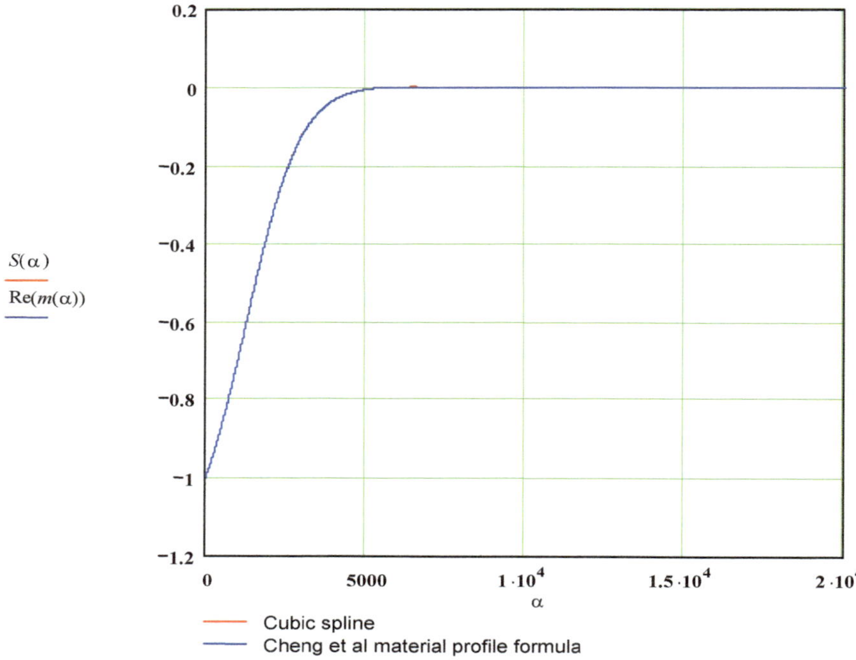

Fig. 7. Spline Approximation $S(\alpha)$ of Cheng Formula $\text{Re}(m(\alpha))$.

The major benefit of this approach is that the cubic polynomial coefficients for all S_j need only be calculated once, making the method very rapid.

3.4 The convergence of the source coil fields

The vector potentials required for evaluating the regions bounding the source coil are derived from equations (11) and (12), which are improper integrals. Since an explicit anti-derivative does not exist for these equations, numerical quadrature was be used. The convergence of these integrals can be studied by considering the form of their integrand for large α, which can be represented in the following way:

$$\int_{a}^{\xi} \frac{d\alpha}{\alpha^n} = \frac{\xi^{1-n} - a^{1-n}}{1-n} \tag{61}$$

If $n > 1$ then $\xi^{1-n} \to 0$ for $\xi \to \infty$. Given this equation (61) is convergent. Stephenson generalises this further by redefining equation (61) as (Stephenson, G. 1974):

$$\int_a^{\zeta} \frac{g(\alpha)}{\alpha^n} \tag{62}$$

where g(α) is some arbitrary function that is bounded and non-zero.

In this instance (62) is said to be convergent if $n > 1$. Given this, it seems reasonable to assume that a large positive value for n is required for a high rate of convergence. An example of this is given for the self inductance L of an air-cored coil (Dodd & Deeds, 1968):

$$L = \pi n_C^2 \mu \int_0^{\infty} \frac{1}{\alpha^5} I(r_2, r_1)^2 \{2\alpha(l_2 - l_1) + 2e^{-\alpha(l_2-l_1)} - 2 + (e^{-2\alpha l_2} + e^{-2\alpha l_1} - 2e^{-\alpha(l_2-l_1)})m\} d\alpha \tag{63}$$

The function shown above has a very rapid rate of convergence due to α being raised to the 5th power. Application of the sensor core boundary equation (28) leads to five core equations for the source coil magnetic flux density $\boldsymbol{B}_S$, which are given below:

$$B_{Sz}^{(1)} = \frac{1}{2} n_c I \mu \int_0^{\infty} \frac{1}{\alpha^2} I(r_2, r_1) J_0(\alpha r) e^{-\alpha z} \{e^{\alpha l_2} - e^{\alpha l_1} - (e^{-\alpha l_2} - e^{-\alpha l_1}) m(\alpha)\} d\alpha \tag{64}$$

$$B_{Sz}^{(2)} = \frac{1}{2} n_c I \mu \int_0^{\infty} \frac{1}{\alpha^2} I(r_2, r_1) J_0(\alpha r)(e^{-\alpha l_1} - e^{-\alpha l_2})(e^{\alpha z} - m(\alpha) e^{-\alpha z}) d\alpha \tag{65}$$

$$B_{Sz}^{(12)} = \frac{1}{2} n_c I \mu \int_0^{\infty} \frac{1}{\alpha^2} I(r_2, r_1) J_0(\alpha r) \{2 - e^{\alpha(z-l_2)} - e^{-\alpha(z-l_1)} + e^{-\alpha z}(e^{-\alpha l_1} - e^{-\alpha l_2}) m(\alpha)\} d\alpha \tag{66}$$

$$B_{Sr}^{(1)} = -\frac{1}{2} n_c I \mu \int_0^{\infty} \frac{1}{\alpha^2} I(r_2, r_1) J_1(\alpha r) e^{-\alpha z} \{e^{\alpha l_2} - e^{\alpha l_1} - (e^{-\alpha l_2} - e^{-\alpha l_1}) m(\alpha)\} d\alpha \tag{67}$$

$$B_{Sr}^{(2)} = -\frac{1}{2} n_c I \mu \int_0^{\infty} \frac{1}{\alpha^2} I(r_2, r_1) J_1(\alpha r)(e^{-\alpha l_1} - e^{-\alpha l_2})(e^{\alpha z} - m(\alpha) e^{-\alpha z}) d\alpha \tag{68}$$

Boundary equations (64) – (68) have a rate of convergence no worse than α^{-2}. Comparison with that of the source coil inductance indicates that the rate of convergence of source coil field vectors is relatively poor.

3.5 Convergence of the basis function fields

The field generated by an ith basis function located on the cylindrical face of the sensor core is given by:

$$A^*_{R\varphi}(r,z) = \mu u_i \int_{z_i-\Delta z}^{z_i+\Delta z} G_R^{(n)}(r,z;r',z')dz' \tag{69}$$

where the total scattered field $A_{R\varphi}(r,z) \cong \sum_{i=1}^{N} A^*_{R\varphi}(r,z)$ and $\Delta_z = |z_{i+1} - z_{i-1}|/4$.

If the axial coordinates of the basis function are $l_{a2} = z_i + \Delta z$ and $l_{a1} = z_i - \Delta z$, with radial coordinate r_a, equation (69) becomes:

$$A^*_{R\varphi} = \frac{r_a}{2}\mu u_i \int_0^\infty \frac{1}{\alpha} J_1(\alpha r_a) J_1(\alpha r)\{2 - e^{\alpha(z-l_{a2})} - e^{-\alpha(z-l_{a1})} + e^{-\alpha z}(e^{-\alpha l_{a1}} - e^{-\alpha l_{a2}})m(\alpha)\}d\alpha \tag{70}$$

Which assumes that field point (r, z) is bounded, with $l_{a2} \le z \le l_{a1}$.

The components of flux density $\boldsymbol{B}^*_{\boldsymbol{R}} = B^*_{Rr}\boldsymbol{a}_{\boldsymbol{r}} + B^*_{Rz}\boldsymbol{a}_{\boldsymbol{z}}$ for equation (70) are:

$$B^*_{Rr} = \frac{r_a}{2}\mu u_i \int_0^\infty J_1(\alpha r_a) J_1(\alpha r)\{-e^{\alpha(z-l_{a2})} - e^{-\alpha(z-l_{a1})} - e^{-\alpha z}(e^{-\alpha l_{a1}} - e^{-\alpha l_{a2}})m(\alpha)\}d\alpha \tag{71}$$

$$B^*_{Rz} = \frac{r_a}{2}\mu u_i \int_0^\infty J_1(\alpha r_a) J_0(\alpha r)\{2 - e^{\alpha(z-l_{a2})} - e^{-\alpha(z-l_{a1})} + e^{-\alpha z}(e^{-\alpha l_{a1}} - e^{-\alpha l_{a2}})m(\alpha)\}d\alpha \tag{72}$$

It is clear that a basis function vector potential has a rate of convergence that is poor, the convergence of it's flux density vector $\boldsymbol{B}^*_{\boldsymbol{R}}$ is even worse. A significant benefit in terms of computational efficiency and accuracy is gained if a method can be found to improve the convergence of these field equations. Note that the fields above and below the basis function, as well as the basis functions on the end faces of the core, have been omitted for brevity.

3.6 The modified free space green's function $G_0(r, z; r', z')$

It is evident that field equations have a poor rate of convergence. Considering only the basis function fields, separate equation (72) into two parts, which are given on the following:

$$B_{Rz}^{*} = \frac{1}{2}\mu u_i \int_0^{\infty} J_1(\alpha r_a) J_0(\alpha r) e^{-\alpha z} \{e^{\alpha l_{a2}} - e^{\alpha l_{a1}}\} d\alpha +$$
$$\frac{1}{2}\mu u_i \int_0^{\infty} J_1(\alpha r_a) J_0(\alpha r) e^{-\alpha z} (e^{-\alpha l_{a1}} - e^{-\alpha l_{a2}}) m(\alpha) d\alpha \qquad (73)$$

Assuming that all other field equations can be treated in the same way, a comparison of the integrands of equation (73) is shown in a normalised form in figure 8, with $m(\alpha) = 1.0$.

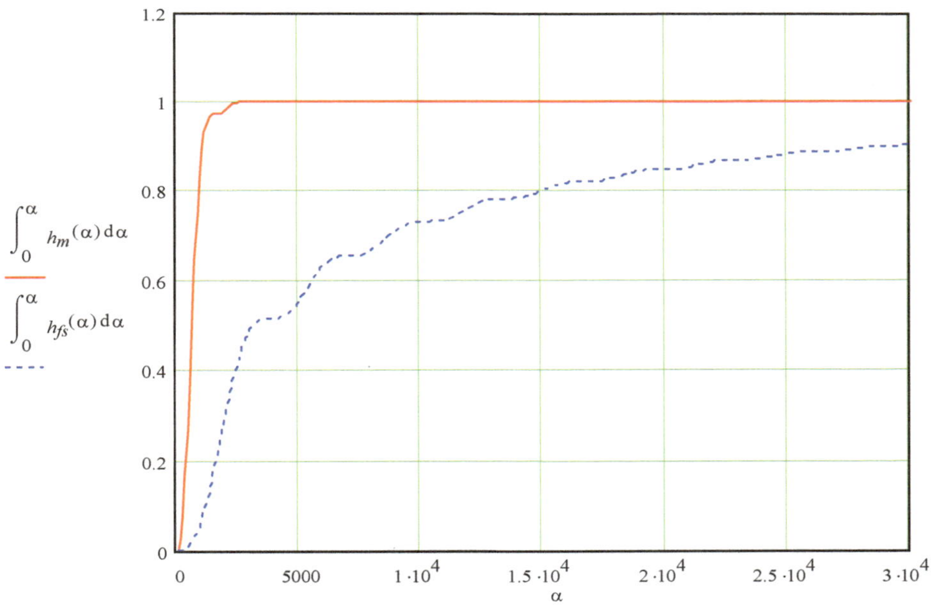

Fig. 8. Free-Space $h_{fs}(\alpha)$ and Material Dependent $h_m(\alpha)$ Convergence.

It is clear from Figure 8 that separating the field equations into two terms, and considering only the material dependent term $h_m(\alpha)$ gives a function with a very rapid rate of convergence. An equation replacing the free-space or material independent term $h_{fs}(\alpha)$ now needs to be determined. Let the delta function coil representing the material independent Green's function $G_0\,(r, z; r', z')$ be formed from discrete current elements IdS. See figure 9.

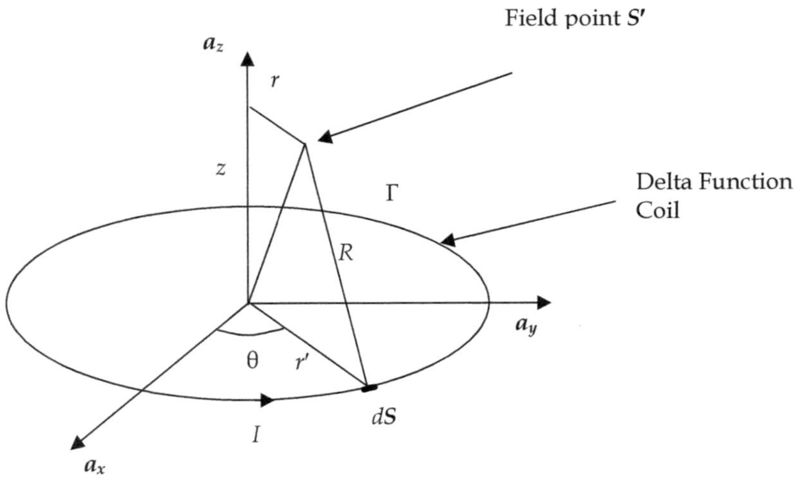

Fig. 9. Current Element *IdS* forming a Delta Function Coil.

The magnetic vector potential generated by *IdS* is given as:

$$A_{R}^{*} = \mu I \int_{\Gamma} \frac{d\boldsymbol{S}}{R}, \tag{74}$$

where μ is media permeability and I is current.

If $\boldsymbol{S} = r'(\cos(\theta)\boldsymbol{a}_x + \sin(\theta)\boldsymbol{a}_y)$, $\boldsymbol{S'} = r\boldsymbol{a}_x + z\boldsymbol{a}_z$ and $R = |\boldsymbol{S'} - \boldsymbol{S}|$, then the free space Green's function $G_0(r, z; r', z')$ is of the following form:

$$G_0(r,z;r',z') = \frac{I}{4\pi} \int_0^{\pi} \frac{\mu \cdot r' \cdot \cos(\theta) d\theta}{\sqrt{(r - r' \cdot \cos(\theta))^2 + (z - z')^2 + (r' \cdot \sin(\theta))^2}} \tag{75}$$

Integration of $G_0(r, z; r', z')$ over r or z gives the relevant equation for the basis functions.

Equation (75) satisfies all the requirements of the Green's function (Sadiku, 1992). Highly accurate calculations of field quantities were found to be possible using this equation.

4. Testing and evaluation

Nearly all eddy current investigations are conducted in the sensor coil region by determining coil impedance. Given this, if the source coil is densely and uniformly wound with a rectangular cross section, having the radial and axial dimensions (r_2, r_1) and (l_2, l_1), the induced voltage V across the coil will be equal to:

$$V = \frac{j\omega 2\pi N_c}{(l_2 - l_1)(r_2 - r_1)} \int_{l_1}^{l_2} \int_{r_1}^{r_2} r(A_{S\varphi}^{(12)}(r,z) + A_{r\varphi}(r,z)) dr dz, \tag{76}$$

where N_c is the number of turns on the source coil. Coil impedance Z can be found by simply dividing V by source current I.

Evaluation of the sensor model proceeded by defining the dimensional and physical properties of the sensor of section 2, which are shown below in table 1. Note that no information for the two pickup coils is given as this is the subject of future work.

Sensor Core		Sensor Source Coil	
Sensor Lift-off:	0.50 mm	Source Coil r_1:	1.45 mm
Core Radius:	0.99 mm	Source Coil r_2:	3.175 mm
Core Length:	6 mm	Source Coil l_1:	3.005 mm
Core Permeability	1000	Source Coil l_2:	3.845 mm
		Source Coil Turns N_c:	294

Table 1. Sensor Properties.

Matrix equation (40) was solved for coefficients u_i and v_i using Mathcad, version 11.0a and source coil impedance determined from equation (76). It was found empirically that 80 collocation points spread evenly along the core-air interface C provided good results.

Source coil self inductance L and resistance R was calculated for differing sensor lift off over steel with the following properties: relative permeability μ_r = 95.6 and conductivity σ = 8.4x10^6 S/m. As a comparison, the sensor model of table 1 was also simulated using the commercial FEM solver MagNet, version 6.25. The results of this, displayed in the form of a normalised impedance plane diagram, are shown in figure 10.

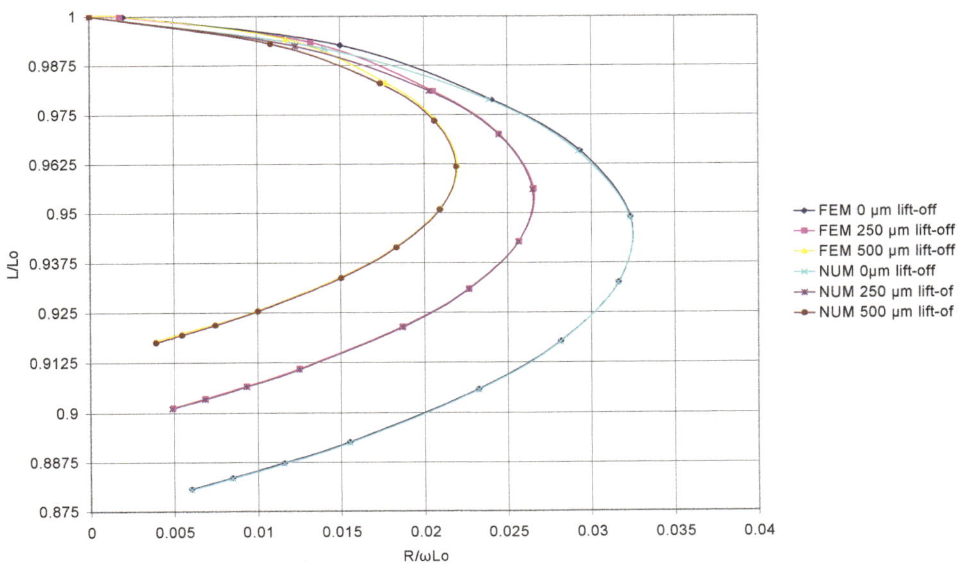

Fig. 10. Normalised Impedance Plane Diagram of Sensor.

Changes to the value of core permeability μ_r were also simulated for the sensor held in free space, positioned above a solid copper substrate (f = 30 kHz) and finally above 100 μm of copper plating on steel (f = 10 kHz). The results of this simulation are displayed in figure 11.

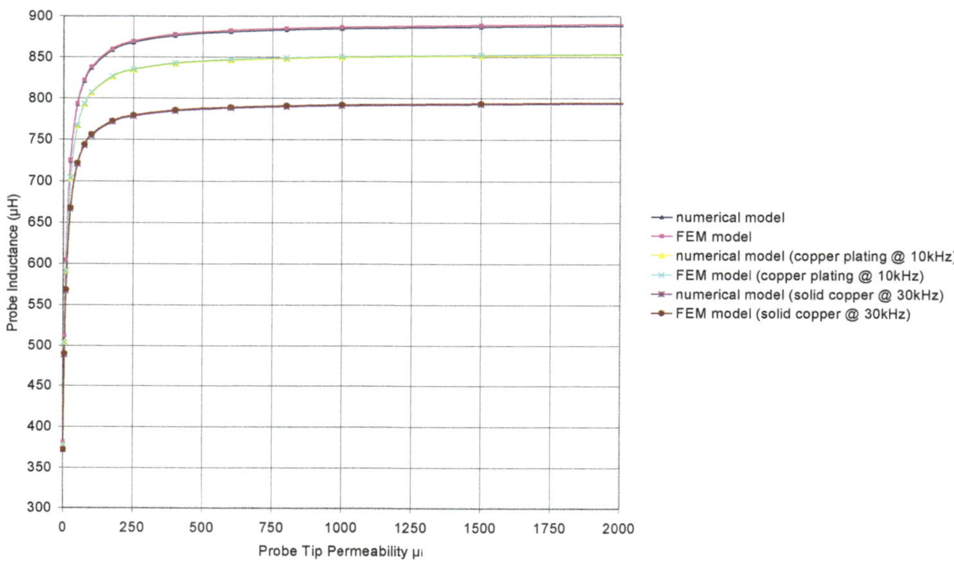

Fig. 11. Source Coil Inductance as a Function of Core Permeability.

5. Conclusion

The benefits of the magnetic moment method developed in this work are:

- Only points on the closed surface bounding the sensor core are discretised.
- The use of a spline function to replace the material dependent matrices $V(M,1)_{12}/V(M,1)_{22}$ of Cheng, Dodd and Deeds (Equation 7), allows for a potentially infinite number of stratified layers to be used to represent non-linear material profiles, with no penalty in terms of computation time or accuracy.
- Separating the basis functions into free-space static and substrate dependent dynamic terms, allows for more efficient computation of magnetic fields on the equivalent boundary surface. Static field components (free space components) only need to be computed once for any given simulation session.
- For a given lift-off and material profile function $m(\alpha)$, probe tip permeability μ_i can be varied without the need to recalculate basis function fields.

Future work entails implementing a Galerkin method of weighted residuals to replace the current collocation method and conducting detailed tests on non-linear material profiles.

6. References

Balanis, C. A. (1989). *Advanced Engineering Electromagnetics,* John Wiley & Sons, ISBN 0-471-50316-9, New York

Becker, R.; Dobmann, G.; & Rodner, C. (1988) *Quantitative eddy current variants for micromagnetic microstructure multiparameter analysis 3MA.* In: *Review of Progress in Quantitative Non-Destructive Evaluation 7B,* edited by Thompson, D.O. & Chimenti, D. E. pp.1703 - 1707, Plenum Press, New York

Blitz, J. (1991). *Electrical and Magnetic Methods of Non-destructive Testing,* 1st edition, Adam Hilger, ISBN 0-7503-0148-1

Bowler, N. (2006). *Frequency dependence of relative permeability in steel.* In: Review of Progress in Quantitative Non-Destructive Evaluation, Vol. 25, edited by Thompson, D.O. & Chimenti, D. E. pp.1269-1276, CP820, AIP, ISBN 0-7354-0312-0

Burden, R. L. Faires, J. D. (1989). *Numerical Analysis,* 4th Edition, Boston: PWS-KENT Publishing Company, ISBN: 0-534-98059-7

Cheng, C. C.; Dodd, C. V. & Deeds, W. E. (1971). *General Analysis of Probe Coils Near Stratified Conductors,* International Journal of Non-destructive Testing, Vol. 3, pp. 109-130

Dodd, C. V. & Deeds, W. E. (1968) 'Analytical Solutions to Eddy-Current Probe-Coil Problems' Journal of Applied Physics, Vol. 39, No. 6. pp 2829-2838

Glorieux C.; Moulder, J.; Basart, J. & Thoen, J. (1999). *The Determination of Electrical Conductivity Using Neural Network Inversion of Multi-frequency Eddy Current Probe Data,* Journal of Physics D. Vol. 32, pp. 616-622

Harrison, D. J.; Jones, L. D. & Burke, S. K. (1996). *Benchmark problems for defect size and shape determination in eddy-current non-destructive evaluation,* Journal of Non-Destructive Evaluation, Vol. 15, No. 1, pp. 21-34

Langhill, T. J. (1999). *Painting Over Hot-Dip Galvanised Steel,* Materials Performance, pp. 44-49, December 1999

Moulder, J.; Uzal, E. & Rose, J. H. (1992). *Thickness and Conductivity of Metallic Layers from Eddy Current Measurements,* Review of Scientific Instruments, Vol. 63, No. 6, pp 3455-3465.

Norton, S. J. & Bowler, J. R. (1993). *Theory of Eddy Current Inversion,* Journal Of Applied Physics, Vol. 73, No. 2, pp. 501-512

Sadiku, M. N. O. (1992). *Numerical Techniques in Electromagnetics,* Florida: CRC Press, ISBN: 0-8493-4232-5

Stephenson, G. (1973). Mathematical Methods for Science Students' 2nd Edition, Published London Longman

Uzal, E.; Moulder, J. C.; Mitra, S. & Rose, J. H. (1993). *Impedance of Coils over Layered Metals with Continuously Variable Conductivity and Permeability: Theory and Experiment,* Journal of Applied Physics, Vol. 74, No. 3, pp. 2076-2089

Yildir, Y. B.; Klimpke, B. W. & Zheng, D. (1992). *A computer program for 2D/RS eddy current problem based on boundary element method,* Available from: http://www.integratedsoft.com/papers/techdocs/tech_2ox.pdf

10

Finite Element Method Applied to the Modelling and Analysis of Induction Motors

M'hemed Rachek and Tarik Merzouki
University Mouloud Mammeri of Tizi-Ouzou
Algeria

1. Introduction

During the past decades, the development of solution methods and the growth of computer capacities have made its possible to solve more and more involved magnetic field problems. Thus, numerical techniques essentially based on the Finite Elements Method (FEM) have been used and has gradually become a standard in electrical machine modelling-design, analysis and optimisation. Electrical machines are electromagnetic devices with combined constrains such as complex geometries and several physical phenomena's. To model them, we must solve the magnetic field non-linear Partial Differential Equation (PDE) derived from the Maxwell's equations combined to the materials properties, and their coupling with phenomena that exist in electromagnetic structures, such as electric circuits, and mechanical motional equations. (Arkkio, 1987; Benali, 1997).

Induction Motor (IM) is an electromagnetic-mechanical actuator where strongly interacts several phenomena such as magnetic field, electrical circuits, mechanical motion. The aim of this chapter is to present an implementation of the finite element method for the modelling of rotating electrical machines, especially the squirrel cage three-phase induction motors. The generalized model consists firstly on strong coupling between the partial differential equation of the magnetic field diffusion and the electric circuits equations obtained from Kirchhoff laws. The model integrates as well realistic geometries, and the non-linear properties of the magnetic materials, as voltage supply of the stator windings. Secondly, the mechanical equation including the rotor movement effects is coupled to the electromagnetic phenomenon through the magnetic force responsible of the rotor motion.

The governing magnetic field time-dependent equation derived from Maxwell formalism is expressed in term of Magnetic Vector Potential (MVP) with only z-direction component for the cases of two dimensional (x,y) cartesian coordinates. The induction motors stator windings are usually in star or delta connection, then the source term of the magnetic field is explicitly an applied line voltage or implicitly the magnetizing current. The squirrel rotor cage is formed by massive conductive bars short-circuited at their ends through massive and conductive end-rings. Mathematically, the squirrel rotor cage appear as a polyphases circuits modelled by the same way that the stator windings but with affecting a zero voltage for each adjacent bars with theirs end-rings portion.

Generally, permittivity and conductivity can be considered as constants, however the magnetic reluctivity of the core ferromagnetic materials depend on the magnetic flux density intensity which is implicitly fixed by the voltage excitation or currents level at each step time of the motor operating. This magnetic flux density-reluctivity non linear dependence is take into account in the model by the classical iterative Newton-Raphson method. The finite element formulation of the non-linear transient coupled magnetic field-electric circuits of induction motors model leads to an algebraic differential equations system. The solution process requires firstly a major loop concerning the time-discretization using the effectiveness Cranck-Nicholson scheme, and secondly for each step time we have to unsure the minor loop convergence of the Newton-Raphson algorithm for determining the appropriate magnetic reluctivity values.

The time stepping finite magnetic field-electric circuits coupled model is sequentially coupled with the mechanical equation of the rotor motion. The interaction between stator and rotor flux densities generate an electromagnetic torque responsible of the motion. Since the physical position of the moving part of the induction motors will change at each time step, the finite element coefficients matrix are consequently changes. The unknown mechanical position of the moving part can be found after solving the mechanical motional equation by the fourth order Rung-Kutta method. To take the movement into account, several strategies have been proposed, the boundary integral method, the air-gap-element, and the connecting meshes through the sliding line, moving band, and the Lagrange multipliers or nodal interpolation techniques (Dreher, et al., 1996).

A particularly elegant and accurate method is that due to (Abdel-Razek, et al., 1982) named the Air-Gap Element (AGE). The air gap element consist on the coupling between the meshes of the stator and rotor through the unmeshed air-gap band. The air-gap appears such as a multi-nodes finite element (Macro-Element) where it corresponding Laplace equation solution leads to an analytical expression of the magnetic vector potential. The combination between the magnetic vector potential of the air gap interfaces leads to a macro-element matrix. At each displacement step the rotor movement is simulated through only new computation of the air-gap element matrix, then the rotor implicitly moved without any changes on the motor mesh topologies. In addition, since the magnetic vector potential is derived from the field analytical solution in the air-gap, the magnetic flux density can be directly deduced permitting an accurate calculation of the electromagnetic torque using the Maxwell stress tensor method.

The magnetic-electric model obtained from the strong coupling of electric circuit equations of stator windings and polyphases rotor squirrel cage and the magnetic vector potential diffusion equations of the magnetic field, are solved using the nodal based finite element method with step-by-step algorithm. Finally, the magnetic vector potential, stator windings currents and bars voltages differences are the unknown variables. The studied simulation concerns different operating modes such as electrical transients where the speed is constant for no-load and nominal conditions, and the general no-load and loaded electro-magneto-mechanical transient mode. Despite complex mathematical background, the simply and detailed presentation of the model offers an important aid for students, teachers and industrial employers for understanding the basis in simulation of electrical machines and particularly induction motors.

2. Magnetic field native equations

The theory of electromagnetic electrical machines modeling is described by the time-space differential Maxwell's equations where the displacement current are neglected because of the low frequency oft he supply source (Joao, et al., 2003; Binns, et al.,1994; Arkkio, 1987):

$$\nabla \times H = J \tag{1}$$

$$\nabla \times E = -\frac{\partial B}{\partial t} \tag{2}$$

$$\nabla \cdot B = 0 \tag{3}$$

Moreover the electric and magnetic fields quantities are related with the material properties expressed by the following constitutive relations:

$$H = \upsilon\left(B^2\right) \cdot B \tag{4.a}$$

$$J = \sigma \cdot E \tag{4.b}$$

Where $\upsilon\left(B^2\right)$ ist he magnetic reluctivity, σ the electric conductivity, H is the magnetic field, and J the conduction current density.

In the frequency domain and time-dependence with taking into account the eddy current, through (2) and (3) the electric field E and the magnetic flux density B are expressed using the magnetic vector potential A and scalar electric potential U^r, such as :

$$E = -\frac{\partial A}{\partial t} + \nabla U^r \tag{5}$$

$$B = \nabla \times A \tag{6}$$

Two types of conductors are considered in the field model parts. A solid conductor corresponds to a massive part of conductive material in the computational domain, whereas a stranded conductor models and thereby assumes the current to be homogeneously distributed along the cross-section of the coil. The positive or negative direction of the current is fixed by the unit vector $d = \pm 1$, as follow.

$$J = \begin{cases} d\dfrac{N_{cn} I_n^s}{S_n} & \text{Stranded stator conductors} \\ -\sigma\dfrac{\partial A}{\partial t} + \sigma\left(\nabla U_m^r\right) & \text{Solid conductors rotor bars} \end{cases} \tag{7}$$

To formulate the magnetic field problem, we consider a two-dimensional domain partitioned into electrically conducting and non-conducting regions as shown in Fig.1. This domain represents for instance the cross-section of an induction motor with length L_δ. The conducting regions are the cross-sections of stranded stator windings conductors Ω_s and solid conductors Ω_b of the rotor bars, the non-conducting ferromagnetic region Ω_{core}, and the air gap region by Ω_{air}.

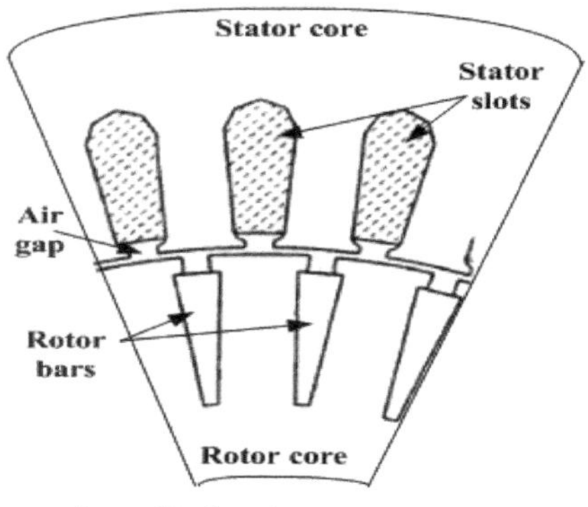

Fig. 1. Geometrical configuration of induction motors

To develop the mathematical model in term of magnetic vector potential in the three-phase induction motor, it is assumed that the magnetic field lies in the cross-sectional two-dimensional (x, y) plane. Hence, only the z-component of the induced current and the magnetic vector potential can be considered. It also assumes that magnetic material of the motor cores is non-linearly isotropic. The magnetic property of the laminated iron cores is modelled by Marrocco approximation of the reclutivity (Brauer, et al, 1985; Hecht, et al., 1990), which is a single-valued nonlinear function of the flux density B, thus exclude the effect of magnetic hysteresis from the analysis.

The fundamental equations obtained from (1)-(6) and describing the time-space variation of the magnetic vector potential with the component $A = (0, 0, A_z(x, y, t))$ has the following form

$$\frac{\partial}{\partial x}\left(\upsilon(B^2)\frac{\partial A_z(x,y,t)}{\partial x}\right) + \frac{\partial}{\partial y}\left(\upsilon(B^2)\frac{\partial A_z(x,y,t)}{\partial y}\right) = -d\frac{N_{cn}I_n^s}{S_n} + \sigma\left(-\frac{\partial A_z(x,y,t)}{\partial t} + \frac{U_m^r}{L_\delta}\right) \quad (8)$$

In the model of an electrical machine, the magnetic field due to the currents in the coils. However, it is often more appropriate to model the feeding circuit as a voltage source, which leads to the combined solution of the magnetic field and circuit equations. The stator phase windings are generally modelled as filamentary conductors, and the rotor bars are modelled as solid conductors with eddy currents.

3. Electric circuits equations model

The computational model of the induction motors can be greatly improved by coupling the circuit equations of the stator and rotor windings with the two-dimensional field equation (8). In the circuit equations, the dependence between current and voltage is solved and the circuit quantities are coupled with the magnetic field by means of flux linkage. Also, the end-windings outside the core region are modelled by including an additional inductance in the circuit model (Hecht, et al., 1990; Kanerva, 2005; Piriou, et al., 1990).

3.1 Electric circuits equations of the stator windings

The delta and star connection (see Fig. 2 and Fig.3) are the two commonly used ways to connect the stator windings. In the delta connection the potential differences induced in the stator windings are equal to the line voltages. In the star connection with neutral point, the potential differences of the stator windings are equal to the phase voltages.

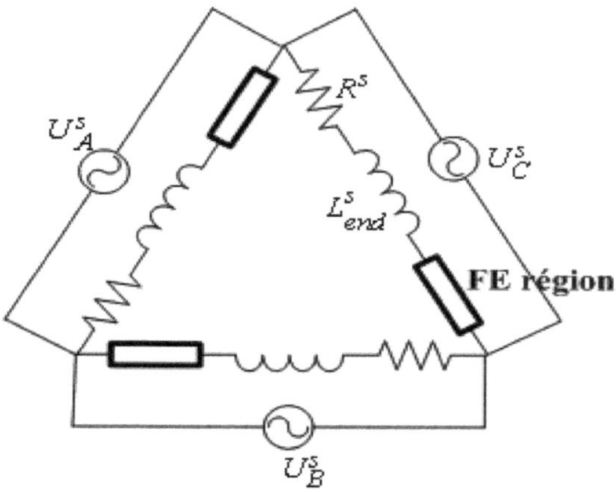

Fig. 2. Stator windings in delta connection.

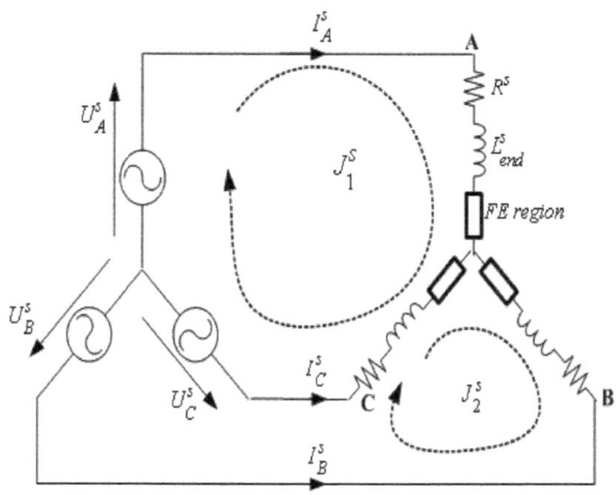

Fig. 3. Stator windings in star connection.

The three phases stator circuit equations are in matrix form:

$$U^s(t) = E^s(t) + R^s I^s(t) + L^s_{end} \frac{dI^s(t)}{dt} \tag{9}$$

$$E_n^s(t) = N_s\left(\frac{L_\delta N_{cn}}{S_n}\right)\sum_{n=1}^{N_{c_n}}\left(\iint_{\Omega_{s_n}^+}\left(\frac{\partial A}{\partial t}\right)d\Omega - \iint_{\Omega_{s_n}^-}\left(\frac{\partial A}{\partial t}\right)d\Omega\right) \qquad n = A,B,C \tag{10}$$

Where A,B,C denote the three stator phase, Ω_s^+ and Ω_s^- are respectively, the cross-sectional areas of the "go" and "return" side of the phase conductors. The column vectors of the potential differences of the stator windings with their currents and electromotive force are detailed as follows:

$$U^s(t) = \begin{Bmatrix} U_A^s(t) \\ U_B^s(t) \\ U_C^s(t) \end{Bmatrix}, \quad E^s(t) = \begin{Bmatrix} E_A^s(t) \\ E_B^s(t) \\ E_C^s(t) \end{Bmatrix}, \quad I^s(t) = \begin{Bmatrix} I_A^s(t) \\ I_B^s(t) \\ I_C^s(t) \end{Bmatrix},$$

$$R^s = \begin{pmatrix} R_A^s & 0 & 0 \\ 0 & R_B^s & 0 \\ 0 & 0 & R_C^s \end{pmatrix}, \quad L_{end}^s = \begin{pmatrix} L_{end_A}^s & 0 & 0 \\ 0 & L_{end_B}^s & 0 \\ 0 & 0 & L_{end_C}^s \end{pmatrix}$$

When the stator windings has star connection with non-connected neutral star point (see Fig. 3), only two from the three phase currents are independent variables, and the third is determined by an additional constraint which unsure a zero sequence of the phases currents $I_C^s = -I_A^s - I_B^s$. For this reason the connectivity matrix is formed:

$$[K] = \begin{bmatrix} 1 & 0 & -1 \\ 0 & 1 & -1 \end{bmatrix} \tag{11}$$

The line voltages V^s and loops currents $J_{1,2}^s$ containing the two independent currents, are formed in the following way:

$$K\begin{Bmatrix} U_{AN}^s \\ U_{BN}^s \\ U_{CN}^s \end{Bmatrix} = \begin{bmatrix} 1 & 1 & 0 \\ 0 & 1 & 0 \end{bmatrix}\begin{Bmatrix} U_{AB}^s \\ U_{BC}^s \\ U_{CA}^s \end{Bmatrix} = Q^s\{V^s\} \tag{12.a}$$

$$\{I^s\} = [K]^{tr}\begin{Bmatrix} J_1^s \\ J_2^s \end{Bmatrix} = [K]^{tr}\begin{Bmatrix} I_A^s \\ I_B^s \end{Bmatrix} = [K]^{tr}\{J_{1,2}^s\} \tag{12.b}$$

3.2 Electric circuits equations of the rotor cage

A network of the non-skewed rotor cage is shown in Fig. 4. For normal operating frequencies (50 or 60 Hz), the inductive component of the inter-bar impedance can be neglected. Two adjacent bars are connected by the end-ring resistances and inductances (Arkkio, 1987; Benali, 1997; Ho, et al., 2000).

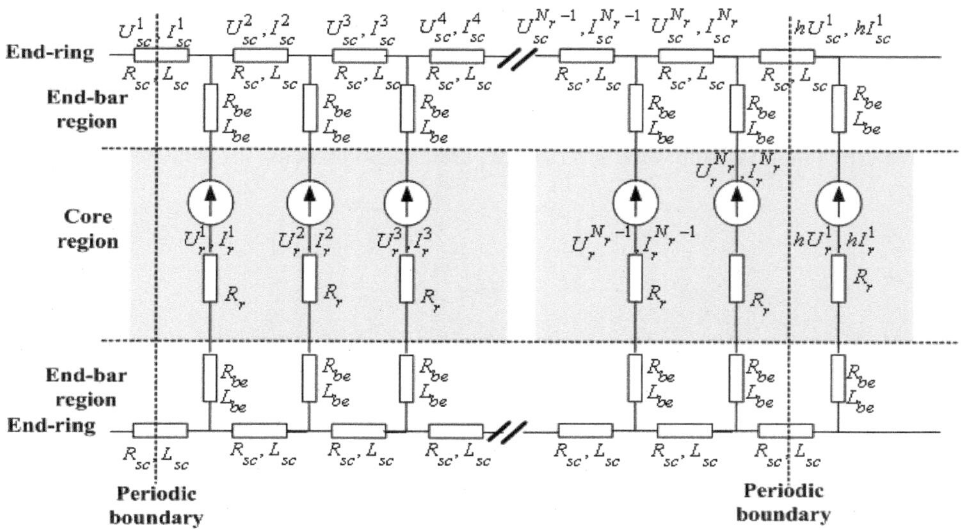

Fig. 4. Electric circuit configuration of squirrel rotor cage.

In a cage rotor, each rotor bar requires its own equation. In time variation, the potential difference induced in the mth rotor bar is given by:

$$U_m^b = R_r I_m^r + R_r \iint_{\Omega} \beta_m^r \sum_{j=1}^{N_d} \left(N_j \frac{\partial A_j}{\partial t} \right) d\Omega \qquad m = 1,\dots,N_b \tag{13}$$

$$\beta_m^r = \begin{cases} 1 & \text{if } (x,y) \text{ belongs to rotor bar } m \\ 0 & \text{Otherwise} \end{cases} \tag{14}$$

Where R_r is $(N_b \times N_b)$ unit matrix.

Integration of the current density in a rotor bar over its cross section S_b gives the total currents of the mth bar. When constant conductivity and uniform cross section area S_b are assumed in the bar and the end-bar, $(N_b \times N_b)$ unit end-bars self inductance L_{be}, and resistance R_{be} matrices are included, the above equation (13) for each bar can be expressed by (15) and (17). All the rotor bars are connected by short-circuit rings in both ends of the rotor core (16). This is taken into account by defining the end-ring unit resistance matrix and the end-ring unit inductance matrix.

$$U_m^r = R_{be} I_m^b + L_{be} \frac{dI_m^b}{dt} + U_m^b \tag{15}$$

$$U_m^{sc} = R_{sc} I_m^{sc} + L_{sc} \frac{dI_m^{sc}}{dt} \tag{16}$$

$$I_m^r = \sigma\left(\iint_{\Omega_b}\left(-\frac{\partial A}{\partial t} + \frac{U_m^r}{L_\delta}\right)d\Omega\right) \tag{17}$$

From Kirchhoff's second law applied to the rotor cage electric circuit (Fig.4), a relation between the potential difference and currents of bars and end-ring are obtained such as:

$$2U^{sc} = M \cdot U^b \tag{18.a}$$

$$I^b = M^{tr} \cdot I^{sc} \tag{18.b}$$

Where M ist he rotor cage connection matrix, and h is the periodicity factor h (+1 if periodic and -1 if non-periodic).

$$M = \begin{bmatrix} 1 & 0 & . & . & . \\ 0 & 1 & . & . & . \\ . & . & . & . & . \\ . & . & . & . & 0 \\ . & . & . & 0 & 1 \end{bmatrix} + \begin{bmatrix} 0 & 0 & . & . & . \\ -1 & 0 & . & . & . \\ . & -1 & . & . & . \\ . & . & . & . & 0 \\ . & . & . & -1 & 0 \end{bmatrix} + \begin{bmatrix} 0 & 0 & . & . & -h. \\ 0 & 0 & . & . & . \\ . & . & . & . & . \\ . & . & . & . & 0 \\ . & . & . & 0 & 0 \end{bmatrix} \tag{19}$$

4. Magnetic field – Electric circuits coupling

4.1 Time stepping finite element formulation of the non-linear magnetic field model

In the electric machine model, the magnetic field in the iron core, windings and air gap is solved by the two-dimensional finite element code and coupled with the voltage equations of the stator and rotor windings. The model is based on the direct coupling, which means that magnetic field equations and electrical circuit equations are solved simultaneously by time-stepping approach with handling magnetic non-linearities using Newton-Raphson iterative algorithm.

In the time stepping formulation, the derivative of the vector potential, stator windings currents and bars voltages are approximated by first-order difference ratios:

$$\frac{\partial}{\partial t}\begin{Bmatrix} A \\ I_n^s \\ U_m^s \end{Bmatrix} = \frac{1}{\Delta t}\left(\begin{matrix} A \\ I_n^s \\ U_m^r \end{matrix}\Bigg|_{k+1} - \begin{matrix} A \\ I_n^s \\ U_m^r \end{matrix}\Bigg|_{k}\right) \tag{20}$$

The time discretization is performed by using the Crank-Nicholson sheme as:

$$\begin{Bmatrix} A \\ I_n^s \\ U_m^r \end{Bmatrix}_{k+1} = \frac{1}{2}\left\{\begin{matrix} \frac{\partial A}{\partial t} \\ \frac{\partial I_n^s}{\partial t} \\ \frac{\partial U_m^r}{\partial t} \end{matrix}\Bigg|_{k+1} + \begin{matrix} \frac{\partial A}{\partial t} \\ \frac{\partial I_n^s}{\partial t} \\ \frac{\partial U_m^r}{\partial t} \end{matrix}\Bigg|_{k}\right\}\Delta t + \begin{Bmatrix} A \\ I_n^s \\ U_m^r \end{Bmatrix}_{k} \tag{21}$$

Several methods can be used for the numerical solution of the magnetic field equation (8), such as reluctance networks, the boundary element method, the finite difference method or the finite element method. In this work, the numerical analysis is based on the finite element method. The two-dimensional geometry is covered by a finite element mesh, consisting of first-order triangular elements. If possible, the cross section of the electrical machine is divided in N_s symmetry sectors, from which only one is modelled by the finite element method and symmetry constraints are set on the periodic or anti-periodic boundary (Nougier, 1999; Binns, et al., 1994; Ho, et al., 1997). The magnetic vector potential can be approximated as the sum of the element shape functions times and nodal potential values:

$$A_z(x,y,t) = \sum_{j=1}^{N_{nodes}} N_j(x,y) \cdot A_{zj}(x,y,t) \tag{22}$$

Where N_{nodes} is the total nodes number of the finite element mesh, $N_j(x,y)$ the shape function, and $A_{zj}(x,y,t)$ is the magnetic vector potential of the node j .

The numerical field equation is derived by Galerkin's method, where (8) is multiplied by shape functions and integrated over the whole finite element mesh with substituting the magnetic vector potential approximation (22). The last line integral term of the formulation (23) correspond to the air-gap contribution due to the rotor movement.

$$\iint_\Omega \sum_{j=1}^{N_{nodes}} \left[\upsilon(B^2)(\nabla N_i)\cdot(\nabla N_j)\{A_j\} + \left(\sigma N_i N_j \frac{\partial\{A_j\}}{\partial t} \right) \right] d\Omega +$$

$$\iint_\Omega \left\{ \sum_{j=1}^{N_{nodes}} \left(-\sigma N_i \left(\beta_m^s U_m^r\right) - N_i \left(\beta_n^s I_n^s\right) \right) \right\} d\Omega = \tag{23}$$

$$\oint_\Gamma \upsilon_o \left(\frac{\partial \sum_{j=1}^{nnt} N_j^{AGE} A_j^{AGE}}{\partial n} \right) d\Gamma$$

The problems in the analysis of the electrical machine are almost non-linearly isotropic due to the presence of ferromagnetic materials. The magnetic permeability is non-homogeneous and will be a function of the local magnetic field which is unknown at the start of the problem. The permeability is low at very low flux densities, rises quickly as the flux density increases and then decreases in the saturation region. As the permeability is unavoidably contained in all of the element stiffness matrices, an iterative process must be used to keep correcting the permeability until it consistent with the field solution. The Newton-Raphson iterative technique is used for the analysis of the non-linear problem (Brauer, et al., 1985; Joao, et al.,2003; Neagoe, et al., 1994).

At the beginning, an unsaturated value of permeability is assigned for each element of the mesh. When solving the problem, the magnitude of the flux density in each element is computed and the magnetic reluctivities are corrected to be consistent with the computed values of the flux density. The problem is then solved again using the new values. This process is continued till a satisfactory result is obtained when the difference between the actual solution and the previous one is smaller than a pre-specified value. The equations for the time-stepping simulation are derived by adding the equations from two successive steps together and replacing the derivatives with expressions (20) and (21). Using this approach, the magnetic vector potential integrals formulation (23) are formed for each node in the finite element mesh. Following, a residual vector $\Re^f$ is obtained after the finite element discretization, and the ith element of the residual vector is:

$$\begin{aligned}
\Re_i^f\left(A_{k+1},U_{n_{k+1}}^r,I_{nk+1}^s\right) = & \iint_\Omega \left\{\sum_{j=1}^{N_n}\left(\nu(A_{k+1})(\nabla N_i)(\nabla N_j)+\frac{2\sigma}{\Delta t}N_iN_j\right)A_j\Big|_{k+1}\right\}d\Omega \\
& -\iint_\Omega \left\{N_i\frac{\sigma}{L_\delta}\sum_{m=1}^{N_b}\beta_m^r\left(U_m^r\Big|_{k+1}\right)+N_i\sum_{n=A,B,C}\beta_j^s\left(I_n^s\Big|_{k+1}\right)\right\}d\Omega \\
& +\iint_\Omega \left\{\sum_{j=1}^{N_n}\left(\nu(A_k)(\nabla N_i)(\nabla N_j)-\frac{2\sigma}{\Delta t}N_iN_j\right)A_j\Big|_{k+1}\right\}d\Omega \\
& -\iint_\Omega \left\{N_i\frac{\sigma}{L_\delta}\sum_{m=1}^{N_b}\beta_m^r\left(U_m^r\Big|_k\right)+N_i\sum_{n=A,B,C}\beta_n^s\left(I_n^s\Big|_k\right)\right\}d\Omega - \oint_\Gamma \upsilon_o\frac{\partial}{\partial n}\left(\sum_{j=1}^{nnt}N_j^{AGE}\cdot A_j^{AGE}\right)d\Gamma
\end{aligned} \tag{24}$$

In matrix form, equation (24) can be written as follows:

$$\begin{aligned}
\Re_i^s\left(A_{k+1},U_{nk+1}^r,I_{nk+1}^s\right) = & \left[S(A_{k+1})+M(A_{k+1})+S^{AGE}(A_{k+1})\right]\{A_{k+1}\}+\left(D^r\right)^T U_{k+1}^r \\
& +\left(D^{sT}K^T\right)\left(J_{1,2}^s\right)_{k+1}+\left[S(A_k)+M(A_k)+S^{AGE}(A_k)\right]\{A_k\}+\left(D^r\right)U_k^r+\left(D^{sT}K^T\right)\left(J_{1,2}^s\right)_k
\end{aligned} \tag{25}$$

4.2 Time stepping finite element formulation of the stator windings equations

The same approximation (20), and (22) is also applied to the winding equations (9) and (10). The resulting equations of the average value of the potential difference at the time steps k and $k+1$ is used to approximate the true potential difference as:

$$U_n^s\Big|_{k+1} = \left(R^sI_n^k+L_{end}^s\frac{dI_n^s}{dt}\right)\Bigg|_{k+1} + N_sL_z\int_\Omega \beta_n^s\frac{\partial A}{\partial t}\Bigg|_{k+1} d\Omega \tag{26.a}$$

$$U_n^s\Big|_k = \left(R^sI_n^s+L_{end}^s\frac{dI_n^s}{dt}\right)\Bigg|_k + N_sL_z\int_\Omega \beta_n^s\frac{\partial A}{\partial t}\Bigg|_k d\Omega \tag{26.b}$$

$$\beta_n^s = \frac{N_{cn}}{S_n}\begin{cases}-1 & \text{Negatively oriented coil}\\ +1 & \text{Positively orriented coil}\\ 0 & \text{Otherwise}\end{cases} \tag{27}$$

After the substitution of the approximation (21) in (26), the voltages equations of the nth phase of the stator windings becomes:

$$\frac{U_n^s\big|_{k+1}+U_n^s\big|_k}{2}=N_sL_z\iint_{\Omega}\beta_n^s\sum_{j=1}^{N_d}N_j\left(\frac{A_j\big|_{k+1}+A_j\big|_k}{\Delta t}\right)d\Omega+\frac{R^s}{2}\left(I_n^s\big|_{k+1}+I_n^s\big|_k\right)+L_{end}^s\frac{I_n^s\big|_{k+1}+I_n^s\big|_k}{\Delta t} \tag{28}$$

The voltages equations (28) are expressed in matrix form as follow:

$$\begin{aligned}\Re_i^s\left(A_{k+1},I_{nk+1}^s\right)= &\left(KD^s\right)A_{k+1}+\left(-\frac{R^s\Delta t+2L_{end}^s}{2N_sL_\delta}\right)KK^T\left(J_{1,2}^s\right)_{k+1}-\\ &\left(KD^s\right)A_k+\left(-\frac{R^s\Delta t-2L_{end}^s}{2N_sL_\delta}\right)KK^T\left(J_{1,2}^s\right)_k+\left(\frac{\Delta t}{2N_sL_\delta}\right)Q^s\left[\left(V_n^s\right)_{k+1}+\left(V_n^s\right)_k\right]\end{aligned} \tag{29}$$

Equation (29) can be written under this following form:

$$\begin{aligned}\Re_i^s\left(A_{k+1},I_{nk+1}^s\right)=&\left(KD^s\right)A_{k+1}+\left(G^sKK^T\right)\left(J_n^s\right)_{k+1}-\left(KD^s\right)A_k\\ &+\left(H^sKK^T\right)\left(J_{1,2}^s\right)_k+\left(C^s\right)\left[\left(V_n^s\right)_{k+1}+\left(V_n^s\right)_k\right]\end{aligned} \tag{30}$$

The different matrix components of (30) are:

$$D_{ij}^s=-\iint_{\Omega^e}\left(\beta_n^s\cdot N_j\right)d\Omega^e \tag{31.a}$$

$$\left[G^s\right]_{(3\times3)}=-\left(\frac{R^s\Delta t+2L_{end}^s}{2N_sL_\delta}\right) \tag{31.b}$$

$$\left[H^s\right]_{(3\times3)}=-\left(\frac{R^s\Delta t-2L_{end}^s}{2N_sL_\delta}\right) \tag{31.c}$$

$$\left[C^s\right]_{(3\times3)}=\left(\frac{\Delta t}{2N_sL_\delta}\right)Q^s \tag{31.d}$$

4.3 Time stepping finite element formulation of the rotor cage equations

By undertaking the same way as the stator windings, after applying Crank-Nicholson scheme, equations (13), (16) and (17) of the voltage equations of the rotor cage becomes:

$$\frac{1}{2}\left(U_m^b\big|_{k+1}+U_m^b\big|_k\right)=\frac{1}{2}R_r\left(I_m^r\big|_{k+1}+I_m^r\big|_k\right)+R_r\int_{\Omega}\beta_m^r\sigma\left\{\sum_{j=1}^{N_d}\frac{A_j\big|_{k+1}-A_j\big|_k}{\Delta t}\right\}d\Omega \tag{32}$$

$$\frac{1}{2}\left(U_m^r\Big|_{k+1}+U_m^r\Big|_k\right)=\frac{R_r}{2}\left(I_m^r\Big|_{k+1}+I_m^r\Big|_k\right)+L_{be}\frac{\left(I_m^r\Big|_{k+1}+I_m^r\Big|_k\right)}{\Delta t}+\frac{1}{2}\left(U_m^b\Big|_{k+1}+U_m^b\Big|_k\right) \quad (33.a)$$

$$\frac{1}{2}\left(U_m^{sc}\Big|_{k+1}+U_m^{sc}\Big|_k\right)=\left(\frac{R_{sc}}{2}+\frac{L_{sc}}{\Delta t}\right)\left(I_m^{sc}\Big|_{k+1}+I_m^{sc}\Big|_k\right) \quad (33.b)$$

The combination of the expressions (32) and (33) with the end-rings voltages and currents (18), lead to the matrix form of the unified loops voltages equations in the rotor cage expressed as:

$$\Re\left(A_{k+1},U_{mk+1}^r\right)=\left(D^r\right)A_{k+1}+\left(C^r\right)\left(U_m^r\right)_{k+1}+\left(-D^r\right)A_k+\left(C^r\right)\left(U_m^r\right)_k+\left(G^r\right)\left(I_m^r\right)_k \quad (34)$$

$$\left[C^r\right]_{(N_b\times N_b)}=\frac{\Delta t}{2L_\delta R_b}\times\left\{I_{N_b\times N_b}+\frac{R_b}{2}\left[\left(R_{sc}+2\frac{L_{sc}}{\Delta t}\right)I_{(N_b\times N_b)}+\left(R_{be}+2\frac{L_{be}}{\Delta t}\right)M_b\right]^{-1}M_b\right\} \quad (35.a)$$

$$\begin{aligned}G^r=\frac{\Delta t}{2L_\delta}I_{N_b\times N_b}-\frac{\Delta t}{2L_\delta}&\left[\left(R_{sc}+2\frac{L_{sc}}{\Delta t}\right)I+\left(R_{be}+2\frac{L_{be}}{\Delta t}\right)M_b\right]^{-1}\\ \times&\left[\left(R_{sc}-2\frac{L_{sc}}{\Delta t}\right)I+\left(R_{be}-2\frac{L_{be}}{\Delta t}\right)M_b\right]\end{aligned} \quad (35.b)$$

$$D_{ij}^r=-\frac{\sigma}{L_\delta}\int_\Omega\left(\beta_i^r\cdot N_j\right)d\Omega \quad (35.c)$$

Where $M_b=\left(M^{tr}\right)M$ is the auxiliary connection matrix.

4.4 Full magnetic field – Electric circuits coupling model

Combining equation (25), (30) and (34) a system of coupled equations is obtained:

$$\begin{aligned}&\begin{bmatrix}[S+M](A_{k+1}^{q+1})+S^{AGE} & \left[D^r\right]^{tr} & \left[D^s\right]^{tr}KK^{tr}\\ \left[D^r\right] & \left[C^r\right] & 0\\ K\left[D^s\right] & 0 & \left[G^s\right]\end{bmatrix}\times\begin{bmatrix}A^{q+1}\\ \left(U_m^r\right)^{q+1}\\ \left(J_{1,2}^s\right)^{q+1}\end{bmatrix}_{k+1}=\\ &-\begin{bmatrix}[S-M]\left(A_k^q\right)+S^{AGE} & \left[D^r\right]^{tr} & \left[D^s\right]^{tr}KK^{tr}\\ \left[D^r\right] & \left[C^r\right] & 0\\ K\left[D^s\right] & 0 & \left[H^s\right]\end{bmatrix}\begin{bmatrix}A^q\\ \left(U_m^r\right)^q\\ \left(I_n^s\right)^q\end{bmatrix}-\begin{bmatrix}0\\ \left(\left[G^r\right]I_m^r\right)_k\\ \left[C^s\right]\left[\left(V_n^s\right)_{k+1}+\left(V_n^s\right)_k\right]\end{bmatrix}\end{aligned} \quad (36)$$

Because of the non linearity of the core material, the stiffness matrix [S] depends on the nodal values of the magnetic vector potential. After applied the Newton Raphson iteration method, a final algebraic system of equations for the nonlinear time-stepping simulation of the electrical machine is obtained such as:

$$\begin{bmatrix} P(A_{k+1}^{q}) & \left[D^{r}\right]^{tr} & \left[D^{s}\right]^{tr} KK^{tr} \\ \left[D^{r}\right] & \left[C^{r}\right] & 0 \\ K\left[D^{s}\right] & 0 & \left[G^{s}\right] \end{bmatrix} \begin{bmatrix} \Delta A_{k+1}^{q+1} \\ \Delta\left(U_{m}^{r}\right)_{k+1}^{q+1} \\ \Delta\left(J_{1,2}^{s}\right)_{k+1}^{q+1} \end{bmatrix} = \begin{bmatrix} \Re^{f}\left(A_{k+1}^{q},\left(U_{m}^{r}\right)_{k+1}^{q},\left(J_{1,2}^{s}\right)_{k+1}^{q}\right) \\ \Re^{r}\left(A_{k+1}^{q},\left(U_{m}^{r}\right)_{k+1}^{q}\right) \\ \Re^{s}\left(A_{k+1}^{q},\left(J_{1,2}^{s}\right)_{k+1}^{q}\right) \end{bmatrix} \tag{37}$$

Where P is the Jacobian matrix system expressed through the following matrices elements (Joao, et al., 2003; Arkkio, 1987; Benali, 1997):

$$S_{ij} = S^{AGE} + \iint_{\Omega} \nu(A_{k+1})(\nabla N_i)(\nabla N_j)d\Omega \tag{38.a}$$

$$P_{ij} = S^{AGE+}S_{ij} + J_{ij} = S^{AGE} + S_{ij} + \iint_{\Omega}\sum_{j=1}^{N_n}\left(\frac{\partial \upsilon\left(A_j\right)_k^q}{\partial A_j}\right)(\nabla N_i)\cdot(\nabla N_j)d\Omega \tag{38.b}$$

$$M_{ij} = \iint_{\Omega} \frac{2\sigma}{\Delta t} N_i N_j \, d\Omega \tag{38.c}$$

5. Mechanical model and movement simulation

5.1 Movement simulation technique

For modelling the movement in rotating electrical machines using the Air-Gap-Element (AGE), the space discretised domain is commonly split up into two subdomains, a stator Ω_{stator}, and rotor Ω_{rotor} meshed domains, and the unmeshed air gap with Γ^{AGE} boundary (Fig. 5).

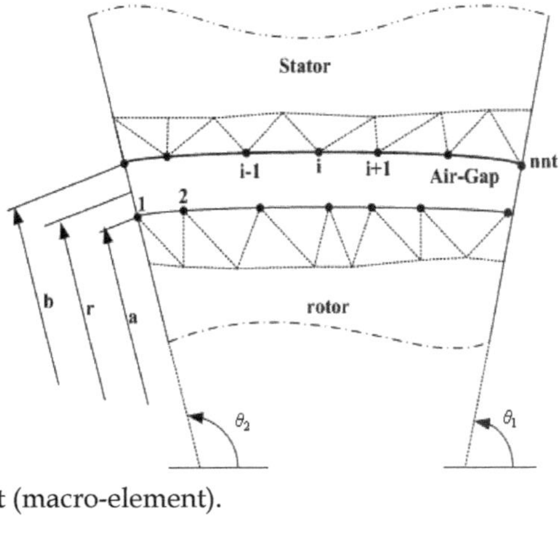

Fig. 5. Air-Gap Element (macro-element).

In a general step-by-step solution of the magnetic field in rotating electrical machines, the stator and rotor magnetic field equations are expressed in their own coordinate systems. The solutions of both fields equations are matched to each other in the air-gap. The rotor is rotated at each time step by an angle corresponding to the mechanical angular frequency, this means that a new finite element mesh in the air gap has to be constructed. The basic form of the air gap element matrix general terms is given by the expression:

$$A(a,b,r,\theta)=\sum_{j=1}^{nntl} A_i\left[\frac{1}{2}a_{0j}+\sum_{r=1}^{\infty}\left[a_{rj}\cos(\lambda_r\theta)+b_{rj}\sin(\lambda_r\theta)\right]\right] = \sum_{j=1}^{nnt}\left(N_j^{AGE}(a,b,r,\theta)\right)\cdot A_j^{\Gamma_{AGE}} \quad (39)$$

Where $\left(a_{0j},a_{rj},b_{rj},\lambda_r\right)$ are the Fourier's expansion coefficient which depends on the air-gap nodes coordinates locations, nnt is the total numbers of the air-gap nodes.

The movement simulation is take into account through the expression (39) while computing the associated matrix of the air-gap element given by the line integral term defined in the formulation (24). The air-gap-element matrix is given as follow:

$$\oint_{\Gamma^{AGE}} \upsilon_0 \sum_{j=1}^{nnt}\left\{A_j^{AGE}\right\}\frac{\partial}{\partial n}\left(N_j^{AGE}(r,\theta)\right)d\Gamma^{AGE} = \left[S^{AGE}\right]\left\{A_j^{AGE}\right\} \quad (40)$$

For a complete development of the air gap element, the reader are invited to detailed implementation given in (Abdel-Razek, et al., 1982; Joao, et al., 2003).

5.2 Mechanical equations and torque computation

In a general case the magnetic field and electric circuits equations are coupled to the rotor mechanical equation through the electromagnetic torque. This includes the interaction between mechanical and electromagnetic quantities (Ho, et al., 2000). The mechanical differential system equations of speed and angular displacement is given as follow:

$$\frac{d}{dt}\begin{bmatrix}\omega\\ \theta\end{bmatrix}=\begin{bmatrix}-\dfrac{f}{J_m} & 0\\ 1 & 0\end{bmatrix}\begin{bmatrix}\omega\\ \theta\end{bmatrix}+\begin{bmatrix}\dfrac{1}{J_m}\left(C_{em}(t)-C_{load}\right)\\ 0\end{bmatrix} \quad (41)$$

At each step time, the computed electromagnetic torque is introduced in the mechanical model (41) solved using the fourth order Rung- Kutta method to get the rotor angular displacement and speed. The electromagnetic torque is computed from Maxwell's Tensor as a function of radial and tangential components of the magnetic flux density:

$$C_{em}=\frac{pr^2L_\delta}{\mu_0}\int_{\theta_1}^{\theta_2} B_r B_\theta d\theta \quad (42)$$

Where r is the rotor external radius, and p the number of poles pairs. The magnetic flux density components $\left(B_r,B_\theta\right)$ are computed in the air gap boundaries through the derivatives expression of the analytical magnetic vector potential shape functions (39).

6. Simulation results and discussions

6.1 Algorithm of Non-linear step by step finite element solver

The theory of the previous paragraphs is applied for the simulation of an induction motor in different operating modes. Two cases are studied, the first one concern the electric transient state simulation of induction motor while considering constant speed, and the second one treat the general electromagnetic-mechanical transient state simulation. The solution process of the electromagnetic-mechanical non-linear transient model is summurised by the following chart (see Fig.6).

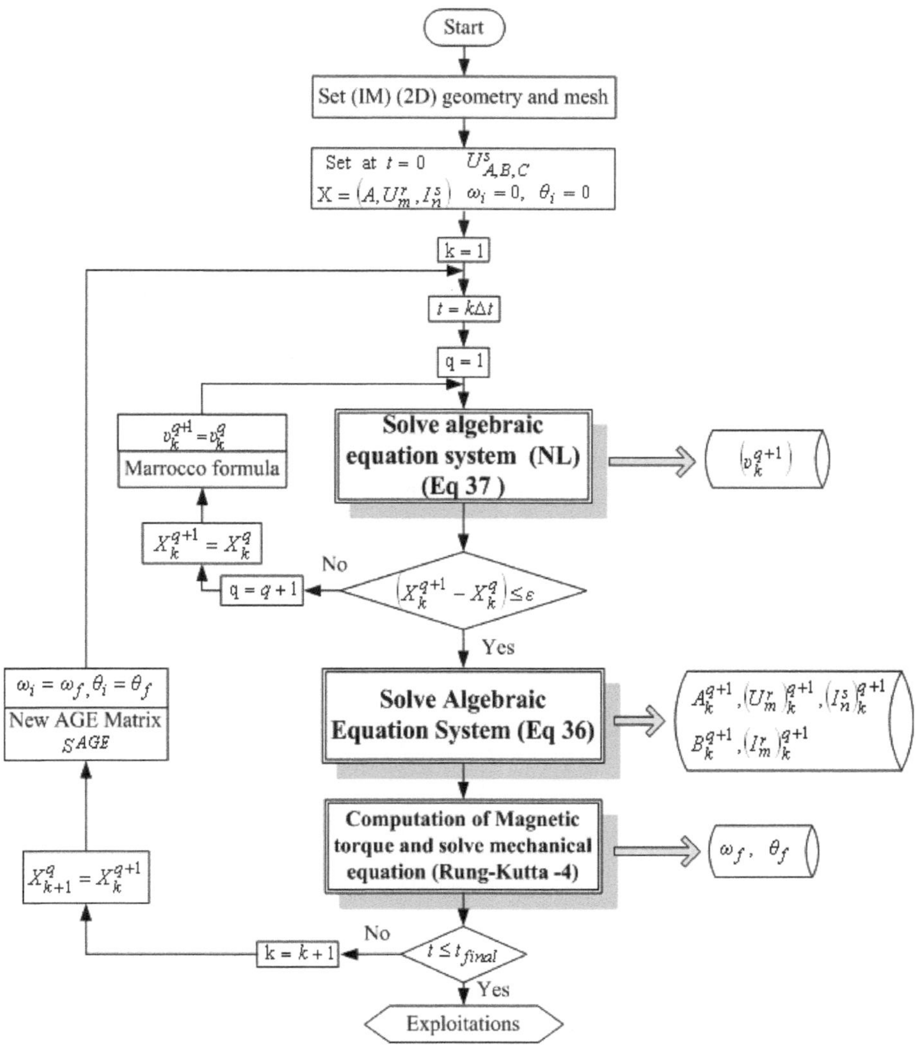

Fig. 6. General algorithm for induction machine numerical analysis.

6.2 Presentation of the studied induction motor

The considered simulated system to apply the model of the present works is an three phase induction motor Leroy Sommer (Mezani, 2004). The poles number is four, the rated power, the efficiency, and voltage are respectively 5.5KW, 84.26% and 380V. The stator windings and the rotor cage (bars-end rings) are made respectively with cooper and aluminium materials. More detailed characteristics of the motor are presented in Table. 1.

Geometrical components	Values (mm)	Physical components	Values
Stator core external diameter	168.0	Rated nominal current	11.62A
Stator core internal diameter	130.0	Nominal torque	37 N.m
Rotor core external diameter	109.2	Power factor	0.865
Rotor core inernal diameter	66.4	Moment of inertia	0.014 (Kg.m2)
Axial length	160.0	Friction coefficient	0.011 (1/ms)
Stator conductors per slot	19	Slip	4.13%
Number of stator slot	48 slots	Stator phase resistance	1.4Ω
Number of rotor slot	24 bars	Stator phase inductance	0.2 mH

Table 1. Studied induction motor geometrical and physical datas.

The stator and rotor slots geometrical dimensions are detailed in the following Fig. 7. To reduce the computational time due to nodes number of the finite element mesh and the geometrical complexity, usually the electrical machines models are created on the smallest symmetrical part of the machine. The (Fig. 8) shows the finite element mesh of the motor studied domain where only ¼ of the motors. The mesh containing 3204 nodes and 5623 first order triangular element is obtained using the Matlab PdeTool mesh automatic generator. We note that the air-gap between the stator and rotor meshes is not meshed and coupled together through the air-gap matrix (40). Homogeneous Dirichlet boundary condition is imposed for the external and internal motor radius, and anti-periodic ones at the other boundaries.

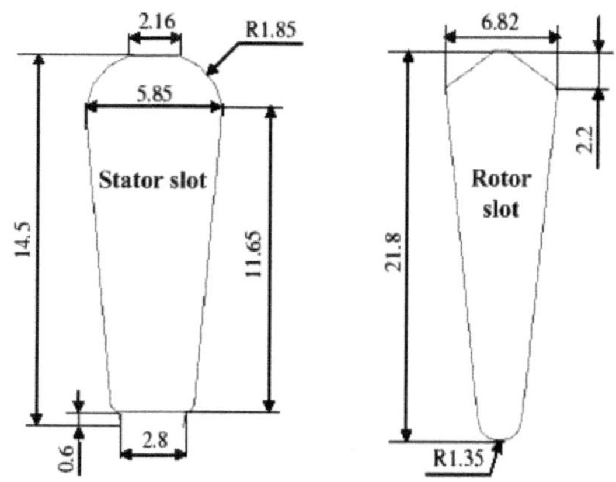

Fig. 7. Gometrical dimensions of the motor slots.

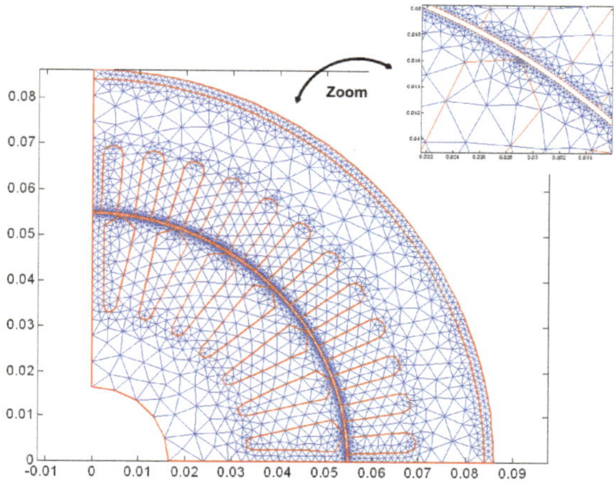

Fig. 8. Finite Element Mesh with (AGE).

6.3 Electromagnetic transient operating condition with constant speed

Results of this part concern the transient electromagnetic state simulation, at no load and nominal operation modes while the speed is considered constant. The electromagnetic transient is simulated while considering that the motor is operating in steady state with constant speed. Stator and rotor meshes are coupled by air-gap-element matrix. The constant speed value is equal to 1495 tr/mn and 1348 tr/mn respectively for the no-load and nominal modes. Since the mechanical phenomena is not considered, the speed is constant and the rotor displacement is not taken into account. The air-gap-element matrix is calculated only once. At each step time, the algebraic system (37) corresponded to Newton-Raphson algorithm is iteratively solved in order to get the magnetic permeability value. The latest is then used to establish the algebraic system (36) which the solution lead to the values of the magnetic vector potential, stator windings currents and the rotor bars voltages.

The stator windings currents wave forms corresponded to the electromagnetic transient state of the no-load and nominal conditions are respectively shown in Fig. 9 and Fig. 10. We note that, in agreement with the theory a high starting currents are obtained which decreases quickly because of the small electromagnetic durations. Electromagnetic torque for the no-load and nominal electromagnetic transient operations is given by the Fig. 11. After a brief transient duration the torque is stabilized at 4.5 N.m and 37.3 N.m values respectively for no load and nominal conditions.

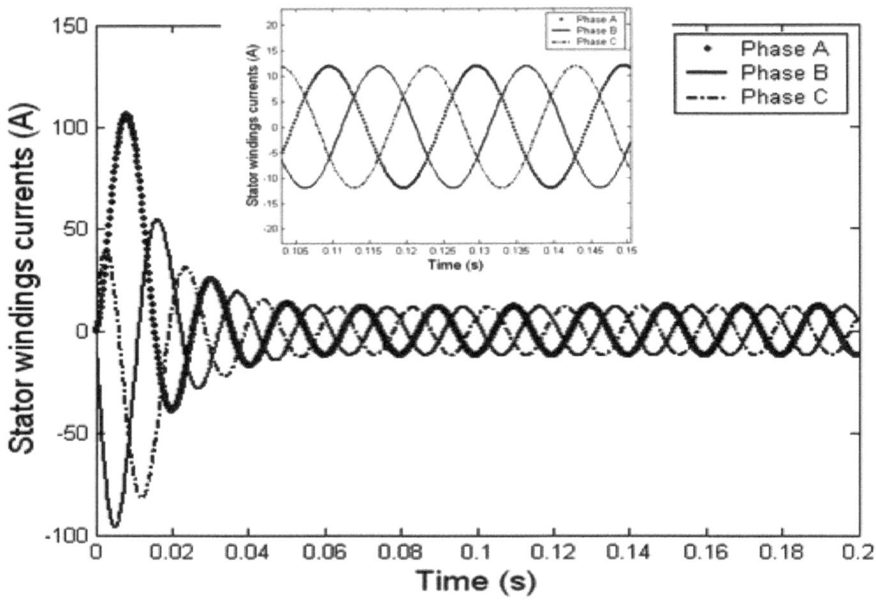

Fig. 9. Stator currents in no-load mode.

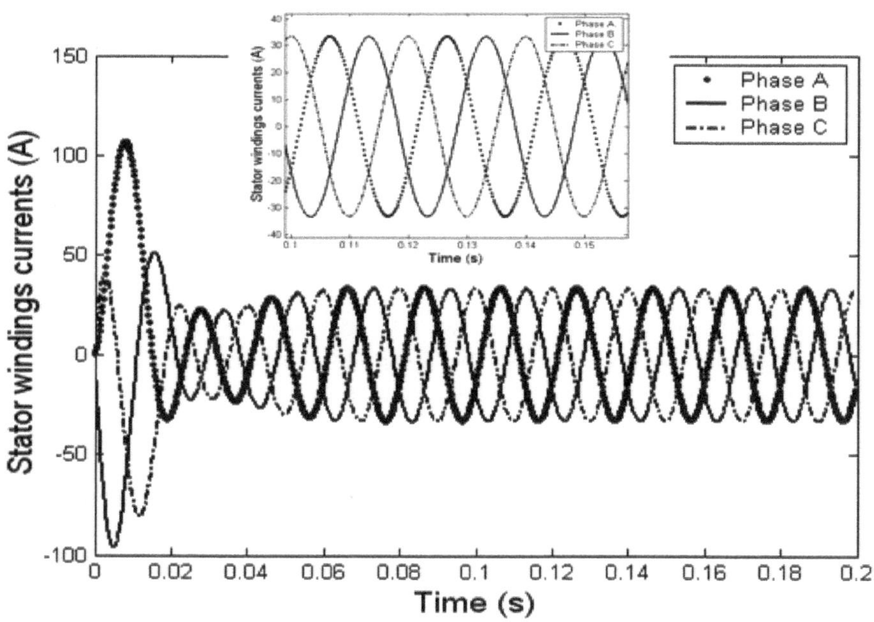

Fig. 10. Stator currents nominal mode.

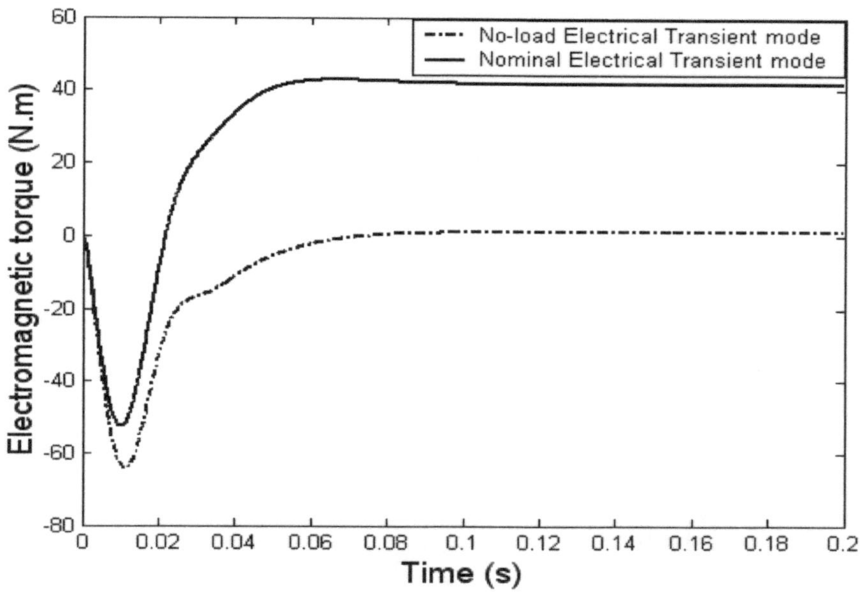

Fig. 11. Electromagnetic torque at no-load and nominal modes.

6.4 Electromagnetic-mechanical transient operating with direct start

Results of this part concern the transient electromagnetic-mechanical general simulation, at no load and load direct start operation modes. The solution process detailed by the Fig.6 is summarized by the following steps. Firstly, the algebraic equation system (37) is solved to get the magnetic reluctivity associated to the voltage level at each step time. Secondly, the algebraic equation system (36) is solved, which permit us to know the stator windings currents, the rotor bar voltages, and the magnetic vector potential which lead to deduce the magnetic flux density, and permit the computation of the electromagnetic torque. The latest is introduced in the mechanical model, which solution leads to the speed and angular displacement of the rotor. Since the mechanical phenomena is considered, the rotor displacement is taken into account, the air-gap-element matrix corresponded to each rotor position is calculated at each displacement step.

The motor simulations concerns a direct start loaded condition with load torque of 10 N.m. The stator windings currents wave forms are given by the Fig. 12 and Fig. 13. Motor angular speed, and electromagnetic torque are given by the Fig. 14, and Fig. 15, respectively.

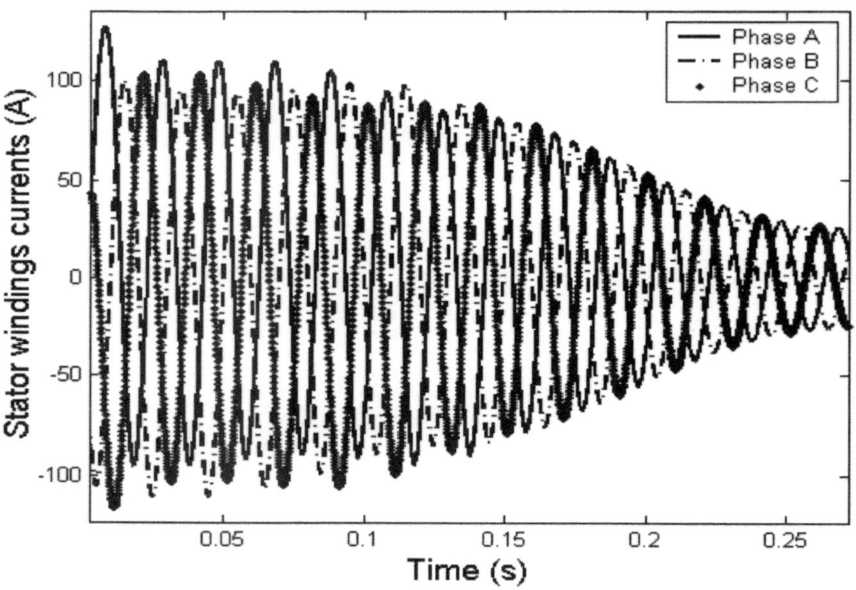

Fig. 12. Stator windings currents for direct start loaded motor.

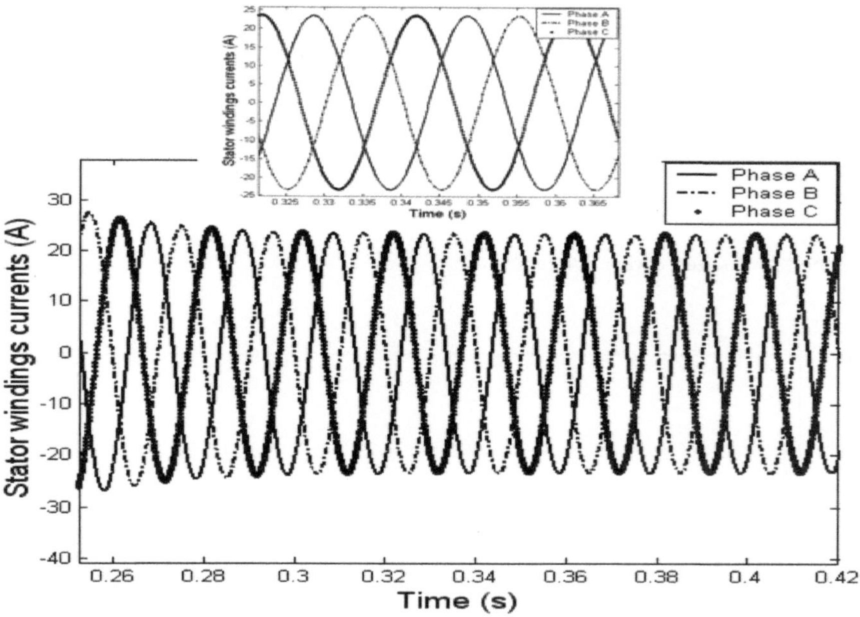

Fig. 13. Steady state stator windings currents for loaded direct start motor.

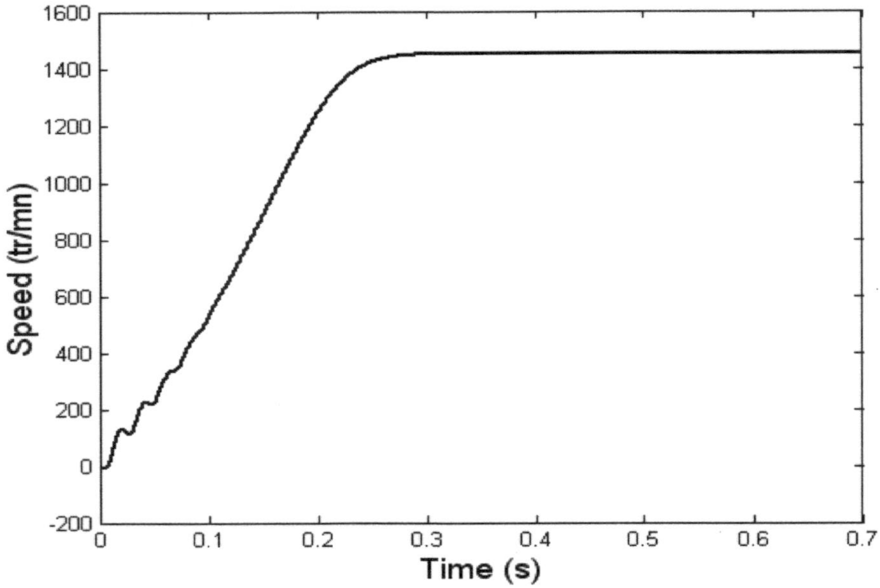

Fig. 14. Speed for loaded direct start motor.

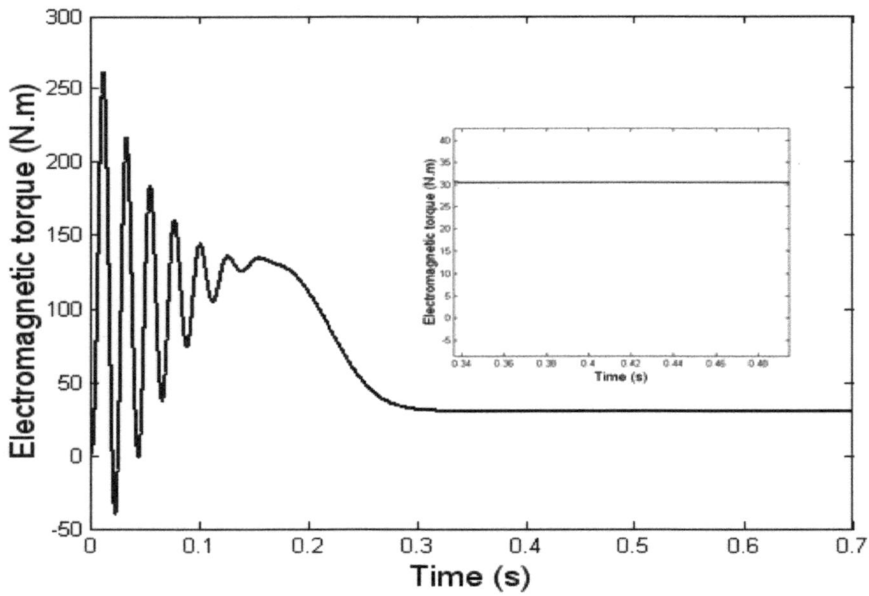

Fig. 15. Electromagnetic torque for loaded direct stat motor.

From the transient stator currents results given by the Fig. 12 we note a high starting currents which reach to nominal steady state values after average half periods such as given by the Fig. 13. The several ascillation of the currents transients behavior is due to the strong electromagnetic and mechanical interaction through theirs corresponded time constants.For the rotor speed given by the Fig.14, we note that after some modulations at the motor start, gthe speed increase linearely till it steady state values according to mechanical first order differential equation. The Fig. 15 of electromagnetic torque show that after some periods oft he magnetic duration the torque reach to it steady state value of average 37.3 N.m.

7. Conclusion

This chapter goal is to present a detailed finite element method use to solved partial differential equation of electromagnetic phenomena occurred in induction motor. The magnetic field equation expressed in term of magnetic vector potential strongly coupled with the electric circuits equations of the stator windings and rotor cage are solved using the nodal based finite element process. The resulting nonlinear time-dependent algebraic differential equations system obtained from the finite element formulation is solved using step-by-step numerical integration based on the Crank-Nicholson scheme, combined to the Newton-Raphson iterative process for handling the magnetic material non-linearity. The electromagnetic and mechanic interaction is considered firstly by the computation of the electromagnetic torque by the Maxwell stress tensor responsible of the rotor displacement, and secondly by solving the mechanical motional equation to get the new rotor angular position. Since the motor is meshed only once, the rotor movement is taking into account by the macro-element method which lead to an air-gap matrix of the movement. The validation of the model is performed through simulation of an induction motor in no-load and loaded direct start operating modes. The numerical results are in good agreement with corresponding results appearing in the recent literature. The contribution of this work can be applied to analyze a large class of electrical machines, and offers an important support for students, teachers and industrial employers for understanding the basis of numerical modelling of electrical machines.

8. References

Abdel-Razek, A.; Coulomb, J.L.; Féliachi, M. & Sabonnadière, J.C. (1982), Conception of an Air-Gap Element for the Dynamic Analysis of the Electromagnetic Field in Electric Machines, *IEEE Transaction On Magnetics,* Vol.18, No.2, (March 1982), pp. 655-659, ISSN 0018-9464.

Arkkio, A. (1987), Analysis of induction motors based on the numerical solution of the magnetic field and circuit equations. PhD Dissertation, Helsinki University of Technology, Sweeden.

Benali, B. (1997), Contribution à la modélisation des systèmes électrotechniques à l'aide des formulations en potentiel : Application à la machine asynchrone, Doctorat thesis, University of Sciences and Technology of Lille, France.

Binns, K.J.; Lawrenson, P.J, & Trowbridge, C.W. (1994), The Analytical and Numerical Solution of Electrical and Magnetic Fields, In: Wiley, (Ed.), ISBN 0471924601, Chichester, England.

Brauer, J.R. ; Ruehl, J.J, & Hirtenfelder, F. (1985), Coupled nonlinear electromagnetic and structural finite element analysis of an actuator excited by an electric circuit. *IEEE Transaction On Magnetics,* Vol.31, No.3, (May 1985), pp. 1861-1864, ISSN 0018-9464.

Dreher, T.; Perrin-Bit, R.; Meunier, G. & Coulomb, J.L. (1996), A 3D finite element modelling of rotating machines involving movement and external circuit, *IEEE Transactions on Magnetics,* Vol.32, No.4, (April 1996), pp. 1070-1073, ISSN 0018-9464.

Hecht, F.; Marrocco, A.; Piriou, F. & Abdel-Razek,A. (1990), Modélisation des systèmes électrotechniques par couplage des équations électriques et magnétiques, *Revue de Phyique Applique,*Vol.25, (July 1990), pp. 649-659, ISSN 0018-9464.

Ho, S.L.; Li, H.L.; Fu, W.N. & Wong, H.C. (2000), A novel approach to circuit-field-torque coupled time stepping finite element modelling of electrical machines. *IEEE Transactions on Magnetics,* Vol.36, No. 4, (July 2000), pp. 1886-1889, ISSN 0018-9464.

Ho, S.L., & Fu, W.N. (1997), A comprehensive approach to the solution of direct-coupled multi-slice model of skewed motors using time stepping eddy-current FEM. *IEEE Transactions on Magnetics,*Vol.33, No.3, (May 1997), pp. 2265-2273, ISSN 0018-9464.

Joao, P.; Bastos, A. & Sadowski, N. (2003), Electromagnetic Modeling by Finite Element Methods, Marcel Dekker Inc, (Ed.), ISBN 0824742699, New York, United States.

Kanerva, S. (2005), Simulation of electrical machines, circuits and control systems using finite element method and system simulation, PhD. Dissertation, Helsinki University of Technology, Sweden.

Mezani, S. (2004), Modélisation électromagnétique et thermique des moteurs à inductions en tenant compte des harmoniques d'espaces, Doctorat thesis, Polytechnical institut of Loraine, Nancy, France.

Neagoe, C. & Ossart, F. (1994), Analysis of convergence in non linear magnetostatics finite element problems. *IEEE Transactions on Magnetics,* Vol.30, No.5, (September 1994), pp. 2865-2868, ISSN 0018-9464.

Nougier, J.P. (1999), Methodes de Calcul Numeriques, In: Masson, (Ed.), Oxford University press, ISBN 0-19-511767-0, New York, United state.

Piriou, F. & Abdel-Razek, A. (1990), A model for coupled magnetic-electric circuits in electric machines with skewed slots. *IEEE Transactions on Magnetics,* Vol.26, No.2, (March 1990), pp. 1096-1100, ISSN 0018-9464.

Part 3

Mechanics and Materials

11

Numerical Modelling of Steel Deformation at Extra-High Temperatures

Marcin Hojny and Miroslaw Glowacki
AGH University of Science and Technology, Krakow
Poland

1. Introduction

Due to the globalized energy crisis and high consciousness of environmental protection in last year's, more and more products and new technology that put stress on energy preservation and environmental protection are being developed. The integrated casting and rolling technologies are newest and very efficient ways of hot strip production. Only few companies all over the world are able to manage such processes. Among them one can mention the plant located in Cremona Italy which develops the Arvedi Steel Technologies - new methods of steel strip manufacturing. They are called ISP (Inline Strip Production) and AST (Arvedi Steel Technologies) processes and are characterized by very high temperature allowed at the mill entry. The instant rolling of slabs which leave the casting machine allows the utilization of the heat stored in the strips during inline casting.

The rolling equipment for the Inline Strip Production process consist of cast rolling machine, liquid core reduction equipment, high reduction mill, inductive heater, Cremona coiling station, descaler, traditional finishing mill and the cooling zone. The initial mould strip thickness is 74 mm and is reduced to 55 mm during liquid core reduction process. The region of maximum strip temperature for a high reduction mill is placed in the strip centre and varies from 1220°C to 1375°C depending on the casting speed. The main benefits of the technology are:

- very low investment costs,
- compact rolling equipment layout - total rolling line is only 170 m long,
- no need of tunnel furnace,
- good product quality - 1 mm strip with best shape and microstructure,
- very low level of heating energy consumption which drops to 0 when the casting speed is over around 0.14 m/s,
- up to 20 times lower water consumption,
- inverse (in comparison to traditional rolling) temperature gradient in the cross-section of the strip, which is very useful for the rolling process,
- low level of installed mill power in high reduction mill (3 rolling stands with 0.5, 0.6 and 0.8 MW) providing reduction from 55 to 12.5 mm by the strip width of 1300 mm.

The AST (Arvedi Steel Technologies) technology is a result of further development of ISP to a real endless process. The main difference between these two technologies is the absence of the heating equipment in case of AST. The whole reduction process is running in one rolling mill consisting of 5 or 7 stands, which can reduce the strip thickness from 55÷70 mm to 0.8 mm. AST is the most compact hot strip production process using oscillating mould technology with excellent efficiency and cost. The total equipment length is 70 to 80 m including casting machine at the front and final coolers at the rear end. The maximal temperature of the strip occurs in central region of its cross-section and varies from 1340°C to 1420°C according to the casting speed. It suggests that the central region of the strand subjected to the rolling is still mushy.

Both the technologies mentioned above ensure huge reduction of rolling forces, very high product quality and low investment costs and their details are usually classified. The main goal of the mentioned new technologies is to significantly lower the rolling forces and to reach very favourable temperature field inside the plate in comparison with traditional processes. However certain problems specific to such a metal treatment arise. The central part of the material is still mushy. This results in changes in material density and occurrence of characteristic temperatures, which have great influence on plastic behaviour of the material. A vital problem is also the lack of data regarding material's thermal and mechanical properties and significant changes of density.

The material behaviour above the solidus line is strongly temperature-dependent. There are a few characteristic temperature values between solidus and liquidus. The Nil Strength Temperature (NST) is the temperature level at which material strength drops to zero while the steel is being heated above the solidus temperature. Another temperature associated with NST is the Strength Recovery Temperature (SRT), which is specific to cooling. At this temperature the material regains strength greater than 0.5 N/mm^2. Nil Ductility Temperature (NDT) represents the temperature at which the heated steel loses its ductility. The Ductility Recovery Temperature (DRT) is the temperature at which the ductility of the material (characterised by reduction of area) reaches 5% while it is being cooled. Over this temperature the plastic deformation is not allowed at any stress tensor configuration.

Very important for plastic behaviour of steel is also its density. It varies with temperature and depends on the cooling rate. The solidification process causes non-uniform density distribution in the controlled volume resulting in non-uniform deformation and heat conduction. There are three main factors causing density changes: solid phase formation, thermal shrinkage and movement of liquid particles inside the solid skeleton. The density plays an important role in both mechanical and thermal solutions.

The most important steel property having crucial influence on metal flow paths is the strain-stress relationship. It is not easy to run the isothermal tests that could be the source of the computation of yield stress function parameters for temperature range close to solidus line. Keeping constant temperature during the whole experiment course is difficult. There are also some difficulties with interpretation of tests results. Lack of good methods of particular metal flow simulation and significant inhomogeneity in strain distribution in the deformation zone lead to weak accuracy of standard FEM solutions.

The mathematical and experimental modelling of mushy steel deformation is an innovative topic regarding the very high temperature range deformation processes. Tracing the related papers published in the past ten years, one can find many papers regarding experimental results (Kang & Yoon, 1997; Koc et al., 1996; Kop et al., 2003) and modelling (Modigell et al., 2004; Hufschmidt et al., 2004) for non-ferrous metals tests. The papers deal mainly with tixotrophy. The first results regarding steel deformation at extra high temperature were presented in the past few years (Jing et al., 2005; Jin et al., 2002). The situation is caused by the very high level of steel liquidus and solidus temperatures in comparison with non-ferrous metals. The deformation tests for non-ferrous metals are much easier. The rising abilities of thermo-mechanical simulators enable steel sample testing and as a result both computer simulation and improvement of new rolling technologies, similar to ISP and AST processes.

The chapter sheds light on these problems. It focuses on the axial-symmetrical computer model, which ensures the possibility of its experimental verification with the help of physical simulation.

2. Thermo-mechanical model of steel deformation in semi-solid state

The numerical analysis of deformation of samples having liquid phase in their central parts shows extremely high strain inhomogeneity requiring application of hybrid analytical-numerical solution of the problem (Hojny & Glowacki, 2008, 2009, 2011). A coupled thermal-mechanical mathematical model was developed for simulation of plastic behaviour of such a species. The model is dedicated to modelling processes, which require high accuracy of resulting parameter fields. Analytical solutions of both incompressibility and mass conservation (for the mushy zone) conditions are important parts of the model. Analytical condition eliminates problems with unintentional specimen volume changes caused by application of numerical methods. The existing, physical changes of steel density in the mushy zone have influence on real variations of controlled volume. On the other hand numerical errors can be a source of volume loss which cause interference with real changes. This effect is very undesirable in modelling of thermal-mechanical behaviour of steel in temperature range characteristic for the transformation of state of aggregation.

The mathematical model of the process consists of two main parts - mechanical and thermal - both of them supported by density changes model. The mechanical part is responsible for the strain, strain rate and stress distribution in a controlled volume. The stress is substantial due to shrinkage and plastic deformation. It can cause cracks when the stress exceeds the ultimate tensile strength, which is very low within discussed temperature range.

2.1 Thermal model

Thermal solution has crucial influence on simulation results, since the temperature has strong effect on remaining variables, especially if the specimen temperature is close to solidus line. Plastic flow of solid and mushy materials, stress distribution and density changes are relevant to the temperature field particularly for deformation of body which consist of both solid and semi-solid regions. The temperature field is a result of solution of Fourier-Kirchhoff equation. The combined Hankel's boundary conditions have been adopted for the model (Glowacki, 2005). The Fourier-Kirchhoff equation in cylindrical coordinate system is written as follows:

$$\frac{1}{r}\frac{\partial}{\partial r}\left(rk_r\frac{\partial T}{\partial r}\right)+\frac{1}{r}\frac{\partial}{\partial \theta}\left(\frac{1}{r}k_\theta\frac{\partial T}{\partial \theta}\right)+\frac{\partial}{\partial z}\left(k_z\frac{\partial T}{\partial z}\right)+Q=\rho c_p\frac{\partial T}{\partial \tau} \tag{1}$$

where r, θ, z are cylindrical coordinate system, T is the temperature distribution in the controlled volume, τ is the time variable, k denotes the heat conduction coefficient (or coefficients matrix in case of thermal inhomogeneity), Q represents the rate of heat generation (or consumption) due to the transformation of the aggregation state, plastic work done and due to electric current flow in case of resistance heating of the sample. Finally, c_p describes the specific heat. The solution of equation (1) has to satisfy appropriate boundary conditions. The Hankel's boundary conditions can be written in form of differential equation (Glowacki, 2005):

$$k\frac{\partial T}{\partial n}+\alpha(T-T_0)+q=0 \tag{2}$$

In equation (2) T_0 is the distribution of the border temperature, q describes the heat flux through the boundary of the deformation zone, α is the heat transfer coefficient, n is a vector which is normal to the boundary surface.

2.2 Mechanical model with analytically controlled compressibility

As mentioned above the mechanical part is responsible for calculation of the strain, strain rate and stress distribution in the deformation zone which consist of solid and semi-solid regions. The analysis of mathematical models which can be applied for strain field calculations for semi-solid steel deformation process has proved good predictive ability of a rigid-plastic model of metal flow. The model is completed with numerical solution of Navier stress equilibrium equations in order to satisfy all the requirements leading to stress field. Rigid-plastic model was selected due to its very good accuracy with reference to strain field during the hot deformation and sufficient correctness of calculated deviatoric part of stress field. Moreover, the elastic part of stress tensor components is very low at temperatures close to solidus line and practically can be neglected in calculations of strain distribution. Classical rigid-plastic solutions are based on optimisation of following power functional:

$$J^*\left[v(r,z)\right]=W_\sigma+W_\lambda+W_t \tag{3}$$

where W_σ is the plastic deformation power, W_λ the penalty for the departure from either incompressibility or mass conservation conditions and W_t the friction power. In presented solution the second part of functional (3) is missing and both incompressibility and mass conservation conditions are given in analytical form constraining the velocity field components. The functional takes the following shape:

$$J^*\left[v(r,z)\right]=W_\sigma+W_t \tag{4}$$

where v describes the velocity field distribution in the deformation zone. In case of functional (4) the optimisation procedure is much more convergent than the one concerning functional (3), because numerical solution of both the mentioned conditions generates a lot of local minima and leads to wide flat neighbourhood of the global optimum. The accuracy of the proposed hybrid solution is also much better because of negligible volume loss caused by numerical errors which is very important for materials with changing density. Fully numerical solution shows lower accuracy contrary to the proposed analytical-numerical one. For solid regions of the sample the incompressibility condition is satisfactory and in cylindrical coordinate system it has been described with an equation:

$$\frac{\partial v_r}{\partial r}+\frac{v_r}{r}+\frac{\partial v_z}{\partial z}=0 \tag{5}$$

where v_r and v_z are the radial and longitudinal velocity field components in cylindrical coordinate system r, θ, z. For the mushy zone equation (5) is replaced by the mass conservation condition, which takes a form:

$$\frac{\partial v_r}{\partial r}+\frac{v_r}{r}+\frac{\partial v_z}{\partial z}-\frac{1}{\rho}\frac{\partial \rho}{\partial \tau}=0 \tag{6}$$

where ρ is the temporary material density and τ is the time variable.

2.3 Density changes

Density distribution is one of the most important properties of the mushy steel which undergo the deformation. Its changes have influence on both the mechanical and thermal parts of the presented model. The knowledge concerning effective density distribution is very important for modelling of deformation of porous and mushy materials. Density changes of liquid, solid-liquid and solid materials are ruled by three phenomena:

- solid phase formation,
- laminar liquid flow through porous material and
- thermal shrinkage.

Total density changes can be calculated according to the Darcy method which can be formulated in a form of differential equation:

$$\frac{\partial \rho}{\partial \tau}=\left(\rho_s X_s+\rho_l X_l\right)\left(\frac{\rho_s}{\rho_l}-1\right)\frac{\partial X_l}{\partial \tau}+\rho_l X_l \operatorname{div} v+\left(\beta_s \rho_s X_s+\beta_l \rho_l X_l\right)\frac{\partial T}{\partial \tau} \tag{7}$$

where X and β are fraction and linear expansion coefficients, respectively. Indexes l and s denote the liquid and solid phases, τ is the time variable, T is the temperature distribution in the controlled volume. Solution of equation (7) requires further time and computer memory resources. Nevertheless another way of taking density into consideration is possible due to temperature dependency of this quantity. In order to avoid additional problems with solution of equation (7) density changes were calculated according to an empirical model taking into consideration the experimental data. The model is slightly less accurate but such a method makes the solution much easier. The density changes of the investigated steels were computed using commercial JMatPro software. Figure 1 presents an example graph of density versus temperature dependency drawn for steel having 0.41% carbon content.

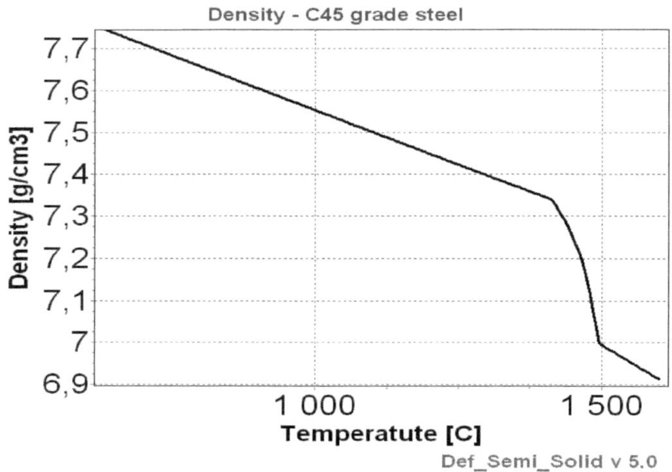

Fig. 1. The density versus temperature curve for C45 grade steel.

More details about presenting thermo-mechanical model, numerical methods and techniques will be described in the next chapter of this book.

3. Numerical systems – Def_Semi_Solid and JMatPro

In aim to allow easy working with the Gleeble® 3800 simulator a user friendly system called Def_Semi_Solid v.5.0 was developed in the Department of Computer Science and Modeling of the Faculty of Metals Engineering and Computer Science in Industry, AGH. The numerical part of the program was developed in FORTRAN/C++ language, which guaranties fast computation and the graphical interface was written using visual version of C++ language, taking advantage of its object oriented character. This approach has sufficient usability both in Windows and Linux based systems. The newest version of Def_Semi_Solid system is equipped with full automatic installation unit (Figure 2) and new graphical interface. It allows the computer aided testing of mechanical properties of steels at very high temperature using Gleeble® 3800 physical simulators to avoid problems which arise by traditional testing procedures. The first module allows the establishment of new projects or working with previously existing ones. The integral parts of each project are: input data for a specific compression/tension test as well as the results of measurements and optimization. In the current version of the program the module permits application of a number of database engines (among other standard MSAcces, dBASE IV and Paradox 7-8 for PC-based systems) and allows the implementation of material databases and procedures of automatic data verification. The next module (the solver) gives user the possibility of managing the working conditions of the simulation process. The inverse analysis can be turned off or on using this part of the system.

Fig. 2. View of DSS program setup window (local version).

The last module (DSS/Post module) is dedicated to the visualisation of the numerical results and printing the final reports. In the current version the possibility of visualization was significant improved. The main are: shading options using OpenGL mode (2D and 3D) and possibility make a full contour map (2D and 3D) as shown in Figure 3.

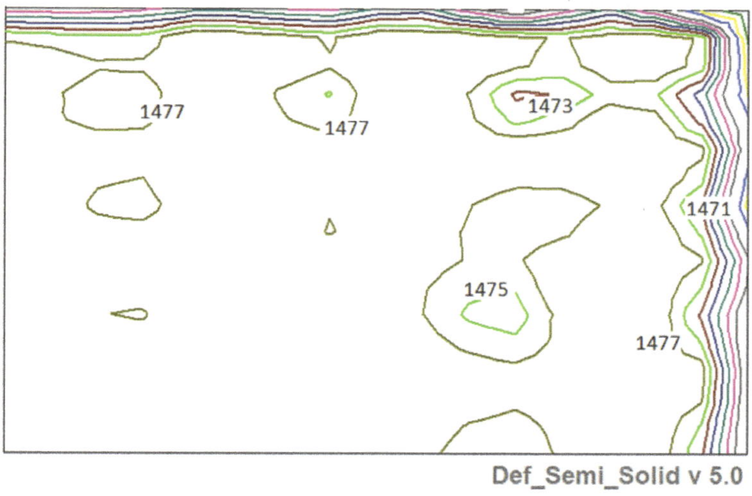

Fig. 3. The Post-processor of the newest version of Def_Semi_Solid system (contour option).

The less visible but powerful heart of the system is of course the solver. The finite element code dedicated to the axial-symmetrical compression/tension tests has been developed. The solution is based on the thermo-mechanical approach with density changes described in the previous section. The second software used during theoretical work was commercial code JMatPro. This program calculates a wide range of materials properties for alloys and steels. Using JMatPro we can make calculations for stable and metastable phase equilibrium, solidification behaviour and properties, thermo-physical and physical properties, phase transformations. JMatPro includes a Java based user interface, with calculation modules using C/C++, and will run under any Windows and under Linux system.

4. Computer aided experimental procedure

The steel grades subjected to series of experiments in Institute for Ferrous Metallurgy in Gliwice, Poland using Gleeble 3800® thermo-mechanical simulator (Figure 4) were the C45 (0.45% C) and S355J2G3So (0.11% C).

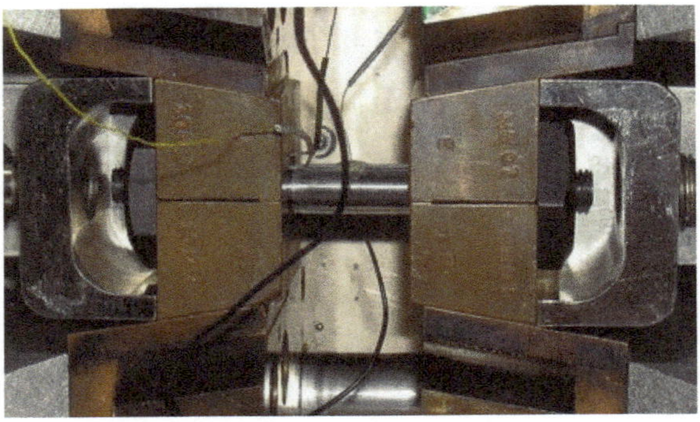

Fig. 4. The standard Gleeble® equipment allowing deformation is semi-solid state.

The essential aim of the investigation was reconstruction of both temperature changes and strain evolution on specimen exposed to simultaneous deformation and solidification. The analysis of metal flow in subsequent regions of the sample deformation zone requires adequate methods. Classical techniques of interpretation of results of compression testing procedures fail due to significant barrelling of the sample which is inevitable at any temperature close to solidus level. The developed user friendly dedicated FEM solution (Figure 5) with variable density based on the hybrid model described in previous sections is the core of strain-stress curves calculation system.

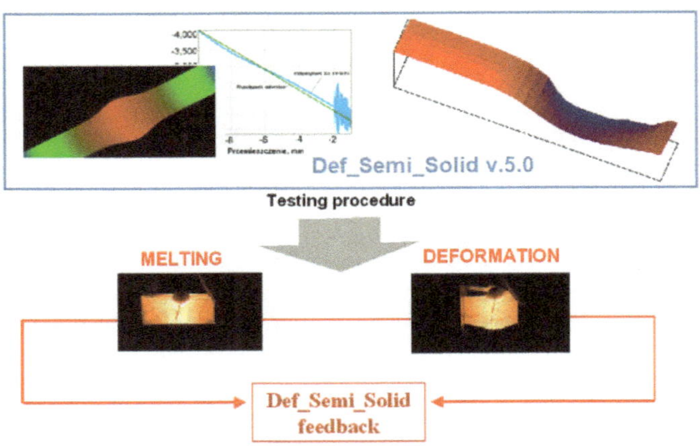

Fig. 5. The Def_Semi_Solid system as a feedback unit with Gleeble® 3800 simulator.

In all cases, experiments ran according to schedule:

stage 1: the sample was prepared (e.g. mounting thermocouples, die selection),
stage 2: melting procedure of the sample was realized,
stage 3: at the end the deformation process was done.

Figure 6 shows the shape of the testing samples and locations of thermocouples used during experiments.

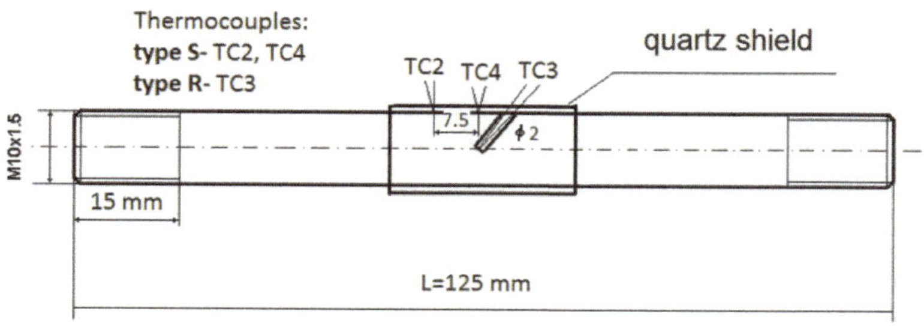

Fig. 6. Samples used for the experiments. TC2, TC3 and TC4 thermocouples.

Material tests in the semi-solid state should be carried out in as isothermal conditions as possible due to the very high sensitivity of material rheology on small changes of temperature (Hojny et al., 2009). The basic reason for uneven temperature distribution inside samples on the Gleeble® simulator is their contact with cooper handles during resistance heating (Figure 7).

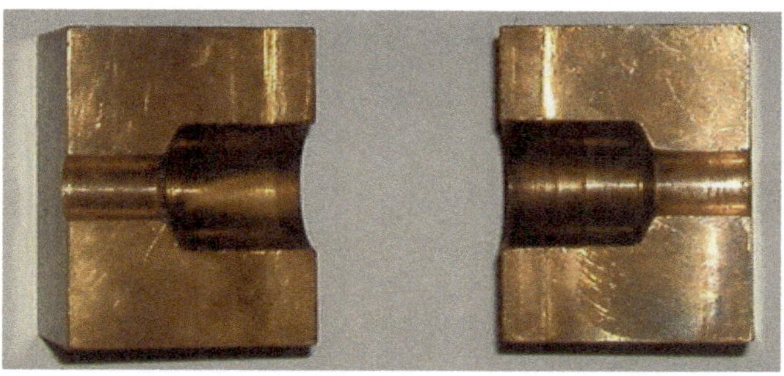

Fig. 7. The handle with short contact zone (sample - handle) used during experiments (called "hot handle").

The estimated liquidus and solidus temperature of the investigated steels are: 1495 °C and 1410 °C, respectively for C45, 1513 °C and 1464 °C, respectively for S355J2G3So. In the next section the example results of the melting and deformation procedure are presented.

4.1 Melting procedure

Thermal solution has crucial influence on simulation results, since the temperature has strong effect on remaining. The resistance heating processes cause non-uniform distribution of temperature inside heated materials especially in longitudinal section of the sample. In the case of the semi-solid steels, such distribution gives significant differences in the microstructure and rheology. The thermo-physical properties of this steel, necessary in calculations, were determined using *JMatPro* software. This software determines this properties on the basis of the chemical composition. The example main properties of C45 grade steel used in calculations are presented in Figure 8 and Figure 9.

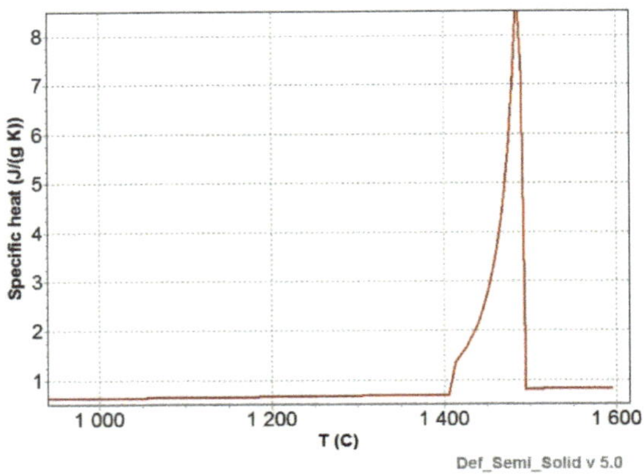

Fig. 8. Specific heat versus temperature for the investigated steel (C45).

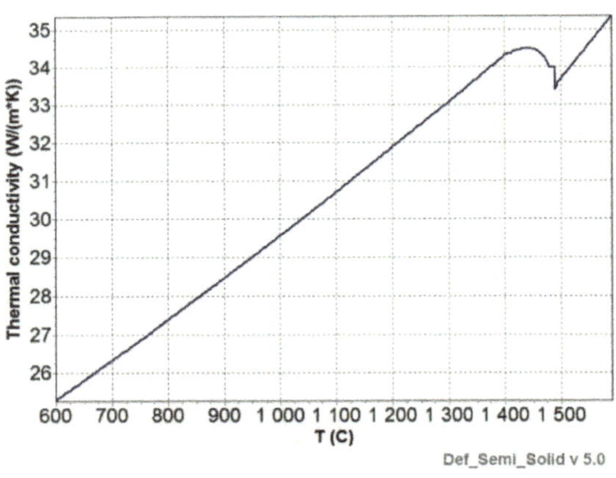

Fig. 9. Thermal conductivity versus temperature for the investigated steel (C45).

In case of each physical and computer simulation samples were heated to 1430 °C and after holding at constant temperature the sample was cooled to nominal deformation temperature. The heat generated (Q) due to direct current flow was calculated using inverse approach. The objective function (F) was defined as a norm of discrepancies between calculated (T_c) and measured (T_m) temperatures (only for indication steering thermocouples TC4) according to the following equation:

$$F(Q)=\int_{\tau_0}^{\tau_1}\left[\left(T_c(Q,r,z,T)-T_m(r,z,T)\right)^2\right]dT \quad (8)$$

where τ is the time variable.

In Figure 10 the comparison between experimental and theoretical temperature versus time curves is presented for steering thermocouple (mounting locations of thermocouples shown in Figure 6),

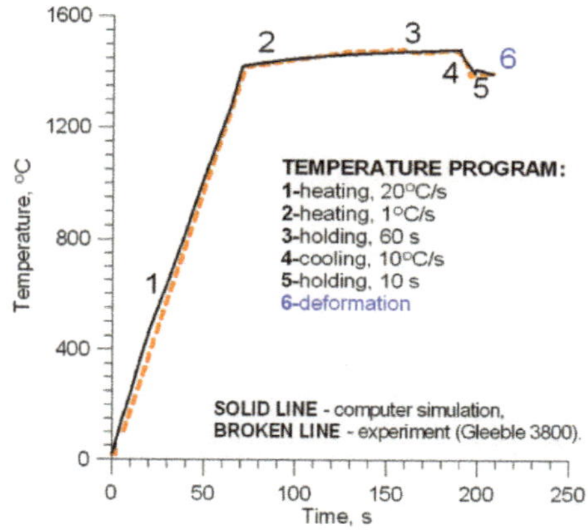

Fig. 10. Comparison between the experimental and theoretical time-temperature curves during initial heating and final compression at 1400°C.

In the final stage of physical simulation for different holding time, the temperature difference between core (TC3 thermocouple) of the sample and the surface (TC4 thermocouple) was analyzed. In all cases the core temperature was higher than surface temperature, for example, difference between core of the sample and surface was about 40 °C for test at 1380 °C (Figure 11).

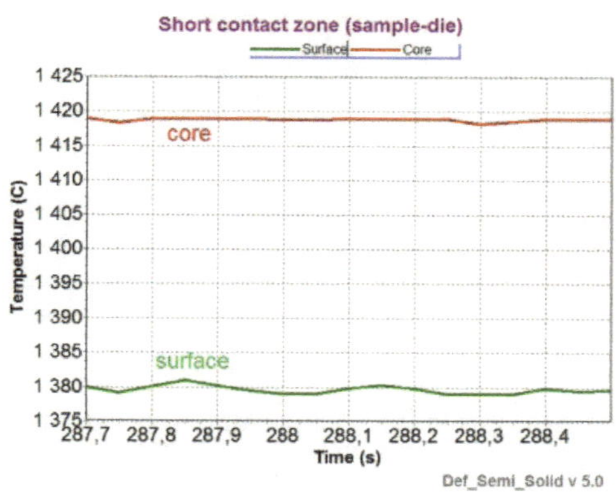

Fig. 11. The temperature change versus time for hot handle (final stage of physical simulation right before deformation at 1380 °C).

The numerical simulation confirmed results obtained during experimental parts. In the Figure 12 temperature distributions in the cross section of the sample tested at temperature 1380 °C are presented for 3 and 6 seconds of heating and final distribution right before deformation.

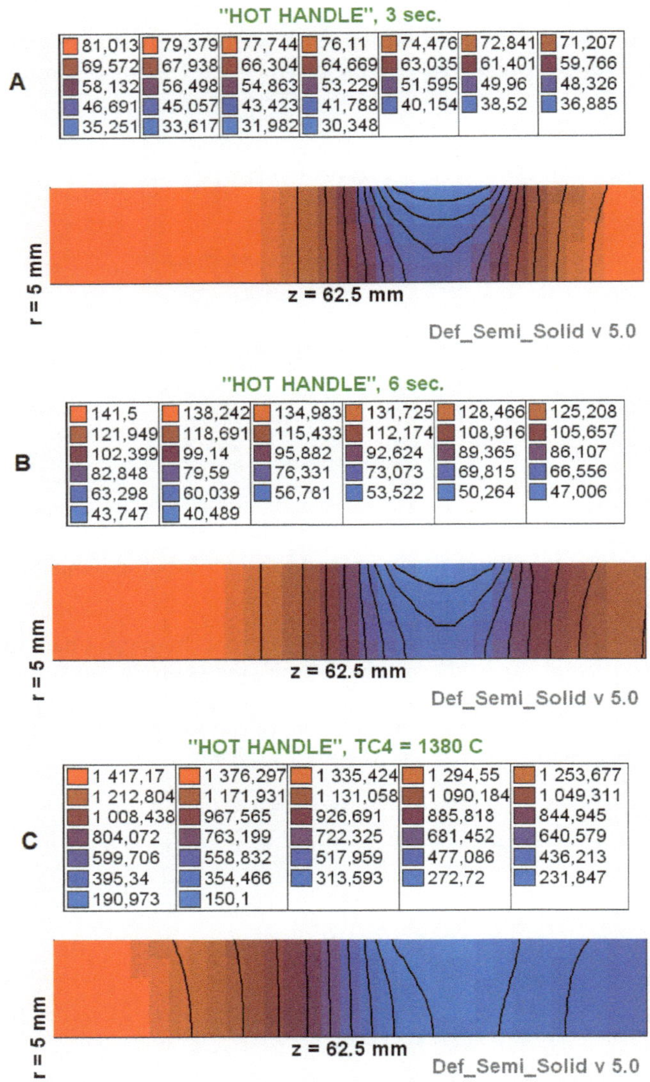

Fig. 12. Distribution of temperature in the cross section of the sample tested at temperature 1380 °C after time heating a) 3 seconds, b) 6 seconds, c) right before deformation.

The one can observe, major temperature gradient between contact surface die-sample. The difference between experimental and theoretical core temperatures for hot handles was 3°C (calculated core temperature equal 1417°C, measured core temperature equal 1420°C.) Finally, the micro and macrostructure of the tested samples was investigated. Figure 13 show example microstructure of the tested samples right before deformation for middle and border of the heating zone.

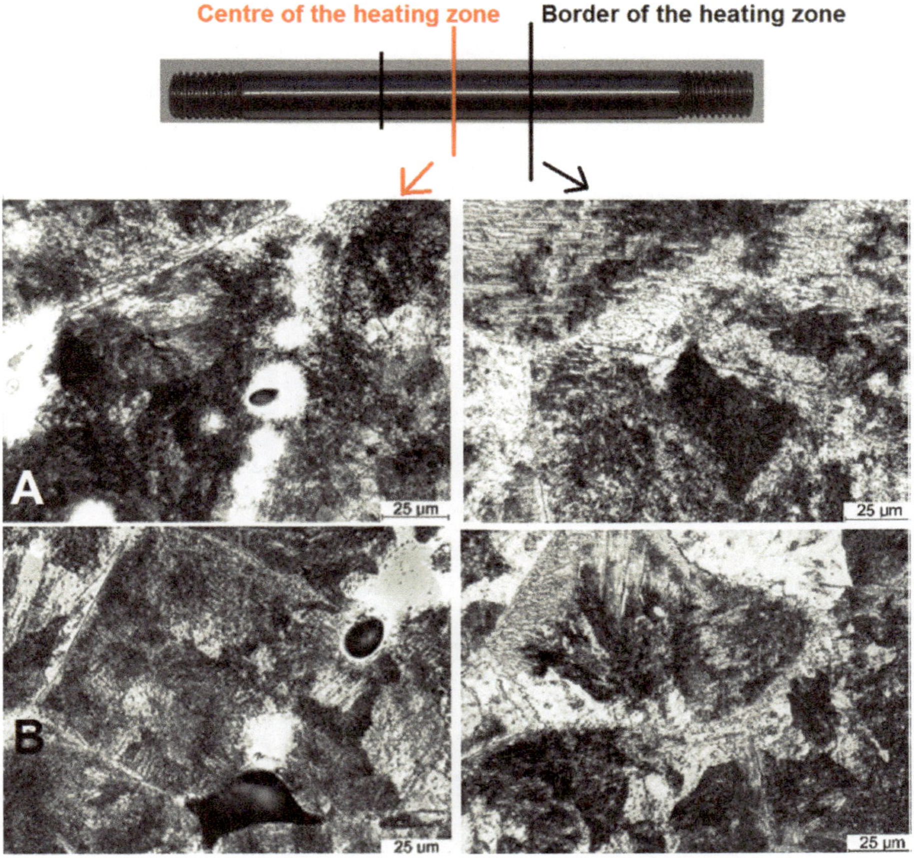

Fig. 13. Microstructure of the middle/border of the sample right before deformation. Variant with hot handle. Magnification: 400x.

Microstructure of the cooled samples consist of pearlite (the darkest phase), bainit (grey phase mainly near the borders of grains) and the bright ferrite (Figure 13). It is result of such phase composition the wide zone of melting and almost twice smaller speed of cooling in the case of samples warmed in „hot" handles. Figure 14 show example macrostructure in the cross-sections of the tested samples right before deformation and calculated core temperature. One can observe that for analyzed temperatures liquid phase particle exist in the central part of the sample. The comparison between experimental results and numerical show that mathematical model of resistance heating right reflect back the physical simulation of resistance heating of samples using Gleeble® 3800 physical simulator.

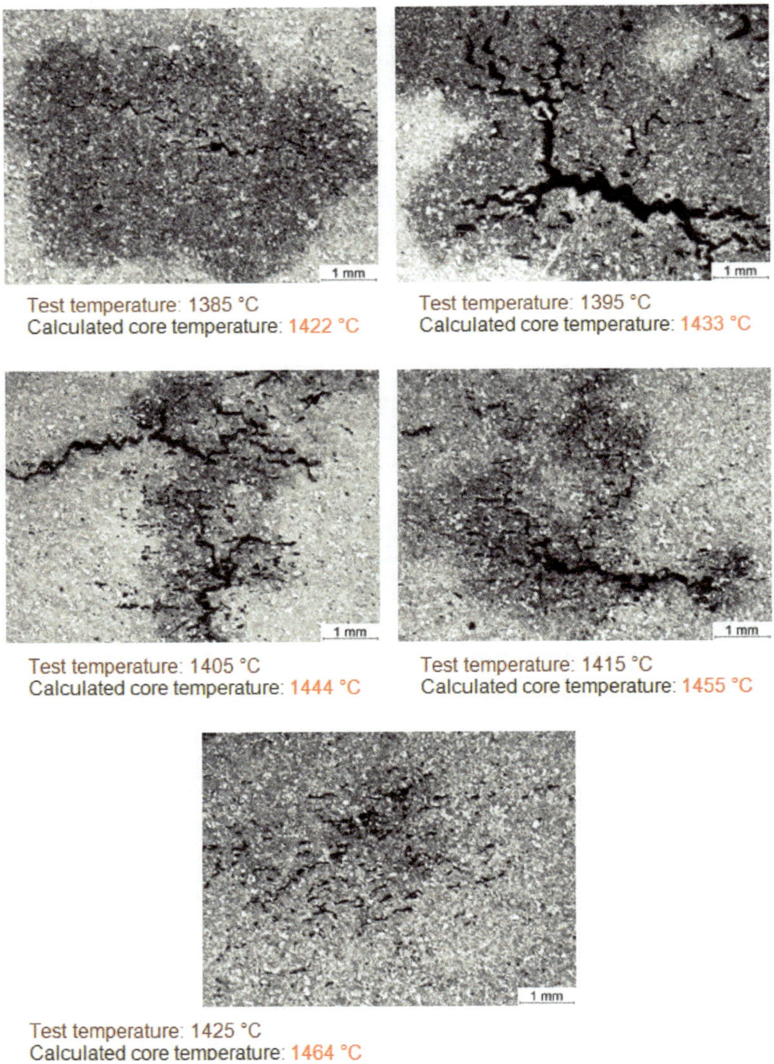

Fig. 14. Macrostructure of the middle of the sample right before deformation. Variant with hot handle. Magnification: 10x.

4.2 Deformation procedure

In the next stage of the experimental part the compression tests were done. During experiments die displacement, force and temperature changes in the heating zone are recorded. Parallel, the computer simulations were realized in order to obtain optimal value parameters of process. The strain-stress curves were described by following equation:

$$\sigma_p = A\varepsilon^n \dot{\varepsilon}^m \exp(-BT) \tag{9}$$

where A, B, n, m are material constant, T - temperature, ε - strain and $\dot{\varepsilon}$ - strain rate. It is not easy to construct isothermal experiments for temperatures higher than 1300ºC. Several serious experimental problems arise. First of all, keeping so high temperature constant during the whole experimental procedure is extremely difficult. There are also severe difficulties concerning interpretation of the measurement results. The significant inhomogeneity in the strain distribution in the deformation zone and distortion of the central part of the sample lead to poor accuracy of the stress field calculated using traditional methods, which are good for lower temperatures. The only possibility to have good coefficients of strain-stress formula is the inverse method (Glowacki & Hojny, 2009). The long calculation time, which is usual by this kind of analysis requires sometimes parallel computation. The application such a method is considered for the future application.

The objective function was defined as a norm of discrepancies between calculated (F_c) and measured (F_m) loads in a number of subsequent stages of the compression according to the following equation:

$$\varphi(x) = \sum_{i=1}^{n} \left[F_i^c - F_i^m \right]^2 \tag{10}$$

The theoretical forces F_c were calculated with the help of sophisticated FEM solver facilitating accurate computation of strain, stress and temperature fields for materials with variable density. The example comparison between the calculated and measured loads are presented in Figure 15-17, showing quite good agreement between both loads.

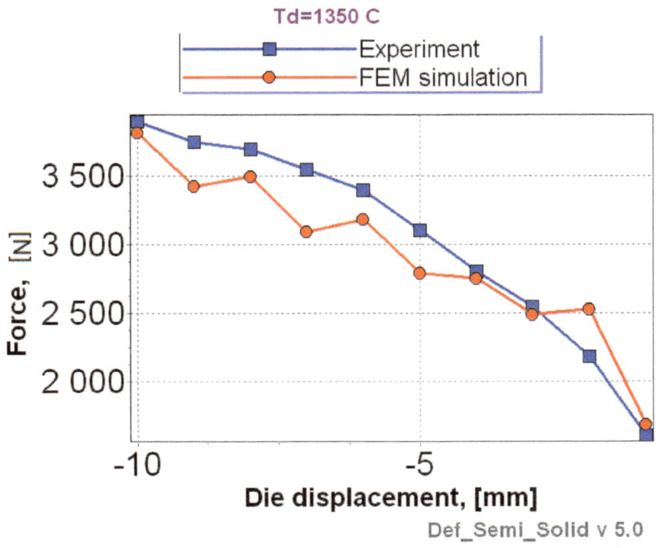

Fig. 15. Comparison between measured and predicted loads at temperature 1350ºC.

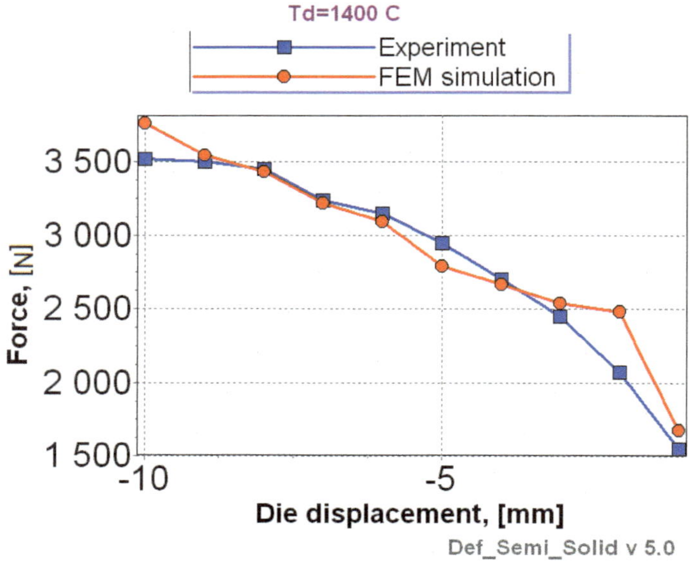

Fig. 16. Comparison between measured and predicted loads at temperature 1400°C.

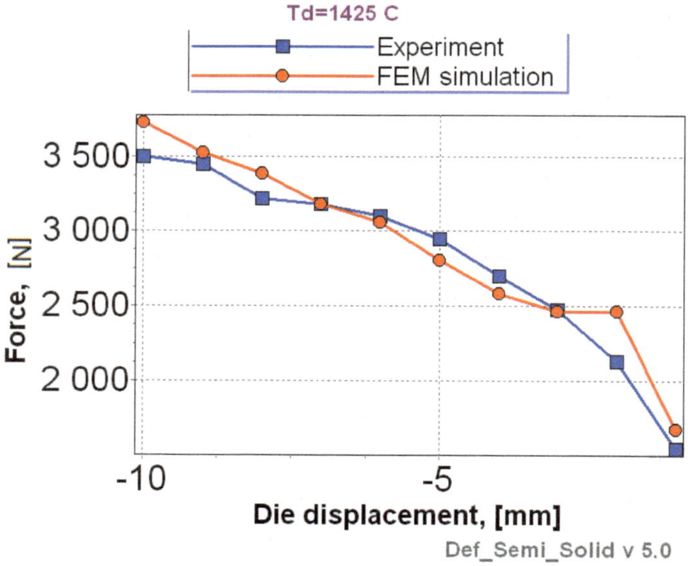

Fig. 17. Comparison between measured and predicted loads at temperature 1425°C.

The coefficients obtained during inverse analysis allow the construction of stress-strain curves, which are presented in Figure 18.

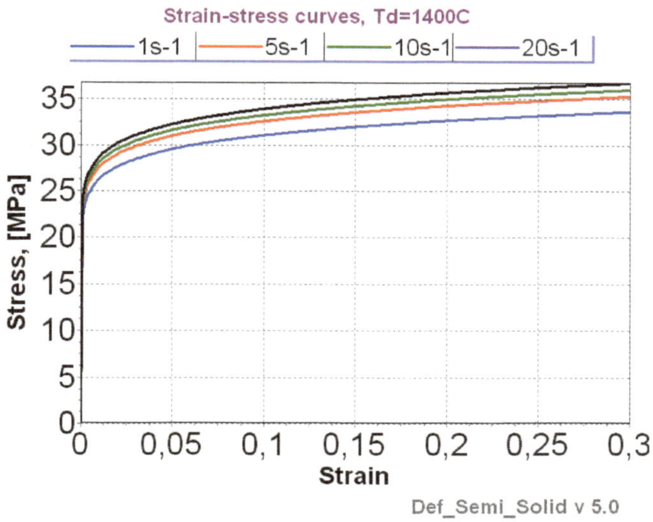

Fig. 18. Stress-strain curves at several strain rate levels for temperature 1400°C.

4.3 Verification of the computer system – Example results

Using previously developed curves, example simulations of compression of cylindrical samples with mushy zone have been performed. The results of the tests demonstrate the possibilities of the computer system. For all series of tests the simulations were done using short contact zone between the sample and simulator jaws. The deformation zone had the initial height of 62.5 mm. The radius of the sample was 5 mm. An example specimen was melted at 1430°C, and then after cooling temperature deformed at demanding temperature. The first variant at 1430°C, the second variant at 1425°C. During the tests each sample was subjected to 10 mm reduction of height. The final temperature for variant no. 1 and no. 2 is shown in Figure 19. The temperature distribution for the both variants is similar. Taking into account the value of core temperature for variant no. 2 one can state the existence of the mushy zone in the sample centre. The analysis of the effective strain fields for specimens no. 1 and no. 2, which are presented in Figure 20 show that influence of density variations on the metal flow scheme is not very significant, although small differences are clearly visible. The analysis of the strain shows maximal values of strain in the central region of the sample. In Figure 21 mean stress distribution is presented. The initial temperature distribution has great influence on the stress field in the deformation zone. The inhomogeneity of the strain field also leads to stress generation. The analysis of the mean stress (Figure 21) shows the stress concentration near the upper surface and in the centre of the sample. In accordance with the earlier conjectures for sample no. 1 the stress level is significantly higher than for sample no. 2. The material properties which have been used in the model are not very accurate because of difficulties in experiment. The existing inhomogeneity of deformation causes problems concerning the calculation of right stress values. The good analysis of the results of experiments needs computer programs to simulate plastic deformation of steel samples. The created model, which takes into account the variable density, can be helpful for the discussed purposes.

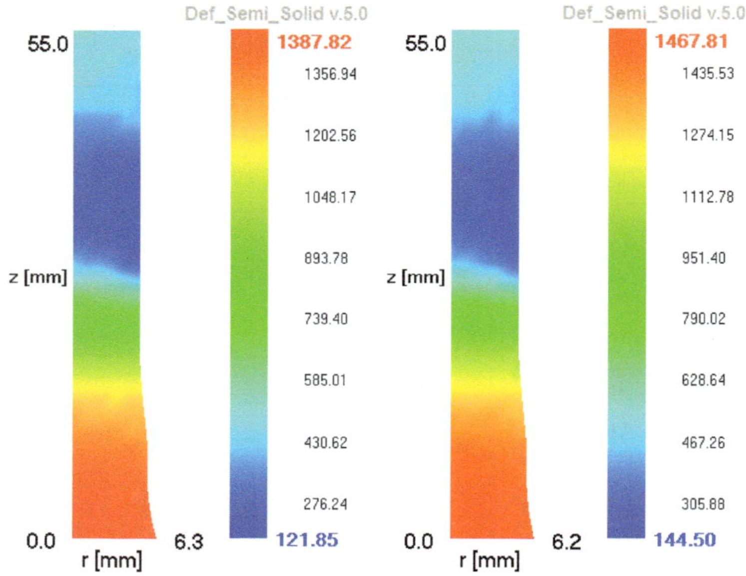

Fig. 19. Example results of computer simulation of deformation at 1350°C (left) and 1425°C (right): final temperature distribution in the cross-section of the sample.

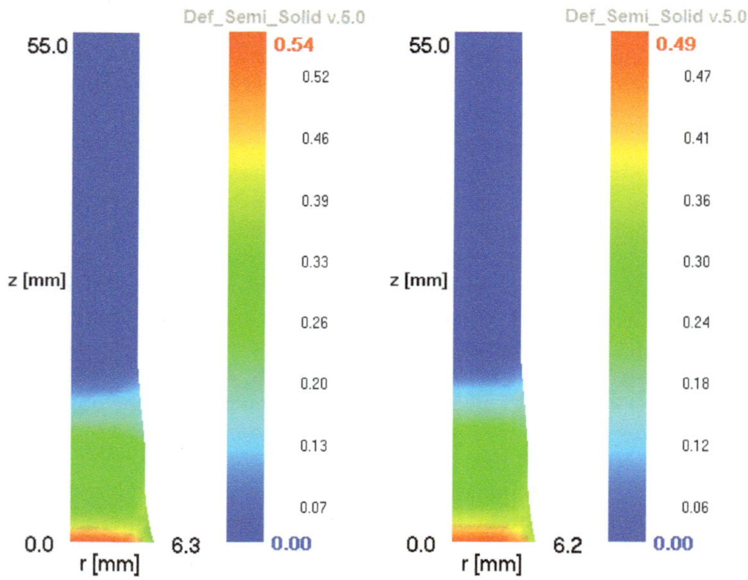

Fig. 20. Example results of computer simulation of deformation at 1350°C (left) and 1425°C (right): final strain distribution in the cross-section of the sample.

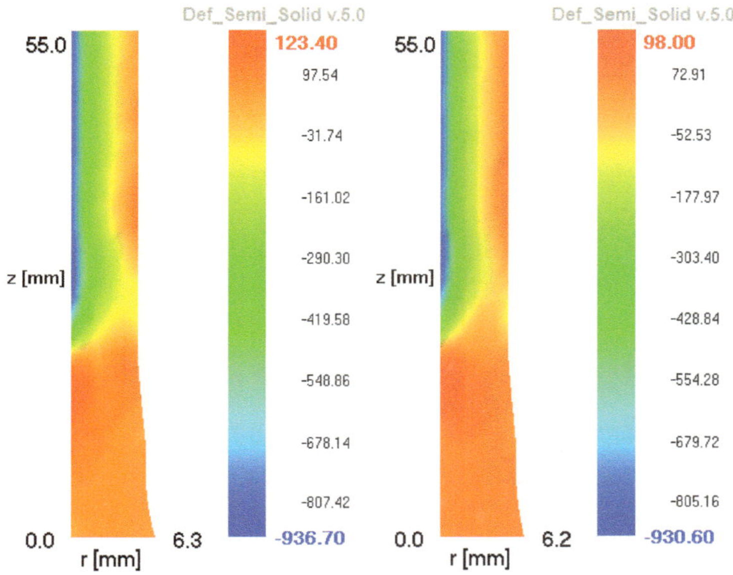

Fig. 21. Example results of computer simulation of deformation at 1350°C (left) and 1425°C (right): final mean stress distribution in the cross-section of the sample.

The calculated and experimental (Figure 22) shapes of the sample allow a rough verification of rheological model. For verification of the computer system two comparative criteria were used:

- comparison of the maximum measured and calculated diameters of the sample,
- comparison between the measured and calculated lengths of zone which was not subjected to the deformation.

Figures 23 and 24 show example application of the 1st and 2nd criterion, respectively. The figures confirm quite good agreement between theoretical and experimental results.

Fig. 22. Final shape of the sample after deformation at 1350°C, 1400°C and 1425°C.

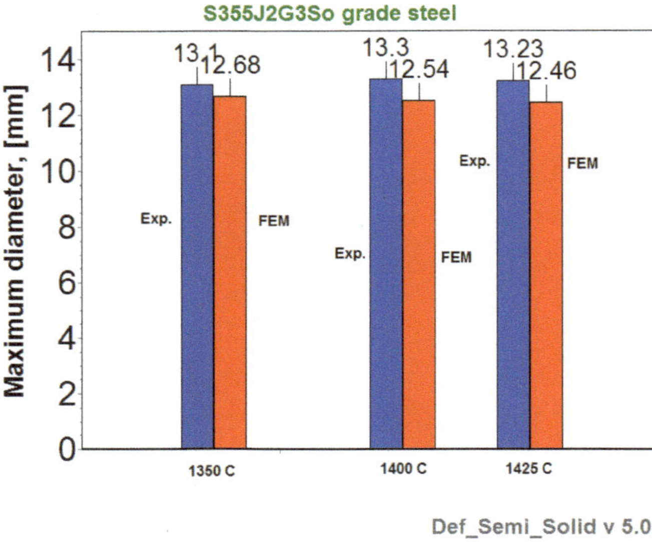

Fig. 23. The comparison of the measured and calculated maximal diameters of the sample - experiments between 1350°C and 1425°C.

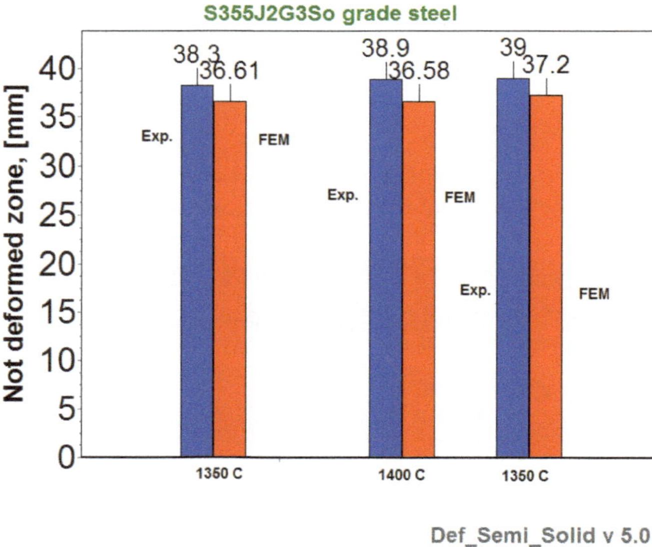

Fig. 24. The comparison between the measured and calculated length of zone which was not subjected to the deformation - Experiments between 1350°C and 1425°C.

5. Conclusions

Computer aided testing of steel samples deformation at coexistence liquid and solid phase requires resolving a number of problems. One of them is the difficulty in determination of material thermal and mechanical properties, such as: coefficients of heat transfer and other thermal properties, diagrams of density changes, which is dependent on temperature, etc. The main problem is the interpretation of compression tests results leading to strain - stress curves. The presented model with incompressibility condition in analytical form allows the simulation of the deformation of material with mushy zone avoiding volume loss, which cause problems with density. The presented Def_Semi_Solid program is a unique tool, which can be very helpful and may enable the right interpretation of results of very high temperature tests. The paper has shown its predictive ability regarding: temperature, shape and size of the deformation zone. The focus of attention were mechanical properties of investigated steel and specific character of theoretical model applied to the analysis. One can observe that the compression tests interpretation was possible only due to application of right model and implementation of the inverse analysis.

6. Acknowledgments

The work has been supported by the Ministry of Science and Higher Education Grant N N508 410637 and partially N N508 585539.

7. References

Glowacki M. (2005). The mathematical modelling of thermo-mechanical processing of steel during multi-pass shape rolling, *Journal of Materials Processing Technology*, Vol. 168, No.2, pp. 336–343, ISSN: 0924-0136

Glowacki M.; Hojny M. (2009). Inverse analysis applied for determination of strain-stress curves for steel deformed in semi-solid state, *Inverse Problems in Science and Engineering*, Vol. 17, No.2,pp. 159–174, ISSN: 1741-5977

Hojny M.; Glowacki M. (2008). Computer modelling of deformation of steel samples with mushy zone, *Steel Research International*, Vol. 79, No. 11, pp. 868-874,ISSN: 1611-3683

Hojny M.; Glowacki M. (2009). The physical and computer modeling of plastic deformation of low carbon steel in semisolid state, *Journal of Engineering Materials and Technology*, Vol. 131, No. 4, pp. 041003-1–041003-7, ISSN: 0094-4289

Hojny M.; Glowacki M.; Malinowski Z. (2009). Computer aided methodology of strain-stress curve construction for steels deformed at extra high temperature, *High Temperature Materials and Processes*, Vol. 28, No. 4, pp. 245–252, ISSN: 0334-6455

Hojny M.; Glowacki M. (2011). Modeling of Strain-Stress Relationship for Carbon Steel Deformed at Temperature Exceeding Hot Rolling Range, *Journal of Engineering Materials and Technology*, Vol. 133, No. 2, pp. 021008-1–021008-7, ISSN: 0094-4289

Hufschmidt M.; Modigell M.; Petera L. (2004). Two-Phase Simulations as a Development Tool for Thixoforming Processes. *Steel Research International*, Vol. 75, No.3, pp. 513–518, ISSN: 1611-3683

Jing Y.L.; Sumio S.; Jun Y. (2005). Microstructural evolution and flow stress of semi-solid type 304 stainless steel. *Journal of Materials Processing Technology*, Vol. 161, No. 3, pp. 396-406, ISSN: 0924-0136

Jin S. D.; Hwan O.K. (2002). Phase-field modelling of the thermo-mechanical properties of carbon steels. *Acta Materialia*, Vol. 50, No. 9, pp. 2259-6454, ISSN: 1359-6454

Kang, C.G.; Yoon, J.H. (1997). A finite-element analysis on the upsetting process of semi-solid aluminium material. *Journal of Materials Processing Technology*, Vol. 66, No.3, pp. 76-84, ISSN: 0924-0136

Koc, M.; Vazquez V.; Witulski T.; Altan T. (1996). Application of the finite element method to predict material flow and defects in the semi-solid forging of A356 aluminium alloys. *Journal of Materials Processing Technology*, Vol. 59, No. 2, pp. 106-112, ISSN: 0924-0136

Kopp R.; Choi J.; Neudenberger D. (2003). Simple compression test and simulation of an Sn–15% Pb alloy in the semi-solid state. *Journal of Materials Processing Technology*, Vol. 135, No. 3, pp. 317-323, ISSN: 0924-0136

Modigell M.; Pape L.; Hufschmidt M. (2004). The rheological behaviour of metallic suspensions. *Steel Research International*, Vol. 75, No.3, pp. 506–512, ISSN: 1611-3683

12

Numerical Evaluation of Product Development Processes

Nadia Bhuiyan
Concordia University
Canada

1. Introduction

The management of new product development (NPD) processes is a continual challenge facing organizations that develop complex, innovative products. While market trends are forcing shorter product development times in order to meet time-to-market (TTM) goals, companies are trying to develop mechanisms to streamline their NPD processes. One approach that has provided much success towards achieving shorter TTM is concurrent engineering (Winner et al., 1988; Clark and Fujimoto, 1991; Blackburn, 1991; Wheelwright and Clark, 1992; Smith and Reinersten, 1991). Concurrent engineering (CE) can broadly be defined as the integration of inter-related functions at the outset of the product development process in order to minimize risk and reduce effort downstream in the process, and to better meet customers' needs (Winner et al., 1988). Multi-functional teams, concurrency of product/process development, integration tools, information technologies, and process coordination are among the elements that enable CE to improve the performance of the product development process (Blackburn, 1991). The traditional NPD process suffers many setbacks. This process evolves in a sequential fashion, where phases follow one another serially, each one dominated by a single functional role. There is little or no cross-communication among various functions, and information generated from one activity gets handed off to the next only after its completion. The commonly encountered problems with this type of process are increased downstream effort, process span time, i.e., the start to finish time of the process, and costs.

In order to study and evaluate the performance of CE and sequential NPD processes, a new approach is used based on an existing mathematical technique called the expected payoff method, which is the basis of decision theory. Under this framework, the mathematics which describe the micro-processes, such as information sharing between team members and overlapping of activities, and their relationships with the macro-process performance in terms of expected payoff (where the macro-process is the overall development process), are described. Network diagrams are presented as a formalism for expressing product development processes. The fundamental concept of the model is based on the premise that team members make decisions or choose actions that maximize the payoff (utility or usefulness) that their actions bring to the team. Team members must obtain, process, and communicate information to one another to make decisions that will optimize their performance.

This chapter is organized as follows. Section 2 discusses the existing literature, and highlights the contributions of the research. Section 3 explains the characteristics of NPD processes. In Section 4, the expected payoff method is described and the results of the mathematical analysis are presented. The results are detailed in Section 5, and in Section 6, conclusions and paths for future research are presented.

2. Literature review

In this section, a review of the relevant theoretical and analytical research is presented. Krishnan et al. (1997) developed a deterministic model based on properties of the design process that help to determine when and how two development activities should be overlapped (Figure 1). These properties are defined as 'upstream information evolution' and 'downstream iteration sensitivity'. The former is the rate at which upstream information converges to a final solution, and the information is modeled as an interval that gets refined over time. Sensitivity describes how vulnerable the downstream activity is to any changes in the upstream information, and is defined by the time needed by the downstream activity to incorporate the changes, which represents rework. Different patterns of information exchange between two activities, represented by the arrows in the diagram, are studied.

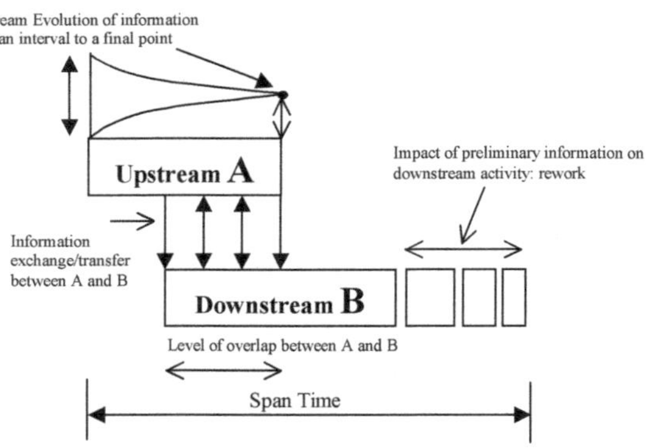

Fig. 1. Krishnan et al.'s model.

The authors address the overlapping problem by studying how values of the two properties determine the extent to which overlapping is appropriate between the dependent activities, A and B, and consequently how the span time is affected. Various overlapping policies between the upstream and downstream activities are examined based on varying the values of these two properties, and an integer program is developed to minimize span time.

Loch and Terwiesch (1998) have developed an analytical model of CE that considers the overlapping of two sequentially dependent activities, an upstream product design activity, and a downstream process design activity. The authors study the trade-offs between the downstream activity using upstream preliminary information to overlap activities, and the corresponding delay this might cause in terms of downstream rework. They suggest that when engineering changes (EC) arise during the product design, this poses the risk of redoing the overlapped work of the downstream activity, and this can be significant if the dependency between the two activities is high. They propose that communication during overlapping can reduce rework effects, but at the cost of communication time. They also use the concepts of evolution and sensitivity.

Yassine et al. (1999) have studied the CE problem of overlapping activities through a decision analytic framework. Using a probabilistic model consisting of an upstream activity and a downstream activity, their methodology finds the optimal overlapping policy based on the study of independent, dependent, and interdependent activities, described as the information structure of a process. A schedule of when to transfer information based on the information structures can fall under one of three categories: sequential, partial overlapping, and concurrent. Sequential transfer of information takes place for dependent activities. Partial overlapping can take place for either dependent or interdependent activities. In both cases, however, the information exchange/transfer must appropriately minimize the risk of downstream rework in the event of a change in the upstream activity. A concurrent schedule can take place when the activities are independent; since neither requires information from the other to proceed, they may be executed in parallel.

Ha and Porteus (1995) developed a simple model that proposes the optimal policy for the frequency and timing of progress reviews in an overlapped process. The authors study two overlapped, interdependent activities, an upstream design activity and a downstream process activity. In contrast to sequentially dependent activities, the nature of interdependent activities requires team members to communicate frequently. They develop a dynamic program that shows that, in order for overlapped activities to be beneficial, the design activity must be accompanied by progress reviews to minimize the risk of downstream rework and thus span time, and to improve quality. However, these gains are only achieved at the expense of the time and cost spent on communication. Therefore, the frequency of communication or progress reviews must be balanced with the value gained from having them. The optimal policy of reviews minimizes span time by providing sufficient information at the right time, helping to identify potential design problems early.

Different methods to address the problem of overlapping have been suggested in the literature. This research contributes to the existing work by introducing a methodology based on decision theory to study the performance of processes.

3. NPD processes

An information processing view of organizations, and thus of product development, is assumed in this research. From this perspective, the product development process must go through a set of decision-making processes to transform information inputs into information outputs, which are used to develop tangible outputs, i.e., the end product(s) (Clark and Fujimoto, 1991; Galbraith, 1973). Therefore, the focus of the models is on the flow of information as it evolves from the beginning to the end of the development process, making the relationships between development activities more readily apparent.

Product development can be defined as the process of undertaking all the activities and processing the information required to develop a concept for a product up to the product's market introduction. NPD processes may vary from one organization to the next, and as such, there is no one standard process agreed to by all. However, the general steps required in a product development process are fundamentally similar (Ulrich and Eppinger 2011). The NPD process defined in this study is a generic one which outlines the major steps in product development. It is a summary of the common phases and activities used in many instances in the literature as well as in the case study, and as such, it is a reasonably accepted approach to representing the product development process (Schilling and Will, 1998; Nihtila, 1999; Eastman, 1980).

The NPD process, shown in Figure 2, begins with the development of a concept for a marketable product (Phase A). In this phase, market requirements are determined, new ideas are generated, screened for economic and technical feasibility, and one is selected. In Phase B, 'Definition', a set of specifications to make the product is defined, and the product architecture is developed. Phase C, 'Development', consists of detailed design, physical prototyping, and testing. Finally, in Phase D, 'Implementation', the product volume is ramped up in manufacturing and launched onto the market.

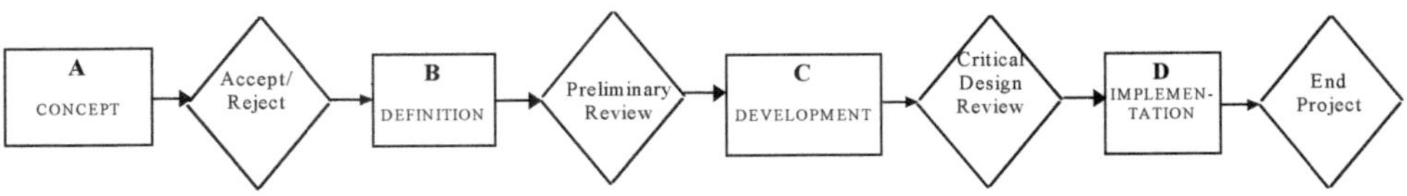

Fig. 2. A schematic diagram for a general stage-gate process with Phases A, B, C, D.

Figure 2 is an example of the traditional NPD process, where the phases are performed sequentially one after the other. Between phases, a one-way dependence is assumed, that is, the downstream phase depends on information generated by the upstream phase, but not vice-versa. This is represented by the uni-directional arrows between phases.

In an NPD process, the relationship between product and process design is mutually interdependent (Tian et al., 1998). This means that the information generated by one or more functions poses contingencies for others, thus, the parameters of the product and the process should be considered simultaneously (Adler 1995). Therefore a higher degree of coordination is required to manage more people collaborating on interdependent activities. In a sequential process, this interdependence is ignored; a dependent relationship is assumed, and this leads to unplanned coordination downstream. While better management of interdependencies does lead to shortened span time as compared to the sequential process, the price is higher cost of upstream effort.

CE uses two main mechanisms to reduce the span time for NPD processes: 1) increased information sharing from the start of a project (functional participation), and 2) overlapping of phases and activities. In a CE process, functional participation takes place through the formation of a team consisting of a representative from each of the functions that contribute to the development of a product. The goal is to make downstream activities easier to perform by releasing preliminary information to them early in the process to allow for overlap of activities. However, due to uncertainty in the early stages of an NPD process, the release of incomplete information to downstream functions may potentially introduce the need for rework should there be a change in upstream information. Thus, potential risks must be carefully examined to ensure that added time and effort are kept to a minimum (Krishnan et al., 1997).

Compared to a sequential approach, CE can decrease span time at the expense of increased interdependencies between activities (sequential to reciprocal). To handle the increased interdependencies, close intensive coordination is required through functional participation. However, this may increase effort.

Figure 3 shows an overlapped CE process. Note that information flows are more frequent than in the sequential case, and they are also bi-directional. Major milestones exist at the same gates as before, and each phase is made up of activities, not shown in the figure.

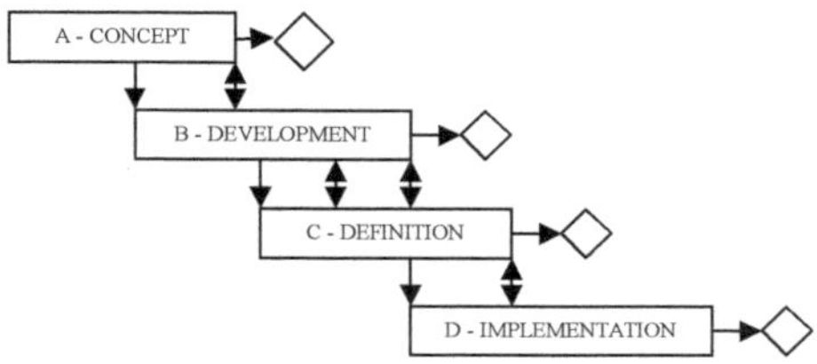

Fig. 3. CE development process.

4. Expected payoff method

In this section, a mathematical approach is described to measure the performance of processes, namely, sequential and CE processes, through the study of macro- and micro-variables. The macro-variable is the expected payoff, while the micro-variables are team interaction and level of overlap. The concepts of information processing and decision-making are presented as the basis of this framework. An information processing view is assumed, so that processes are studied through the way in which several team members perform activities such as acquiring, communicating, and processing information in order to make decisions, which in turn organizes the way activities are executed. Processes and corresponding team activities are modeled via networks of interconnected elements. These elements transform inputs into outputs, and represent people, machines, or other real-world objects. Each network realizes an output which is a measure of process performance, and is used to evaluate and compare processes. The methodology is based on the expected payoff method, a technique used in decision theory. It is applied in the calculation of a simple model of both a sequential and a CE process, and the results are compared.

The principle of the expected payoff method has been applied mainly in the field of economics, management science, and in certain areas of artificial intelligence, with respect to decision-making. In this field, economists study 'the best use of available (limited) resources' (Marschak and Radner, 1972). There has been no use of this method in the evaluation of CE in new product development processes. In an organizational environment, teams are also concerned with making the best use of alternatives or limited resources. The interested reader can find several readings in the literature on the principle of utility theory and its various applications (Marschak and Radner, 1972; Fishburn, 1970; Marschak, 1959; Marschak, 1954; von Neumann and Morgenstern, 1943). The framework developed in this part of the research will compare a simple model of a sequential process to a CE process, and evaluate the two in terms of the total expected payoff.

4.1 Methodology

The approach assumes that individuals in a team work towards achieving common goals with common interests and beliefs, within the constraints of their work, all of which guide their behavior. Given the complexities of such a situation, the problem is allocating appropriate information at the right time, such that team members can make the 'right' decisions which serve to accomplish their common goals (Marschak and Radner, 1972). This chapter will describe the means by which the activities of teams can be described, as well the mathematical analysis which can evaluate team performance, namely, through the use of the expected payoff method. The expected utility or payoff of an action measures the usefulness that an action brings to a person. By combining this with probability theory, decision theory helps a person determine that the action which maximizes his or her expected payoff over all possible actions (from this point forward, for simplicity, the term 'his' will be understood to include 'his/her'). The development of the expected payoff function will be described in detail. Processes can be explicitly represented through network diagrams that illustrate the activities that team members must perform, the inter-relatedness of activities through information requirements, and the communication required among team members (Figure 4). A network realizes a response function, or outcome function, which is based on the actions of the individuals in the organization, and these actions affect the outcome or expected payoff. Among various possible network configurations, the network with the greatest expected payoff is considered optimal.

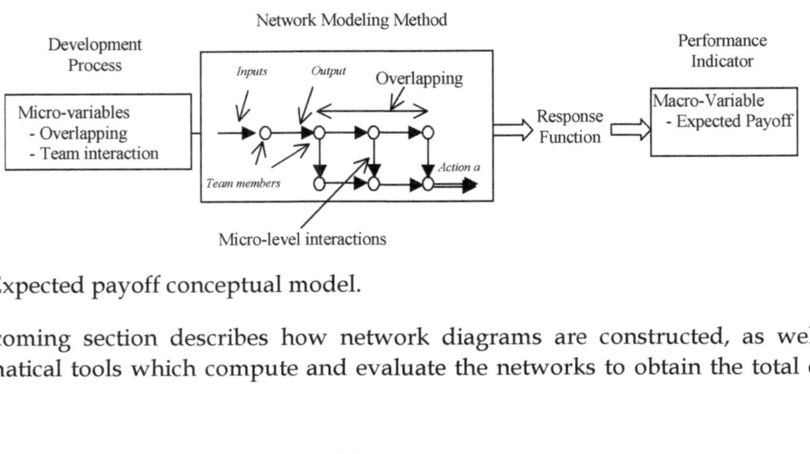

Fig. 4. Expected payoff conceptual model.

An upcoming section describes how network diagrams are constructed, as well as the mathematical tools which compute and evaluate the networks to obtain the total expected payoff.

4.2 Mathematical model: Definition of fundamental quantities

In the following sections, the fundamental quantities of the expected payoff model are defined mathematically.

4.2.1 Actions and outcomes

Faced with a set of alternatives, the decision made by the decision-maker is called his action, a. An action, or decision, can have more than one outcome (or result or consequence). This is denoted as $r = \rho(a)$, where ρ is the outcome function of the action taken. The possible outcomes also depend on external factors out of the decision-maker's control, which can be called the environment, represented by the variable x. Since an outcome depends on both the action taken and the environment, the outcome function can now be expressed as $r = \rho (x, a)$. Because x is uncertain, the outcome variable r given a is also said to be uncertain. The decision problem now is made up of a set X of alternative states of the environment x, a set A of all possible actions a, a set R of all possible outcomes r, and an outcome function ρ from X x A to r, giving the outcome of each state-action pair, $r = \rho (x, a)$.

4.2.2 Decision rules

The problem of choosing among alternative actions can be generalized by saying that individuals choose among rules of actions or strategies, rather than from a set of possible actions alone. In an organization, rules of action play a very big role in contingency planning, where team members must decide in advance how they will respond to incoming information. This is obviously important in making economic decisions because individuals must be ready to act as soon as they can (e.g. stock brokers). In the context of an engineering firm, if a task is to design and develop a new product and get it to market as quickly as possible, the designers make use of incoming information as soon as they receive it, and they must additionally decide upon how much of it should be transferred to downstream functions, and when.

An action can now be described as $a = \alpha (y)$, where α is the decision function, and y is the information which will be obtained in the future. It should be noted that the information y is not the same as the variable x, the state of the environment, which describes information already received. The expression says that an action a depends upon the information y received.

4.2.3 Information

Information can be generated and obtained by team members through various means, such as through observation, communication, and/or computation. There are two sets of information available to the decision-maker: one is the set X of all possible states of the environment, and the other is the set Y of all possible information signals. An information signal y is a partition of the environment X. The information structure η is the partitioning of X into different signals of y. Therefore, a signal y will correspond to each x in X. An information structure is thus defined as $y = \eta (x)$. Any partition of X can be viewed as a way to describe the states of the environment. As an example, suppose a marketing manager can offer a customer a product in small, medium, or large. The customer wants either small or medium. The marketing manager must make a decision about which size to choose, which will impact the design of the product, and the information relevant to his decision is the size small or medium. The set X of all possible sizes is thus partitioned into one subset, small, and another subset, medium, and this partitioning defines the information structure, η.

4.3 Expected payoff

The expected payoff method is based on the premise that every individual has preferences as to how to prioritize a list of alternatives due to personal beliefs or interests (assuming that the individual is consistent). Preferences can be described by the ranking of alternatives according to some subjective probability distribution, consistent with what the person believes will happen. Under uncertainty, each of the alternatives is an action which may result in one or more outcomes, as discussed.

In this sense, the term 'utility' refers to the usefulness an action brings to an individual. For the order of preference given to all actions, each position can be assigned a single number representing the utility of each, or a person's desirability of the occurrence of an event, thus capturing his preferences. The probability of each action occurring is represented by the subjective probability assignments. The expected utility of an action is therefore the sum of the utilities of its various possible outcomes, weighted by the probability of each outcome's occurrence.

Given these basic definitions described in the previous section, for a set R of alternative outcomes r1… rN, if Zi(a) denotes the event that an action a results in the outcome ri (since ri = ρ (x, a)), then the "expected utility" Ω for an action a is:

$$\Omega\ (a; \rho, \pi, \upsilon) = \Sigma\ \upsilon(ri)\ \pi\ [Zi(a)] \tag{1}$$

where:

π = subjective probability function
υ = utility function.

The left-hand side of the expected utility function in (1) shows that the expected utility depends only on the decision-maker's action, given the functions ρ, π, and υ, which describe the factors which are out of his control. The individual's actions are under his control, and his goal is to choose the action which maximizes the corresponding expected utility. In the utility function υ (r) in (1), r can be replaced to obtain the new payoff function ω: υ (r) = υ [ρ (x, a)] $\equiv$ ω (x, a). The expression in (1) can be further simplified by stating that, given the set X of alternative states of the environment x, the probability of the state x can be written as Φ (x) = π ({x}), where Φ is the probability density (or mass) function, and x is assumed to be a random variable that is normally distributed. This expression is, in other words, the probability of the set X consisting of the single element x denoted by {x}. The expected utility function in (1) can be re-written as:

$$\Omega\ (a; \omega, \Phi) \equiv E\ \omega\ (x, a) = \Sigma\ \omega\ (x, a)\ \Phi\ (x) \tag{2}$$

The expression in (2) can now be called the expected payoff of the action a, where the expected utility depends on the decision-maker's action only, and where ω and Φ describe the factors uncontrolled by the decision-maker. Though the utility and the probability functions may be thought of as being controllable, it is assumed for simplicity that they are not, and that they are treated as givens of the problem. By replacing the actions by decision rules, and introducing the information structure into the equation, the payoff function can be re-written as:

$$\omega\ (x, a) = \omega\ [x, \alpha\ (y)] = \omega\ [(x, \alpha\ (\eta\ (x))] \tag{3}$$

From this, the expected payoff becomes:

$$U = \Sigma\ \omega\ [(x, \alpha\ (\eta\ (x))]\ \Phi\ (x) \equiv \Omega\ (\eta, \alpha; \omega, \Phi) \tag{4}$$

The expected payoff now depends on the decision function α and the information structure η, and on the factors over which the decision-maker has no control, namely ω, and Φ. The information structure η is assumed to be under the control of the team member; each member has the ability to observe and partition the information into the subsets needed for his activity. The individual has more than one pair (η, α) available, and he will choose the one that maximizes U. This expression is the measure that describes the performance of the various processes through the evaluation of actions under uncertainty. It is used to evaluate the process network diagrams to be developed in upcoming sections. The optimal process structure will be that which maximizes the expected payoff of the network under certain conditions, given the probability distribution of the states of the environment.

4.3.1 Expected payoff as a quadratic function

The payoff function can be expressed as a quadratic function of the team action variables. Although this is an approximation, it is useful. The quadratic function is one that has been used to describe many real-life phenomena, such as in economics for the law of diminishing returns. The concave quadratic function describes the expected payoff function in that there is a point that is optimum, i.e., the maximum point, and before and after this point, the value of the payoff decreases. The use of functions of orders higher than two is very complex and difficult to solve, and a linear function is neither sufficient nor appropriate to describe the present phenomena in detail since it is not expected that the payoff function continuously increases or decreases. Also, since the goal here is to make comparisons between two process structures, the relative comparisons do not require the payoff function to be exact. Taking the case of two members in a team, 1 and 2, where each must make a decision, then the quadratic payoff function can be chosen as:

$$u = -a12 - a22 + 2Q\ a1\ a2 - 2\eta 1\ (x)\ a1 - 2\eta 2\ (x)\ a2 \quad (5)$$

(the use of functions of x will be suppressed for simplicity in the future).

This particular form of the quadratic function is similar to the one used by Marschak and Radner (1972), with some of the coefficients chosen to simplify calculations. In the above expression, Q measures the interaction between a1 and a2, the action variables of team members 1 and 2, respectively, and must be between zero and one. The interaction is one of the micro-variables of the process model. For M action variables, if the second derivative of the expected payoff function exists, then a measure of the interaction between the action variables i and j is $\partial 2\omega\ /\partial ai\partial aj$. In other words, it measures "the degree to which a change in action j influences the effect of a change in action i on the payoff for given values of the other action variables and of x" (Marschak and Radner, 1972, p.101). The functions η1 (x) and η2 (x) are related to the information structure.

The assumption that the payoff is quadratic gives meaning to the variances and the correlations of the information variables. Normal distributions are fully described by their means and variances. The variance can help gauge uncertainty, as it takes the difference between the maximum and minimum values of x. For multivariate distributions, the correlation coefficient describes the degree of statistical interdependence between variables (above, r describes the correlation between two variables). Due to the interdependencies in processes, it is often important to understand how one variable affects another. Whether this is the correlation between action variables or the environment variables, it is reasonable that these correlations may affect the information structure and/or probability of occurrence chosen by the organizer. Another simplification is the normalizing assumption, where each variable is considered to be measured from its mean, so that m = 0, and E (m) = 0. There is no loss of meaning since this is simply a coordinate transformation. In this case, the correlation coefficient becomes:

$$r = r12 = Ex1x2/s1s2, \text{ or}$$

$$Ex1x2 = r * s1s2.$$

These properties of probability distributions will be useful in solving the expected payoff functions and determining the macro-variables of interest, discussed in upcoming sections.

4.3.2 Multi-person teams

For n members in a team, then there will also be n information structures and n decision rules. Each member i chooses an action ai from set Ai of all possible alternatives. The payoff function can be written as:

$$u = \omega\ (x, a1\ ,a2\ ...)$$

where u is now the utility to the team (and to each of its members). Although ai is the action variable controlled by the ith member, ai itself can be an m-tuple of many distinct variables, each controlled by the ith member. If there is no interaction among the action variables, then the payoff is said to be additive, and the form of the expression becomes:

$$\omega\ (x, a) = \Sigma\ \omega\ i\ (x, ai)$$

If, however, there are interactions among action variables, then the quadratic function must include an extra term to express this interaction, namely Q, as before.

4.3.3 Team decision functions and information functions

In a single person team, the person's action is related to the decision function through a = α(y). For a multi-person team, there are now n decision functions, α = (α1 … αn) and ai = α1(yi). The same decision rules as before can be applied for a team. The joint action of the team members is a = (a1,… an), and y = (y1 , ... yn) is the team information, so there are n decision functions, and the team decision rule can then be denoted as α = (α1, α2,…. αn). The same expression for an action a = α(y) for a single person team is also applicable for teams, keeping in mind what each term means individually. The information structure for each team member can be expressed as yi = ηi (x), and for the team, the information structure is η = (η1,... η n). Then, for y = η (x), and a = α (y), a = α [η (x)] applies for the team action. The payoff of the team can be written, as before:

$$u = \omega\ (x, a1\ ,a2\ ,\ldots) = \omega\ (x, \alpha\ 1\ [\eta 1\ (x)], \ldots\ \alpha\ n\ [\eta n\ (x)]) = \omega\ (x, \alpha\ [\eta\ (x)],$$

and the expected payoff of the team is:

$$E\ (u) = \Omega\ (\eta, \alpha) = E\ (\omega\ (x, \alpha\ [\eta\ (x)]) \tag{6}$$

4.3.4 Consideration of time

All the discussion up until now has involved the static case of the team decision problem, but time can be incorporated into the various concepts. If one team member's action at time t (t = 1,...T) is ai(t), and x(t) is the state of the world at time t, then the team action variable becomes:

$$a = [a1(1),\ldots an(1),\ldots an(T)],$$

and the state of the world is:

$$x = [x(1),\ldots x(T)].$$

For yi (t) = ηi (x, t), the action variable becomes:

$$ai(t)= ai[\eta i\ (x,t\ ),t\ ].$$

An important assumption of this situation is that for actions that are spaced apart in time, the larger the time difference, the less the interaction between those actions. Therefore, it is assumed for simplicity that the actions that are distant in time need less coordination than those that are closer together. The payoff function with no interaction is thus additive in time, and can be expressed as:

$$\omega\ (x, a) = \Sigma\ \omega t\ [x, a\ (t)]\ (\text{for } t=1,\ldots T).$$

4.4 Design of network models

Networks can be used as a powerful tool to represent and evaluate the structure of a process, and more specifically, the structure of information flow and work patterns in a team. A network can be defined as a system of interconnected elements, all of which work together to produce a desired output. A network consists of the following basic components:

- element (represented by a circle): the component which has the function of transformation of information. An element can represent a human being, a machine, a communication tool, etc., in the process of performing an activity.
- input(s) (represented by an arc into the element): required for each element. These inputs are various types of information (e.g. information or actions coming from the previous element's output, noise from the environment, team members' personal knowledge or expertise, etc.).
- output(s) (denoted by an arc leaving the element): the result of each element. This can be in the form of 1) processed information, 2) a simple relay or distribution of information, or 3) an action being sent out as a result of the transformation process, either to another element or to the environment.

Each element in a network has an input which is transformed into or transferred as an output; the message of this output then feeds into one or more downstream elements. Elements are connected to one another through the input and output arcs, which carry information messages to and from elements. Messages coming from the environment (i.e., external to the organization) are called observations. Messages from one element to another are communication, while messages going out into the environment are called actions. In the context of an organization, networks can be used to represent processes from an information processing point of view. Once all intermediate elements have been completed, a final action(s) is issued, which signals project completion. Networks can be organized according to time structure.

4.4.1 Connections between elements

The connections between elements in a network can be described in the form of a square array. For each element i and j, the set of all possible messages that can be sent from i to j, is denoted by Bij. Any messages that come from outside, that is, from the environment, are described by the set Zi, and Ei which denotes the set of all possible values of noise coming from the environment to element i. This noise can be information that is observed from outside the organization, such as customer input, best practices, etc. The messages sent out to the environment are defined as the action variables, a = (a1, … an), for n actions, where a is the team action variable.

The set Bi0 denotes the set of all possible messages from element i to the environment. This set will consist either of the Cartesian product of some sets Aj, where for each j, Aj is the set of all possible values that action variable aj can take, or it will be empty since not all elements will have an action as an output. The set B0i is symmetric to this set, and it represents the set of all possible messages from the environment outside to an element i, which is the Cartesian product of Zi and Ei. Therefore, the set Bi of possible alternative output messages of element i is denoted by Bi = Π Bij (j=0...m). For m elements, the set B́i of combined messages from other elements to i is given by B́i = Π Bki (k=0...m). The transformation of each element i is expressed through the task function βi = (βi0,..., βim), which transforms each input message into an output message. The set Bii is empty as it is assumed that messages will not be sent from an element to itself. Figure 5 below shows an example of a simple network diagram. The corresponding square array consisting of the sets Bij in Table 1 illustrates message transfers between elements in the figure. The symbol Φ denotes an empty set.

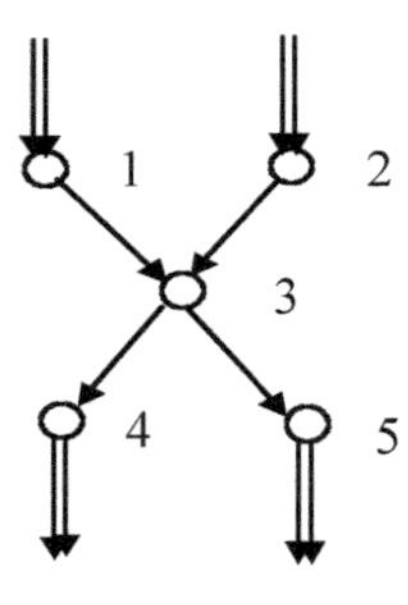

Fig. 5. Simple network diagram.

	0	1	2	3	4	5
0	□	B01	B02	□	□	□
1	□	□	□	B13	□	□
2	□	□	□	B23	□	□
3	□	□	□	□	B34	B35
4	B40	□	□	□	□	□
5	B50	□	□	□	□	□

Table 1. Information dependencies.

The time distribution and spatial distribution of members in a team must be separated. For teams in a dynamic environment, networks are divided into time periods by several elements based on the structure of information flow, which is illustrated by elements broken down into intermediate stages or actions. Figures 6 and 7 show the network diagrams of possible sequential and CE processes.

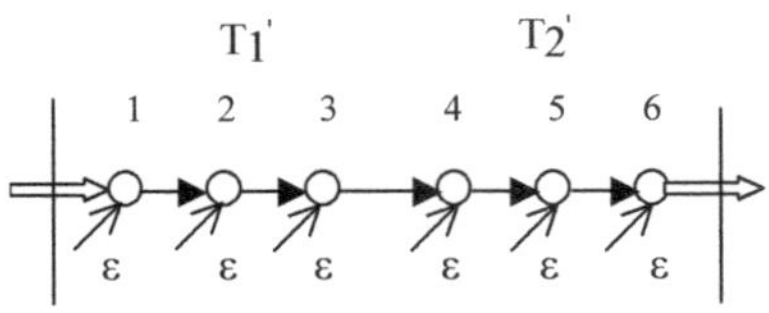

Fig. 6. Sequential network diagram.

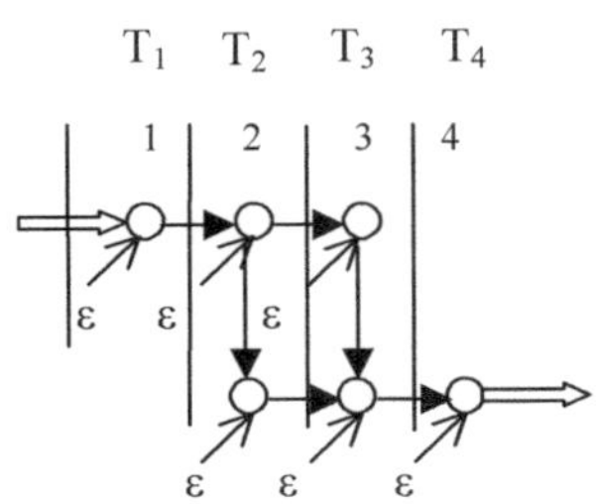

Fig. 7. CE network diagram.

In the context of NPD processes, as an example, an element i can be a designer who receives information ε from the environment, say from the customer. An action a can be the release of design specifications from the designer to another element, say the manufacturing resource. This resource then uses information from this action as well as information from observations through personal experience and/or company databases for example, transforms the new combined information, and takes an action, such as manufacturing the product. This action is sent out to the environment, i.e., the customer.

4.5 Application to models

A simple model is designed in this section using network diagrams and its evaluation using the expected payoff method will be studied. The main purpose here is to compare the relative differences between process structures in terms of the expected payoff. In the evaluation that follows, the expected payoff function is equation (5), repeated here:

$$u = - a12 - a22 + 2Q\ a1\ a2 - 2\eta 1\ a1 - 2\eta 2\ a2 \qquad (5)$$

Recall that the coefficient "Q" in (5) denotes the interaction, specifically in this case between two overlapping activities occurring in the same time period. This function evaluates every step of performing work, and also evaluates the different processes as a whole (i.e., each intermediate step is evaluated using this function, as well as the overall structure of the network).

4.5.1 Assumptions of the model

In order to compare the two processes at the same level, some assumptions must be made to ensure consistency. First, both of the processes begin with the same input information variable ηi which is a random variable dependent upon x. Thus, the first member of each process begins by observing the same information that is coming from the environment. Another assumption in the model is that the members inside the organization not only receive information from other sources (i.e., other elements or the environment), they also contribute to the processing of their work through the use of their own expertise, which is denoted in the models by ε as an input into each element (Howard, 1966). However, this is considered as being a special state of the set X of information from the environment despite the fact that it comes from the element itself. Therefore, during the evolution of the activity, not only does the information that a member receives get processed, but also because each member is contributing his own knowledge and expertise, this pooled information adds value to the activity, which results in an increase in the expected payoff.

Earlier, it was mentioned that the choice of the payoff function as a quadratic equation is appropriate since a quadratic function has a maximum point. This assumption is important and must be re-stated. Furthermore, since the expected payoff is the measure being used to compare relative process performance, an absolute measure is unnecessary, so the problem of defining a specific and accurate form of the function can be avoided.

The cost of a network is not considered in the models. Marschak and Radner (1972) did not include this important factor explicitly in their decision functions, although they acknowledge its importance. There is a cost associated with decision-making, with how information is obtained, with team organization, etc. In the context of this research, cost was not chosen as a parameter of the models, however, since cost can help in assessing the trade-offs of one process design over another.

The sequential and CE process networks are created as sequences of single-period decision problems (see Figures 9 and 10), where the interaction between periods is assumed to be zero, i.e., Q=0, as previously discussed in Section 4.4.4. Thus, interaction is assumed to be zero across periods, though there is interaction within time periods. Therefore, it is assumed that the total payoff function is additive in time. In each time period, optimal decisions are made. This difference in interactions addresses the case for which the sum of the maximum expected payoffs is equal to maximum of the sum of expected payoffs. In other words:

$$\text{Max}\ \Sigma\ E\ \omega\ (x, a) = \Sigma\ \text{Max}\ E\ \omega\ (x, a) = \Sigma\ E\ \omega\ \text{max}\ (x, a).$$

Thus the maximum expected payoff is calculated in each period, which are the added up to give the total maximum expected payoff. In other words, the expected value of the maximum payoffs is equal to the total maximum expected payoffs for each period.

Rework is not modeled in either the sequential or CE process. Though this is a simplifying assumption not characteristic of most NPD processes, the model is presented in basic form, with the intent of bringing out some essential features of the expected payoff method.

The simple network shown below in Figure 8 will be defined here to illustrate how a network diagram is evaluated in terms of its gross expected payoff ('gross' since the cost of a network is not considered). The network is assumed to be in one time period, reflecting a single action. In cases when there is more than one element, each element in the network can be evaluated separately in terms of its expected payoff, and the total expected payoff is simply the sum of the individual ones. Actions taken at different times can be considered to be corresponding to different team members.

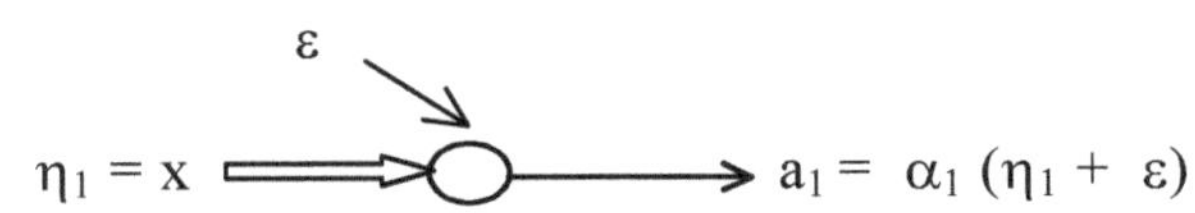

Fig. 8. Action taken in one time period.

Figure 8 illustrates an action. Element 1 has $\eta 1 = x$ as input variable, where $\eta 1$ is a random variable dependent upon the state of nature x (which is suppressed). The state variable observed by the team member at element 1 is processed, which also receives some information ε from outside (qualified as information such as team member's personal expertise), which, for simplicity, is considered to be a constant. This information is processed, and an action a1 is taken, which is a function of the inputs to the element. The information ε combined with the information $\mu 1$ is additive. With a single action, the payoff function is chosen as a quadratic in one input variable, in the form:

$$\omega = -a12 + 2a1x \tag{6}$$

Taking the derivative of (6) with respect to a1 and setting it equal to zero gives:

$\omega'(a1) = -2a1 + 2x = 0$, and solving for a1gives the best decision function, denoted by $\dot{\alpha}$:

$$\dot{\alpha}(x) = x \tag{7}$$

which is the optimal decision. The second derivative of (6) is negative (-2), ensuring a maximum point, so plugging (7) back into (6) and taking the expected value of the payoff gives the following expected value of the maximum payoff:

$$\Omega = E(x + \varepsilon)2 = s2 \tag{8}$$

where s2 is the variance of x. The decision function in equation (7) has a distribution of possible decisions, which implies that multiple choices can be made. Assuming that this distribution is normal, then, equation (8) shows that the payoff is equal to the variance. This means that in making a decision, i.e., reducing possible choices to a single value, the payoff is equal to the value of the reduction of uncertainty of information. It is reasonable to conclude that the larger the variance, i.e., the more uncertain the decision, then, the more benefit (payoff) there is in making a decision.

4.5.2 Sequential engineering network diagram

The sequential engineering network diagram is illustrated in Figure 9, and consists of six time periods, T1 to T6, which represent the division of the sequential work done by six different functional team members. Team member 1 receives complete information represented by the state variable x, and then uses this information, along with his own expertise represented by ε, to complete his activity. At the end of his activity, he sends complete information to the downstream activity, which is again processed by the second team member. The output of this activity is a message sent to the next team member, etc.

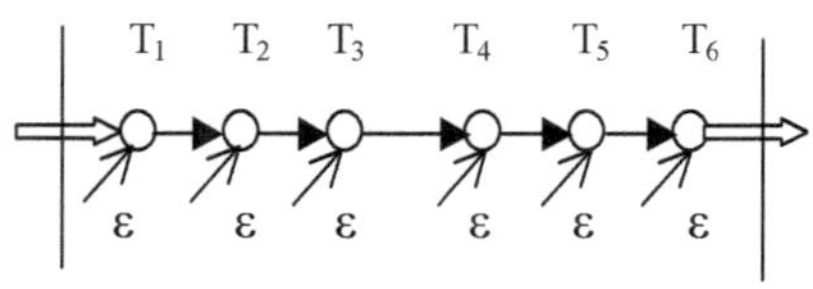

Fig. 9. Sequential network diagram.

Because of the assumption of no cross-functional communication in a sequential process, there is no interaction between team members, and the communication of information is assumed to be 'over-the-wall', thus even if there is some interaction, it is assumed to be so weak that it is negligible.

Evaluation of the Network

1. First Period T1:

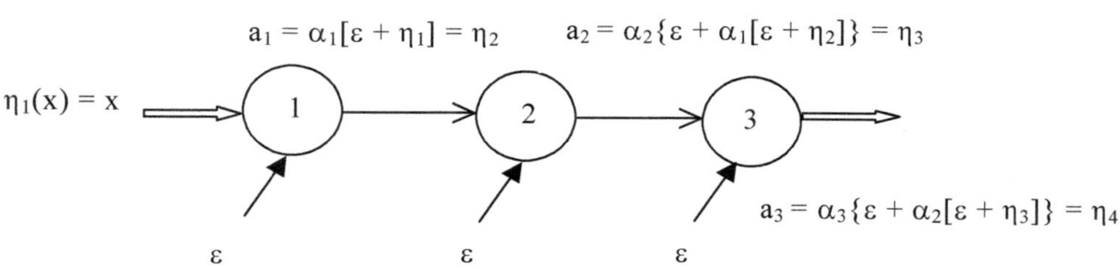

Similar to the example above, team member 1 receives complete information, and every member within the network contributes his special technical knowledge to the flow, denoted by ε. As before, the payoff function is (6) and the expected payoff is (9):

$$\omega = -a12 + 2a1x \tag{6}$$

$$\Omega1 = s02 \tag{9}$$

where s02 = E(x + ε) 2.

2. Second Period T2:

As in time period 1, member 2 receives output x from element 1, giving the payoff:

$$\Omega\, 2 = s12 \tag{10}$$

where s12 = E(x + ε + ε) 2.

A similar procedure as above applies to each time period, up until time period six. The total Expected Value of the Maximum Payoff:

$$\Omega = \Sigma\ (i = 1\ldots6)\ \Omega\, i = \Omega\, 1 + \Omega\, 2 + \Omega\, 3 + \Omega\, 4 + \Omega\, 5 + \Omega\, 6$$

$$= s02 + s12 + s22 + s32 + s42 + s52$$

If it is assumed for simplicity that the variance for each information structure is the same, then the total expected payoff becomes:

$$\Omega\ TOT = s02 + s12 + s22 + s32 + s42 + s52 = 6s2 \tag{11}$$

4.5.3 Concurrent engineering network diagram

The concurrent engineering diagram shown in Figure 10 is an appropriate modification of the sequential engineering network. It takes into account the two teams of three members each, but this time with a few added features. The two teams' activities are now overlapping in time periods T2 and T3. These two teams now are also communicating with each other through the transfer of information denoted by the arrows between overlapped activities. There is interaction between the two members from both teams in the same two time periods. Since rework is not modeled, overlapping part of the six time periods in the sequential process gives the resulting four time periods in the CE process. The main comparison of interest at this point is the difference between expected payoffs.

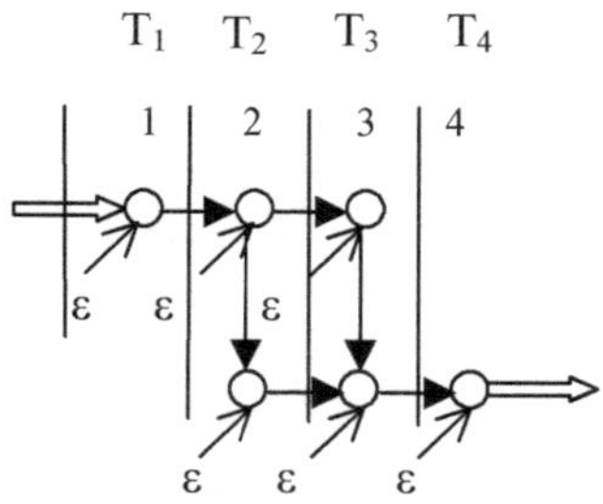

Fig. 10. CE network diagram.

Evaluation of the Network

1. First Period T1:

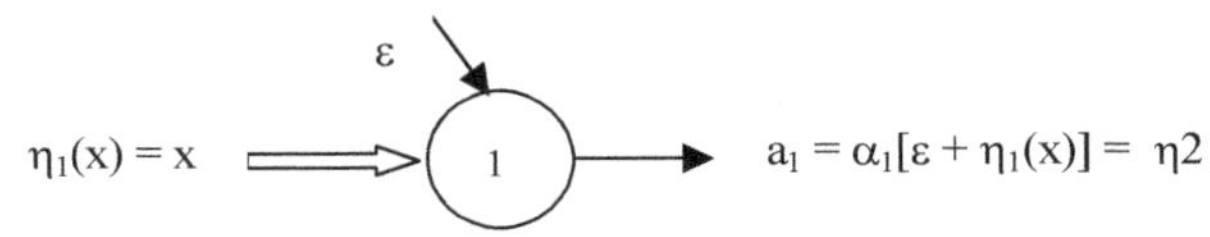

Again, it is assumed that member 1 obtains complete information x, that is the information structure $\eta 1(x)=x$. Also, every member within the network contributes his special technical knowledge to the processing and transferring of information. The output $\alpha[x] = a1$ is determined as before. Choosing $u1= \omega (x, a1) = -2a12 + 2a1[x + \varepsilon]$, the best decision function is:

$$\dot{\alpha}[x] = x + \varepsilon$$

and the expected value of the maximum payoff for the first time period is:

$$\Omega 1 = s\,02 \tag{12}$$

where $s02 = E[x + \varepsilon]$

2. Second Period T2:

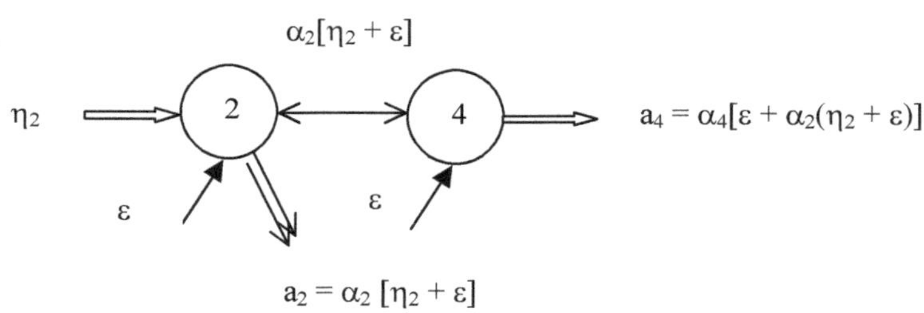

The payoff function is:

$$\omega(x, a) = -a22 - a42 + 2Q\, a2\, a4 - 2\,\eta\, 2\, a2 - 2\,\eta 4\, a4 \tag{13}$$

Taking the first derivative of the payoff function first with respect to a2 and a4 setting each equal to zero:

$$\delta\omega/\delta a2 = -2a2 + 2Q\, a4 - 2\,\eta\, 2 = 0 \tag{14}$$

$$\delta\omega/\delta a4 = -2a4 + 2Q\, a2 - 2\,\eta\, 4 = 0 \tag{15}$$

gives the following system of equations:

$$-\, a2 + Q\, a4 = \eta\, 2 \tag{16}$$

$$Q\ a2 - a4 = \eta\ 4 \quad (17)$$

Solving for a2 and a4 yields the following best decision functions for T2 period actions:

$$\dot{\alpha}2 = [-1/(1\text{-}Q2)] * \eta\ 2 + [-Q/(1\text{-}Q2)] * \eta\ 4$$
$$= [-1/(1\text{-}Q2)] * [x + \varepsilon + \varepsilon] + [-Q/(1\text{-}Q2)] * [x + \varepsilon + \varepsilon + \varepsilon] \quad (18)$$

$$\dot{\alpha}4 = [-Q/(1\text{-}Q2)] * \mu\ 2 + [-1/(1\text{-}Q2)] * \eta\ 4$$
$$= [-Q/(1\text{-}Q2)] * [x + \varepsilon + \varepsilon] + [-1/(1\text{-}Q2)] * [x + \varepsilon + \varepsilon + \varepsilon] \quad (19)$$

Plugging (18) and (19) into (13) gives the payoff for time period T2:

$$\omega = [\eta\ 22 + 2Q\ \eta\ 2\ \eta\ 4 + \eta\ 42]/\ [1\text{-}Q2], \quad (20)$$

and the expected payoff is:

$$\Omega\ 2 = E\omega = [s12 + 2Qr12\ s1\ s2 + s22]/\ [1\text{-}Q2] \quad (21)$$

where:

$$s12 = E\ [x + \varepsilon + \varepsilon]\ 2$$

$$s22 = E\ [x + \varepsilon + \varepsilon + \varepsilon]2$$

$$r12 = [E\ (x + \varepsilon + \varepsilon)(x + \varepsilon + \varepsilon + \varepsilon)]/\ s1\ s2$$

where r12 is the correlation coefficient and Q is the interaction.

3. Third Period T3:

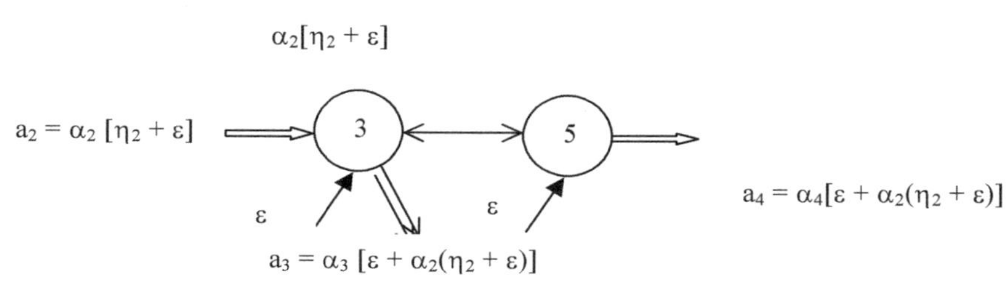

The payoff function is:

$$\omega = -a32 - a52 + 2Q\ a3\ a5 - 2\ \eta\ 3\ a3 - 2\ \eta\ 5\ a5 \quad (22)$$

where:

$$\eta\ 3 = \varepsilon + \dot{\alpha}\ 2 = [-1/(1\text{-}Q2)] * [x + \varepsilon + \varepsilon] + [-Q/(1\text{-}Q2)] * [x + \varepsilon + \varepsilon + \varepsilon] + \varepsilon \quad (23)$$

$$\eta\ 5 = \varepsilon + \dot{\alpha}\ 2 = [-Q/(1\text{-}Q2)] * [x + \varepsilon + \varepsilon] + [-1/(1\text{-}Q2)] * [x + \varepsilon + \varepsilon + \varepsilon] + \varepsilon \quad (24)$$

Performing the same calculations as in T2 gives the following best decision functions:

$$\acute{\alpha}3 = [-1/(1-Q2)] * \eta\, 3 + [-Q/(1-Q2)] * \eta\, 5$$

$$= [-1/(1-Q2)]\ \{-1/(1-Q2)] * [x + \varepsilon + \varepsilon] + [-Q/(1-Q2)] * (x + \varepsilon + \varepsilon + \varepsilon)+ \varepsilon\} + \quad (25)$$

$$[-Q/(1-Q2)]\ \{-Q/(1-Q2)] * [x + \varepsilon + \varepsilon] + [-1/(1-Q2)] * (x + \varepsilon + \varepsilon + \varepsilon) + \varepsilon\}$$

$$\acute{\alpha}5 = [-Q/(1-Q2)] * \eta\, 3 + [-1/(1-Q2)] * \eta\, 5$$

$$= [-Q/(1-Q2)]\ \{-1/(1-Q2)]\ [x + \varepsilon + \varepsilon] + [-Q/(1-Q2)] * (x + \varepsilon + \varepsilon + \varepsilon) + \varepsilon\ \} + \quad (26)$$

$$[-1/(1-Q2)]\ \{-Q/(1-Q2)]\ [x + \varepsilon + \varepsilon] + [-1/(1-Q2)] * (x + \varepsilon + \varepsilon + \varepsilon) + \varepsilon\ \}$$

The expected payoff is:

$$\Omega\, 3 = [s32 + 2Qr34s3\ s4 + s42]/\ [1-Q2] \quad (27)$$

where:

$$s32 = E[-1/(1-Q2)\ [x + \varepsilon + \varepsilon] + [-Q/(1-Q2)] * (x + \varepsilon + \varepsilon + \varepsilon) + \varepsilon]2$$

$$s42 = E[-Q/(1-Q2)\ [x + \varepsilon + \varepsilon] + [-1/(1-Q2)] * (x + \varepsilon + \varepsilon + \varepsilon) + \varepsilon]2$$

$$r34 = E[\eta 3\ \eta\ 5]/\ s3\ s4$$

4. Fourth Period T4:

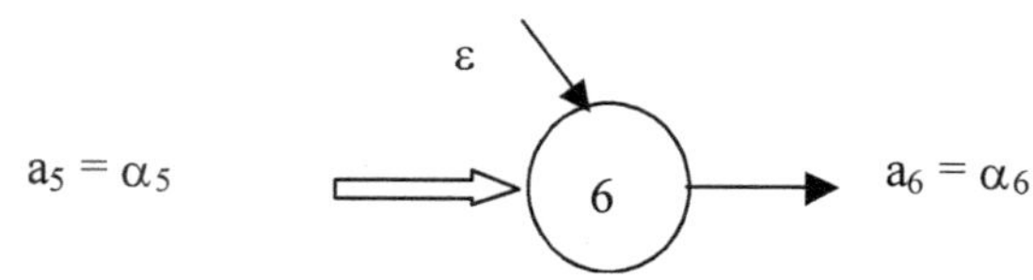

The payoff function is chosen as:

$$\omega = -a62 + 2a6\ \eta\ 6 \quad (28)$$

where:

$$\eta\ 6 = \varepsilon + \acute{\alpha}\ 5$$

$$= (-Q/(1-Q2)\ \{[-1/(1-Q2)]\ [x + \varepsilon + \varepsilon] + [-Q/(1-Q2)]\ (x + \varepsilon + \varepsilon + \varepsilon + \varepsilon) + \varepsilon] + \varepsilon$$

$$+ (-1/(1-Q2)\ \{[-Q/(1-Q2)]\ [x + \varepsilon + \varepsilon] + [-1/(1-Q2)]\ (x + \varepsilon + \varepsilon + \varepsilon) + \varepsilon] + \varepsilon\} \quad (29)$$

The best decision function is:

$$\acute{\alpha}\ 6 = \eta\ 6 \quad (30)$$

Therefore the expected value of the maximum payoff is:

$$\Omega\ 4 = s52 \tag{31}$$

where:

$$s52 = E\ [\varepsilon + \dot{\alpha}\ 5]2$$

Total Expected Value of the Maximum Payoff:

$$\Omega = \Sigma\ (i = 1...4)\ \Omega\ i = \Omega\ 1 + \Omega\ 2 + \Omega\ 3 + \Omega\ 4$$

$$= s02 + [s12 + 2Qr\ s1\ s2 + s22]/\ [1\text{-}Q2] + [s32 + 2Qr34s3\ s4 + s42]/\ [1\text{-}Q2] + s52$$
$$= [s02 + s52] + [s12 + s22 + s32 + s42]/\ [1\text{-}Q2] + [2Q(r12\ s1\ s2 + r34s3\ s4]/\ [1\text{-}Q2] \tag{32}$$

$$\Omega\ TOT = A + B/[1\text{-}Q2] + CQ/[1\text{-}Q2]$$

where the coefficients A, B, C are:

$$A = s02 + s52$$

$$B = s12 + s22 + s32 + s42$$

$$C = 2(r12s1\ s2 + r34s3\ s4)$$

5. Results

From the calculations in the previous section, the total expected payoffs are summarized below for each of the two processes:

Sequential: $$\Omega\ TOT = s02 + s12 + s22 + s32 + s42 + s52 \tag{11}$$

CE: $$\Omega\ TOT = A + B/[1\text{-}Q2] + CQ/[1\text{-}Q2] \tag{32}$$

where the coefficients A, B, and C are as before. The equation for the expected value of the maximum payoff for the sequential process is a constant with respect to Q, while for the CE process it is polynomial in Q. If it is assumed for simplicity that all variances are equal and the correlation coefficients are equal to zero, i.e., the information variables are independent, then Figure 11 depicts the resulting curves for each process.

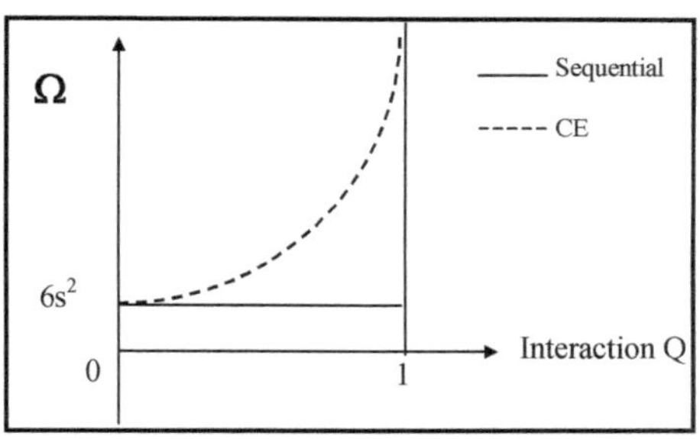

Fig. 11. Expected payoff vs interaction.

This analysis shows that a CE process is always better than a sequential one in terms of expected value of the maximum payoff. This is contradictory to practical observations of both processes. This is due to the fact that the analytical model oversimplifies the sequential process, whereby it is assumed that there is virtually no interaction between phases, and that information is 'thrown over the wall' from one function to another. Under this assumption, there is no interaction, which naturally results in an expected payoff that is independent of the interaction, Q, thus giving a constant. Additionally, it is also assumed in the modeling process that the contribution of each member's specialized information is the same for both the sequential and CE processes. This results in the total expected payoff for the sequential process being always lower than that of CE. Again, this assumption is not consistent with practical observations. In order to make the analysis more meaningful, some further assumptions should be made with regards to team members in a sequential process as compared to a CE process.

In some practical situations, a sequential process can be better than a CE process (Krishnan et al., 1997). When this is true, in the modeling process it is reasonable to assume that for a sequential process, every team member's knowledge and information is sufficient to allow him to finish his activity independently. In fact, it may even be argued that in a sequential process, the amount and types of information that functional members must possess is greater than members in a cross-functional team, which allows them to finish their activity independently. They must possess not only information about their own specialization, but they must also have, to some extent, information about other functions as well. After all, a designer will not design a product which requires milling if the company does not own a milling machine. In contrast, in a CE process, it can be assumed that members on a cross-functional team do not need to possess as much information about other functions since sharing of information will occur naturally as a consequence of teamwork, in which case it is reasonable to assume that more work is required to obtain information. Therefore, the variance of knowledge and information measured by s2 is assumed to be larger for members in a sequential process than for the same members who would work in the overlapped periods in a CE process. This implies that the lack of information or knowledge by members in a CE process can be compensated by the exchange of information in the overlapped periods.

Given this assumption, the straight line in Figure 11 would move up the y-axis, while the CE curve would remain the same. This would create a point of intersection between the two curves, indicating that, for a given point of interaction, one process will be superior to the other in terms of expected payoff. For simplicity, it is assumed that for the CE process, all variances are equal to 1, and that the correlation coefficients are equal to 0. It can be further assumed that team members 2, 3, 4, and 5 in a sequential process have a variance that is slightly higher than the same members in a CE process, who, as explained above, exchange information during the overlapped periods. For simplicity, the variance for the sequential members' information is taken to be one-quarter higher than that of the CE members' information i.e., si2 (sequential) = 1.25 si2 (CE). Plugging these values back into 11 and 32, the total expected payoffs are:

Sequential: Ω TOT = 8.25

CE: Ω TOT = 2 + 4/[1-Q2]

This analysis is now illustrated in Figure 12.

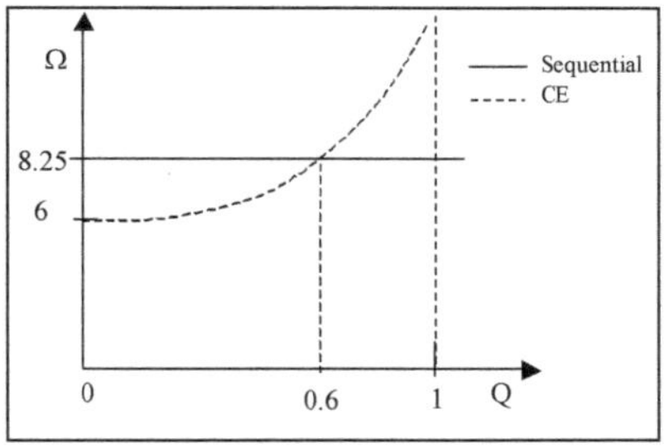

Fig. 12. Expected payoff vs interaction.

The curve for the CE process shows how the expected value of the maximum payoff changes with interaction, showing that as team interaction increases, the expected payoff increases as well. For this particular case, it was found that a sequential process has a higher expected payoff when the interaction is lower than 0.6, and a CE process has a higher payoff for values of interaction greater than 0.6. In other words, the results show that when actions in a CE process highly influence one another, i.e., the interaction is higher than 0.6, then a CE process is more valuable in terms of expected value of the maximum payoff. If the interaction between action variables is not strong, i.e., less than 0.6, then the sequential process is sufficient, and superior in terms of expected payoff.

In conclusion, the expected payoff method from decision theory provided some initial results in the comparison of a sequential and CE process. From the mathematical derivation presented in this chapter, comparing equation (11) to (32) shows that a CE process is always more valuable than a sequential process in terms of expected payoff. In most instances in reality, however, a sequential process has some benefit. Under the conditions when this holds true, the sequential process has a higher total expected payoff when the interaction intensity is low, while CE is better than a sequential process for high interaction.

6. Conclusion

The expected payoff method is presented here as a very simple introduction to studying NPD processes. A more elaborate and detailed development is in progress (Kong and Thomson, 2001), the results of which are expected to provide a major contribution to the existing body of work in studying organizational processes and their coordination.

Some avenues for future research are now discussed. In the comparison of sequential and CE processes, it is assumed that there is no interaction between team members in the sequential process, thus emulating the 'over-the-wall' approach, where team members throw information over an invisible wall. In practice however, there exist interactions among members (or departments) of a team, though they may be very weak. Future work should consider this. It was stated that the expected payoff method assumes that individuals in a team work towards achieving common goals with common interests and beliefs within the constraints of their work, all of which guide their behavior. For a CE team, this is conceivable in the sense that any 'team' usually works together to achieve some goal, and a cross-functional team, ideally, works towards the common project goals of being on time, and within budget. However, in practice there is tension between meeting project and functional goals, as team members have project-specific goals, but also have departmental obligations to fulfill. The same assumption is debatable for a sequential process, where functional teams in different activities tend to have differing goals. For example, in isolation, a designer's goal is to create a product design without much concern for the production process that will build it. Similarly, a marketing manager's goal is to get customers to buy the company product without much concern for how the product will be made. This is partly due to the fact that functional goals are tied to functional rewards. Taking into account this divergence of beliefs would require further analysis into economic and organization theory where individuals' actions are based on self-interest. A more detailed description of the influence of time on the payoff function must be developed. Presently, it is assumed that interaction between action variables at different times is weaker the farther apart they are in time. However, if there is interaction between actions at different times, the payoff function will not be additive in time. The sequential process will have constraints which link actions that are distant in time, and can no longer be evaluated as a series of single-period problems, in which interaction is so weak that it does not exist. Most activities are not deterministic in a product development process. In fact, many situations arise where a stochastic relation between activities apply. A commonly occurring phenomenon is the failure of one or more activities, which consequently require rework. Rework loops in the network diagrams must be expressed to incorporate this very important characteristic of development processes. The function of rework in a network is to prevent the expected payoff from reaching a maximum when an activity is reworked, though a maximum can be reached after a few iterations but at a greater cost. The measure of the expected utility of a network has always been considered in its gross form, that is, without any consideration for the cost of the network. In reality, obtaining information can be very costly, and though one network may be superior to another in terms of the expected payoff, the cost of that network may not justify its use. This concept should also be incorporated into the models.

7. Acknowledgment

The author is grateful to Dr. Vincent Thomson and Dr. Linghua Kong for their guidance and constructive advice during this research.

8. References

Adler, P.S. (1995). Interdepartmental Interdependence and Coordination: The Case of the Design/Manufacturing Interface, *Organizational Science*, Vol.6, No.2, pp. 147-167.

Blackburn J. (1991) New Product Development: The New Time Wars, in J. Blackburn (ed.), Time-Based Competition: The Next Battleground in *American Manufacturing*, (pp. 121-163), Business One Irwin, ISBN 13: 9781556233210, Homewood, Ill.

Clark, K.B. and Fujimoto, T. (1991). *Product Development Performance*, Harvard Business School Press, ISBN-13: 978-0875842455, Boston, Massachusetts.

Eastman, R.M. (1980) Engineering information release prior to final design freeze. *IEEE Transactions on Engineering Management*, No.27, pp. 37-41.

Fishburn, P.C. (1970). Utility Theory for Decision-Making, Publications in *Operations Research*, No. 18. John Wiley and Sons, New York.

Galbraith, J. (1973). Designing Complex Organizations, Addison-Wesley, ISBN:0201025590, Massachusetts.

Ha, A.Y and Porteus, E.L. (1995). Optimal Timing of Reviews in Concurrent Design for Manufacturability, *Management Science*, Vol.41, No.9, pp. 1431-1447.

Howard, R. (1966). Information Value Theory. *IEEE Transactions on Systems Science and Cybernetics*, Vol.SSC-2, No.1, pp. 22-27.

Krishnan, V., Eppinger, S.D., and Whitney, D.E. (1997). A Model-Based Framework to Overlap Product Development Activities, *Management Science*, Vol.43, No.4, pp. 437-451.

Loch. C. and Terwiesch, C. (1998). Communication and Uncertainty in Concurrent Engineering, *Management Science*, Vol.44, No.8, pp. 1032-1047.

Marschak, J. and Radner, R. (1972) *Economic Theory of Teams*, Yale University Press, ISBN 13: 9780300012798, New Haven and London.

Marschak, J. (1959). Efficient and viable organizational forms, in *Modern Organization Theory*, M. Haire, ed, Wiley, New York.

Marschak, J. (1954). Towards an economic theory of organization and information, In R.M. Thrall, *Decision Processes*, C.H. Coombs, and R.L. Davis, eds. Wiley, New York, 1954.

Nihtila, J. (1999). R&D-Production integration in the early phases of new product development projects, *Journal of Engineering and Technology Management*, Vol.16, pp. 55-81.

Schilling, M.A. and Will, C.W. (1998). Managing the New Product Development Process: Strategic Imperatives, *IEEE Engineering Management Review*, Vol.26, No.4, pp. 55-68.

Smith, P. G. and Reinertsen. (1991). *Developing Products in Half the Time*, Van Nostrand Reinhold Publishers, ISBN 0-442-00243-2, New York.

Tian, H. et al., "A Review Upon Concurrent Engineering," *9th IFAC Symposium on Information Control and Manufacturing*, France, June 1998, p. 514.

Ulrich, K.T., and Eppinger, S.D. (2011). *Product Design and Development*, McGraw-Hill, ISBN-13: 9780073404776, New York, 2011.

von Neumann, J. and Morgenstern, O. (1953). *Theory of Games and Economic Behaviour*, Princeton University Press, ISBN 978-0-691-13061-3, Princeton, N.J., 3d. ed.

Wheelwright, S.C., and Clark, K.B. (1992). Revolutionizing Product Development, Quantum Leaps in Speed, Efficiency and Quality, The Free Press, ISBN-13: 978-0029055151, New York.

Winner, R. I., J. P. Pennell, H. E. Bertrand, and M. M. G. Slusarezuk (1988). *The Role of Concurrent Engineering in Weapons System Acquisition*, Institute for Defense Analyses, Alexandria, VA, USA, IDA Report R-338.

Yassine, A.; Chelst, K.R. and Falkenburg D. R. (1999). A Decision Analytic Framework for Evaluating Concurrent Engineering, *IEEE Transactions on Engineering Management*, Vol.46, No.2, pp. 144-157.

13

Distinct Element Method Applied on Old Masonry Structures

Marwan Al-Heib
Ineris – Ecole des Mines de Nancy, Parc de Saurupt
France

1. Introduction

Masonry structures have specific aspects and different numerical approaches are available for studying their behavior. The analysis of masonry constructions is a complex task (Lourenco, 2002), especially under special loads and when the soil-structure interaction becomes essential for studying the real behavior. Usually, salient aspects are:

- Difficult and expensive characterization of the mechanical properties of the materials used;
- Large variability of mechanical properties, due to workmanship and use of natural materials;
- Significant changes in the core and constitution of structural elements, associated with long construction periods;
- Unknown construction sequence;
- Unknown existing damage in the structure.

In addition, under the different loading conditions, many experimental studies have shown that joints or interfaces are the weakest zones of masonry structures. Figure 1 shows some masonry failure modes, according to Sutcliffe et *al.*, 2001.

Several methods and computational tools are available (Massart et al, 2005) for the assessment of the mechanical behavior of old constructions. The empirical approaches and the Eurocode (6) recommendations are generally satisfactory for engineers. The methods resort to different theories or approaches, resulting in: different levels of complexity (from simple graphical methods and hand calculations to complex mathematical formulations and large systems of non-linear equations), different availability for the practitioner (from readily available in any consulting engineer office to scarcely available in a few research-oriented institutions and large consulting offices), different time requirements (from a few seconds of computer time to a few days of processing) and, of course, different costs. Three approaches (Figure 2) are generally employed by engineers and researchers to model the masonry element: equivalent medium, discontinuous medium using continuous numerical approach (finite element and boundary element methods) and discontinuous medium using distinct element approach (distinct element method). The distinct element code will be employed herein to model masonry structures.

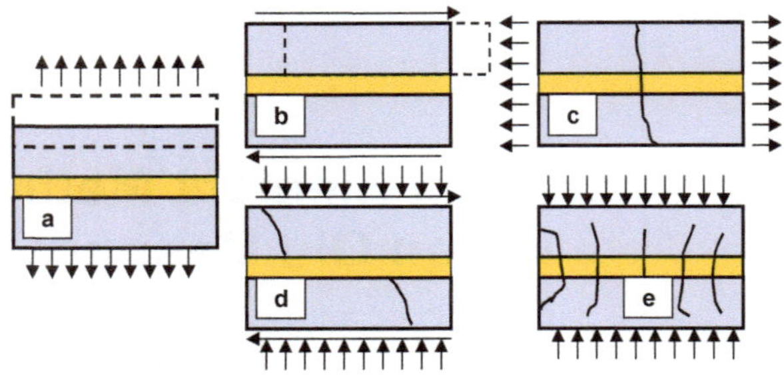

Fig. 1. Masonry failure modes a- direct tensile cracking of joint, b-sliding along joint c-cracking of unit and joint, diagonal tensile cracking of units e-compressive failure due to mortar militancy (Idris et al, 2009).

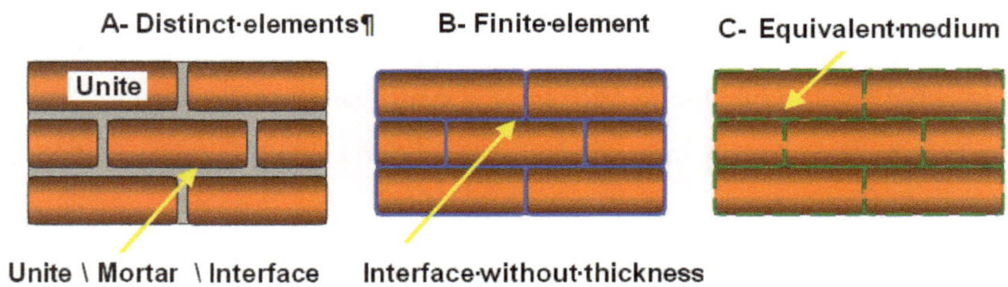

Fig. 2. Different approaches to model the behavior of masonry structures (Idris et al, 2009).

Two case studies will be presented in this chapter. The first case study concerns the simulation of the behavior of an underground structure of old tunnel supported by masonry of stone elements. The second case study concerns particularly the behavior of a masonry wall under the effect of an underground excavation (tunnel, mine, soil settlement, etc.).

2. Distinct element method

2.1 Description and background

A numerical model must represent two types of mechanical behavior in a discontinuous system: (1) behavior of the discontinuities; and (2) behavior of the solid material. In addition, the model must recognize the existence of contacts or interfaces between the discrete bodies that comprise the system. Numerical methods are divided into two groups according to the way in which they treat behavior in the normal direction of motion at contacts. In the first group (using a soft-contact approach), a finite normal stiffness is taken to represent the measurable stiffness that exists at a contact or joint.

The distinct element method was presented for the first time by Cundall in Nancy (1971), it considers the medium as an assembly of distinct rigid blocs that are linked together by joints. One can distinguish between rigid blocs and deformable blocs. Deformable blocs can be studied using the difference element method.

A discontinuous medium is distinguished from a continuous medium by the existence of interfaces or contacts between the discrete bodies that comprise the system. Discrete methods can be categorized both by the way they represent contacts and by the way they represent the discrete bodies in the numerical formulation.

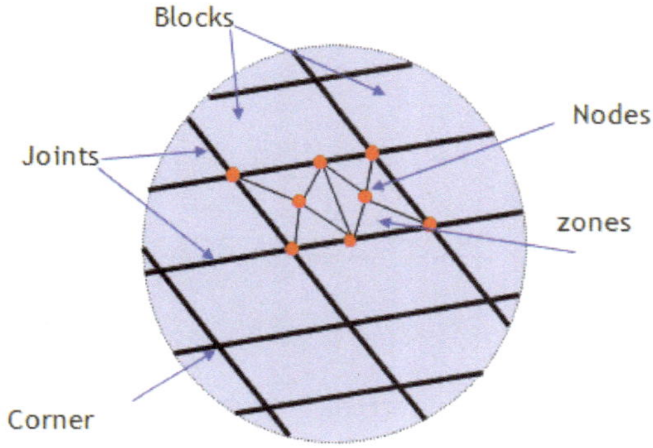

Fig. 3. Distinct element method and principal aspects.

In the second group (using a hard-contact approach), interpenetration is regarded as nonphysical, and algorithms are used to prevent any interpenetration of the two bodies that form a contact. The discrete (or distinct) element methods fall within the general classification of discontinuous analysis techniques. Originally used to model jointed and fractured rock masses (Tzamtzis et al, 2004), they were developed for the analysis of structures composed of particles or blocks and are especially suitable for problems in which a significant part of the deformation is accounted for by relative motion between blocks. *Masonry provides a natural application for these techniques*, as the deformation and failure modes of these structures are strongly dependent on the role of the joints. This approach is well suited for collapse analysis, and may thus provide support for studies of safety assessment, namely of historical stone masonry structures under earthquakes.

Two main features of the discrete element method (DEM) led to its use for the analysis of masonry structures. One is the allowance for large displacements and rotations between blocks, including their complete detachment. The other is the automatic detection of new contacts as the calculation progresses. Block material may be assumed rigid or deformable. Concerning masonry blocks, they are generally bonded by a lime or cement mortar. The model does not take the thickness of the mortar into account. Many numerical works have been performed for modeling masonry structures with the discrete elements methods (Verdel, 1994, Lemos, 1998). These studies looked essentially to the dynamic solicitation on dams and historic buildings.

2.2 Masonry joint modeling

In discrete element models, the representation of the interface between blocks relies on sets of point contacts (Figure 4 and Figure 6). Adjacent blocks can touch along a common edge segment or at discrete points where a corner meets an edge or another corner. At each contact, the mechanical interaction between blocks is represented by a force (stress), resolved into a normal (F_n or σ_n) and a shear (F_s or τ) component. Contact displacements are defined as the relative displacement between two blocks at the contact point. In the elastic range, contact forces and displacements are related through the contact stiffness parameters (normal and shear).

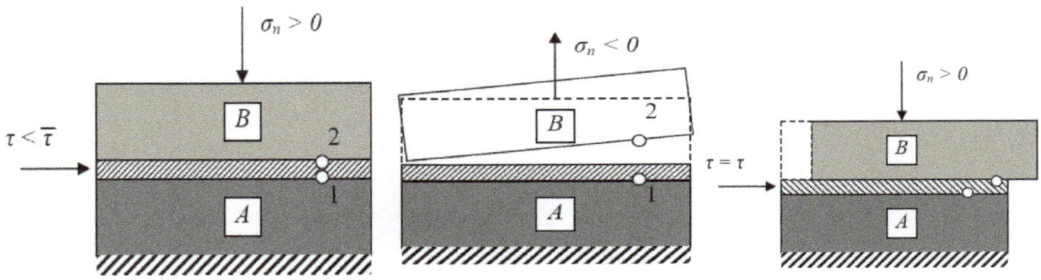

Fig. 4. Coulomb slip model with residual strength (shear and normal behavior).

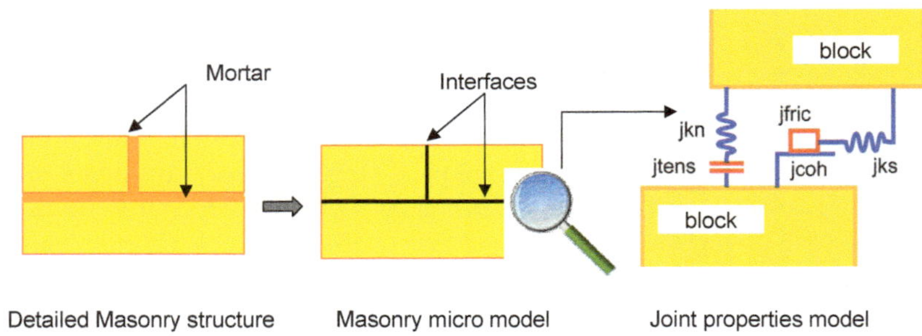

Fig. 5. Interface model code (jkn: joint normal stiffness, jks joint shear stiffness, jcoh: joint cohesion, jfric: joint friction angle and jtens joint tensile strength) (Idris et al, 2009).

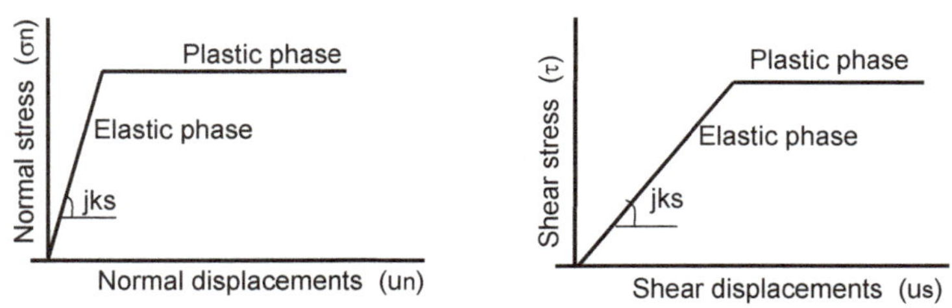

Fig. 6. Joint behavior (respectively) under normal and shear loads.

The mechanical behavior of joints is described as follows, (Itasca, 2000):

- The response to normal loading is expressed by the normal stiffness, jkn and normal displacement Δ un:

$$\Delta\sigma n = jkn\ \Delta un \tag{1}$$

- The shear stress increment is calculated as:

$$\Delta\tau = jks\ \Delta us \tag{2}$$

Where jks, jkn are joint shear stiffness and normal stiffness and Δ us and Δ un are shear displacement and normal displacement of joint. The value of jkn will depends on the contact area ratio between the two joint surfaces and the relevant properties of the joint filing material, if present (Souley, 1993). The value of jks depends on the roughness of the joint surface, which can be determined by the distribution, amplitude, and inclination of the asperities on the friction along the joint, the cohesion due to interlocking, and the strength of the filing material, if present. Figure 5 shows the evolution of joint behavior under normal and shear loads. Figure 6 resume a joint proprieties model for the distinct elements method.

The following parameters are used to define the mechanical behavior of the contacts: the normal stiffness (kn), shear stiffness (ks), friction angle (ϕ), cohesion (c) and tensile strength (R_t). To approximate a displacement-weakening response, the Coulomb slip model with residual strength (Figure 7) is used.

$$\tau = coh + \sigma n \tan fric \tag{3}$$

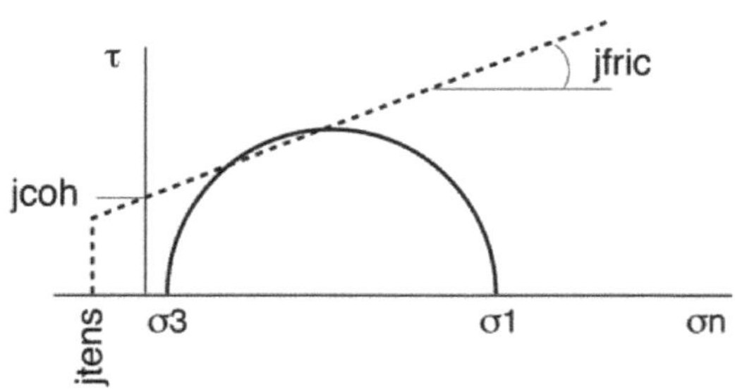

Fig. 7. Elasto-plastic Mohr-Coulomb joint model, code (jcoh: joint cohesion, jfric: joint friction angle and jtens joint tensile strength) (Itasca, 2000, UDEC).

2.3 Calculation method and algorithm

The algorithm of the calculation method for the discrete element method (DEM) must take into account the fact that the geometry of the system, as well as the number and type of contacts between the discrete bodies, may change during the analysis. In the discrete element method the structural analysis, both static and dynamic, is based on explicit algorithms. Among the most important capabilities of DEM that make it very suitable for masonry structures could be mentioned: the ability to simulate progressive failure associated with crack propagation; the capability of simulating large displacements/rotations between blocks; the fact that contact points are updated automatically as block motion occurs and the fact that the problem of interlocking is overcome by automatically rounding the corners.

The calculations performed in DEM alternate between applications of a force and displacement law at all contacts and Newton's second law at all blocks. The force-displacement law is used to find contact forces from known (and fixed) displacements.

Newton's second law gives the motion of the blocks resulting from the known (and fixed) forces acting on them. If the blocks are deformable, motion is calculated at the grid points of the triangular finite-strain elements within the blocks. Then, the application of the block material constitutive relations gives new stresses within the elements. Figure 8 schematically shows the calculation cycle for the distinct element method.

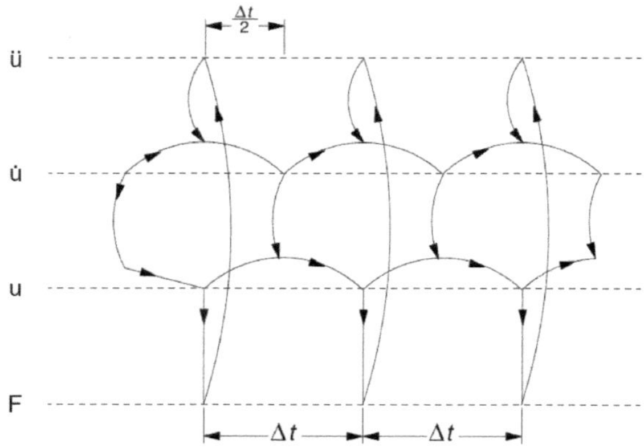

Δt: time step, U: Displacement, Ü: Acceleration, U: Velocity, F: Force

Fig. 8. Calculation cycle used in distinct element formulation (Itasca, 2000).

The motion of an individual block is determined by the magnitude and the direction of resulting out-of-balance moment and forces acting on it. Consider the one-dimensional motion of a single mass acted on by a varying force, *F(t)*. Newton's second law of motion can be written in the form

$$\frac{d\dot{u}}{dt} = \frac{F}{m} \tag{4}$$

where $\dot{u}$ = velocity; t = time; and m = mass.

The central difference scheme for the left-hand side of Eq. (4) at time t can be written as

$$\frac{d\dot{u}}{dt} = \frac{\dot{u}^{(t+\frac{\Delta t}{2})} - \dot{u}^{(t-\frac{\Delta t}{2})}}{\Delta t} \tag{5}$$

Substituting Eq. (5) in Eq. (4) and rearranging yields

$$\dot{u}^{(t+\frac{\Delta t}{2})} = \dot{u}^{\left(t-\frac{\Delta t}{2}\right)} + \frac{F^{(t)}}{m}\Delta t \tag{6}$$

With velocities stored at the half-time step point, it is possible to express displacement as

$$u^{(t+\Delta t)} = u^t + \dot{u}^{\left(t+\frac{\Delta t}{2}\right)\Delta t} \tag{7}$$

Because the force depends on displacement, the force/displacement calculation is done at a one-time instant. Figure 1.4 illustrates the central difference scheme with the order of calculation indicated by the arrows. The central difference scheme is "second-order accurate" – i.e., first-order error terms vanish from the solution. This is an important characteristic that prevents long-term drift in a distinct element simulation. For blocks in two dimensions that are acted upon by several forces as well as gravity, the velocity equations become:

$$\dot{u}_i^{(t+\frac{\Delta t}{2})} = \dot{u}_i^{(t-\frac{\Delta t}{2})} + (\frac{\sum F_i^{(t)}}{m} + g_i)\,\Delta t \tag{8}$$

$$\dot{\theta}_i^{(t+\frac{\Delta t}{2})} = \dot{\theta}_i^{(t-\frac{\Delta t}{2})} + (\frac{\sum M_i^{(t)}}{I})\,\Delta t \tag{9}$$

where $\dot{\theta}$ = angular velocity of block about centroid;
I = moment of inertia of block;
ΣM = total moment acting on the block;
$\dot{u}_i$ = velocity components of block centroid; and
g_i = components of gravitational acceleration (body forces).

In Eq. (9) and those that follow, indices i denote components in a Cartesian coordinate frame, and summation is implied for repeated indices in an expression. The new velocities in Eq. (9) are used to determine the new block location according to:

$$x_i^{(t+\Delta t)} = x_i^{(t)} + \dot{u}_i^{(t+\frac{\Delta t}{2})}\Delta t \tag{10}$$

$$\theta_i^{(t+\Delta t)} = \theta_i^{(t)} + \dot{\theta}_i^{(t+\frac{\Delta t}{2})}\Delta t \tag{11}$$

Where θ = rotation of block about centroid; and
x_i = coordinates of block centroid.

In summary, each timestep (Δt) produces new block positions that generate new contact forces. Resulting forces and moments are used to calculate linear and angular accelerations of each block. Block velocities and displacements are determined by integration over increments in time. The procedure is repeated until a satisfactory state of equilibrium or continuing failure results. Mechanical damping is utilized in the equations of motion to provide both static and dynamic solutions. Gridpoint forces are obtained as a sum of three terms:

$$F_i = F_i^z + F_i^c + F_i^l \tag{12}$$

F_i^l is the external applied loads.

F_i^c is resulted from the contact forces and exists only for gridpoints along the block boundary. Finally, the contribution of the internal stresses in the zones adjacent to the gridpoint is calculated as:

$$F_i^{(z)} = \int_c \sigma_{ij}\, n_j\, d_s \tag{13}$$

where σij is the zone stress tensor; and nonlinear and post-peak strength models are readily incorporated into the code in a direct way without recourse to devices such as equivalent stiffness or initial strains. In an explicit program, however, the process is much simpler –

after each time step, the strain state of each zone is known. The program then needs to know the stress in each zone in order to proceed to the next time step. The stress is uniquely defined by the stress-strain model whether it is a linearly elastic relation or a complex, nonlinear and post-peak strength model.

The basic failure model for blocks is the Mohr-Coulomb failure criterion with a non-associated flow rule. Other nonlinear plasticity models available in *Distinct Element codes (UDEC, 3DEC)* are the Drucker-Prager failure criterion, the ubiquitous joint model and strain-softening models for both shear and volumetric (collapse) yield.

2.4 Data limited problems and numerical modeling recommendations

It is necessary when data are limited to measure the quality and the quantity of available data in order to help the understanding of the problem to be solved. In soil-structure interaction and in many other branches of engineering geology, this is a category with limited data available. In this case, it is necessary before starting to be clear on why we are building the model; a good conceptual model can lead to savings in time and money on field tests that are better designed, and then a first model to identify with realistic data and then analyzing the mechanism of the problem, the visualization and the analysis help to understand the behavior of the model. Once we have learned all we can from the simple model or models, then a more complex model can be used. We apply this methodology in this paper.

3. Tunnel masonry structure support proposed models

The first case study (Idris et al, 2009) concerns the simulation of the behavior of an underground structure (Figure 9): old tunnel supported by masonry of stone elements, through this example, we insist on the importance of discontinuities behavior and their characterization (friction angle and cohesion). A majority of the world's tunnels are currently more than 100 years old; these would all be considered as ancient infrastructure. Old tunnels are often supported by a masonry structure. The type of masonry support or lining depends upon utilization of the high compressive strength in the stones, which explains the vaulted section shape of old tunnels supported by masonry.

Old underground constructions, especially tunnels, display specific characters regarding behavioral evolution over time. Infrastructure environment, surrounding ground and used construction materials all contribute to this evolution. Apart from the environment and evolution in surrounding soil and in the absence of an effective isolation system for such underground structures, subsoil water can easily penetrate the masonry joints and circulate within. Over time and in the presence of other aggressive ambient factors, several physical, chemical and biological processes may develop inside the masonry structure; this phenomenon and its impact are collectively called the tunnel-ageing phenomenon. One impact is the alteration in mechanical properties of construction materials (masonry structure composed of blocks and mortar). As a result, various types of disorders appear inside old tunnels (Figure 9); these would include: longitudinal or transverse structural cracks, convergence and partial masonry collapse. The instability of old tunnels depends on the interaction between soil, tunnel support (blocks and mortar).

The study focuses on the evolution of masonry support joint mechanical behavior in built tunnels over time. This study is carried out with the help of the experimental design strategy and numerical modeling by the well-known Universal Distinct Element Code (UDEC).

A tunnel masonry structure is a discontinuous medium consisting of blocks bonded to each other by mortar; in addition, such a structure forms an interface with the surrounding soil. The Distinct Element Method (DEM) is a suitable technique for modeling these structures. By means of the Universal Distinct Element Code (UDEC), a simplified micro-model of an ancient tunnel has been derived (Idris et al., 2009), (Figure. 10). The representative model is positioned at a shallow depth of 20 m. The masonry-supporting section consists of a regular rectangular and square limestone blocks (Figure. 10). Masonry blocks are bonded by lime mortar. The masonry support thickness is 80 cm and the sidewall height amounts to 3 m. By taking into account model section symmetry, only half of each set-up needed to be modeled.

Fig. 9. Illustration and examples of degradations of masonry tunnels.

The soil surrounding the tunnel consists of a homogeneous mix of clay and sand (Verdel and Bigarre, 1999). Table 1 lists the basic mechanical properties assigned to the surrounding soil, masonry and masonry joints, based on the work by Verdel and Bigarre (1999), Hoek (2000) and Janssen (1997).

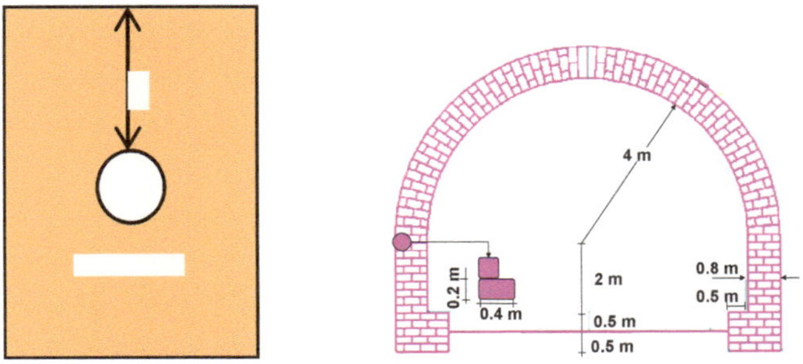

Fig. 10. Old tunnel model using Universal Distinct Element Code UDEC.

Surrounding ground			Masonry			Masonry joints		
Param.	*Unit*	*Value*	*Param.*	*Unit*	*Value*	*Param.*	*Unit*	*Value*
M	Kg/m^3	1900	*M*	Kg/m^3	2000	*jkn*	GPa/m	150
E	MPa	200	E	MPa	6000	jks	GPa/m	69.7
ν		0.3	ν		0.2	jcoh	MPa	1.2
C	MPa	0.50	C	MPa	3	jfric	°	25
φ	°	20	φ	°	30	jtens	MPa	0.4
Tr	MPa	0.10	T	MPa	1			

M: Volumic mass; E: Young modulus; ν: Poisson's ratio; C: Cohesion; φ: Friction angle; T: Tensile strength;
jkn, jks: Normal, tangential joint stiffness; jcoh: Joint cohesion; jfric: joint friction angle; jtens: joint tensile strength

Table 1. Mechanical properties choice of surrounding ground and masonry.

Calculations were carried out in plane strain: the soil and masonry follow a perfect elasto-plastic Mohr–Coulomb plasticity criterion. The calculation step was proceeded by two main stages: model consolidation in the initial stress condition prior to tunnel excavation; and tunnel excavation and simultaneous installation of masonry support. The calculation could then be continued until reaching model equilibrium.

3.1 Ageing simulation of masonry joints in built tunnels

This part of our study sought to understand the evolution in masonry joints behavior over time and evaluate the influence of certain masonry joint mechanical properties on the behavior of masonry old tunnels and the surrounding ground. To simplify the simulation of the complex ageing phenomenon, the strategy has consisted of utilizing experimental designs and response surfaces in combination with various data analyses in order to identify the most powerful experimental factors influencing joint masonry structure behavior over time. A factorial experiment entails a statistical study in which each observation is categorized according to more than one factor. Such an experimental set-up makes it possible to study the effect of each factor on the response variable, while requiring fewer observations than when conducting separate experiments for each factor independently. It also allows studying the effect of the interaction between factors on the response variable (Barrentine, 1999).

In this study, two different experimental designs were proposed to simulate the evolution of mechanical properties of joints. The first experimental design explains the evolution of joints filling material properties and the second explains the evolution of ratio (normal stiffness/ shear stiffness).

3.2 First proposed experimental design

Many factors may influence joint mechanical behavior parameters. In order to predict the 'potential behavior' of the joints under loading, three distinct joint parameters must be introduced into the analysis (Goodman et *al.*, 1968):

- The unit stiffness across the joint, jkn, which characterizes the elastic phase behavior;
- The unit stiffness along the joint, jks, which characterizes the elastic phase behavior;

- Joint fill material proprieties (cohesion, tensile strength, friction angle).

The first experimental design represents the evolution of joint cohesion, joint tensile strength, and joint friction angle. To evaluate the influence of each chosen factor on masonry structure behavior, it proved necessary to observe significant changes in model behavior once factor values had been changed.

Two significant response factors were detected; herein is the cumulated length of open joints and the cumulated length of joints at limiting friction (slip joints). Open joint means that the induced tension stress is greater than joint tension strength and slip joint means the shear strength is less than the induced shear stress using Mohr-Coulomb criteria.

For this purpose, a complete factorial design was proposed; this three-level design is written as K^n factorial design (with K = 3: the studied factor number, n: level number). This nomenclature means that three factors are considered, each one at three distinct levels (Barrentine, 1999). Consequently, a complete factorial design with 27 experiments was proposed. Table 2 contains all of the experimental results (i.e. changed experimental factors and observed responses). In all simulations, soil and masonry blocks properties have been given the unchanged values shown in Table 2.

Surrounding ground			Masonry			Masonry joints		
Param.	Unit	Value	Param.	Unit	Value	Param.	Unit	Value
M	Kg/m3	1900	M	Kg/m3	2000	jkn	GPa/m	5
E	MPa	100	E	MPa	10000	jks	GPa/m	2
ν		0.3	ν		0,3	jcoh	MPa	1
C	MPa	0.1	C	MPa	6	jfric	°	40
φ	°	30	φ	°	60	jtens	MPa	0
Tr	MPa	0.10	Tr	MPa	3			

M: Volumic mass; E: Young modulus; ν: Poisson's ratio; C: Cohesion; φ: Friction angle; T: Tensile strength; jkn, jks: Normal, tangential joint stiffness; jcoh: Joint cohesion; jfric: joint friction angle; jtens: joint tensile strength

Table 2. Characterization of soil, blocs and joints of masonry.

Figure 11 and Figure 12 provide some selected results, which show the influence of joint parameters on the length of open and slip joint evolution on mechanical behavior of masonry support structure.

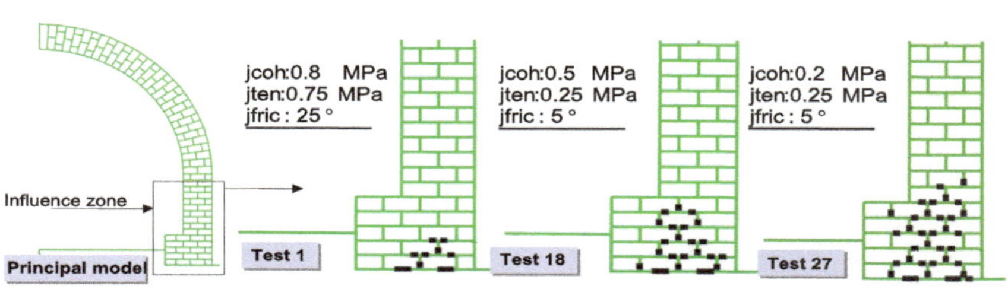

Fig. 11. A sample of numerical simulation results; the observed response is the total length of open joints (σ_n<JR_T).

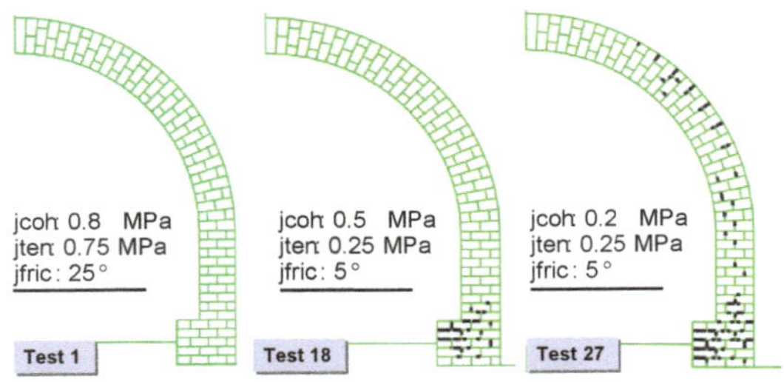

Fig. 12. A sample of numerical simulation results, the observed response is the total length of joints at limiting friction (slip joints).

3.3 Response surface analysis

To summarize the experimental design results, 3D graphical response surfaces were generated, where the predicted responses (cumulated length of open joints and cumulated length at limiting friction) were indicated by a plane surface distance that relates every pair of modified factors (Kresic, 1997). These response surfaces yield a graphical indication of the reliability of results obtained; they also make it possible to compare dual influences from the studied factors and to observe possible interactions between them (Figure. 13 and Figure. 14).

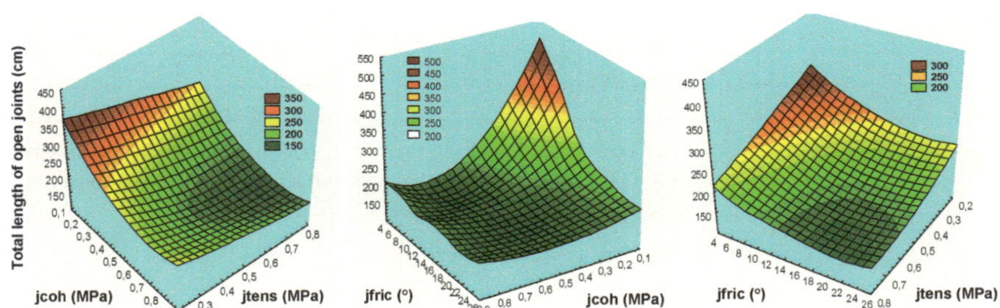

Fig. 13. Response surfaces for relating the total length of open joints with the data set (jcoh: joint cohesion, jtens: joint tensile strength and jfric: joint friction angle).

Response surface (Figure. 14) analysis highlights a number of key points:

The joint cohesion has a higher influence than joint tensile strength on the cumulated length of open joints.

- The evolution in two factors together (joint cohesion and joint friction angle) exerts remarkable influence on masonry block mechanical behavior (i.e. on the cumulated length of open joints) which means that these two parameters have an important interaction influence on the observed response.
- The difficulty is in comparing the influence of joint cohesion with that of joint friction angle on the mechanical behavior of masonry.

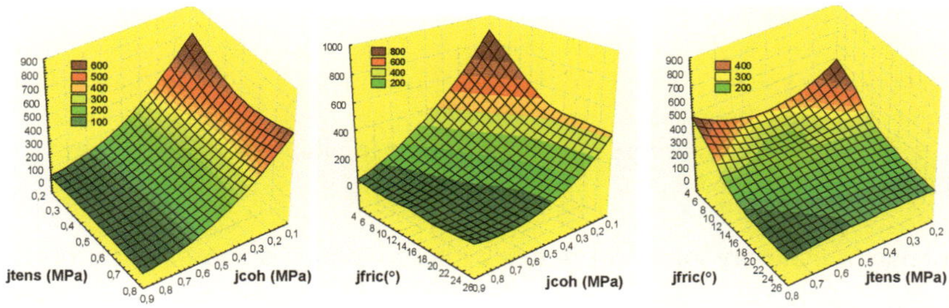

Fig. 14. Response surfaces for relating the total length of joints at limiting friction with the data set (jcoh: joint cohesion, jtens: joint tensile strength and jfric: joint friction angle).

3.4 Second proposed experimental design

The determination of jkn and jks is a complex operation and many authors propose experimental studies. The second experimental design expresses just the evolution of the ratio (jkn/jks) to evaluate its influence on masonry structure behavior, where:

- jkn is joint normal stiffness;
- jks is joint shear stiffness.

The (jkn/jks) ratio may exceed 100 (Souley, 1993), so this experimental design suggests that the ratio (jkn/jks) changes between 2 and 100, where jkn remains constant and only jks changes its value. It proved necessary to observe significant changes in model behavior once the (jkn / jks) ratio had been changed. The observed response in this step of the study is the total length of shear displacements detected on the tunnel masonry support section after every modeling test. Table 3 provides the detailed experimental design and the obtained results. This experimental design contains 12 modeling tests. Figure 15 provides some selected results, which show the influence of (jkn/jks) ratio evolution on mechanical behavior of masonry support structure.

Figure 16 shows the relations between (jkn/jks) ratio and observed response of the cumulated length of joint shear displacements. On figure 16, we can distinguish a remarkable rapid increase in the total length of joint shear displacements when the (jkn/jks) ratio varies between 2 and 20.

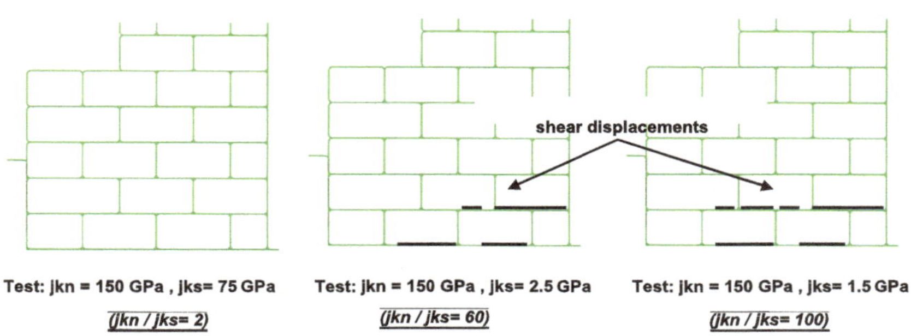

Fig. 15. Some numerical simulation results examples.

After that, the relation between the (jkn/jks) ratio and the cumulated length of joint shear displacements seems to be a linear relation and the cumulated length of joint shear displacements seems to increase slightly with the rise of the (jkn/jks) ratio more than 20. We can extract from the previous analysis that the (jkn /jks) ratio increases and directly influences the behavior of masonry support structure in old tunnels by the increase of shear displacement along affected joints.

3.5 Influence of masonry joint proprieties evolution on the surrounding ground of tunnel

The evolution of mechanical joint parameters (jcoh, jten, jfric, jkn and jks) did not have any significant influence on the mechanical behavior of the surrounding soil. The significant evolutions concern only the masonry joint behavior. Generally, in old tunnels, the use of the masonry support is strong enough, the models inspired from real built tunnels, with 80 cm of masonry support which have very strong support. Masonry structure is mainly loaded in compression due to its vaulted section shape. This massive support may explain the absence of block mechanical parameters that influence the behavior of surrounding soil (Idris et *al.*, 2008).

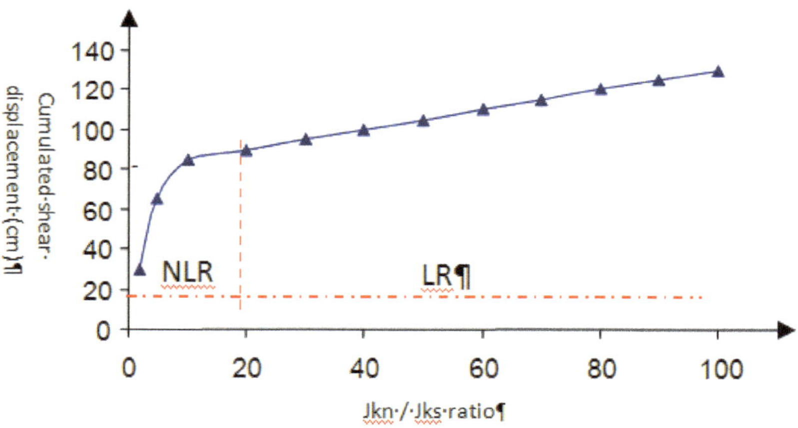

NLR: Non Lenar Relation LR: Linear Relation, Jkn : Joint normal stiffness, Jks : Joint shear stiffness

Fig. 16. Graphical presentation of shear displacements as a function of the jkn/jks ratio.

3.6 Conclusion

The study concerned the behavior of masonry tunnel structures due to the ageing phenomena by using numerical modeling and experimental design. A first experimental design was proposed to simulate ageing effects of old tunnel behavior; a complete factorial experimental design, which expresses the evolution of three selected masonry joints mechanical properties, was then forwarded. The factors selected for the present study were: masonry joint cohesion, joint tensile strength and joint friction angle. All experimental design tests were modeled by means of the distinct element method. Two significant responses were detected, they are respectively: the total length of open joints; the total length of joints at limiting friction (slip joints).

An analysis of results indicated that the three studied masonry joint mechanical factors have significant influence on masonry mechanical behavior expressed by the total length of open joints. Joint cohesion is the most important factor, then joint tensile strength and finally joint friction angle. Only the interaction of joint cohesion and joint friction angle have significant influence on the total length of open joints. For the total length of joints at limiting friction, only joint cohesion and joint friction angle have a significant influence and joint cohesion remains the most important factor. These two factors have a significant interaction influence on the mechanical behavior of tunnel masonry support.

The second proposed experiment design expressed the evolution of the ratio jkn/jks. The detected significant response was the total length of shear displacements along masonry joints. Results showed that shear displacements increase according to jkn/jks ratio increase. The total length of shear displacements evolution follows two types of relations as a function of the jkn/jks ratio; firstly it follows a non-linear relation according to certain jkn/jks values (jkn/jks=[2-20]), then it changes (increases slightly) behavior according to a linear model.

4. Impact of underground movement on a masonry wall

The second case study concerns in particular the behavior of a masonry wall under the effect of an underground excavation (tunnel, mine, soil settlement, etc.). The main objective is to verify and to improve the comprehension of masonry structure behavior using the numerical modeling approach by distinct element method.

The excavation of tunnels and underground mines modifies the initial stress distribution and induces displacements on the ground surface (Peck, 1964, Standing, 2008, Al Heib, 2008). Figure 17 diagrammatically shows the surface subsidence trough above an advancing tunnel. The surface subsides and structures can be damaged due to the induced strains (Figure 17). For 'Greenfield sites', i.e. those without the presence of buildings or subsurface structures, the shape of this trough transverse to the axis of the tunnel closely approximates to a normal Gaussian distribution curve - an idealization which has considerable mathematical advantages. The subsidence trough consists of vertical and horizontal displacements.

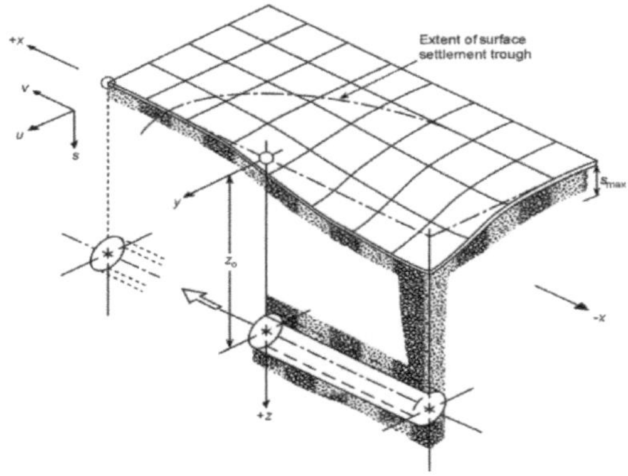

Z_0: depth of the tunnel, S_{max}: maximum subsidence

Fig. 17. Surface settlement above an advancing tunnel (Standing and Burland, 2008).

The amount of damage caused by subsidence depends upon the magnitude and the type of ground movements, structural factors and geological factors (Burland, 1995, Standing and Burland, 2008). The magnitude of ground movements on the structure are governed by its location and orientation in relation to the underground workings and the depth of underground cavities (Figure 18).

Fig. 18. Damages induced by compression strain due to soil subsidence (Deck et al., 2003).

4.1 Damage due to horizontal strain

The different types of movements can affect structures in different ways (Figure 19). Vertical subsidence may affect tall buildings (local tilt) and long buildings (deferential settlements). Horizontal extension and compression strain, tilt and curvature are the causes of the most commonly seen type of subsidence damages. Damage to buildings is generally caused by differential horizontal movements (horizontal strain) and the concavity and convexity of the subsidence profile. The extension horizontal strain is characterized by the fracturing of the masonry and the compression strain is characterized by squeezing-in of voids (doors and windows). Unlike settlements, there are fewer case histories where horizontal movements have been measured. The maximal horizontal displacement depends on the soil behavior and the geometry of the excavations; it generally equals 40% of the maximal vertical displacement (Lake et al, 1998). Horizontal displacements can be differentiated to give the horizontal strain ε_h at any location on the ground surface.

4.2 Building damage assessment and soil-structure interaction

It is clear from the above, considering tunnel construction and underground excavation, even with Greenfield conditions, that precise prediction of ground movements due to ground excavation is not realistic. However, it is possible, *for non-stiff buildings*, to make reasonable estimates of the likely range of movements provided excavation is carried out under the control of suitably qualified and experienced engineers and highlighted by numerical and physical modeling (Standing and Burland, 2008).

Building deformation and its potential damage caused by subsidence in urban areas has a major impact on the planning and construction process of any underground excavation. The use of Greenfield subsidence parameters to determine the level of the damages appears as a conservative approach; it can lead to costly projects. Potts and Addenbrooke (1997) and others (Dimmok et al, 2008) introduce design charts for tunnels to consider the influence of the buildings own stiffness, thus leading to more realistic predictions of induced deformations. The approach introduced by Potts and Addenbrooke was based on continuous and simplified 2D numerical modeling. The building was modeled by weightless, elastic beams. Franzius (2004) presented the results of an extensive parametric study using 3D FE analysis. The relative stiffness expressions were introduced to relate the stiffness of a surface structure to the stiffness of the ground; they defined the relative bending and axial stiffness respectively:

$$\rho^* = \frac{EI}{E_s(\frac{B}{2})^4} \text{ and } \alpha^* = \frac{EA}{E_s(\frac{B}{2})} \tag{14}$$

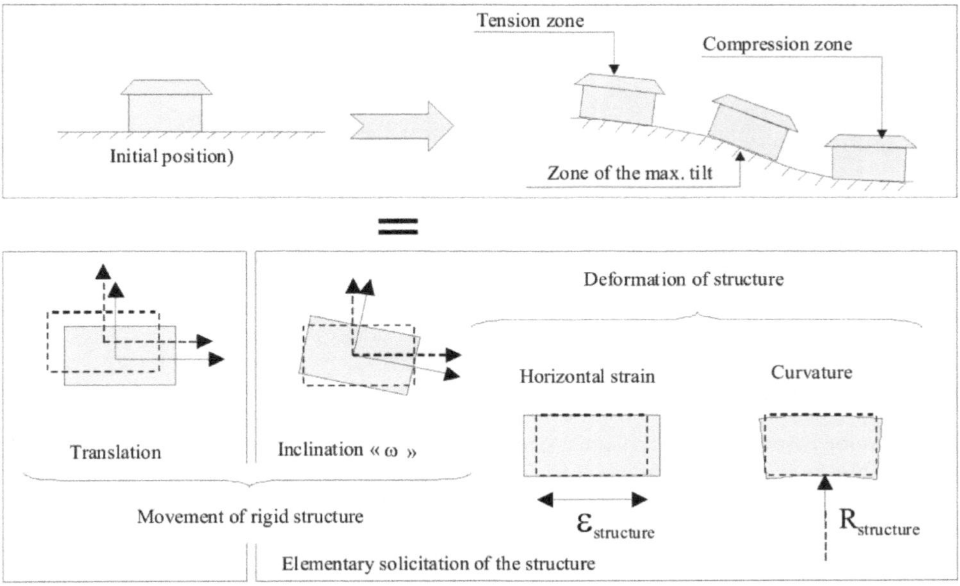

Fig. 19. Transfer surface movement to the surface structure (Deck et al, 2003).

Son and Cording (2007) had evaluated the stiffness for a masonry building using the distinct element code UDEC, their studies were limited to evaluating the influence of Young modulus and shear modulus for different value parameters and taking into account the presence of windows and doors. Giorgia Giardina et al (2008), studied masonry wall damaging due to tunneling; they confirm, using finite elements, that the Greenfield approach is too conservative, and that the total interaction (coupling) approach is more realistic but needs more time and energy to obtain the results. The masonry wall is generally very vulnerable due to the compression strain and shear strain as shown in the pictures. The failure generally appears along the mortar joints around openings corresponding to weakness zones of the walls. To study the effect of soil-structure interaction and the role of the stiffness due to the horizontal strain, we consider the numerical modeling approach.

4.3 Numerical model description

The present work is focused on the behavior of a masonry wall under horizontal strain. In addition to the evaluation of the stiffness of the masonry wall, the objective of the numerical modeling is to quantify the movement transfer from the soil to the masonry structure. The class of the wall damage depends on the characterization of the wall: geometry and proprieties. The 3D numerical models use the distinct elements code (3DEC). The model presents an individual wall of unreinforced masonry; the wall is under only the effect of the horizontal displacements and dead load. The horizontal displacements were applied on the boundary of soil (model). The masonry wall dimensions are: 10 m length, 5 m high and 25 cm thick. The wall consists of masonry units of 50 cm * 25 cm * 25 cm (reference). Two configurations were studied with and without windows. The model has two parts: the soil and the wall, the soil behavior is considered as elastic-plastic, homogenous and isotropic. The plastic criterion (failure) is the Mohr-Coulomb defined by cohesion and friction angle,

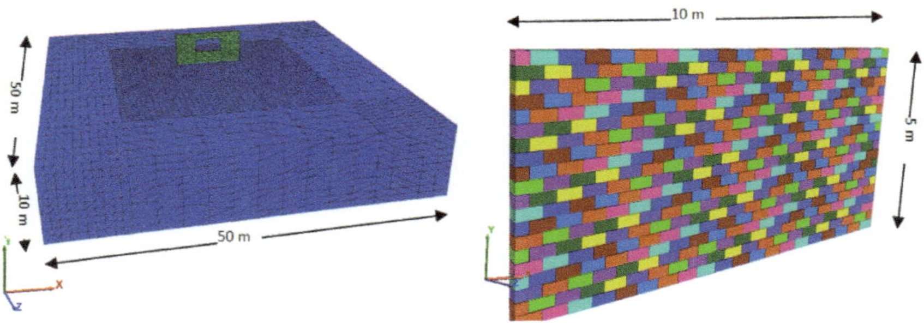

Fig. 20. 3DEC numerical model using distinct element method for ISS.

The behavior assumption of masonry units is as elastic and isotropic material. The mortar between blocks corresponds to vertical and horizontal joints. The behavior of the joints is considered as elastic-plastic and the plastic criterion is Mohr-Coulomb which is defined by the friction angle, the cohesion and the tensile strength. The table presents the priorities of soil, blocs and joints of masonry. Table 1 lists the basic mechanical properties assigned to the soil, the masonry and the masonry joints; they are based on the estimations from different international publications. The characterization of soil and building are very close to those used by Potts and Addenbrooke. The calculation was divided into two main stages: the stage of the consolidation to obtain the initial stress condition and the application of the horizontal displacements on the lateral boundary surface. The calculation could then be continued until reaching model equilibrium. Two directions of displacement were studied. The first one corresponds to in-plane solicitation and the second one corresponds to out-plane. The boundary condition of the horizontal displacement induced in the soil a compression horizontal strain in Greenfield equal to 8 mm/m.

4.4 Evaluation of relative stiffness of the wall

Two assumptions were employed to quantify the relative stiffness according to the relation introduced by Potts and Addenbrooke (1996): the first one considers the wall as a continuous single bloc and the second assumption considers the wall as an assembly of masonry units (Figure 21). The following relations (Figure C) determine the equivalent Young modulus (E_1, E_2) and the equivalent shear modulus (G_{12}) taking into account the geometrical and mechanical parameters of discontinuities for the second assumption. The second assumption reduces the stiffness of the wall.

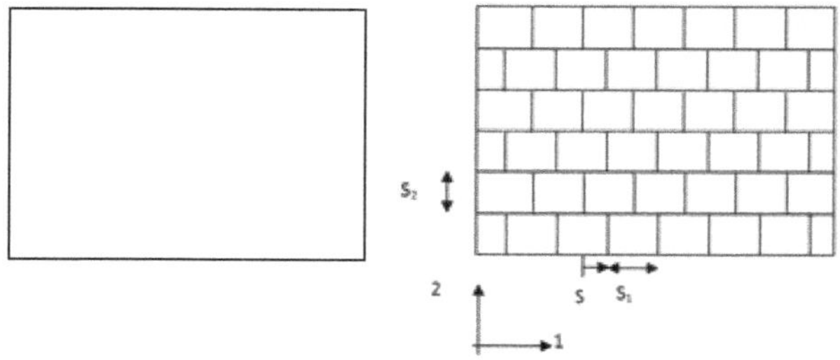

a- Single bloc Masonry blocs

$$E_1 = \frac{E}{1 + \frac{B_{n1} E}{S_1 \; K_{n1}}} \qquad \nu_1 = \frac{\nu}{1 + \frac{B_{n1} E}{S_1 \; K_{n1}}}$$

$$E_2 = \frac{E}{1 + \frac{E}{S_2 \; K_{n2}}} \qquad \nu_2 = \frac{\nu}{1 + \frac{E}{S_2 \; K_{n2}}}$$

$$\frac{1}{G_{12}} = \frac{1}{G} + \frac{B_{t1}}{S_1 \; K_{S1}} + \frac{B_{t2}}{S_2 \; K_{S2}}$$

With

$$B_{t1} = \left[1 + \frac{K_{n2}}{K_{s1}} \frac{S}{S_2} \left(1 - \frac{S}{S_1} \right) \right]^{-1}$$

$$B_{n1} = \left[1 + \frac{K_{s2}}{K_{n1}} \frac{S}{S_2} \left(1 - \frac{S}{S_1} \right) \right]^{-1}$$

E_1, E_2 : Young modulus, ν_1, ν_2: Poisson ratio,
G_{12}: Shear modulus, S1,S2: horizontal and vertical spacing.

c- equivalent elastic parameters

Fig. 21. Assumptions for determining the relative stiffness of the masonry wall.

Table 3 lists the parameters assigned to the wall and the estimated axial and flexion stiffness of the masonry wall.

Continue Wall			Masonry wall		
Parameter	*Unit*	*Value*	*Parameter*	*Unit*	*Value*
I	***m^4***	***32e-4***	***I***	***m^4***	***32e-4***
A	***m^2***	***0.125***	***A***	***m^2***	***0.125***
E	MPa	10000	*E*	MPa	2000
EI	MNm^2	32	EI	MNm^2	2156
Es	MPa	100	Es	MPa	100
B	m	10	B	m	10
ρ*	°1/m	83	ρ*	°1/m	17
α*		100	α*		21.56

I: Volumetric mass; E: Young modulus;

Table 3. Wall stiffness calculation according to the Potts and Addenbrooke approach.

Observing the charts which were introduced by Potts and Addenbrooke (Figure 22), the studying wall is stiff enough and the transfer of soil deformations to the wall is small, may be equal to zero for all structure positions from the position of the underground excavation (e/B). The adopted approach by Potts and Addenbrooke is very simplified for a masonry wall and the realistic behavior can be suspected as completely different from the above conclusions according to in-situ observations.

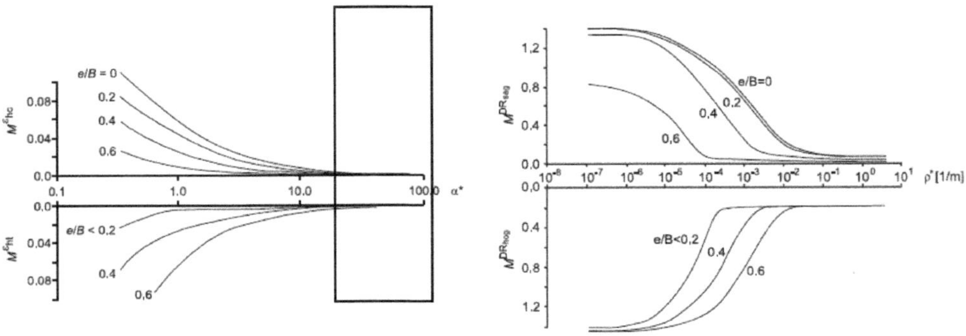

Fig. 22. Chart for determining the strain (ε and D) transfer from soil to structures function of relative axial α and flexion ρ stiffness.

4.5 Results analysis

The analysis of results will be limited to the behavior of the masonry wall due to the effect of the horizontal displacements (horizontal strain). We looked for the state of joints (mortar): elastic, open or sliding; and the horizontal and vertical displacements of the wall. We compare the horizontal strains of the masonry wall to the horizontal strain of the soil under Greenfield conditions.

The first configuration corresponds to in-plane solicitation without windows. The Figure 23 presents the horizontal and the vertical displacements of the wall. The horizontal displacement direction corresponds to the direction of the solicitation and the vertical displacement direction is oriented toward the top. The distribution of displacements depends on the localization of the cracks due to the failure of vertical joints. A principal discontinuity was formed near the second part of the wall; the principal discontinuity is associated with three or four cracks (joints). The dip of discontinuity is about 45° to 60°. The variation of the horizontal strain decreases from the bottom to the top of the wall. The maximum horizontal strain is 1.4 mm/m. The transfer of the horizontal strain from soil to the wall is equal to 17.5%.

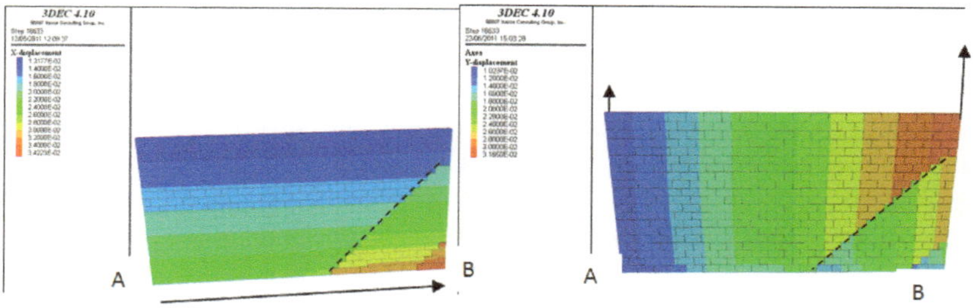

Fig. 23a. Horizontal and vertical displacements of the wall.

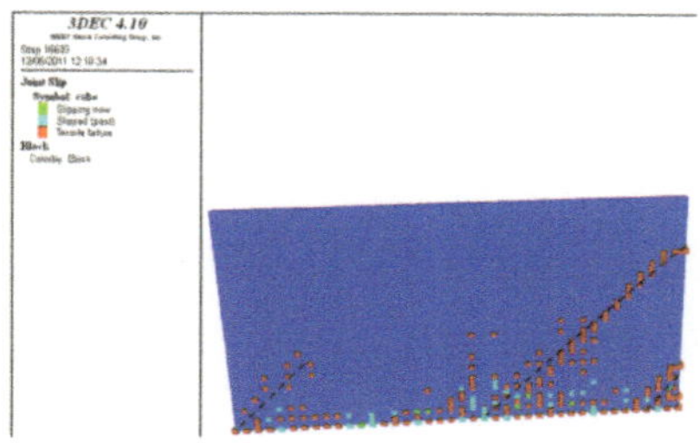

Fig. 23b. Failure of the joints creating inclined and vertical cracks.

The second configuration corresponds to the introduction of a window in the centre of the masonry wall, which surface corresponds to 16% of the wall surface. The presence of a void in the wall modifies the distribution of the displacements and the localization of the cracks (Figure 24). The horizontal strain increases and it is equal to 2.65 mm/m instead of 1.4 mm/m, the transferred strain is equal to 33% of soil strain. The stiffness of the wall decreases due to the window, the density of cracks increases also, we observe new cracks in the upper part of the wall above the window zone due to tension stresses. The numerical results are very close to the in-situ observations due to the subsidence.

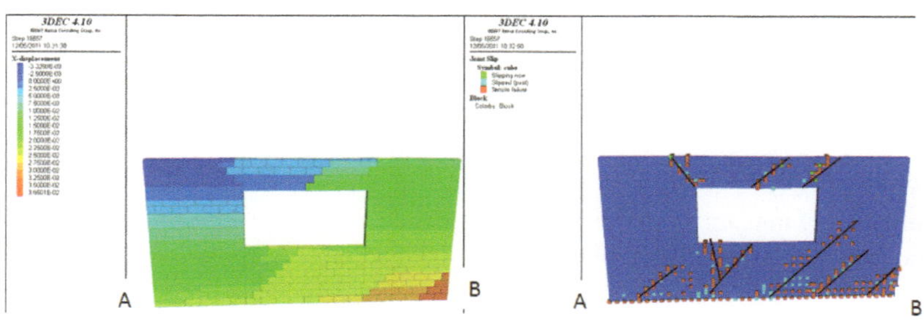

a- Horizontal displacement b- Failure localization

Fig. 24. Behavior of masonry wall under the effect of ground subsidence due to the underground excavation - results obtained by 3DEC numerical model.

The third and the fourth configurations correspond to the horizontal displacement applied perpendicularly to the wall (out-plane). Figure 25 presents the mechanism of the wall deformation, the displacement of the bottom part of the wall is the opposite of the displacement of the upper part. The behavior of the wall is influenced by the direction of the solicitation and the stiffness of the wall. A horizontal open crack was created by the impact of horizontal displacement point C), the localization of this crack is at 1 m from the base of the horizontal strain between points A and C. It is equal to 100 mm/m, it is much greater than the applied horizontal strain (8 mm/m) on the ground, and in the second part of the wall, and the strain is less than the soil horizontal strain.

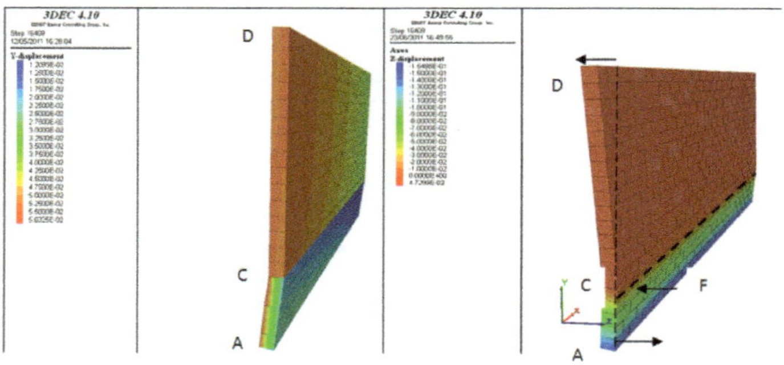

a- Horizontal displacement b- Failure localization

Fig. 25. Behavior of masonry wall under the effect of ground subsidence due to the underground excavation - results obtained by 3DEC numerical model.

The fourth configuration takes into account the presence of the window. The presence of the window increases the number of cracks very strongly. Some blocks lose their contacts because of the free surface and can fall.

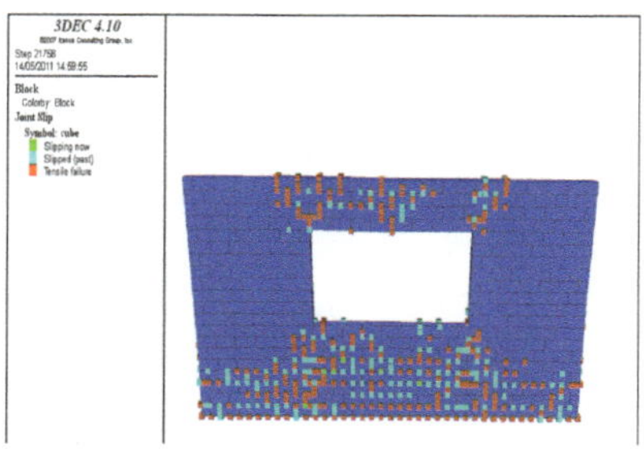

Fig. 26. Behavior of masonry wall under the effect of ground subsidence due to the underground excavation - results obtained by 3DEC numerical model.

Table 4 resumes and compares the values of horizontal displacements (Ux and Zu) and vertical displacement (Uy) and the localization damages according to the behavior of vertical and horizontal joints. Figure 27 presents an interpretation of the wall behavior under the influence of the horizontal strain. The wall is influenced by the ground displacements. This solicitation of the wall by the horizontal strain reveals traction failures in the joints of masonry blocks. These failures can form oblique cracks in the wall along vertical and horizontal mortar.

A ground horizontal strain, in Greenfield conditions, equaling 8 mm/m can induce severe damage on structures according to different prediction methods without the effect of soil-structure interaction (Deck et al., 2003, table 5).

	Ux (mm)	Uy (mm)	Uz (mm)	Localization of damages
In-plane	34	31	4.9	One principal inclined shear crack and many incomplete shear cracks associated with open cracks, with the direction of horizontal displacement.
Out-plane	18	56	-154	Horizontal crack at 1.5 m from the base and distortion and tilt of transversal section
In-plane - window	36		7	Inclined shear cracks associated with open cracks, with the direction of horizontal displacement. Up the window, inclined cracks in the opposite direction
Out-plane - window	11.5	62	144	Distortion of the wall associated to the fall of certain blocks, increase in the number of cracks

Table 4. Displacements and damage localization of masonry wall.

The damages can be determined by the chart of Boscardin and Cording (1989) and Potts and Addenbrooke (1996) and Table 5. In this case, damage must be very severs. The results of numerical modeling using 3D distinct elements code highlights this point and shows this evaluation must be improved taking into acount the interaction between soil and structure and the behavior of joints. The numerical results are strongly close to the physical modeling results obtained by Cox (Cox, 1980) and in situ observations (Deck, 2002).

Limits horizontal strain ε mm/m	< 0.5 for negligible degradations. 0.5 to 0.75 for very light damage. 0.75 to 1.5 for light damage. 1.5 to 3 for moderate damage. > 3 for severe damage.

Table 5. Classes of degradation according to horizontal strain.

5. Conclusion

The study focused on the behavior of the masonry wall due to horizontal deformation of the soil; the study also focused on soil-structure interaction and the movement of transfer of soil structure using numerical modeling. A 3D numerical model of the masonry wall was made to simulate the behavior. The study examined the observance of the level and location of damage due to horizontal deformation of the soil; particular attention was granted to the location of cracks based on the direction of solicitation and the existing window. Different configurations have been calculated and analyzed horizontal and vertical movements.

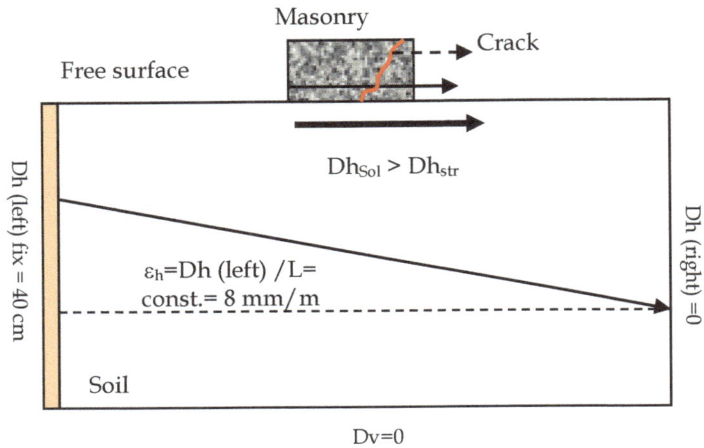

D_h: horizontal displacement, D_v: vertical displacement, ε_h: horizontal deformation

Fig. 27. Analyze of soil structure interaction due to the ground movement.

The resulting analysis indicated that a masonry wall formed by masonry units is more sensitive to horizontal strain than a continuous structure idealized by a beam. We can conclude that the theory of beams and the Potts and Addenbrooke chart underestimate the impact of horizontal strain. The out-plane horizontal strain can seriously damage the masonry wall, with the introduction of a main horizontal crack, the level of damage in this case is more important than the in-plane horizontal solicitation. The presence of a window decreases the stiffness and increases the damage to the wall.

In conclusion, numerical modeling using the distinct element method is an original tool to improve the comprehension of wall damaging; the obtained results are very close to in-situ observations and can supplement an advancing progress for the evaluation of masonry structures.

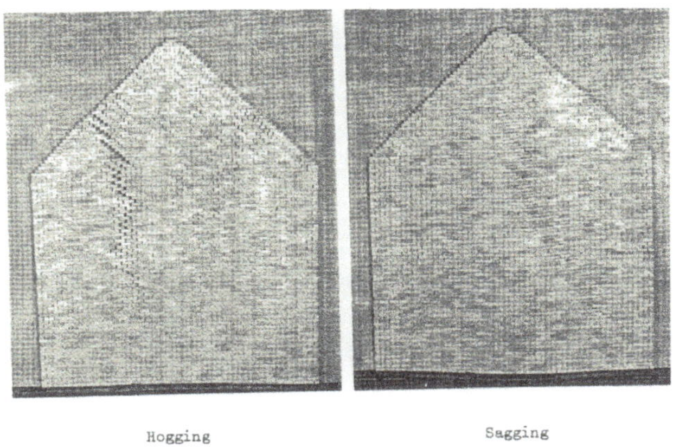

Fig. 28. Effect of horizontal strain on masonry wall using physical modeling (Cox, 1980).

6. References

Al Heib M. (2008). State Of The Art Of The Prediction Methods Of Short And Long-Term Ground Movements (Subsidence And Sinkhole) For The Mines In France. In: Coal Geology Research Progress. ISBN: 978-1-60456-596-6. Editor: Thomas Michel and Hugo Fournier. 2008 Nova Science Publishers, Inc. pp 53-76.

Barrentine, Larry B., (1999). An introduction to design of experiments: a simplified approach. Experiments with Three Factors. The American Society for Quality, pp. 27–35 (Chapter 3).

Boscardin, M.D. and Cording, E.G. (1989). Building response to excavation-induced settlement, *Journal of Geotechnical Engineering*, ASCE, 115(1): 1-21.

Burland, J.B. (1995). Assessment of risk of damage to buildings due to tunneling and excavations. Invited special lecture, in: Proc. 1st Int. Conf. Earthquake Geotechnical Engineering, IS-Tokyo, 1189-1201.

Cox D. W. (1980).- Modeling stochastic behaviour using the friction table with examples of cracked brickwork and subsidence. Proc. of the 2nd international conference on ground movements *and structures*, Cardiff (Avril 1980), Edité par GEDDES J. D., Pentech press, pp. 307-328.

Cundall, P. A., (1971), A computer model for simulating progressive large scale movements in blocky rock systems', in Proc. of the Symposium of the International Society of Rock Mechanics (Nancy, France, 1971), 1, Paper No. II-8.

Deck O., Al Heib M., and Homand F. (2003). Taking the soil-structure interaction into account in assessing the loading of a structure in a mining area. Engineering Structure 25:435-448.

Dimmock, P.S. and Mair R.J. (2008). Effect of building stiffness on tunneling-induced ground movement, Tunneling and Underground Space Technology, 23: 438-450.

Eurocode 6, (1996). Calcul des ouvrages en maçonnerie – Partie 4: Méthode de calcul simplifies pour les ouvrages non armée, NBN EN (1996-3) (ANB).

Franzius, J.N., Potts, D.M., Burland, J.B., (2006). The response of surface structures to tunnel construction, Proc. Inst. Civ. Eng. Geotech. Eng., 159 (1), 3–17.

Giorgia Giardina, Max A.N. Hendriks, Jan G. Rots (2008). Numerical analyses of tunnel-induced settlement damage to a masonry wall. 7th fib PhD Symposium in Stuttgart, Germany

Goodman, R. E., Taylor, R. L., and Brekke, T. L., (1968). A Model for the Mechanics of Jointed Rock, Journal of the Soil Mechanics and Foundations Div., ASCE, Vol. 94, No SM3, pp. 637-659.

Itasca, 2000. UDEC Universal Distinct Elements Code Manual. Continuously yielding joint model. Itasca Consulting Group Inc., pp. 1–16.

Idris J., Al Heib M., Verdel T. (2009). Masonry joints mechanical behaviour evolution in built tunnels. Analysis by numerical modelling and experimental design, Tunnelling and Underground Space Technology, 24 (2009) 617–626.

Janssen, H.J.M., (1997). Structural masonry. In: Balkema, A.A. (Ed.), Numerical studies with UDEC. Centre for civil engineering research and codes. Roterdam, Netherlands, ISBN 90 5410680 8, pp. 96–106.

Hoek, E., (2000). Practical Rock engineering. Rock Mass Properties. rocscience.com on line, Chapter 11, pp. 161–203.

Lourenço, Paulo B., (2002). Guidelines for the analysis of historical masonry structures. University of Minho, Guimarães, Portugal

Kresic, N., (1997). Quantitative Solution in Hydrology and Groundwater Modelling, Inverse Distance to a Power Method. Lewis Publishers, pp.121–123.

Lake L.M., Rankin W.J. and Hawley J. (1992). Prediction and effects of ground movements caused by tunnelling in soft ground beneath urban areas. CIRIA funders report CP/5 129p.

Massart T.J., Peerlings R.H.J., Geers M.G.D., Gottcheiner S., (2005). Mesoscopic modeling of failure in brick masonry accounting for three-dimensional effects. Engineering Fracture Mechanics 72 (2005) 1238–1253

Lemos, J. V. (1998). Discrete Element Modeling of the Seismic Behavior of Stone Masonry Arches," in Computer Methods in Structural Masonry 4 (4th International Symposium, Florence, Italy, September 1997), pp. 220-227, G. N. Pande et al., Ed. London: E&FN Spon.National Coal Board 1975. Subsidence Engineers Handbook. National Coal Board Production Dept., U.K.

Peck, R. (1969). Deep excavations and tunnelling in soft ground, Proceedings of the 7th International Conference on Soil Mechanics Foundation Engineering, Mexico, 3:225-290.

Potts, D.M. and Addenbrooke, T.I. (1997). A structure's influence on tunneling-induced ground movements. Proc. Instn. Civil Engrs., Geotechnical Engineering, 125 : 109-125.

Souley, M., (1993). Modelling of jointed rock masses by distinct element method, influence of the discontinuities constitutive laws upon the stability excavation. Doctoral thesis of (Institut national polytechnique de Lorraine), pp.75-136.

Son M. and Cording E. J. (2007). Evaluation of building stiffness for building response analysis to excavation-induced ground movements. Journal of Geotechnical engineering. ASCE/August 2007 (995-1002).

Standing, J. and Burland, J.B. (2008). Impact of underground works on existing infrastructure, Invited lecture in Post-Mining, France, 1-39.

Sutcliffe D.J., Yu H.S., Page A.W., (2001). Lower bound limit analysis of unrienforced masonry shear walls, Computers and structures, pp. 1295-1312.

Tzamtzis A.D. and Asteris P.G., (2004). FE Analysis of Complex Discontinuous and Jointed Structural Systems (Part 1: Presentation of the Method - A State-of-the-Art Review). Electronic Journal of Structural Engineering, pp. 75-92.

Verdel T., (1994). La méthode des éléments distincts (DEM) : un outil pour évaluer les risques d'instabilités des monuments en maçonnerie. Applications à des cas égyptiens. Actes du Septième Congrès de l'Association Internationale de Géologie de l'Ingénieur, Lisbonne, Septembre 1994, pp 3551-3560, Ed BALKEMA (ISBN 90 54 10 503 8)

Verdel, T. and Bigarre, P. (1999). Modélisation de tunnels anciens avec le logiciel UDEC, Rapport INERIS, Société SIMECSOL, pp. 1-12.

14

Inverse Analysis Applied to Mushy Steel Rheological Properties Testing Using Hybrid Numerical-Analytical Model

Miroslaw Glowacki
AGH University of Science and Technology
Poland

1. Introduction

Integrated casting and rolling technologies are most recent and very efficient way of hot strip production. More and more companies all over the world are able to manage such processes. The mentioned technologies ensure huge reduction of rolling costs, very high product quality and low investment costs. Computer simulation is of vital importance to the development of "know how" theory for these processes. The lack of publications concerning mechanical properties and behaviour of steels simultaneously subjected to both plastic deformation and solidification was the inspiration for the investigation. This also necessitated the development of an appropriate mathematical model of mushy-steel deformation. The contribution summarizes the results of the author's recent theoretical research concerning the computer simulation of mushy steel published in recent years in well-known journals and book chapters [Glowacki, 2006; Glowacki at al., 2010; Glowacki & Hojny, 2006, 2009; Hojny & Glowacki, 2008, 2009a, 2009b, 2011; Hojny at al., 2009].

As an example of a company providing the integrated casting and rolling technologies one can mention the plant located in Cremona, Italy which develops the new methods of steel strip manufacturing. They are called Inline Strip Production (ISP) and Arvedi Steel Technology (AST) processes and are characterized by very high temperature allowed at the mill entry. The instant rolling of slabs which leave the casting machine allows for the utilization of the heat stored in the strips during inline casting. Both the mentioned technologies ensure huge reduction of rolling forces and their details are usually classified.

The development of "know how" theory for the semi-solid steel rolling technology requires numerical modelling. The development of appropriate mathematical models is limited by the lack of thermal and mechanical properties concerning mushy steels deformation in temperature range which is close to solidus line. The work presented in the current contribution is an attempt to cover the gap providing a proposition of a hybrid numerical-analytical model of semi-solid steel deformation. The mathematical modelling of steel deformation in semi-solid state, as well as experimental work in this field, are innovative topics regarding the very high temperature range deformation processes. Tracing the related papers published in the past 10 years one can find many dealings with experimental results for non-ferrous metals tests (Kang & Yoon, 1997; Koc at al., 1996; Kopp at al., 2003; Sang-

Yong at al., 2001; Zhao at al., 2006). The first results regarding steel deformation at extra high temperature were presented during last few years (Li, 2005; Seol, 1999, 2002). Most of the problems concerning semi-solid steel testing are caused by the very high level of steel liquidus and solidus temperatures in comparison with non-ferrous metals. The deformation tests for non-ferrous metals are much easier. The rising abilities of thermo-mechanical simulators enable investigation of steel samples and as a result both computer simulation and the development of new, very high temperature rolling technologies like Arvedi ISP and AST processes. The lack of mathematical models describing the steel behaviour in the last phase of solidification with simultaneous plastic deformation was the inspiration of the investigation described in the proposed book chapter.

The main goal of the chapter is to present problems of theoretical work leading to the development of a methodology of very high temperature testing of steel samples while their central parts are still mushy. In such conditions the deformation of samples is strongly inhomogeneous and all the well-known methods of yield stress curve examination fail due to significant barrelling of the sample. Although the investigation concerned both physical tests and dedicated simulation system, the author sacrifices the contribution to the hybrid model which is the heart of the system. With the help of inverse analysis it allows for the right interpretation of deformation tests providing data regarding the mushy steel rheological properties.

2. Physical basis and characteristic features of steel deformed at very high temperature

The rolling equipment for the ISP process allows for reduction of initial mould strip thickness from 74 mm to 55 mm during liquid core reduction process. The region of maximum strip temperature for a high reduction mill is located in the strip centre and varies from 1220 °C to 1375 °C depending on the casting speed. The main benefits of the technology are: inverse temperature gradient, good product quality, very low level of heating energy consumption, up to 20 times lower water consumption in comparison to traditional rolling, low level of installed mill power, compact rolling equipment layout, no need for tunnel furnace and very low investment costs. The AST technology is a result of further development of ISP into a real endless process and the benefits of its application are even greater. The whole reduction process is running in one rolling mill consisting of 5 or 7 stands, which can reduce the strip thickness from 55÷70 mm to 0.8 mm. The maximum temperature of the strip occurs in central region of its cross-section and varies from 1340 °C to 1420 °C depending on the casting speed. This suggests that the central region of the strand subjected to the rolling is still mushy.

The main benefits of the new very high temperature technologies are significantly lower rolling forces and very favourable temperature field inside the steel plate. However, certain problems arise which are specific for this kind of metal treatment. The central parts of slabs are mushy and the solidification is not yet finished while the deformation is in progress. This results in changes in material density and occurrence of characteristic temperatures having great influence on the plastic behaviour of the material (Senk, 2000; Suzuki, 1988). The nil strength temperature (NST), strength recovery temperature (SRT), nil ductility temperature (NDT) and ductility recovery temperature (DRT) have effect on steel plastic behaviour and limit plastic deformation. The Nil Strength Temperature (NST) is the temperature level at which material strength drops to zero while the steel is being heated above the solidus temperature. Another temperature associated with NST is the Strength Recovery Temperature (SRT). At this temperature the cooled material regains strength greater than 0.5 N/mm2. Nil Ductility Temperature (NDT) represents the temperature at which the heated steel loses its ductility. The Ductility Recovery Temperature (DRT) is the temperature at which the ductility of the material (characterised by reduction of area) reaches 5% while it is being cooled. Over this temperature the plastic deformation is not allowed at any stress tensor configuration.

Significant changes of density and lack of data regarding material's thermal and mechanical properties are vital problems of the modelling. They have great influence on steel rheology and heat transfer. An issue of great importance is the lack of strain-stress relationships, which in the temperature range above 1400 °C strongly depend on the density and are very temperature sensitive. It is not easy to run isothermal tests that could be the source of the computation of yield stress function parameters for such high temperatures. There are also some problems with the interpretation of tests results.

Density is very important for plastic behaviour of mushy steel plates. It varies with temperature and depends on the cooling rate. The solidification process causes non-uniform density distribution in the controlled volume resulting in non-uniform deformation and heat conduction. There are three main factors causing density changes: solid phase formation, thermal shrinkage and movement of liquid particles inside the solid skeleton. The density plays an important role in both mechanical and thermal solutions.

The contribution sheds some light on the physical problems but it focuses on the axial-symmetrical computer model, which ensures the right simulation of mushy steel samples deformation reflecting the physical requirements. The presented model fills the gap in modelling of plastic behaviour of semi-solid steels.

3. Hybrid numerical-analytical model of mushy steel deformation

Testing of steels at temperature higher than 1400 °C is difficult due to deformation instability and risk of sample damage during experiment. Such experiments do not assure the strain homogeneity and cannot be interpreted using traditional methods. Appropriate interpretation of the results is possible only with the help of a computer aided engineering system. The contribution reports a new model underlying such a system developed by the author's team. Together with GLEEBLE physical simulator equipped with high temperature module the code allows for investigation of properties of semi-solid steel.

The numerical solver is the less visible yet very powerful kernel of the system. It is based on a thermal-mechanical model with variable density. The mechanical part of the model is a hybrid variational solution with analytical mass conservation condition constraining the velocity field components. The accuracy of the proposed solution is very good due to negligible volume loss guaranteed by the analytical form of the mass conservation condition. This is important for materials with variable density and is not captured by classical solutions. Analytical condition eliminates problems with unintentional specimen volume changes caused by application of numerical methods. The existing, physical changes of steel density in the mushy zone have influence on real variations of controlled volume.

On the other hand numerical errors can be a source of volume loss which interferes with real changes. This effect is very undesirable in modelling of thermal-mechanical behaviour of steel in temperature range characteristic for the (transformation of state of aggregation).

The mentioned mechanical and thermal parts of the mathematical model of the process are supported by a third one, i.e. the density changes model. The mechanical part is responsible for the strain, strain rate and stress distribution in a controlled volume.

4. Thermal part of the model

Heat exchange between solid metal and environment, and its flow inside the metal is controlled by a number of factors. During phase change two additional phenomena have to be taken into account. Note that in the process of deformation of steel at temperature of liquid to solid phase transition there are two sources of heat changes. On the one hand heat is generated due to the state transformation. On the other hand it is secreted as a result of plastic deformation. In addition, steel density variations also cause changes of body temperature.

Thermal solution has a major impact on simulation results, since the temperature has strong effect on remaining variables. This is especially evident if the specimen temperature is close to solidus line when the body consist of both solid and semi-solid regions. In such case the affected phenomena are: plastic flow of solid and mushy materials, stress evolution and density changes. The theoretical temperature field is a solution of Fourier-Kirchhoff equation with appropriate boundary conditions.

The most general form of the Fourier-Kirchhoff equation in any coordinate system can be written in operator form as follows:

$$\nabla^T(\mathbf{\Lambda}\,\nabla T) + Q = c_p\rho\left(\boldsymbol{v}^T\nabla T + \frac{\partial T}{\partial \tau}\right) \tag{1}$$

where T is the temperature distribution in the controlled volume and $\mathbf{\Lambda}$ denotes the symmetrical second order tensor called heat transformation tensor. In case of thermal inhomogeneity the whole tensor has to be considered. Q represents the rate of heat generation (or consumption) due to the phase transformation, due to plastic work done and due to electric current flow (resistance heating of the sample is usually applied). Finally c_p describes the specific heat, ρ the steel density, $\boldsymbol{v}$ the velocity vector of specimen particles and τ the elapsed time. The heat transformation tensor consists of a set of anisotropic heat transformation coefficients and can be given in a form:

$$\mathbf{\Lambda} = \begin{pmatrix} \lambda_{xx} & \lambda_{xy} & \lambda_{xz} \\ \lambda_{yx} & \lambda_{yy} & \lambda_{yz} \\ \lambda_{zx} & \lambda_{zy} & \lambda_{zz} \end{pmatrix} \tag{2}$$

In the case of anisotropic bodies, the solution is carried out locally, and the axes of coordinate system are oriented in accordance with the principal directions of the thermal conductivity. In this case all off-diagonal components of the heat transformation tensor are zeros ($\lambda_{ij} = 0, i \neq j$) and equation (2) becomes:

$$\mathbf{\Lambda} = \begin{pmatrix} \lambda_{xx} & 0 & 0 \\ 0 & \lambda_{yy} & 0 \\ 0 & 0 & \lambda_{zz} \end{pmatrix} \tag{3}$$

Furthermore for a thermally isotropic material $\lambda_{xx} = \lambda_{yy} = \lambda_{zz} = \lambda$ and tensor of the heat transformation can be written in the index notation can be as:

$$\Lambda_{ij} = \lambda\delta_{ij} \tag{4}$$

where δ_{ij} is the Kronecker delta.

The temperature of samples compressed in axially-symmetric process can be determined by solving the appropriate form of Fourier-Kirchhoff equation. Here the equation will be expressed in the cylindrical coordinate system, which is a natural choice for the cylindrically-shaped samples. It takes following differential form:

$$\frac{1}{r}\frac{\partial}{\partial r}\left(r\lambda_r\frac{\partial T}{\partial r}\right) + \frac{1}{r}\frac{\partial}{\partial \theta}\left(\frac{1}{r}\lambda_\theta\frac{\partial T}{\partial \theta}\right) + \frac{\partial}{\partial z}\left(\lambda_z\frac{\partial T}{\partial z}\right) + Q = \rho c_p\frac{\partial T}{\partial \tau} \tag{5}$$

The assumption of axial symmetry can be considered appropriate for the tensile and compression tests of steel in semi-solid state in all physically stable cases. It is invalid only for failed experiments. The symmetry simplifies the model by implying identical temperature distribution at any axial sample cross-section. This results in the equation:

$$\frac{\partial T}{\partial \theta} = 0 \tag{6}$$

Equation (5) can be further simplified if the heat properties of the medium are assumed isotropic. By calculating the differentials in equation (5) and using equation (6) we get the following form of Fourier-Kirchhoff equations for isotropic, axially-symmetric heat flow:

$$\lambda\left(\frac{\partial^2 T}{\partial r^2} + \frac{1}{r}\frac{\partial T}{\partial r} + \frac{\partial^2 T}{\partial z^2}\right) + Q = \rho c_p\frac{\partial T}{\partial \tau} \tag{7}$$

Equation (7) needs to be solved with appropriate initial and boundary conditions. The initial conditions relate to cases of non-stationary heat exchange. Most solutions use Cauchy condition which assume the known a priori temperature distribution at time τ_0: $T_\Omega(\tau_0) = f_\Omega$. In a particular (but often adopted) case the temperature is assumed to be constant throughout the considered area $T_\Omega(\tau_0) = T_0 = const$.

Boundary conditions have more complex nature and relate to all cases of heat transfer and describe the spatial aspect of the heat exchange. The considered continuous medium changes its temperature though convection, radiation, conduction, or a combination of these phenomena. Theoretical solutions of the problem are generally subject to one or more boundary conditions. Combined Hankel's boundary conditions have been adopted for the presented model. The conditions for axially-symmetrical problem can be written in form of a differential equation:

$$\lambda r \frac{\partial T}{\partial n} + \alpha(T - T_0) + q = 0 \tag{8}$$

In equation (8) T_0 is the distribution of border temperature, q describes the heat flux through the boundary of the deformation zone, α is the heat transfer coefficient and n is a vector which is normal to the boundary surface. More details concerning the problem can be found in (Glowacki, 1996). Equation (7) subject to condition (8) defines the problem of temperature evolution during the whole process of heating and deformation of the samples.

Note that (7) is a spatiotemporal equation. The solution of such equations is difficult because in general case the temperature is a function of both location (r, z) and time τ.

$$T = T(r, z, \tau) \tag{9}$$

In addition, the used boundary conditions, appropriate for the cooling or heating of the sample are also described by differential equation. For that reason equation (7) is solved in a two-step process (Zienkiewicz at al., 2005):

- the corresponding steady-state equation is solved. After FEM discretization this yields a matrix algebraic equation,
- the solution obtained in the first step is then adapted to non-steady-state conditions using a transient discretization of the time variable.

4.1 Thermal model for steady-state heat flow process

The Fourier-Kirchhof equation (7) for the steady heat flow can be written as:

$$\lambda \left(\frac{\partial^2 T}{\partial r^2} + \frac{1}{r} \frac{\partial T}{\partial r} + \frac{\partial^2 T}{\partial z^2} \right) + Q = 0 \tag{10}$$

Application of finite element method for solving problems of heat flow requires a functional. Equation (10) together with the boundary conditions given by equation (8) needs to be expressed in a variational setting.

Consider the problem of optimizing the general form of the heat flux power functional.

$$\chi = \int_V f(r, z, T, T_r, T_z)\ dV + \int_S \left(qT + \tfrac{1}{2}\alpha(T - T_0)^2 \right)\ dS \tag{11}$$

where f is a function of position, temperature and temperature gradient:

$$T_r = \frac{\partial T}{\partial r}; \qquad T_z = \frac{\partial T}{\partial z} \tag{12}$$

This function is specified in the relevant domain V with the boundary S. Let us consider a small variation of the functional (11):

$$\delta\chi = \int_V \left(\frac{\partial f}{\partial T}\delta T + \frac{\partial f}{\partial T_r}\delta T_r + \frac{\partial f}{\partial T_z}\delta T_z \right) dV + \int_S [q\delta T + \alpha(T - T_0)\ \delta T] dS \tag{13}$$

that can be rewritten as:

$$\delta\chi = \int_V \delta T \left[\frac{\partial f}{\partial T} - \frac{\partial}{\partial r}\left(\frac{\partial f}{\partial T_r}\right) - \frac{\partial}{\partial z}\left(\frac{\partial f}{\partial T_z}\right)\right] dV + \int_S \delta T \left(q + \alpha(T - T_0) + l_r \frac{\partial f}{\partial T_r} + l_z \frac{\partial f}{\partial T_z}\right) dS \quad (14)$$

where l_r and l_z are the direction cosines of normal to the outer surface with respect to Or and Oz axes, respectively.

A necessary condition for the functional (11) to reach extreme value for a given function is for the variation $\delta\chi$ to be equal to 0. Since equation (14) must be satisfied for any variation δT, the expressions in brackets have to be zero at an extreme:

$$\frac{\partial}{\partial r}\left(\frac{\partial f}{\partial T_r}\right) + \frac{\partial}{\partial z}\left(\frac{\partial f}{\partial T_z}\right) - \frac{\partial f}{\partial T} = 0 \quad (15)$$

for the entire volume V and

$$l_x \frac{\partial f}{\partial T_x} + l_z \frac{\partial f}{\partial T_z} + q + \alpha(T - T_0) = 0 \quad (16)$$

for its boundary S. Can therefore be concluded that if one satisfy the equations (15) and (16) than the functional (11) reaches an optimum. Both of these formulations are equivalent. The above reasoning is the solution of so called Euler problem. In the presented particular case the appropriate form of the function f is as follows:

$$f = r\left[\frac{1}{2}\lambda(T_r^2 + T_z^2) - QT\right] \quad (17)$$

where T_r and T_z are given by relationships (12). In this case the equations (15) and (16) can be written as follows:

$$\begin{aligned} \lambda\left(\frac{\partial^2 T}{\partial r^2} + \frac{1}{r}T_r + \frac{\partial^2 T}{\partial z^2}\right) + Q = 0 \\ \lambda r \frac{\partial T}{\partial n} + q + \alpha(T - T_0) = 0 \end{aligned} \quad (18)$$

The presented reasoning shows that the assumption of steady-state heat flow leads to equations (18). The first of them is identical with the equation (10), and the second to boundary condition (8). Thus, according to the Euler reasoning, the solution of equation (10) satisfying the boundary condition (8) is the functional extremal:

$$\chi = \int_V r\left\{\frac{1}{2}\lambda\left[\left(\frac{\partial T}{\partial r}\right)^2 + \left(\frac{\partial T}{\partial z}\right)^2\right] - QT\right\} dV + \int_S \left(qT + \frac{1}{2}\alpha(T - T_0)^2\right) dS \quad (19)$$

Optimization of the functional (19) in the domain of discrete functions is based on replacement of the continuous real function of the temperature distribution $T(r,z)$ by their discrete counterparts. In the proposed solution the finite element method was used for that purpose. The discretization of the control volume was done accordingly. The temperature distribution function was discretized according to the formula:

$$T(r,z) = \mathbf{n}^T(r,z)\,\mathbf{T} \tag{20}$$

where $\mathbf{n}(r,z)$ is a vector of the shape function and $\mathbf{T}$ is a nodal temperature vector. After substituting (20) and its derivatives to (19) it takes the discrete form:

$$\chi = \int_V r\left\{\frac{1}{2}\lambda\left[\left(\frac{\partial \mathbf{n}^T}{\partial r}\mathbf{T}\right)^2 + \left(\frac{\partial \mathbf{n}^T}{\partial z}\mathbf{T}\right)^2\right] - Q\mathbf{n}^T\mathbf{T}\right\}dV + \int_S\left(q\mathbf{n}^T\mathbf{T} + \frac{1}{2}\alpha(\mathbf{n}^T\mathbf{T} - T_0)^2\right)dS \tag{21}$$

From the mathematical point of view, equation (21) no longer defines a functional, but a function of many variables. Nevertheless hereinafter it still will be referred to as a functional. Its derivative with respect to $\mathbf{T}$ is given as follows:

$$\frac{\partial \chi}{\partial \mathbf{T}} = \int_V r\left[\lambda\mathbf{T}^T\left(\frac{\partial \mathbf{n}}{\partial r}\frac{\partial \mathbf{n}^T}{\partial r} + \frac{\partial \mathbf{n}}{\partial z}\frac{\partial \mathbf{n}^T}{\partial z}\right) - \mathrm{Q}\mathbf{n}^T\right]dV + \int_S (q\mathbf{n}^T + \alpha(\mathbf{T}^T\mathbf{n} - T_0)\mathbf{n}^T)dS \tag{22}$$

Equation (22) one can written in matrix form as:

$$\mathbf{HT} + \mathbf{p} = \mathbf{0} \tag{23}$$

where matrix $\mathbf{H}$ and vector $\mathbf{p}$ have shapes:

$$\begin{aligned} \mathbf{H} &= \int_V r\lambda\left(\frac{\partial \mathbf{n}}{\partial r}\frac{\partial \mathbf{n}^T}{\partial r} + \frac{\partial \mathbf{n}}{\partial z}\frac{\partial \mathbf{n}^T}{\partial z}\right)dV + \int_S \alpha\mathbf{n}\mathbf{n}^T dS \\ \mathbf{p} &= -\int_V rQ\mathbf{n}\; dV - \int_S (\alpha T_0 - q)\,\mathbf{n}\; dS \end{aligned} \tag{24}$$

The system of linear equations (23) can be solved using standard methods of linear algebra. This yields the discrete temperature vector $\mathbf{T}$.

4.2 Thermal model for non-steady-state heat flow

For the non-steady-state heat flow equation (7) has to be used instead of equation (10). A derivation similar to the one for the steady-state flow and the same space discretization lead to formulation of discrete form of functional equivalent to equation (7). It is analogous to functional (21).

$$\begin{aligned} \chi = \int_V r\left\{\frac{1}{2}\lambda\left[\left(\frac{\partial \mathbf{n}^T}{\partial r}\mathbf{T}\right)^2 + \left(\frac{\partial \mathbf{n}^T}{\partial z}\mathbf{T}\right)^2\right] - \left[Q - \rho c_p\frac{\partial}{\partial \tau}(\mathbf{n}^T\mathbf{T})\right]\mathbf{n}^T\mathbf{T}\right\}dV + \\ + \int_S\left(q\mathbf{n}^T\mathbf{T} + \frac{1}{2}\alpha(\mathbf{n}^T\mathbf{T} - T_0)^2\right)dS \end{aligned} \tag{25}$$

Differentiation of functional (25) with respect to $\mathbf{T}$ leads to relation which is similar to (22).

$$\frac{\partial \chi}{\partial \mathbf{T}} = \int_V r\left[\lambda \mathbf{T}^T\left(\frac{\partial \mathbf{n}}{\partial r}\frac{\partial \mathbf{n}^T}{\partial r} + \frac{\partial \mathbf{n}}{\partial z}\frac{\partial \mathbf{n}^T}{\partial z}\right) - \left(Q - \rho c_p \frac{\partial \mathbf{T}^T}{\partial \tau}\mathbf{n}\right)\mathbf{n}^T\right] dV + \\ + \int_S (q\mathbf{n}^T + \alpha(\mathbf{T}^T\mathbf{n} - T_0)\mathbf{n}^T) dS \tag{26}$$

The system (26) can be written in a matrix form analogous to equation (23):

$$\mathbf{HT} + \mathbf{C}\frac{\partial \mathbf{T}}{\partial \tau} + \mathbf{p} = \mathbf{0} \tag{27}$$

where $\mathbf{H}$ and $\mathbf{p}$ are matrices given by (24), and $\mathbf{C}$ can be expressed as:

$$\mathbf{C} = \int_V \rho c_p \mathbf{n}\, \mathbf{n}^T dV \tag{28}$$

An assumption of linear temperature change in very short time interval $\Delta\tau$ and application of weighted Galerkin's residual method leads to an equation which is a discrete (with respect to time) counterpart of equation (27).

$$\overline{\mathbf{H}}\mathbf{T}_{i+1} + \overline{\mathbf{p}} = \mathbf{0} \tag{29}$$

Matrix $\overline{\mathbf{H}}$ and vector $\overline{\mathbf{p}}$ in equation (29) are described by the following relations:

$$\overline{\mathbf{H}} = \left(2\mathbf{H} + \frac{3}{\Delta\tau}\mathbf{C}\right) \\ \overline{\mathbf{p}} = \left(\mathbf{H} - \frac{3}{\Delta\tau}\mathbf{C}\right)\boldsymbol{T}_i + 3\mathbf{p} \tag{30}$$

Equation (29) can be used to compute the vector of nodal temperatures $\mathbf{T}_{i+1}$ after a time step $\Delta\tau$ (i.e. at $\tau = \tau_{i+1} = \tau_i + \Delta\tau$) provided that initial value $\mathbf{T}_i$ for $\tau = \tau_i$ is known.

5. Mechanical model

A mathematical model of the compression process is based on the theory of plastic flow (Chakrabarty, 2006). The principle of the upper assessment (Bower, 2010), calculus of variations (Adhikari, 1998), approximation theory and optimization methods (Findaeisen at al., 1980 ; Nocedal & Wright 2006) and numerical methods for solving partial differential equations (Evans 1988; Polyanin, & Zaitsev, 2004; Pinchover & Rubinstein, 2005), including the finite element method (Zienkiewicz at al., 2005) were used. The following assumptions were established:

- deformation and stress state are axial-symmetrical,
- deformed material is isotropic but inhomogeneous,
- the material behaviour is rigid-plastic - the relationship between the stress tensor and strain rate tensor is calculated according to the Levy-Mises flow law, which is given as:

$$\sigma_{ij} - \frac{1}{3}\sigma_{kk}\delta_{ij} = \frac{2}{3}\frac{\sigma_p}{\dot{\varepsilon}_i}\dot{\varepsilon}_{ij} \tag{31}$$

Rigid-plastic model was selected due to its very good accuracy at the strain field during the hot deformation and sufficient correctness of calculated deviatoric part of the stress field. Moreover, the elastic part of each stress tensor component is very low at temperatures close to solidus line and can in practice be neglected in calculations of strain distribution. The limits for plastic metal behavior are defined according to Huber-Mises-Hencky yield criterion:

$$\sigma_{ij}\sigma_{ij} = 2\left(\frac{\sigma_p}{\sqrt{3}}\right)^2 \tag{32}$$

In equations (31) and (32) σ_{ij} denotes the stress tensor components, σ_{kk} represents the mean stress, δ_{ij} is the Kronecker delta, σ_p indicates the yield stress, $\dot{\varepsilon}_i$ is the effective strain rate, and $\dot{\varepsilon}_{ij}$ denotes strain rate tensor components. The components are given by an equation:

$$\dot{\varepsilon}_{ij} = \frac{1}{2}\left(\nabla_i v_j + \nabla_j v_i\right) \tag{33}$$

In cylindrical coordinate system $Or\theta z$ the solution is a vector velocity field defined by the distribution of three coordinates $\boldsymbol{v} = (v_r, v_\theta, v_z)$. The field is a result of optimization of a power functional, which can be written in general form as the sum of power necessary to run the main physical phenomena related to plastic deformation. Due to the axial-symmetry of the sample the velocity field the circumferential component of the velocity field can be neglected and the functional is usually formulated as:

$$J[\boldsymbol{v}] = \dot{W} = \dot{W}_\sigma + \dot{W}_\lambda + \dot{W}_f \tag{34}$$

Component $\dot{W}_\sigma$ occurring in equation (34) represents the plastic deformation power, $\dot{W}_\lambda$ is the power which is a penalty for the departure from mass conservation condition, $\dot{W}_f$ denotes the friction power and $\boldsymbol{v} = (v_r, v_z)$ describes the reduced velocity field distribution.

Rigid-plastic formulation of metal deformation problem requires the condition of mass conservation in the deformation zone. In case of solids and liquids with a constant density, this condition can be simplified to the incompressibility condition. Such a condition is generally satisfied with sufficient accuracy during the optimization of functional (34). In most solutions a slight, but noticeable loss of volume is observed. The loss is caused by incomplete fulfilment of the incompressibility condition imposed on the solution in numerical form. It is negligible in case of traditional computer simulation of deformation processes although in some embodiments more accurate methods are used to restore the volume of metal subjected to the deformation. Unlike this case the density of semi-solid materials varies during the deformation process and these changes result in a physically reasonable change in the volume of a body having constant mass. The size of the volume loss due to numerical errors is comparable with changes caused by fluctuation in the density of the material.

A further problem specific to the variable density continuum is power $\dot{W}_\lambda$, which occurs in functional (34). It is used in most solutions and has a significant share of total power. Even when the iterative process approaches the end, this power component is still significant, especially if the convergence of the optimization procedures is insufficient. In case of discretization of the deformation area (e.g. using the finite element method) if one focuses solely on the $\dot{W}_\lambda$ a number of possible optimal solutions appear. They are related to a number of possible directions of movement of discretization nodes providing the volume preservation of the deformation zone. Each of these solutions creates a local optimum for $\dot{W}_\lambda$ power and thus for the entire functional (34). This makes it difficult to optimize because of lack of uniform direction of fall of total power which leads to global optimum. The material density fluctuation causes further optimization difficulties, resulting from additional replacement of incompressibility condition with a full condition of mass conservation.

The proposed solution requires high accuracy in ensuring the incompressibility condition for the solid material or mass conservation condition for the semi-solid areas. This approach stems from the fact that the errors resulting from the breach of these conditions can be treated as a volume change caused by the steel density variation in the semi-solid zone. High accuracy solution is required also due to large differences in yield stress for the individual subareas of the deformation zone. In the discussed temperature range they appear due to even slight fluctuations in temperature. In presented solution the second component of functional (34) is left out and mass conservation condition is given in analytical form constraining the radial (v_r) and longitudinal (v_z) velocity field components. The functional takes the following shape:

$$J[\boldsymbol{v}] = \dot{W}_\sigma + \dot{W}_t \tag{35}$$

In case of functional (35) the numerical optimisation procedure converges faster than the one for functional (34) due to the reduced number of velocity field parameters (only radial components are optimisation parameters) and the lack of numerical form of mass conservation condition. The accuracy of the proposed hybrid solution is higher also due to negligible volume loss caused by numerical errors which is very important for materials with variable density.

As mentioned before the solution of the problem is a velocity field in cylindrical coordinate system in axial-symmetrical state of deformation. Optimization of metal flow velocity field in the deformation zone of semi-variational problem requires the formulation according to equation (35). The radial velocity distribution $v_r(r,\theta,z)$ and the longitudinal one $v_z(r,\theta,z)$ are so complex that such wording in the global coordinate system poses considerable difficulties. These difficulties are the result of the mutual dependence of these velocities. Therefore the basic formulation will be written for the local cylindrical coordinate system $Or\theta z$ with a view to the future discretization of deformation area using one of the dedicated methods. In addition one will find that the deformation of cylindrical samples is characterized by axial symmetry. As demonstrated by experimental studies conducted using semi-solid samples the symmetry may be disturbed only as a result of unexpected leakage of liquid phase.

Such experiments, however, are regarded as unsuccessful and not subject to numerical analysis. Establishment of the axial symmetry, which except in cases of physical instability can be considered valid also for the process of compression or tensile test of semi-solid samples, allows one to simplify the model because of the identical strain distribution at any axial sample cross-section. Considerations will therefore be carried out in Orz coordinates for the sample cross-sectional using one of the planes containing the sample axis. Components of power functional given by (35) have been formulated in accordance with the general theory of plasticity by relevant equations. The plastic power for the deformation zone having volume of V is given by the subsequent relation:

$$\dot{W}_\sigma = \int_V \sigma_i \dot{\varepsilon}_i \, dV \tag{36}$$

where σ_i is the effective stress and $\dot{\varepsilon}_i$ denotes the effective strain. The plastic deformation starts when the rising effective stress reaches yield stress limit σ_p $(\sigma_i = \sigma_p)$ according to yield criterion given by equation (32). Effective strain occurring in equation (36) is calculated on the basis of the strain tensor components $\dot{\varepsilon}_{ij}$ according to following relationship:

$$\dot{\varepsilon}_i = \sqrt{\frac{2}{3}\dot{\varepsilon}_{ij}\dot{\varepsilon}_{ij}} \tag{37}$$

The components are given by equation (33). For axial-symmetrical case the strain has a form:

$$\begin{pmatrix} \frac{\partial v_r}{\partial r} & 0 & \frac{1}{2}\frac{\partial v_r}{\partial z} + \frac{1}{2}\frac{\partial v_z}{\partial r} \\ 0 & \frac{v_r}{r} & 0 \\ \frac{1}{2}\frac{\partial v_r}{\partial z} + \frac{1}{2}\frac{\partial v_z}{\partial r} & 0 & \frac{\partial v_z}{\partial z} \end{pmatrix} \tag{38}$$

The second component of functional (35) is responding for friction. To compute friction power on the boundary S of area V a model given by the subsequent equation was used:

$$\dot{W}_t = \int_S m \frac{\sigma_p}{\sqrt{3}} \|\overline{\boldsymbol{v}}\| \, dS \tag{39}$$

In equation (39) m is the so called friction factor which is usually experimentally selected and $\overline{\boldsymbol{v}}$ is a relative velocity vector of metal and tool $\overline{\boldsymbol{v}} = \boldsymbol{v} - \boldsymbol{v}_t$. In case of tensile test the samples are permanently fixed in jaws of a physical simulator and friction must not be taken into account. However, compression test requires sharing the friction power which is significant.

5.1 The model of sample velocity field

Clearly defined deformation field resulting from the optimal solution of functional (37) cannot be calculated without one of the conditions mentioned before. For the solid zones the incompressibility condition can be described by universal operator equation independently of the mechanical state of the deformation process:

$$\nabla \boldsymbol{v} = 0 \tag{40}$$

Because the semi-solid zone is characterized by density change due to still ongoing progress of steel state of aggregation, the condition of incompressibility is inadequate to reflect changes and was replaced with the mass conservation condition, which describes the following modified operational equation:

$$\nabla \boldsymbol{v} - \frac{1}{\rho}\frac{\partial \rho}{\partial t} = 0 \tag{41}$$

The basis for the optimization of functional (35) is the velocity field determined by appropriate system of velocity functions in the concerned area. These functions are then the source of deformation field and other physical quantities affecting the power functional formulation. Obtaining an accurate real velocity field requires the use of velocity functions depending on a number of variational parameters. The functions should be flexible enough to map the field throughout the whole volume of the deformation zone. Analytical description of each component of the velocity field with a single function in the whole area of deformation is not preferred. This approach creates difficulties especially in areas not subjected to the deformation where the velocity function should remain constant. Therefore, the solution to the problem of semi-solid metal flow was based on the method proposed by Malinowski in (Malinowski, 1986, 1997, 2005). This method involves the breakdown of the elements and the deformation velocity field approximation by polynomials with coefficients different for each element. The method was originally applied to solutions with a constant volume. The author of the current paper has developed a new method for semi-solid materials by adapting the source one to the analysis of materials with variable density.

In the case of deformation of axial-symmetrical bodies the incompressibility condition is given by following differential equation:

$$\frac{\partial v_r}{\partial r} + \frac{v_r}{r} + \frac{\partial v_z}{\partial z} = 0 \tag{42}$$

For the semi-solid area equation (42) is replaced by the mass conservation condition due to existing density changes. The longitudinal velocity has been calculated as an analytical function of radial velocity using this condition. In cylindrical coordinate system the condition has been described with an equation:

$$\frac{\partial v_r}{\partial r} + \frac{v_r}{r} + \frac{\partial v_z}{\partial z} - \frac{1}{\rho}\frac{\partial \rho}{\partial \tau} = 0 \tag{43}$$

Equation (42) is a special case of equation (43) and therefore the proposed solution will consider the dependence (43) as more general. In (43) ρ is the temporary material density and τ is the time variable. The proposed variational formulation makes the longitudinal velocity dependent on the radial one. Condition (43) allows for the calculation of $\partial v_z/\partial z$ derivative as a function of $\partial v_r/\partial r$ after analytical differentiation of radial velocity distribution function $v_r(r,z)$. Hence, the longitudinal velocity is calculated as a result of analytical integration according to following equation:

$$v_z = -\int \left(\frac{\partial v_r}{\partial r} + \frac{v_r}{r} - \frac{1}{\rho}\frac{\partial \rho}{\partial \tau} \right) dz \tag{44}$$

In this case the velocity field depends only on one function - the radial velocity distribution. Both the components (v_r and v_z) satisfy the mass conservation imposed on the velocity field. The functional takes the form of equation (35) and in case of application of one of the methods requiring discretization (FEM, FDM or any meshless method) the number of discrete parameters is significantly reduced (at least by half). Only the right class of the velocity field distribution functions is problematic. The functions must be relevant for description of the material deformation and sufficiently flexible. Hence, the whole control volume is usually divided into sub-areas and the functions are defined in local coordinate systems for each sub-region. It requires the definition of both the local system and transformation from local to global one. The r coordinate acts as an independent variable (abscissa) in global area and varies in the range of r_m to R_m. The z coordinate depends on r and is limited by functions describing both the area boundaries: lower $z_l = f(r)$ and upper $z_u = g(r)$. Considering all the assumptions two linear functions, binding both the systems - global Orz and local one $O\xi\eta$ were defined

$$\begin{aligned} \xi(r,z) &= \frac{r}{R_m} \\ \eta(r,z) &= \frac{2z - g(r) - f(r)}{g(r) - f(r)} \end{aligned} \tag{45}$$

The main assumption of the presented model is the dependence of the longitudinal velocity distribution function $v_z(\xi,\eta)$ on the radial velocity distribution function $v_r(\xi,\eta)$. For this purpose, the form of the function v_r has to be determined on the basis of analysis of the velocity field components distribution in the control area. For further discussion one assumes the following form v_r function:

$$v_r(\xi,\eta) = \frac{1}{2}\frac{r v_0}{g(r) - f(r)}\left(1 + \frac{\partial \psi(\xi,\eta)}{\partial \eta}\right) \tag{46}$$

where v_0 is the GLEEBLE jaw velocity and $\psi(\xi,\eta)$ is a distribution function of velocity field components in local coordinate system. It should be remembered that for the areas in which the steel is in solid state the incompressibility condition given by dependence (42) should be taken into account and for zones with semi-liquid steel mass conservation equation (43) is valid. Linking the longitudinal velocity v_z with the radial one is done precisely through these two conditions. Taking into account the more general equation (43) and assuming a known value of the radial velocity one can be determine the longitudinal one using the following dependence:

$$v_z(\xi,\eta) = \int \left[\frac{1}{\rho}\frac{\partial \rho}{\partial t} - \frac{\partial v_r(\xi,\eta)}{\partial r} - \frac{v_r(\xi,\eta)}{r} \right] dz \tag{47}$$

The consequence of such a conduct is the fact that this condition is imposed on the velocity field in an analytical form. As already mentioned it is of major importance for optimizing the correct flow field for the steel being in semi-solid conditions.

In order to relate both the velocities the derivative of the velocity with respect to the radial coordinate has to be calculated first. Having in mind the dependence of f and g on r and similar one of ψ on ξ and on η one can write:

$$\frac{\partial v_r}{\partial r}=\frac{v_0}{2}\left[\frac{\partial}{\partial r}\left(\frac{r}{g-f}\right)\left(1+\frac{\partial \psi}{\partial \eta}\right)+\frac{r}{g-f}\frac{\partial}{\partial r}\left(1+\frac{\partial \psi}{\partial \eta}\right)\right] \tag{48}$$

After some differentiations and arrangements relationship (48) can be written in a form:

$$\frac{\partial v_r}{\partial r}=\frac{v_0}{2(g-f)}\left(1+\frac{\partial \psi}{\partial \eta}+\xi\frac{\partial^2 \psi}{\partial \xi\,\partial \eta}\right)-\frac{r v_0\left(\frac{\partial g}{\partial r}-\frac{\partial f}{\partial r}\right)}{2(g-f)^2}\left[1+\frac{\partial \psi}{\partial \eta}+\frac{\partial^2 \psi}{\partial \eta^2}\left(\frac{\frac{\partial g}{\partial r}+\frac{\partial f}{\partial r}}{\frac{\partial g}{\partial r}-\frac{\partial f}{\partial r}}+\eta\right)\right] \tag{49}$$

Taking into account equation (49) and relationship (43) one can calculate the derivative of the longitudinal velocity with respect to z.

$$\frac{\partial v_z}{\partial z}=-\frac{\partial v_r}{\partial r}-\frac{v_r}{r}+\frac{1}{\rho}\frac{\partial \rho}{\partial \tau}=\frac{r v_0\left(\frac{\partial g}{\partial r}-\frac{\partial f}{\partial r}\right)}{2(g-f)^2}\left[1+\frac{\partial \psi}{\partial \eta}+\frac{\partial^2 \psi}{\partial \eta^2}\left(\frac{\frac{\partial g}{\partial r}+\frac{\partial f}{\partial r}}{\frac{\partial g}{\partial r}-\frac{\partial f}{\partial r}}+\eta\right)\right]- \\ -\frac{v_0}{2(g-f)}\left(1+\frac{\partial \psi}{\partial \eta}+\frac{1}{2}\xi\frac{\partial^2 \psi}{\partial \xi\,\partial \eta}\right)+\frac{1}{\rho}\frac{\partial \rho}{\partial \tau} \tag{50}$$

After appropriate integration the velocity is given by the following relationship:

$$v=-\frac{v_0}{4}\left\{2\left(\eta+\frac{g+f}{g-f}+\psi\right)+\xi\frac{\partial \psi}{\partial \xi}-\frac{\left(\frac{\partial g}{\partial r}-\frac{\partial f}{\partial r}\right)}{g-f}\left[\eta+(1-r)\psi+r\left(\frac{\frac{\partial g}{\partial r}+\frac{\partial f}{\partial r}}{\frac{\partial g}{\partial r}-\frac{\partial f}{\partial r}}+\eta\right)\frac{\partial \psi}{\partial \eta}\right]\right\}+ \\ +\frac{\eta(g-f)+g+f}{2\rho}\frac{\partial \rho}{\partial \tau} \tag{51}$$

Function $\psi=\psi(\xi,\eta)$ occurring in all the relationships describing the velocity field can be under the Weierstrass theorem approximated by polynomials. Approximation of $\psi(\xi,\eta)$ with the help of one polynomial in the whole deformation zone, although possible in some cases, is impractical and is a source of many problems. On the other hand the division of areas into smaller sub-areas requires continuity. To ensure continuity of the velocity field and strain field in the whole zone of deformation, including the boundaries of the sub-regions, function $\psi(\xi,\eta)$ should be at least of class C^2.

6. Density changes and their influence on remaining models

In the proposed solution one of the most important parameters is the density. Its changes influence the mechanical part of the presented model and strongly depend on the temperature. The knowledge of effective density distribution is very important for modelling deformation of mushy materials. In the presented solution a model of density changes based on empirical data was applied.

Density distribution is one of the most important properties of the mushy steel which is subjected to the deformation. Its changes have influence on both the mechanical and thermal parts of the presented model. On the other hand, the density is strongly dependent on the temperature. Moreover, the solidification process causes non-uniform density distribution in the controlled volume. Since, the knowledge concerning effective density distribution is very important for the behaviour of deformation of porous and mushy materials and the modelling of such species requires good density changes model.

Density variations of liquid, semi-solid and solid materials are ruled by three phenomena:

- solid phase formation,
- laminar liquid flow through porous material and
- thermal shrinkage.

Transient rate of density changes is ruled by an equation:

$$\frac{\partial \rho}{\partial \tau} = \frac{\partial \rho_p}{\partial \tau} + \frac{\partial \rho_f}{\partial \tau} + \frac{\partial \rho_t}{\partial \tau} \tag{52}$$

In (52) the subsequent right hand derivatives of ρ_p, ρ_f i ρ_t with respect to transient time variable τ denote the density changes as a result of three mentioned phenomena One may calculate the density changes due to solid phase formation according to the relationship:

$$\frac{\partial \rho_p}{\partial \tau} = [\rho_s(1 - X_l) + \rho_l X_l]\left(\frac{\rho_s}{\rho_l} - 1\right)\frac{\partial X_l}{\partial \tau} \tag{53}$$

where X_l and X_s are the shares of liquid and solid phases are semi-steel. Changes in density caused by laminar flow of the liquid phase through the porous material are described by the equation:

$$\frac{\partial \rho_f}{\partial \tau} = \rho_l X_l \left(\frac{\partial v_r}{\partial r} + \frac{v_r}{r} + \frac{\partial v_z}{\partial z}\right) \tag{54}$$

In (54) v is the velocity of the metal particles flow. Changes in density due to thermal shrinkage depend on the speed of changes in temperature and coefficients of linear thermal expansion β_s i β_l of both solid and liquid phases:

$$\frac{\partial \rho_t}{\partial \tau} = [\beta_s \rho_s (1 - X_l) + \beta_l \rho_l X_l] \frac{\partial T}{\partial \tau} \tag{55}$$

where T is the temperature on an absolute scale. Issues of density changes mechanisms were the subject of (Glowacki, 2002). Changes in the density as a result of the velocity and temperature of the metal particles substantially complicate the problem of optimizing the metal flow velocity field. Coupled solution of all the problems is difficult and very often an uncoupled model is used.

6.1 Empirical model of density changes

The density changes model is rather complex and its solution is associated with an additional increase in computational complexity of the total solution. Regardless of the solution used the development of a right model is a problem in itself. It requires addressing a number of issues related to the change of state, the flow of the liquid phase in the presence of solid steel frames, etc. This is an important issue - however, it requires commitment of substantial computer resources and long computation times. Hence another way of taking density into consideration is possible due to temperature dependency of this quantity (Glowacki, 1996). In order to avoid additional problems with solution of differential equation, density changes were calculated according to an empirical model taking into consideration experimental data. The model is slightly less accurate but such a method makes the solution much easier. The solution seems to be a good alternative way to predict changes in mushy steel. In proposed approach the density is depending on:

- temperature,
- chemical composition of the material and,
- steel microstructure.

The study published in (Glowacki, 1998), which is result of investigation carried out for steel in the solid state, shows that for typical forming processes impact of a steel grade on change in the density resulting from temperature changes is small. For determination of density in these conditions for both carbon and low-alloy steels it is proposed to apply following empirical equation:

$$\rho = \frac{7850}{(1 + \Delta l)^3}; \left[\frac{\mathrm{kg}}{\mathrm{m}^3}\right] \tag{56}$$

In equation (56) Δl is calculated according to following formula:

$$\Delta l = 0{,}004 \left(\frac{T + 273}{1000}\right)^2$$

Similar equation can be used for austenitic steels:

$$\rho = \frac{7897}{(1 + \Delta l)^3}; \left[\frac{\mathrm{kg}}{\mathrm{m}^3}\right] \tag{57}$$

The Δl parameter from (57) is calculated as:

$$\Delta l = -0{,}00358 + 0{,}00947\frac{T+273}{1000} + 0{,}0103\left(\frac{T+273}{1000}\right)^2 - 0{,}00298\left(\frac{T+273}{1000}\right)^3$$

Similar dependence can be used for high-alloy steels. In this case it is necessary to modify the equation (56) in a manner appropriate for the particular steel grade. Thus, for temperature range which is proper for traditional process of steel hot deformation the calculation of changes in density seems to be pretty simple. Such temperatures are characteristic for certain sample areas.

Otherwise presents itself the problem for higher temperature ranges, where the deformation occurs during the simultaneous metal solidification. Here the density variations may be significant. For purposes of the current mathematical model an approach proposed by Mizukami was used (Mizukami at al., 2002). For carbon steels containing no other elements the density changes are functions of temperature. Steels were tested with a wide range of carbon content, which ranges from 0.005% to 0.56% by mass. The authors develop tests for typical steels having chemical composition expressed in% by mass given in Table. 1.

Left side of Figure 1 shows the change in density for MC1 grade steel as a function of temperature. For steel changing its states of aggregation some plots of density in the various phases of the transformation process has been developed. The right side of Figure 1 shows the course of the changes in the density of liquid phase as a function of ΔT_l - undercooling temperature with respect to the liquidus line. Changes in density are presented in relation to the base density of 7060 [kg/m^3].

Steel	ULC	LC	MC1	MC2	HC
[C]	0.005	0.040	0.110	0.140	0.550
[Si]	0.010	0.040	0.100	0.160	0.150
[Mn]	0.120	0.190	0.480	0.540	0.910
[P]	0.014	0.026	0.020	0.016	0.021
[S]	0.003	0.006	0.008	0.003	0.001

Table 1. Chemical composition (mass %) of typical steels tested by authors of (Mizukami at al., 2002).

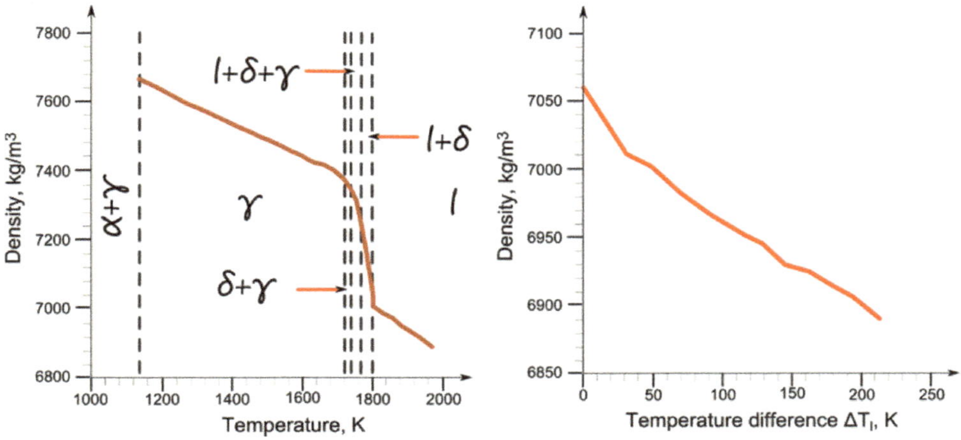

Fig. 1. Density of MC1 grade steel as a function of temperature (left) and density changes of its liquid phase as a function of temperature increase (right). Plots are based on data published in (Mizukami at al., 2002).

Subsequent charts presented in Figure 2 show changes in density of δ i γ phases, respectively. Both of them are functions of undercooling temperature ΔT_δ and ΔT_γ of appropriate phases with respect to solidus temperature.

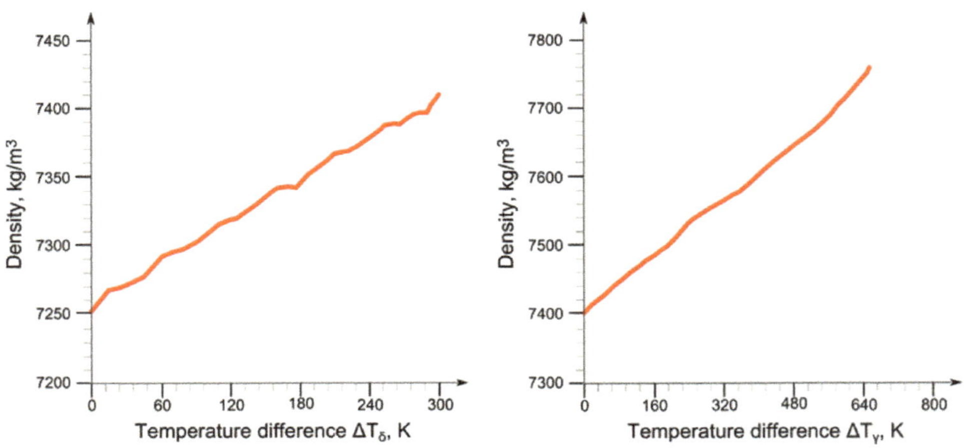

Fig. 2. Density changes of steel phase δ (left) and γ (right) of MC1 steel grade as a function of temperature increase – based on data published in (Mizukami at al., 2002).

The presented graphs were used to develop analytical dependencies, which describe changes in the density of steel during the transformation of state of aggregation. In the region of coexistence of δ and γ phases density was estimated using the additivity rule. The correctness of this approximation was verified by comparing the theoretical results with those which were obtained from the measured values.

The density of carbon steels depends on the temperature and the existing fraction of liquid phase. The effect of alloying elements (except of coal) on the density of each steel phase is small, although the concentration of these components significantly affect the fraction of the phases. The density of each phase is calculated according to the following equations:

$$\begin{aligned} \rho_l &= 7{,}02 - 5{,}50 \cdot 10^{-4}\,\Delta T_l \\ \rho_\delta &= 7{,}27 + 3{,}07 \cdot 10^{-4}\,\Delta T_\delta \\ \rho_\gamma &= 7{,}41 + 4{,}80 \cdot 10^{-4}\,\Delta T_\gamma \end{aligned} \tag{58}$$

In equation (58) ρ_l, ρ_δ and ρ_γ indicate densities of liquid steel and its δ and γ phases, respectively. The density in the regions of occurrence of several phases simultaneously is given by the following equations:

$$\begin{aligned} \rho_{l+\delta} &= \rho_l^0 + \Delta\rho_{l/\delta} \cdot X_\delta \\ \rho_{l+\gamma} &= \rho_l^0 + \Delta\rho_{l/\gamma} \cdot X_\gamma \\ \rho_{l+\delta+\gamma} &= \rho_l^0 + \rho_{l/\delta} \cdot f_\delta + \Delta\rho_{l/\gamma} \cdot f_\gamma \end{aligned} \tag{59}$$

where ρ_l^0 denotes the density of the liquid phase for temperature discrepancy ΔT_l, $\Delta\rho_{l/\delta}$ and $\Delta\rho_{l/\gamma}$ are density differences between δ and γ phases for temperature drop from liquidus to solidus level, X_δ and X_γ are the fractions of δ i γ phases in surrounding liquid phase, respectively. and finally f_δ i f_γ are relative fractions of δ i γ phases. The density of $\delta + \gamma$ phase was estimated according to relationship:

$$\rho_{\delta+\gamma} = \rho_\delta \cdot X_\delta + \rho_\gamma \cdot X_\gamma \tag{60}$$

7. Mushy steel flow stress curves development

The subsequent part of the chapter deals with the computation of mushy steel flow stress curves based on the developed mathematical model which helps to avoid interpretational problems occurring in traditional testing procedures. Proper interpretation of the experimental results is possible with the help of appropriate computer aided testing system. Such a user friendly dedicated computer system with variable density has been developed (Glowacki & Hojny, 2009; Hojny & Glowacki, 2009a). The system codename called *Def_Semi_Solid* is a result of theoretical research conducted in a team lead by the chapter author with the financial support of grants awarded by Polish Committee of Scientific Research. The system in itself is not a subject of the chapter and its details are not discussed. The program was developed using an object oriented technique and is compatible with both Windows and Unix based platforms.

During experiments a few quantities were recorded. Among them the most important are GLEEBLE jaws displacement, force and temperature. This is a start point for the inverse analysis. The system calculates the shape and size of the deformation zone and strain and stress fields as well as optimal values of flow stress curve parameters. The model described in the previous section allows for the comparison of theoretical and experimental results for non-uniform temperature field. Isothermal tests in the temperature range over 1400 °C are impossible even using sophisticated equipment like GLEEBLE simulator. The presented model is a solution to the experimental problems. The analysis of metal flow in subsequent regions of the sample deformation zone requires adequate methods. Classical techniques of interpretation of results of compression testing procedures fail due to significant samples barrelling which is inevitable at any temperature close to solidus level and which requires right analysis of metal flow in subsequent regions of the sample deformation zone.

A number of steel grades were subjected to series of experiments in Institute for Ferrous Metallurgy in Gliwice, Poland using GLEEBLE 3800 simulator. Example results of examination of two steels are reported in the current contribution. The first one is the 18G2A grade steel having 0.16% of carbon and the second was the S355J2G3So grade with 0.11% of carbon content. The essential aim of the investigation was the reconstruction of both temperature changes and strain evolution on specimen exposed to simultaneous deformation and solidification. The inverse procedure has been reported in (Glowacki & Hojny, 2009). Example results of inverse analysis are shortly described in succeeding subsections.

7.1 Characteristic temperature levels

As mentioned before, apart from the liquidus and solidus temperatures, four other temperature levels are characteristic for the mushy steel behaviour. All the levels split the liquidus-solidus range into intervals. The most important for the extra high temperature rolling process design is the nil ductility temperature (NDT). The plastic deformation of a steel specimen is possible only below the NDT temperature. The temperature levels have to be calculated according to results of series of difficult experiments which are not a subject of the current paper. For carbon steels with the carbon content of around 0.1 % the equilibrium liquidus and solidus temperature levels are 1523°C and 1482°C, respectively and the NDT is 1420°C. One must note that the last one is a conventional temperature of a sample surface (indicated during experimental procedure). The maximum and minimum temperatures in the sample's central cross-section may differ by 60- 70 °C. The equilibrium liquidus and solidus temperatures for 18G2A grade steel are 1513°C and 1465°C, respectively. The measured mean value of NDT temperature of the steel falls into the range of 1420°C÷1425°C. The NDT is related to the temperature at which the last liquid phase particles existing in the central part of the sample disappear in static processes. It has been observed that for temperatures higher than NDT a remainder of liquid phase still exist in the central part of the sample (Hojny & Glowacki, 2009a). For dynamic cooling and deformation processes in some regions of the sample the remainder of liquid phase can be observed at temperatures lower than NDT because the difference between sample surface and its central region is higher than for quasi-static processes.

7.2 Yield stress functions

The well-known Voce formula (Voce, 1955) was adopted for the description of the shape of yield stress function. Figure 3 presents four subsequent stages of an example compression test at higher sample surface temperature, i.e. 1425°C for the quasi-static process. One can observe that the experiment was successful (no metal outflow) and the deformation of the sample was realised despite the significant barrelling of the sample.

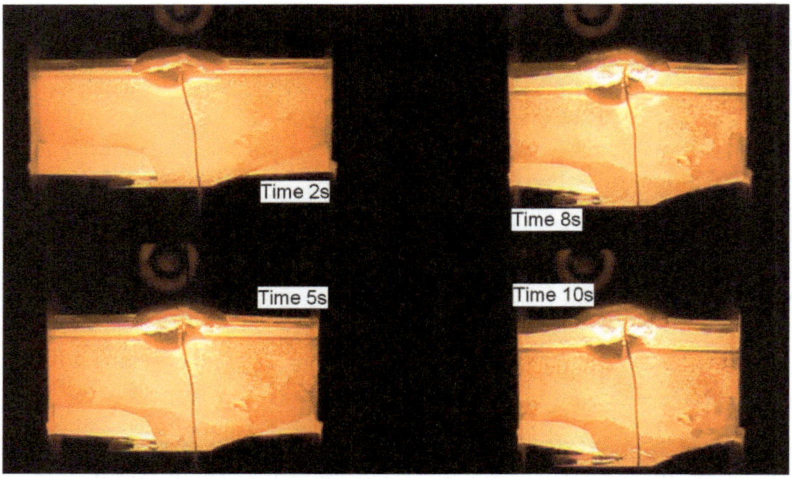

Fig. 3. Four stages of the deformation process ran at temperature 1425°C for the quasi-static deformation process. The figure presents the central part of the sample.

Due to significant strain inhomogeneity inverse analysis is the only method allowing for appropriate calculation of coefficients of yield stress functions at any temperature higher than NDT. The objective function of the analysis was defined as a norm of discrepancies between calculated (F^c) and measured (F^m) loads in a number of subsequent stages of the compression according to the following equation:

$$\varphi(x) = \sum_{i=1}^{n}(F_i^c - F_i^m)^2 \tag{61}$$

where n is the number of subsequent intervals of stress versus strain curve. The theoretical forces F^c were calculated with the help of sophisticated numerical solver being the implementation of the model which was described in this chapter. Due to the very low level of recorded stresses the experimental curves obtained from the GLEEBLE machine are noisy. Before the application of inverse analysis they were smoothed using Fast Fourier Transformation (FFT) algorithm.

The final shape of the curves for 18G2A and S355J2G3So grade steels after interpretation using inverse analysis are presented in figures 8 and 16, respectively. Figure 4 summarise the results of calculation of example coefficients of Voce formula for 18G2A grade steel which was deformed in a quasi-static process. The effective strain inside the deformation zone varied from 0 to 0.6 and the effective strain rate reached its maximum value of 2.9 s^{-1} in final stage of the deformation process. The presented curves are plotted using the calculated coefficients of Voce curve for temperature levels observed in the sam ples' cross-sections and for strain rate equal to 1 s^{-1}.

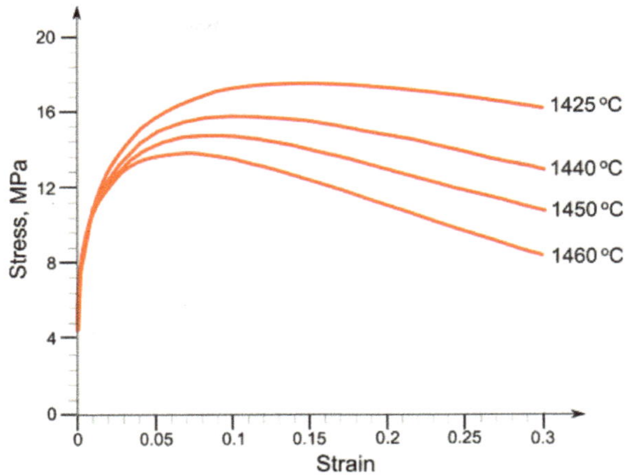

Fig. 4. Flow stress vs. strain at several temperature levels for 18G2A steel grade deformed during quasi-static process - strain rate 1 s-1 (Glowacki & Hojny, 2010).

Example results of investigation of S355J2G3So grade steel are presented in Figure 5. The investigation procedures were analogous to those applied in case of18G2A grade steel.

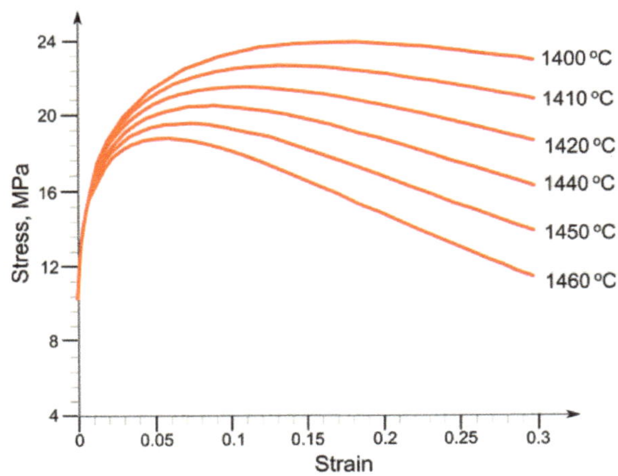

Fig. 5. Stress-strain curves at several temperature levels from the range of 1400-1450°C (S355J2G3So grade steel, quasi-static process - strain rate 1 s-1 (Glowacki & Hojny, 2010).

8. Conclusions

Modelling of deformation of steel samples with mushy zone requires resolving several problems which are characteristic for the temperature range close to the solidus level. Some of the problems are independent of the strain and stress state of the material and are similar for both axial-symmetrical and three dimensional cases. The computation of characteristic temperatures and temperature-dependent sudden changes of steel plastic properties require advanced methods of computer simulation. The most important property for material plastic behaviour is the yield stress function describing strain-stress curve. The proposed analytical model allows computation of such kind relationships.

The chapter has been dedicated to hybrid numerical-analytical model of semi-solid steel behaviour under plastic deformation. Application of inverse analysis and the proposed model allows for the testing of rheological properties of steels at temperature higher than 1400 °C. The results of the research are crucial for a unique computer system allowing for proper interpretation of the results of very high temperature compression tests. The classical interpretation of such results is improper due to strong strain inhomogeneity. The developed system is a tool to overcome many interpretational problems allowing for the computation of the appropriate shape and parameters of yield-stress curves. The curves have crucial influence on the results of computer simulation of semi-solid steel deformation.

The model presented in the current contribution is an axial-symmetrical one. The author have run further investigations leading to the development of a fully three-dimensional model of integrated casting and rolling processes as well as the soft reduction process, that is a part of strip casting technology. Like the model presented in the hereby chapter the spatial one also focuses on three main aspects: thermal, mechanical and density changes. Further intention of the research is the development of fully three dimensional model of mushy steel behaviour during rolling of plates with mushy region. The model will be useful for technologists working on the development of an integrated casting and rolling process. It is the most recent technology of sheet steel production, which is very profitable and requires extremely low energy consumption - very important for steel and automotive industries.

The compression tests carried out have shown good predictive ability of the proposed solution. They show that the flow stress above the NDT is strongly temperature and strain rate dependent. Low carbon steels, having carbon content of 0.11% and 0.16%, have been investigated in wide temperature range and strain rate. Example results of the experimental work were presented delivering a set of equations describing rheological behaviour of the investigated steels. The presented model and experimental procedure requires further investigation leading to the improvement of the solution and modelling additional phenomena accompanying the simultaneous deformation and solidification processes.

9. Acknowledgments

The work has been supported by the Polish Ministry of Science and Higher Education - Grant No. N N508 585539.

10. References

Adhikari S.K. (1998). Variational principles for the numerical solution of scattering problems. Wiley, New York USA, ISBN 0471181935

Bower, A.F. (2010) Applied mechanics and solids, CRC Press - Taylor & Francis Group, New York USA, ISBN 987-1-4398-0247-2

Chakrabarty, J. (2006). Theory of plasticity. Elsevier Butterworth-Heinemann, Oxford UK, ISBN 978-0-7506-6638-2

Evans, L.C. (1998), Partial Differential Equations, American Mathematical Society, ISBN 0821807722.

Findaeisen, W., Szymanowski, J., Wierzbicki, A (1980). Theory and optimization methods. PWN, Warszawa, ISBN 8301009764

Glowacki, M. (1996). Finite element three-dimensional modelling of the solidification of a metal forming charge, *Journal of Materials Processing Technology*, Vol. 60, No. 1-4, pp. 501-504, ISSN 0924-0136

Glowacki, M. (1998). Thermal-mechanical-microstruktural model of shape rolling. *Dissertations and Monographs*, Vol. 76, AGH Publishing, Krakow Poland, ISSN 0867-6631

Głowacki, M. (2002) Possibilities of mathematical modeling of deformation of samples with mushy zone, *Proceedings of 44th Mechanical Working and Steel Processing Conference*, pp. 1151-1162, Orlando USA, September 1, 2002, ISBN 978-1-886362-62-8

Glowacki, M. (2006). Mathematical modelling of deformation of steel samples with mushy zone, In: *Research in Polish metallurgy at the beginning of XXI century, Committee of Metallurgy of the Polish Academy of Science*, K. Swiatkowski, (Ed.), 305-324, Publishing House Akapit, Krakow Poland

Glowacki, M. & Hojny, M. (2006). Development of a computer system for high temperature steel deformation testing procedure, *Proceedings of Simulation, Design and Control of Foundry Processes*, pp. 145-156, Krakow Poland, November 22-24, 2006

Glowacki, M. & Hojny, M. (2009). Inverse analysis applied for determination of strain-stress curves for steel deformed in semi-solid state, *Inverse Problems in Science and Engineering*, Vol.17, No. 2, pp. 159–174, ISSN 1741-5977

Glowacki, M. & Hojny, M. (2010). Investigation of mushy steel rheological properties at temperatures close to solidus level, In: *Polish metallurgy 2006–2010 in time of the worldwide economic crisis , Committee of Metallurgy of the Polish Academy of Science*, K. Swiatkowski, (Ed.), 193-212, Publishing House Akapit, Krakow, Poland, ISBN 978-83-60958-59-9

Glowacki, M., Hojny, M. & Jędrzejczyk, D. (2010). Hybrid analytical-numerical system of mushy steel deformation. In : *Recent studeis in meshless & other novel computational methods*, B. Sarler & S.N. Atluri, (Eds.), pp. 35-54, Tech Science Press, ISBN-10 0-9824205-4-4, USA

Hojny, M. & Glowacki, M. (2008). Computer modelling of deformation of steel samples with mushy zone, *Steel Research International*, vol. 79, No. 11, (2008), pp. 868-874, ISSN 1611-3683

Hojny, M. & Glowacki, M. (2009a) The methodology of strain - stress curves determination for steel in semi-solid state, *Archives of Metallurgy and Materials*, Vol. 54, No. 2, pp. 475–483, ISSN 1733-3490

Hojny, M. & Glowacki, (2009b) The physical and computer modelling of plastic deformation of low carbon steel in semi-solid state, *Transactions of the ASME, Journal of Engineering Materials and Technology*, Vol. 131 No. 4, pp. 041003-1–041003-7, ISSN 0094-4289

Hojny, M., Glowacki, M. & Malinowski Z. (2009), Computer aided methodology of strain-stress curve construction for steels deformed at extra high temperature, *High Temperature Materials and Processes*, Vol. 28, No. 4, pp. 245–252, ISSN 0334-6455

Hojny, M. & Glowacki, M. (2011). Modeling of Strain-Stress Relationship for Carbon Steel Deformed at Temperature Exceeding Hot Rolling Range, *Transactions of the ASME, Journal of Engineering Materials and Technology*, Vol. 133, No. 2, pp. 021008-1–021008-7, ISSN 0094-4289

Kang, C.G. & Yoon, J.H. (1997). A finite-element analysis on the upsetting process of semi-solid aluminum material, *Journal of Materials Processing Technology*, Vol. 66, No. 1-3, pp. 76-84, ISSN 0924-0136

Koc, M., Vazquez, V., Witulski, T. & Altan, T. (1996). Application of the finite element method to predict material flow and defects in the semi-solid forging of A356 aluminum alloys, *Journal of Materials Processing Technology*, Vol. 59, No. 4, pp. 106-112, ISSN 0924-0136

Kopp, R., Choi, J. & Neudenberger D. (2003). Simple compression test and simulation of an Sn–15% Pb alloy in the semi-solid state, *Journal of Materials Processing Technology*, Vol. 135, No. 2-3, pp. 317-323, ISSN 0924-0136

Leader, J. J. (2004). Numerical Analysis and Scientific Computation. Addison Wesley, Boston Massachusetts, ISBN 978-0-201-73499-7

Li, J.Y., Sugiyama, S. & Yanagimoto, J. (2005), Microstructural evolution and flow stress of semi-solid type 304 stainless steel, *Journal of Materials Processing Technology*, Vol. 161, No. 3, pp. 396-406, ISSN 0924-0136

Malinowski, Z. (1986). Analysis of upsetting process based on velocity fields, PhD thesis, AGH Krakow Poland, in Polish

Malinowski, Z. (1997). Effect of heat generation on flow stress deformation based on the axially symmetric compression test. Metallurgy & Foundry Engineering, 23, 1997, 459-467, ISSN 1239-2325

Malinowski Z. (2005). Numerical models in metal forming and heat transfer. Wyd. AGH Publishing, Krakow Poland, in Polish, ISBN: 83-89388-98-7

Mizukami, H., Yamanaka, A. & Watanabe, T. (2002). Prediction of density of carbon steels, *ISIJ International*, Vol. 42, No. 4, pp. 375-384, ISSN 0915-1559

Nocedal J. & Wright S.J. (2006). Numerical Optimization. Springer-Verlag, Berlin Germany. ISBN 0-387-30303-0

Pinchover, Y. & Rubinstein, J. (2005). An Introduction to Partial Differential Equations, New York: Cambridge University Press, ISBN 0521848865.

Polyanin, A.D. & Zaitsev, V.F. (2004). Handbook of Nonlinear Partial Differential Equations, Boca Raton: Chapman & Hall/CRC Press, ISBN 1584883553.

Sang-Yong, L., Jung-Hwan L. & Young-Seon L. (2001). Characterization of Al 7075 alloys after cold working and heating in the semi-solid temperature range. *Journal of Materials Processing Technology*, Vol. 111, No. 1-3, pp. 42-47, ISSN 0924-0136

Seol, D.J., Won, Y.M., Yeo, T., Oh, K.H., Park, J.K. & Yim, C.H. (1999). High Temperature Deformation Behavior of Carbon Steel in the Austenite and δ-Ferrite Regions, *ISIJ International*, Vol. 39, No. 1, pp. 91-98, ISSN 0915-1559

Seol, D.J., Oh, K.H., Cho, J.W., Lee, J.E., Yoon, U.S. (2002). Phase-field modelling of the thermo-mechanical properties of carbon steels, Acta Materialia, Vol. 50, No. 9, pp. 2259-2268, ISSN 1359-6454

Senk, D., Hagemann, F., Hammer, B., Kopp, R., Schmitz, H.P. & Schmitz, W. (2000). Umformen und Kühlen von direktgegossenem, Stahlband, *Stahl und Eisen*, Vol. 120, No. 6, pp. 65-69, ISSN 0340-4803

Suzuki, H.G., Nishimura, S. & Yamaguchi S. (1988). Physical simulation of the continuous casting of steels, *Proceedings of Physical Simulation of Welding, Hot Forming and Con- tinuous Casting*, pp. 166-191, Canmet Canada, May 2-4, 1988

Voce, E. (1955). A Practical Strain Hardening Function, *Metallurgia*, vol. 51, 1955, pp. 219-226, ISSN 0141-8602

Zhao Y.Q., Wu W.L. & Chang H. (2006). Research on microstructure and mechanical properties of a new α + Ti2Cu alloy after semi-solid deformation, *Materials Science and Engineering*, Vol. 416, No. 1-2, pp. 181-186, ISSN 0921-5093

Zienkiewicz, O.C., Taylor, R. L. & Zhu, J.Z. (2005). The Finite Element Method: Its Basis and Fundamentals, Elsevier Butterworth-Heinemann, Oxford UK, ISBN 0-7506-6320-0

15

Numerical Schemes for Fractional Ordinary Differential Equations

Weihua Deng and Can Li
School of Mathematics and Statistics, Lanzhou University, Lanzhou
People's Republic of China

1. Introduction

Fractional calculus, which has almost the same history as classic calculus, did not attract enough attention for a long time. However, in recent decades, fractional calculus and fractional differential equations become more and more popular because of its powerful potential applications. A large number of new differential equations (models) that involve fractional calculus are developed. These models have been applied successfully, e.g., in mechanics (theory of viscoelasticity), biology (modelling of polymers and proteins), chemistry (modelling the anomalous diffusion behavior of Brownian particles), electrical engineering (electromagnetic waves) etc (Bouchaud & Georges, 1990; Hilfer, 2000; Kirchner et al., 2000; Metzler & Klafter, 2000; Zaslavsky, 2002; Mainrdi, 2008). Meanwhile, some rich fractional dynamical behavior which reflect the inherent nature of realistic physical systems are observed. In short, fractional calculus and fractional differential equations have played more and more important role in almost all the scientific fields. One of the most important fractional models is the following initial value problem

$$D_*^\alpha y(t) = f(t, y(t)), \quad y^{(k)}(x_0) = y_0^{(k)}, \quad k = 0, 1, \cdots, \lceil \alpha \rceil - 1, \tag{1}$$

where $\alpha \in (0, \infty)$, $y_0^{(k)}$ can be any real numbers, and D_*^α denotes the fractional derivative in the Caputo sense, defined by

$$D_*^\alpha y(t) = J^{n-\alpha} D^n y(t) = \frac{1}{\Gamma(n-\alpha)} \int_0^t \frac{y^{(n)}(\tau)}{(t-\tau)^{1+\alpha-n}} d\tau, \tag{2}$$

here $n := \lceil \alpha \rceil$ is the smallest integer not less than α, $y^{(n)}(x)$ is the classical nth-order derivative of $y(x)$ and for $\mu > 0, J^\mu$ is the μ-order Riemann-Liouville integral operator expressed as follows

$$J^\mu y(t) = \frac{1}{\Gamma(\mu)} \int_0^t \frac{y(\tau)}{(t-\tau)^{1-\mu}} d\tau.$$

The use of Caputo derivative in above equation is partly because of the convenience to specify the initial conditions. Since the initial conditions are expressed in terms of values of the unknown function and its integer-order derivatives which have clear physical meaning (Podlubny, 1999; 2002). However, from pure mathematical viewpoint, the Riemann-Liouville derivative is more welcome and many earlier research papers use it instead of Caputo derivative (Podlubny, 1999). In general, specifying some additional conditions is necessary to make sure the discussed equation has a unique solution. These additional conditions, in many situations, describe some properties of the solution at the initial time (Heymans & Podlubny, 2005), the fractional derivative does not have convenient used physical meaning (there are already some progress in the geometric and the physical interpretation of the fractional calculus (Podlubny, 2002) and the physical interpretation of the initial conditions in terms of the Riemann-Liouville fractional derivatives of the unknown function has also been discussed in (Podlubny, 2002)).

Just like the classic calculus and differential equations, the theories of fractional differentials, integrals and differential equations have been developing. With the development of the theories of fractional calculus, many research monographs are published, e.g., (Oldham & Spanier, 1974; Podlubny, 1999; Samko et al., 1993). In the literatures, several analytical methodologies, such as, Laplace transform, Mellin transform, Fourier transform, are restored to obtain the analytical solutions of the fractional equations by many authors (Metzler & Klafter, 2000; Podlubny, 1999; Samko et al., 1993; Zaslavsky, 2002; Mainrdi, 2008), however, similar to treating classical differential equations, they can mainly deal with linear fractional differential equation with constant coefficients. Usually, for nonlinear systems these techniques do not work. So in many cases the more reasonable option is to find its numerical solution. As is well known, the difficulty of solving fractional differential equations is essentially because fractional calculus are non-local operators. This non-local property means that the next state of a system not only depends on its current state but also on its historical states starting from the initial time. This property is closer to reality and is the main reason why fractional calculus has become more and more useful and popular. In other words, this non-local property is good for modeling reality, but a challenge for numerical computations. Much effort has been devoted during the recent years to the numerical investigations of fractional calculus and fractional dynamics of (1) (Lubich, 1985; 1986; Podlubny, 1999). More recently, Diethelm et al successfully presented the numerical approximation of (1) using Adams-type predictor-corrector approach (Diethelm & Ford, 2002a) and the detailed error analysis of this method was given in (Diethelm et al., 2004). The convergent order of Diethelm's predictor-corrector approach was proved to be $\min(2, 1+\alpha)$ (Diethelm et al., 2004). Because of the non-local property of the fractional derivatives, the arithmetic complexity of their algorithm with step size h is $O(h^{-2})$, whereas a comparable algorithm for a classical initial value problem only gives rise to $O(h^{-1})$. To improve the accuracy and reduce the arithmetic complexity, some techniques such as the Richardson extrapolation, short memory principle and corresponding mixed numerical schemes are developed. In (Deng, 2007a), we present an improved version of the predictor-corrector algorithm with the accuracy increased to $\min(2, 1+2\alpha)$ and half of the computational cost is reduced comparing to the original one in (Diethelm et al., 2004). Furthermore, we apprehend the short memory principle from a new viewpoint (Deng, 2007b); after using the nested meshes presented in (Ford & Simpson, 2001) and combining the short memory principle and the predictor-corrector approach, we minimize the computational complexity to $O(h^{-1}\log(h^{-1}))$ at preserving the order of accuracy 2.

This chapter briefly reviews the recent development of the predictor-corrector approach for fractional dynamic systems. The plan of this chapter is as follows. In Section 2, we briefly discuss the short memory principle and the nested meshes. In Section 3, the predictor-corrector schemes and its improved versions are presented, meanwhile the convergent order and arithmetic complexity are also proposed. In Section 4, we provide two numerical examples to illustrate the performance of our numerical schemes. Conclusions are given in the last section.

2. Short memory principle

We can see that the fractional derivative (2) is an operator depending on the past states of the process $y(t)$ (see Fig 1). However, for $t \gg 0$ and $0 < \alpha < 1$, the behavior of $y(t)$ far

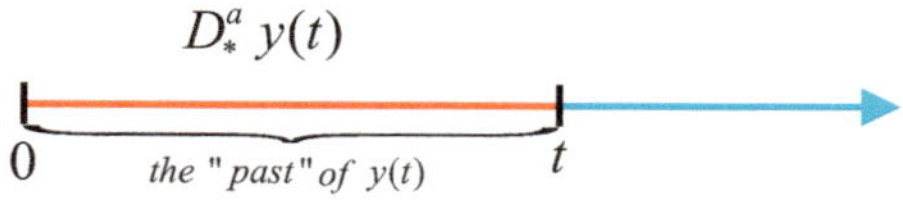

Fig. 1. The fractional derivative operating on the "past" of $y(t)$ (the red part).

from upper terminal suggests that the original fractional derivative can possibly be replaced by the fractional derivative with a moving lower terminal (Podlubny, 1999). This means that the history of the process $y(t)$ can be approached only over a fixed period of recent history. Inspired by this kind of idea, Podlubny in (Podlubny, 1999) introduces a *fixed integral length memory principle* for Riemann-Liouville derivative (see Fig 2). He shows that the truncation

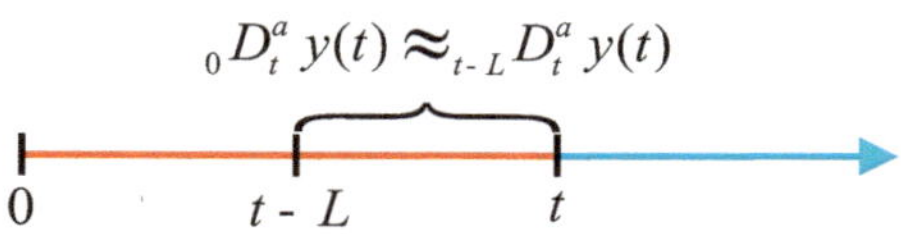

Fig. 2. The "short-memory" principle with "fixed memory length" $L(< t)$, where ${}_0D_t^\alpha$ denote the Riemann-Liouville derivative operator.

error gives $E < ML^{-\alpha}/\Gamma(1-\alpha)$ with a fixed integral length L. To accelerate the computation without significant loss of accuracy, in (Ford & Simpson, 2001) Ford and Simpson present a *short-memory principle* for Caputo derivative. In (Deng, 2007b) we apprehend the short memory principle from a new viewpoint, and then it is closer to reality and extends the effective range of short memory principle from $\alpha \in (0,1)$ to $\alpha \in (0,2)$. In view of the scaling property of fractional integral, the nested meshes are possibly used. And the *nested meshes* can be produced by splitting the integral interval $[0, t_n]$ into

$$[0.t_n] = [0, t_n - p^m\mathcal{T}] \cup [t_n - p^m\mathcal{T}, t_n - p^{m-1}\mathcal{T}] \cup \cdots [t_n - p^2\mathcal{T}, t_n - p\mathcal{T}] \cup [t_n - p\mathcal{T}, t_n], \quad (3)$$

where $\mathcal{T} = \omega h, h \in \mathbb{R}^+, m, \omega, p \in \mathbb{N}$ and $p^m\mathcal{T} \leq t_n < p^{m+1}\mathcal{T}$. Denote $M_h = \{hn, n \in \mathbb{N}\}$ and $l_1, l_2 \in \mathbb{N}, l_1 > l_2$, then $M_{l_2h} \supset M_{l_1h}$.

3. Predictor-corrector schemes

The problem of determining $y(t)$ by means of the information of *initial value* y_0, is called *initial value problem*. As the classic theory of ordinary differential equation, if a function $y(t)$ satisfies the initial value problem (1) pointwisely, then $y(t)$ is called the *solution of the fractional differential equation* (1). For the existence and uniqueness of the solutions of fractional differential equations, one can see (Podlubny, 1999). But the explicit formula of the solution $y(t)$ can't usually be obtained in spite of we can prove the existence of solution. The most reasonable way is to use numerical methods, and the obtained solution is called the *numerical approximation of the solution of the differential equations* (1) and denoted by y_h in the following sections.

3.1 Numerical schemes

In this section, we show the predictor-corrector schemes of (1). Using the Laplace transform formula for the Caputo fractional derivative (Podlubny, 1999)

$$\mathcal{L}\{D_*^\alpha y(t)\} = s^\alpha Y(s) - \sum_{k=0}^{n-1} s^{\alpha-k-1} y^{(k)}(0),\ n-1 < \alpha \leq n. \tag{4}$$

From (1), we have

$$s^\alpha Y(s) - \sum_{k=0}^{n-1} s^{\alpha-k-1} y^{(k)}(0) = F(s, Y(s)), \tag{5}$$

or

$$Y(s) = s^{-\alpha} F(s, Y(s)) + \sum_{k=0}^{n-1} s^{-k-1} y^{(k)}(0), \tag{6}$$

where $F(s, Y(s)) = \mathcal{L}(f(t, y(t)))$. Applying the inverse Laplace transform gives

$$y(t) = \sum_{k=0}^{\lceil\alpha\rceil-1} y_0^{(k)} \frac{t^k}{k!} + \frac{1}{\Gamma(\alpha)} \int_0^t (t-\tau)^{\alpha-1} f(\tau, y(\tau)) d\tau, \tag{7}$$

where the fact

$$\mathcal{L}\{J^\mu y(t)\} = \mathcal{L}\{\frac{1}{\Gamma(\mu)} \int_0^t \frac{y(\tau)}{(t-\tau)^{1-\mu}} d\tau\} = \mathcal{L}\{\frac{t^{\mu-1}}{\Gamma(\mu)} * y(t)\} = s^{-\mu} Y(s),$$

and

$$\mathcal{L}\{t^{\mu-1}\} = s^{-\mu}\Gamma(\mu).$$

are used. The approximation is based on the equivalent form of the Volterra integral equation (7). A fractional Adams-predictor-corrector approach was firstly developed by Diethelm et al (Diethelm et al., 2002b) to numerically solve the problem (7). Using the standard quadrature techniques for the integral in (7), denote $g(\tau) = f(\tau, y(\tau))$, the integral is replaced by the trapezoidal quadrature formula at point t_{n+1}

$$\int_0^{t_{n+1}} (t_{n+1} - \tau)^{\alpha-1} g(\tau) d\tau \approx \int_0^{t_{n+1}} (t_{n+1} - \tau)^{\alpha-1} \tilde{g}_{n+1}(\tau) d\tau, \tag{8}$$

where $\tilde{g}_{n+1}$ is the piecewise linear interpolation of g with nodes and knots chosen at $t_j, j = 0, 1, 2, \cdots, n+1$. After some elementary calculations, the right hand side of (8) gives

$$\int_0^{t_{n+1}} (t_{n+1} - \tau)^{\alpha-1} \tilde{g}_{n+1}(\tau) d\tau = \frac{h^\alpha}{\alpha(\alpha+1)} \sum_{j=0}^{n+1} a_{j,n+1} g(t_j), \tag{9}$$

where the uniform mesh is used and h is the stepsize. And if we use the product rectangle rule, the right hand of (8) can be written as

$$\int_0^{t_{n+1}} (t_{n+1} - \tau)^{\alpha-1} \tilde{g}_{n+1}(\tau) d\tau = \sum_{j=0}^{n} b_{j,n+1} g(t_j), \tag{10}$$

where

$$a_{j,n+1} = \begin{cases} n^{\alpha+1} - (n-\alpha)(n+1)^{\alpha}, & \text{if} \quad j = 0, \\ (n-j+2)^{\alpha+1} - 2(n-j+1)^{\alpha+1} + (n-j)^{\alpha+1}, & \text{if} \quad 1 \le j \le n, \\ 1, & \text{if} \quad j = n+1 \end{cases}$$

and

$$b_{j,n+1} = \frac{h^{\alpha}}{\alpha}[(n+1-j)^{\alpha} - (n-j)^{\alpha}], \quad \text{if} \quad 0 \le j \le n+1.$$

Then the predictor and corrector formulae for solving (7) are given, respectively, by

$$y_h^P(t_{n+1}) = \sum_{k=0}^{\lceil\alpha\rceil-1} \frac{t_{n+1}^k}{k!} y_0^{(k)} + \frac{1}{\Gamma(\alpha)} \sum_{j=0}^{n} b_{j,n+1} f(t_j, y_h(t_j)) \tag{11}$$

and

$$y_h(t_{n+1}) = \sum_{k=0}^{\lceil\alpha\rceil-1} \frac{t_{n+1}^k}{k!} y_0^{(k)} + \frac{h^{\alpha}}{\Gamma(2+\alpha)} f(t_{n+1}, y_h^P(t_{n+1})) + \frac{h^{\alpha}}{\Gamma(2+\alpha)} \sum_{j=0}^{n} a_{j,n+1} f(t_j, y_h(t_j)). \tag{12}$$

The approximation accuracy of the scheme (11)-(12) is $O(h^{\min\{2,1+\alpha\}})$ (Diethelm et al., 2004).

Now we make some improvements for the scheme (11)-(12). We modify the approximation of (8) as (Deng, 2007a)

$$\int_0^{t_{n+1}} (t_{n+1}-\tau)^{\alpha-1} g(\tau) d\tau \approx \int_0^{t_n} (t_{n+1}-\tau)^{\alpha-1} \widetilde{g}_n(\tau) d\tau + \int_{t_n}^{t_{n+1}} (t_{n+1}-\tau)^{\alpha-1} g(t_n) d\tau, \tag{13}$$

where $\widetilde{g}_n$ is the piecewise linear interpolation for g with nodes and knots chosen at $t_j, j = 0, 1, 2, \cdots, n$. Then using the standard quadrature technique, the right hand of (13) can be recast as

$$\int_0^{t_n} (t_{n+1}-\tau)^{\alpha-1} \widetilde{g}_n(\tau) d\tau + \int_{t_n}^{t_{n+1}} (t_{n+1}-\tau)^{\alpha-1} g(t_n) d\tau = \frac{h^{\alpha}}{\alpha(\alpha+1)} \sum_{j=0}^{n} \widetilde{b}_{j,n+1} g(t_j) \tag{14}$$

where

$$\widetilde{b}_{j,n+1} = \begin{cases} \begin{cases} a_{j,n+1}, & \text{if} \quad 0 \le j \le n-1, \\ 2^{\alpha+1} - 1, & \text{if} \quad j = n. \end{cases} & \text{if} \quad n > 0, \\ b_{0,1} = \alpha + 1, & \text{if} \quad n = 0. \end{cases}$$

Hence, this algorithm for the predictor step can be improved as (Deng, 2007b)

$$y_h^P(t_{n+1}) = \sum_{k=0}^{\lceil\alpha\rceil-1} \frac{t_{n+1}^k}{k!} y_0^{(k)} + \frac{h^{\alpha}}{\Gamma(2+\alpha)} \sum_{j=0}^{n} \widetilde{b}_{j,n+1} f(t_j, y_h(t_j)), \tag{15}$$

The new predictor-corrector approach (15) and (12) has the numerical accuracy $O(h^{\min\{2,1+2\alpha\}})$ (the detailed analysis is left in Section 3.2). Obviously half of the computational cost can be reduced, for $0 < \alpha \leqslant 1$, if we reformulate (15) and (12) as

$$y_h^P(t_{n+1}) = \begin{cases} y_0 + \dfrac{h^\alpha}{\Gamma(\alpha+1)} f(y_h(t_0), t_0), & \text{if } n = 0, \\ y_0 + \dfrac{h^\alpha}{\Gamma(\alpha+2)} \cdot (2^{\alpha+1} - 1) \cdot f(y_h(t_n), t_n) & \\ + \dfrac{h^\alpha}{\Gamma(\alpha+2)} \sum\limits_{j=0}^{n-1} a_{j,n+1} f(y_h(t_j), t_j), & \text{if } n \geqslant 1 \end{cases} \tag{16}$$

and

$$y_h(t_{n+1}) = \begin{cases} y_0 + \dfrac{h^\alpha}{\Gamma(\alpha+2)} \left(f(y_h^P(t_1), t_1) + \alpha \cdot f(y_h(t_0), t_0) \right), & \text{if } n = 0, \\ y_0 + \dfrac{h^\alpha}{\Gamma(\alpha+2)} \left(f(y_h^P(t_{n+1}), t_{n+1}) + (2^{\alpha+1} - 2) \cdot f(y_h(t_n), t_n) \right) & \\ + \dfrac{h^\alpha}{\Gamma(\alpha+2)} \sum\limits_{j=0}^{n-1} a_{j,n+1} f(y_h(t_j), t_j), & \text{if } n \geqslant 1. \end{cases} \tag{17}$$

The arithmetic complexity of the above two predictor-corrector schemes ((11)-(12), and (16)-(17)) is $O(h^{-2})$, where h is step size.

For further reducing the computational cost, we understand the short memory principle from a new viewpoint, and then go to design the predictor-corrector scheme (Deng, 2007b). For $\alpha \in (0,1)$ and $\alpha \in (1,\infty)$, we rewrite (7) as, respectively,

$$\begin{aligned} y(t_{n+1}) &= y(t_n) + \frac{1}{\Gamma(\alpha)} \int_{t_n}^{t_{n+1}} (t_{n+1} - \tau)^{\alpha-1} f(\tau, y(\tau)) d\tau \\ &+ \frac{1}{\Gamma(\alpha)} \int_0^{t_n} \left((t_{n+1} - \tau)^{\alpha-1} - (t_n - \tau)^{\alpha-1} \right) f(\tau, y(\tau)) d\tau, \quad \alpha \in (0,1) \end{aligned} \tag{18}$$

and

$$\begin{aligned} y(t_{n+1}) &= \sum_{k=1}^{\lceil \alpha \rceil - 1} \frac{y_0^{(k)}}{k!} (t_{n+1}^k - t_n^k) + y(t_n) + \frac{1}{\Gamma(\alpha)} \int_{t_n}^{t_{n+1}} (t_{n+1} - \tau)^{\alpha-1} f(\tau, y(\tau)) d\tau \\ &+ \frac{1}{\Gamma(\alpha)} \int_0^{t_n} \left((t_{n+1} - \tau)^{\alpha-1} - (t_n - \tau)^{\alpha-1} \right) f(\tau, y(\tau)) d\tau, \quad \alpha \in (1,\infty). \end{aligned} \tag{19}$$

By observation of (18) and (19), we can see that the non-local property of D_*^α induces the term

$$\frac{1}{\Gamma(\alpha)} \int_0^{t_n} \left((t_{n+1} - \tau)^{\alpha-1} - (t_n - \tau)^{\alpha-1} \right) f(\tau, y(\tau)) d\tau. \tag{20}$$

In fact, if $\alpha \in (0,2)$ the integration kernel of (20) fades faster when the time history becomes longer, more concretely,

$$\begin{aligned}
&\frac{1}{\Gamma(\alpha)}\int_0^{t_n}((t_{n+1}-\tau)^{\alpha-1}-(t_n-\tau)^{\alpha-1})f(\tau,y(\tau))d\tau \\
&=\frac{1}{\Gamma(\alpha)(\alpha-1)}\int_0^{t_n}\left(\int_{t_n-\tau}^{t_{n+1}-\tau}z^{\alpha-2}dz\right)f(\tau,y(\tau))d\tau \\
&=\frac{1}{\Gamma(\alpha)(\alpha-1)}\int_{t_{n-1}}^{t_n}\left(\int_{t_n-\tau}^{t_{n+1}-\tau}z^{\alpha-2}dz\right)f(\tau,y(\tau))d\tau \\
&+\frac{1}{\Gamma(\alpha)(\alpha-1)}\int_{t_{n-2}}^{t_{n-1}}\left(\int_{t_n-\tau}^{t_{n+1}-\tau}z^{\alpha-2}dz\right)f(\tau,y(\tau))d\tau \\
&+\cdots+\frac{1}{\Gamma(\alpha)(\alpha-1)}\int_{t_1}^{t_2}\left(\int_{t_n-\tau}^{t_{n+1}-\tau}z^{\alpha-2}dz\right)f(\tau,y(\tau))d\tau \\
&+\frac{1}{\Gamma(\alpha)(\alpha-1)}\int_{t_0}^{t_1}\left(\int_{t_n-\tau}^{t_{n+1}-\tau}z^{\alpha-2}dz\right)f(\tau,y(\tau))d\tau \qquad (21)\\
&=\frac{1}{\Gamma(\alpha)(\alpha-1)}\int_{t_{n-1}}^{t_n}(z_1^*(\tau))^{\alpha-2}f(\tau,y(\tau))d\tau+\frac{1}{\Gamma(\alpha)(\alpha-1)}\int_{t_{n-2}}^{t_{n-1}}(z_2^*(\tau))^{\alpha-2}f(\tau,y(\tau))d\tau \\
&+\cdots+\frac{1}{\Gamma(\alpha)(\alpha-1)}\int_{t_1}^{t_2}(z_{n-1}^*(\tau))^{\alpha-2}f(\tau,y(\tau))d\tau \\
&+\frac{1}{\Gamma(\alpha)(\alpha-1)}\int_{t_0}^{t_1}(z_n^*(\tau))^{\alpha-2}f(\tau,y(\tau))d\tau,
\end{aligned}$$

where $z_m^*(\tau) \in (t_{m-1}, t_{m+1})$, $m = 1,2,\cdots,n$. Obviously, we can see that the integration (20)'s kernel $(t_{n+1}-\tau)^{\alpha-1}-(t_n-\tau)^{\alpha-1}$ decays (algebraically) by the order $2-\alpha$ when $\alpha \in (0,2)$, but in (Ford & Simpson, 2001) the integral kernel $(t_{n+1}-\tau)^{\alpha-1}$ decays with the order $1-\alpha$ when $\alpha \in (0,1)$. This is the main reason why we can extend the range of the *short memory principle* of fractional differential equations from $\alpha \in (0,1)$ to $\alpha \in (0,2)$. For the nested mashes defined by (3), we can take the step length h in the integral $[t_n - p\mathcal{T}, t_n]$ and in the subsequent intervals $[t_n - p^2\mathcal{T}, t_n - p\mathcal{T}], [t_n - p^3\mathcal{T}, t_n - p^2\mathcal{T}], \cdots, [t_n - p^m\mathcal{T}, t_n - p^{m-1}\mathcal{T}], [t_0, t_n - p^m\mathcal{T}]$, step lengths $ph, p^2h, \cdots, p^{m-1}h, p^m h$ are used, respectively. Noting that $(t_n - p^m\mathcal{T}) - 0$ can't be divided by $p^m h$, so the integral in the interval $[0,l]$ $(l = (t_n - p^m\mathcal{T}) - \lfloor (t_n - p^m\mathcal{T})/(p^m h)\rfloor \cdot (p^m h))$ is ignored, it does not destroy the computational accuracy in general. We will pay our attention to the numerical approximation by using *short memory principle* in the integration (20) in the following part.

With the similar approximation of (13), the product trapezoidal quadrature formula is applied to replace the integral of (19), where nodes $t_j, j = n, n+1$ are taken with respect to the weighted function $(t_{n+1} - \cdot)^{\alpha-1}$ for the first integral and nodes $t_j, j = 0,1,\cdots,n$ are used w.r.t. the weighted function $(t_{n+1} - \cdot)^{\alpha-1} - (t_n - \cdot)^{\alpha-1}$ for the second integral, and it yields

$$\begin{aligned}
&\int_{t_n}^{t_{n+1}}(t_{n+1}-\tau)^{\alpha-1}f(\tau,y(\tau))d\tau \approx \int_{t_n}^{t_{n+1}}(t_{n+1}-\tau)^{\alpha-1}\widetilde{f}_{n+1}(\tau,y(\tau))d\tau \\
&=\frac{h^\alpha}{\alpha(\alpha+1)}(\alpha f(t_n,y(t_n))+f(t_{n+1},y(t_{n+1}))), \quad \alpha \in (1,\infty) \qquad (22)
\end{aligned}$$

and

$$\begin{aligned}
&\int_0^{t_n} \left((t_{n+1}-\tau)^{\alpha-1} - (t_n-\tau)^{\alpha-1}\right) f(\tau, y(\tau))d\tau \\
&\approx \int_0^{t_n} \left((t_{n+1}-\tau)^{\alpha-1} - (t_n-\tau)^{\alpha-1}\right) \widetilde{f}_n(\tau, y(\tau))d\tau \\
&= \frac{h^\alpha}{\alpha(\alpha+1)} \sum_{j=0}^{n} \widetilde{a}_{j,n} f(t_j, y(t_j))
\end{aligned} \tag{23}$$

where

$$\widetilde{a}_{j,n} = \begin{cases} (n+1)^{\alpha+1}(\alpha-n) + n^\alpha(2n-\alpha-1) - (n-1)^{\alpha+1}, & \text{if} \quad j=0, \\ (n-j+2)^{\alpha+1} - 3(n-j+1)^{\alpha+1} + 3(n-j)^{\alpha+1} - (n-j-1)^{\alpha+1}, & \text{if} \quad 1 \le j \le n-1, \\ 2^{\alpha+1} - \alpha - 3, & \text{if} \quad j = n. \end{cases} \tag{24}$$

For the first integral of (18) or (19), the product rectangle formula is used

$$\int_0^{t_n} (t_{n+1}-\tau)^{\alpha-1} f(\tau, y(\tau))d\tau \approx \int_0^{t_n} (t_{n+1}-\tau)^{\alpha-1} f(t_n, y(t_n))d\tau = \frac{h^\alpha}{\alpha} f(t_n, y(t_n)). \tag{25}$$

Combing (22)-(25), for $\alpha \in (1,2)$, the new predictor-corrector approach gives

$$\begin{aligned}
y_h^P(t_{n+1}) &= y_0^{(1)} \cdot h + y_h(t_n) + \frac{h^\alpha}{\Gamma(\alpha+1)} f(t_n, y_h(t_n)) \\
&+ \frac{h^\alpha}{\Gamma(\alpha+2)} \sum_{j=0}^{n} \widetilde{a}_{j,n} f(t_j, y_h(t_j)), \quad \alpha \in (1,2)
\end{aligned} \tag{26}$$

and

$$\begin{aligned}
y_h(t_{n+1}) &= y_0^{(1)} \cdot h + y_h(t_n) + \frac{h^\alpha}{\Gamma(\alpha+2)} \left(\alpha f(t_n, y_h(t_n))\right. \\
&\left.+ f(t_{n+1}, y_h^P(t_{n+1}))\right) + \frac{h^\alpha}{\Gamma(\alpha+2)} \sum_{j=0}^{n} \widetilde{a}_{j,n} f(t_j, y_h(t_j)), \quad \alpha \in (1,2).
\end{aligned} \tag{27}$$

For $\alpha \in (2,\infty)$, the predictor-corrector approach is

$$\begin{aligned}
y_h^P(t_{n+1}) &= \sum_{k=1}^{\lceil\alpha\rceil-1} \frac{y_0^{(k)}}{k!} (t_{n+1}^k - t_n^k) + y_h(t_n) \\
&+ \frac{h^\alpha}{\Gamma(\alpha+1)} f(t_n, y_h(t_n)) + \frac{h^\alpha}{\Gamma(\alpha+2)} \sum_{j=0}^{n} \widetilde{a}_{j,n} f(t_j, y_h(t_j)), \quad \alpha \in (2,\infty)
\end{aligned} \tag{28}$$

and

$$\begin{aligned}
y_h(t_{n+1}) &= \sum_{k=1}^{\lceil\alpha\rceil-1} \frac{y_0^{(k)}}{k!} (t_{n+1}^k - t_n^k) + y_h(t_n) + \frac{h^\alpha}{\Gamma(\alpha+2)} (\alpha f(t_n, y_h(t_n)) + \\
& f(t_{n+1}, y_h^P(t_{n+1}))) + \frac{h^\alpha}{\Gamma(\alpha+2)} \sum_{j=0}^{n} \widetilde{a}_{j,n} f(t_j, y_h(t_j)), \quad \alpha \in (2,\infty).
\end{aligned} \tag{29}$$

The predictor and corrector formulae based on the analytical formula (19) is fully described by (28) and (29) with the weighted $\widetilde{a}_{j,n}$ defined by (24). It is apparent that the same term $\sum_{j=0}^{n} \widetilde{a}_{j,n} f(t_j, y_h(t_j))$ exists in both predictor (28) and corrector formulae (29), we just need to compute one times at each predictor-corrector iteration step.

In order to reduce the computational cost, in conjunction with the nested meshes memory principle (3), decompose the integral and still use the product trapezoidal quadrature formula at each subinterval but with different step lengths.

$$\begin{aligned}
&\int_0^{t_n} \left((t_{n+1}-\tau)^{\alpha-1} - (t_n-\tau)^{\alpha-1}\right) f(\tau, y(\tau))d\tau \\
&\approx \left(\int_{t_n-p\omega h}^{t_n} + \sum_{i=1}^{m-1} \int_{t_n-p^{i+1}\omega h}^{t_n-p^i\omega h} + \int_0^{t_n-p^m\omega h}\right)\left((t_{n+1}-\tau)^{\alpha-1} - (t_n-\tau)^{\alpha-1}\right)\widetilde{f}_n(\tau, y(\tau))d\tau \\
&= \frac{h^\alpha}{\alpha(\alpha+1)} \sum_{j=n-p\omega}^{n} b_{j,p^0,n} f(t_j, y(t_j)) \qquad (30) \\
&+ \sum_{i=1}^{m-1} \frac{(p^i h)^\alpha}{\alpha(\alpha+1)} \left(\sum_{j=0}^{(p-1)\omega} b_{j,p^i,n} f(t_n - p^i(\omega+j)h, y(t_n - p^i(\omega+j)h)) \right) \\
&+ \frac{(p^m h)^\alpha}{\alpha(\alpha+1)} \sum_{j=0}^{\lceil n/p^m - \omega \rceil - 1} b_{j,p^m,n} f(t_n - p^m(\omega+j)h, y(t_n - p^m(\omega+j)h)), \qquad (31)
\end{aligned}$$

where

$$b_{j,p^0,n} = \begin{cases} (p\omega+1)^\alpha(\alpha-p\omega)+(p\omega)^\alpha(2p\omega-\alpha-1)-(p\omega-1)^{\alpha+1}, & \text{if } j = n-p\omega, \\ (n-j+2)^{\alpha+1} - 3(n-j+1)^{\alpha+1} + 3(n-j)^{\alpha+1} & \\ -(n-j-1)^{\alpha+1}, & \text{if } n-p\omega+1 \le j \le n-1, \\ 2^{\alpha+1} - \alpha - 3, & \text{if } j = n \end{cases}$$

and for $i = 1, 2, \cdots, m$,

$$b_{j,p^i,n} = \begin{cases} -(1/p^i+\omega)^{\alpha+1} + \omega^{\alpha+1} + (1/p^i+\omega+1)^{\alpha+1} & \\ -(\omega+1)^{\alpha+1} - ((1/p^i+\omega)^\alpha - \omega^\alpha)(\alpha+1), & \text{if } j = 0, \\ (\omega+j-1+1/p^i)^{\alpha+1} - (\omega+j-1)^{\alpha+1} - 2(\omega+j+1/p^i)^{\alpha+1}, & \\ +2(\omega+j)^{\alpha+1} + (\omega+j+1+1/p^i)^{\alpha+1} - (\omega+j+1)^{\alpha+1}, & \text{if } 1 \le j \le (p-1)\omega - 1, \\ (p^i\omega - 1 + 1/p^i)^{\alpha+1} - (p^i\omega-1)^{\alpha+1} - (p^i\omega + 1/p^i)^{\alpha+1} & \\ +(p^i\omega)^{\alpha+1} + ((p^i\omega+1/p^i)^\alpha - (p^i\omega)^\alpha)(\alpha+1), & \text{if } j = (p-1)\omega. \end{cases}$$

Employing above analysis we obtain the following predictor-corrector algorithm

$$\begin{aligned}
y_h^P(t_{n+1}) &= y_0^{(1)} \cdot h + y_h(t_n) + \frac{h^\alpha}{\Gamma(\alpha+1)} f(t_n, y_h(t_n)) + \frac{h^\alpha}{\alpha(\alpha+1)} \sum_{j=n-p\omega}^{n} b_{j,p^0,n} f(t_j, y(t_j)) \\
&+ \sum_{i=1}^{m-1} \frac{(p^i h)^\alpha}{\alpha(\alpha+1)} \left(\sum_{j=0}^{(p-1)\omega} b_{j,p^i,n} f(t_n - p^i(\omega+j)h, y(t_n - p^i(\omega+j)h)) \right) \\
&+ \frac{(p^m h)^\alpha}{\alpha(\alpha+1)} \sum_{j=0}^{\lceil n/p^m - \omega \rceil - 1} b_{j,p^m,n} f(t_n - p^m(\omega+j)h, y(t_n - p^m(\omega+j)h)) \qquad (32)
\end{aligned}$$

and

$$y_h(t_{n+1}) = y_0^{(1)} \cdot h + y_h(t_n) + \frac{h^\alpha}{\Gamma(\alpha+2)}(\alpha f(t_n, y_h(t_n))$$
$$+ f(t_{n+1}, y_h^P(t_{n+1}))) + \frac{h^\alpha}{\alpha(\alpha+1)} \sum_{j=n-p\omega}^{n} b_{j,p^0,n} f(t_j, y(t_j))$$
$$+ \sum_{i=1}^{m-1} \frac{(p^i h)^\alpha}{\alpha(\alpha+1)} \left(\sum_{j=0}^{(p-1)\omega} b_{j,p^i,n} f(t_n - p^i(\omega+j)h, y(t_n - p^i(\omega+j)h)) \right)$$
$$+ \frac{(p^m h)^\alpha}{\alpha(\alpha+1)} \sum_{j=0}^{\lceil n/p^m - \omega \rceil - 1} b_{j,p^m,n} f(t_n - p^m(\omega+j)h, y(t_n - p^m(\omega+j)h)). \quad (33)$$

In this case, the nested meshes predictor-corrector algorithm has the computational cost of $O(h^{-1} log(h^{-1}))$ for $\alpha \in (1,2)$.

3.2 Error analysis and convergent order

In this section, we make the local truncation error and convergent order analysis for the improved predictor-corrector approaches (16)-(17), (26)-(29), and (32)-(33). First we present several lemmas which will be used later.

Lemma 1. *(Diethelm et al., 2004) Suppose $g \in C^2[0,T]$,*

$$\left| \int_0^{t_{n+1}} (t_{n+1}-\tau)^{\alpha-1} g(\tau) d\tau - \int_0^{t_{n+1}} (t_{n+1}-\tau)^{\alpha-1} \tilde{g}_{n+1}(\tau) d\tau \right| \le C_\alpha h^2, \quad (34)$$

where C_α only depends on α.

Proof.

$$\left| \int_0^{t_{n+1}} (t_{n+1}-\tau)^{\alpha-1} (g(\tau) - \tilde{g}_{n+1}(\tau)) d\tau \right|$$
$$\le \frac{\|g''\|_\infty}{2} \sum_{j=1}^{n+1} \int_{t_{j-1}}^{t_j} (t_{n+1}-\tau)^{\alpha-1} (t_j - \tau)(\tau - t_{j-1}) d\tau$$
$$= \frac{\|g''\|_\infty h^{\alpha+2}}{2\alpha(\alpha+1)} \sum_{j=1}^{n+1} \left((n-j+2)^{\alpha+1} + (n-j+1)^{\alpha+1} + \frac{2}{\alpha+2}((n-j+2)^{\alpha+2} - (n-j+1)^{\alpha+2}) \right)$$
$$= \frac{\|g''\|_\infty h^{\alpha+2}}{2\alpha(\alpha+1)} \sum_{j=1}^{n+1} \left((j+1)^{\alpha+1} + j^{\alpha+1} + \frac{2}{\alpha+2}(1 - (n+1)^{\alpha+2}) \right)$$
$$= -\frac{\|g''\|_\infty h^{\alpha+2}}{2\alpha(\alpha+1)} \left(2\int_1^{n+1} \tau^{\alpha+1} d\tau - \sum_{j=1}^{n+1} ((j+1)^{\alpha+1} + j^{\alpha+1}) \right)$$
$$\le \begin{cases} \frac{\|g''\|_\infty h^{\alpha+2}}{24} \sum_{j=1}^{n+1} j^{\alpha-1}, & \text{if } \alpha < 1, \\ \frac{\|g''\|_\infty h^{\alpha+2}}{24} \sum_{j=1}^{n+1} (j+1)^{\alpha-1}, & \text{if } \alpha \ge 1, \end{cases} \le \begin{cases} \frac{\|g''\|_\infty h^{\alpha+2}}{24} \int_0^{n+1} \tau^{\alpha-1} d\tau, & \text{if } \alpha < 1, \\ \frac{\|g''\|_\infty h^{\alpha+2}}{24} \int_2^{n+2} \tau^{\alpha-1} d\tau, & \text{if } \alpha \ge 1, \end{cases}$$
$$\le C_\alpha h^2. \quad (35)$$

With the similar method, we can obtain the following lemma.

Lemma 2. *Suppose* $g \in C^2[0,T]$,

$$\left| \int_0^{t_n} (t_{n+1}-\tau)^{\alpha-1} g(\tau) d\tau - \int_0^{t_n} (t_{n+1}-\tau)^{\alpha-1} \tilde{g}_n(\tau) d\tau \right| \le C_\alpha h^2, \tag{36}$$

where C_α *only depends on* α.

The error of the product rectangle rule is given as

Lemma 3. *Suppose that* $\partial f(\tau, y(\tau))/\partial t \in C[0,t)$, *for some suitable* t, *then we have*

$$\left| \frac{1}{\Gamma(\alpha)} \int_{t_n}^{t_{n+1}} (t_{n+1}-\tau)^{\alpha-1} (f(\tau, y(\tau)) - f(t_n, y(t_n))) d\tau \right| \le C_\alpha h^{\alpha+1}, \tag{37}$$

where C_α *only depends on* α.

Proof.

$$\begin{aligned}
&\left| \frac{1}{\Gamma(\alpha)} \int_{t_n}^{t_{n+1}} (t_{n+1}-\tau)^{\alpha-1} (f(\tau, y(\tau)) - f(t_n, y(t_n))) d\tau \right| \\
&\le \frac{\|\partial f(\tau, y(\tau))/\partial \tau\|_\infty}{\Gamma(\alpha)} \int_{t_n}^{t_{n+1}} (t_{n+1}-\tau)^{\alpha-1} (\tau - t_n) d\tau \\
&= \frac{\|\partial f(\tau, y(\tau))/\partial \tau\|_\infty}{\Gamma(\alpha)} \frac{1}{\alpha(\alpha+1)} h^{\alpha+1} \\
&\le C_\alpha h^{\alpha+1} \quad \text{where} \quad C_\alpha = \frac{\|\partial f(\tau, y(\tau))/\partial \tau\|_\infty}{\Gamma(\alpha+2)}.
\end{aligned} \tag{38}$$

Lemma 4. *Suppose that* $\partial^2 f(\tau, y(\tau))/\partial^2 \tau \in C[0,t)$, *for some suitable* t, *then we have*

$$\left| \frac{1}{\Gamma(\alpha)} \int_{t_n}^{t_{n+1}} (t_{n+1}-\tau)^{\alpha-1} (f(\tau, y(\tau)) - \tilde{f}_{n+1}(\tau, y(\tau))) d\tau \right| \le C_\alpha h^{\alpha+2}, \tag{39}$$

where C_α *only depends on* α.

Proof. According to the property of linear interpolation polynomials,

$$f(\tau, y(\tau)) - \tilde{f}_{n+1}(\tau, y(\tau)) = f[\tau, t_n, t_{n+1}](\tau - t_n)(\tau - t_{n+1}), \tag{40}$$

where $f[\tau, t_n, t_{n+1}]$ is second divided differences. And using the fact

$$\begin{aligned}
&\int_{t_j}^{t_{j+1}} (t_{n+1}-\tau)^{\alpha-1} (\tau - t_j)(\tau - t_{j+1}) d\tau \\
&= \int_{t_j}^{t_{j+1}} (t_{n+1}-\tau)^{\alpha-1} [(t_j - t_{n+1} + t_{n+1} - \tau)(t_{j+1} - t_{n+1} + t_{n+1} - \tau)] d\tau \\
&= \int_{t_j}^{t_{j+1}} [t_{n-j+1} t_{n-j} (t_{n+1}-\tau)^{\alpha-1} - (t_{n-j} + t_{n-j+1})(t_{n+1}-\tau)^{\alpha} + (t_{n+1}-\tau)^{\alpha+1}] d\tau \\
&= \frac{1}{\alpha(\alpha+1)} (t_{n-j+1}^{\alpha+1} t_{n-j} - t_{n-j}^{\alpha+1} t_{n-j+1}) + \frac{1}{(\alpha+1)(\alpha+2)} (t_{n-j}^{\alpha+2} - t_{n-j+1}^{\alpha+2}) \\
&= \frac{2}{\alpha(\alpha+1)(\alpha+2)} (t_{n-j+1}^{\alpha+2} - t_{n-j}^{\alpha+2}) - \frac{h}{\alpha(\alpha+1)} (t_{n-j+1}^{\alpha+1} + t_{n-j}^{\alpha+1}) \quad \text{for all} \quad j \ge 0,
\end{aligned} \tag{41}$$

we have

$$\begin{aligned}
&\left|\frac{1}{\Gamma(\alpha)}\int_{t_n}^{t_{n+1}}(t_{n+1}-\tau)^{\alpha-1}(f(\tau,y(\tau))-\widetilde{f}_{n+1}(\tau,y(\tau)))d\tau\right|\\
&=\left|\frac{1}{\Gamma(\alpha)}\int_{t_n}^{t_{n+1}}(t_{n+1}-\tau)^{\alpha-1}f[\tau,t_n,t_{n+1}](\tau-t_n)(\tau-t_{n+1})d\tau\right|\\
&=\left|\frac{f[\xi,t_n,t_{n+1}]}{\Gamma(\alpha)}\right|\cdot\int_{t_n}^{t_{n+1}}(t_{n+1}-\tau)^{\alpha-1}(\tau-t_n)(\tau-t_{n+1})d\tau\\
&=\left|\frac{f''(\eta,y(\eta))}{2\Gamma(\alpha)}\right|\cdot\int_{t_n}^{t_{n+1}}(t_{n+1}-\tau)^{\alpha-1}(\tau-t_n)(\tau-t_{n+1})d\tau\\
&=\left|\frac{f''(\eta,y(\eta))}{2\Gamma(\alpha)}\right|\cdot\frac{h^{\alpha+2}}{(\alpha+1)(\alpha+2)}\quad\text{(taking } j=n \text{ in (41))}\\
&\le C_\alpha h^{\alpha+2}\quad\text{where}\quad C_\alpha=\frac{\|f''(\eta,y(\eta))\|_\infty}{2\Gamma(\alpha)(\alpha+1)(\alpha+2)}.
\end{aligned}\tag{42}$$

Here $\xi,\eta\in[t_n,t_{n+1}]$ and in the above equalities the second integral mean value theorem and the properties of second divided differences are used.

Lemma 5. *Suppose that $\partial^2 f(\tau,y(\tau))/\partial^2\tau\in C[0,t)$, for some suitable t, then we have*

$$\left|\frac{1}{\Gamma(\alpha)}\int_0^{t_n}((t_{n+1}-\tau)^{\alpha-1}-(t_n-\tau)^{\alpha-1})(f(\tau,y(\tau))-\widetilde{f}_n(\tau,y(\tau)))d\tau\right|\le C_\alpha h^{\min\{\alpha+2,3\}},\tag{43}$$

where C_α only depends on α.

Proof. The idea of this lemma's proof is similar to the above lemmas, namely

$$\begin{aligned}
&\left|\frac{1}{\Gamma(\alpha)}\int_0^{t_n}((t_{n+1}-\tau)^{\alpha-1}-(t_n-\tau)^{\alpha-1})(f(\tau,y(\tau))-\widetilde{f}_n(\tau,y(\tau)))d\tau\right|\\
&\le\frac{\|f''(\eta,y(\eta))\|_\infty}{2\Gamma(\alpha)}\cdot\left|\sum_{j=0}^{n-1}\int_{t_j}^{t_{j+1}}((t_{n+1}-\tau)^{\alpha-1}-(t_n-\tau)^{\alpha-1})(\tau-t_j)(\tau-t_{j+1})d\tau\right|\\
&\le\frac{\|f''(\eta,y(\eta))\|_\infty}{2\Gamma(\alpha)}\cdot\left|\sum_{j=0}^{n-1}\int_{t_j}^{t_{j+1}}(t_{n+1}-\tau)^{\alpha-1}(\tau-t_j)(\tau-t_{j+1})d\tau\right.\\
&\left.-\int_{t_j}^{t_{j+1}}(t_n-\tau)^{\alpha-1}(\tau-t_j)(\tau-t_{j+1})d\tau\right|\\
&=\frac{\|f''(\eta,y(\eta))\|_\infty}{2\Gamma(\alpha)}\cdot\left|\sum_{j=0}^{n-1}\left[\frac{2}{\alpha(\alpha+1)(\alpha+2)}(t_{n-j+1}^{\alpha+2}-t_{n-j}^{\alpha+2})-\frac{h}{\alpha(\alpha+1)}(t_{n-j+1}^{\alpha+1}+t_{n-j}^{\alpha+1})\right.\right.\\
&\left.\left.-\frac{2}{\alpha(\alpha+1)(\alpha+2)}(t_{n-j}^{\alpha+2}-t_{n-j-1}^{\alpha+2})+\frac{h}{\alpha(\alpha+1)}(t_{n-j}^{\alpha+1}+t_{n-j-1}^{\alpha+1})\right]\right|\quad\text{(using(41))}\\
&=\frac{\|f''(\eta,y(\eta))\|_\infty}{2\Gamma(\alpha)}\cdot\left|\sum_{j=0}^{n-1}\left\{\frac{-2(2t_{n-j}^{\alpha+2}-t_{n-j+1}^{\alpha+2}-t_{n-j-1}^{\alpha+2})}{\alpha(\alpha+1)(\alpha+2)}-\frac{h}{\alpha(\alpha+1)}(t_{n-j+1}^{\alpha+1}-t_{n-j}^{\alpha+1})\right\}\right|\\
&=\frac{\|f''(\eta,y(\eta))\|_\infty}{2\Gamma(\alpha)}\cdot\left|\left\{\frac{-2(t_n^{\alpha+2}-t_{n+1}^{\alpha+2}-t_1^{\alpha+2})}{\alpha(\alpha+1)(\alpha+2)}-\frac{h}{\alpha(\alpha+1)}(t_{n+1}^{\alpha+1}+t_n^{\alpha+1}-t_1^{\alpha+1})\right\}\right|
\end{aligned}$$

$$
\begin{aligned}
&= \frac{\|f''(\eta,y(\eta))\|_\infty}{2\Gamma(\alpha)} \cdot \left| \left\{ \frac{-2(t_n^{\alpha+2} - t_{n+1}^{\alpha+2})}{\alpha(\alpha+1)(\alpha+2)} - \frac{h}{\alpha(\alpha+1)}(t_{n+1}^{\alpha+1} + t_n^{\alpha+1}) + \frac{h^{\alpha+2}}{(\alpha+1)(\alpha+2)} \right\} \right| \\
&= \frac{\|f''(\eta,y(\eta))\|_\infty}{2\Gamma(\alpha)} \cdot \left| \left\{ \frac{-h(t_{n+1}^{\alpha+1} + t_n^{\alpha+1} - 2(z^*)^{\alpha+1})}{\alpha(\alpha+1)(\alpha+2)} + \frac{h^{\alpha+2}}{(\alpha+1)(\alpha+2)} \right\} \right| \quad \text{(mean value theorem)} \\
&= \frac{\|f''(\eta,y(\eta))\|_\infty}{2\Gamma(\alpha)} \cdot \left| \left\{ \frac{-h((t_{n+1}^{\alpha+1} - (z^*)^{\alpha+1}) - ((z^*)^{\alpha+1} - t_n^{\alpha+1})}{\alpha(\alpha+1)} + \frac{h^{\alpha+2}}{(\alpha+1)(\alpha+2)} \right] \right\} \\
&= \frac{\|f''(\eta,y(\eta))\|_\infty}{2\Gamma(\alpha)} \cdot \left| \left\{ \frac{-h^2((z^{**})^\alpha - (\tilde{z}^{**})^\alpha)}{\alpha} + \frac{h^{\alpha+2}}{(\alpha+1)(\alpha+2)} \right\} \right| \quad \text{(mean value theorem)} \\
&= \frac{\|f''(\eta,y(\eta))\|_\infty}{2\Gamma(\alpha)} \cdot \left| \left\{ -h^3((z^{***})^{\alpha-1} + \frac{h^{\alpha+2}}{(\alpha+1)(\alpha+2)} \right\} \right| \quad \text{(mean value theorem)} \\
&\le C_\alpha \cdot h^{\min\{\alpha+2,3\}},
\end{aligned}
\tag{44}
$$

where $z^* \in [t_n, t_{n+1}], z^{**} \in [z^*, t_{n+1}] \subset [t_n, t_{n+1}], \tilde{z}^{**} \in [t_n, z^*] \subset [t_n, t_{n+1}], z^{***} \in [\tilde{z}^{**}, z^{**}] \subset [t_n, t_{n+1}]$ and

$$C_\alpha = \frac{\|f''(\eta,y(\eta))\|_\infty}{2\Gamma(\alpha)} \cdot \left| (z^{***})^{\alpha-1} - \frac{1}{(\alpha+1)(\alpha+2)} \right|.$$

Lemma 6. *(Diethelm et al., 2004) Assume that the solution of the initial value problem satisfies*

$$\left| \int_0^{t_{n+1}} (t_{n+1}-\tau)^{\alpha-1} D_*^\alpha y(\tau) d\tau - \sum_{k=0}^{n+1} \tilde{b}_{j,n+1} D_*^\alpha y(t_j) \right| \le C_1 t_{n+1}^{\gamma_1} h^{\delta_1} \tag{45}$$

and

$$\left| \int_0^{t_{n+1}} (t_{n+1}-\tau)^{\alpha-1} D_*^\alpha y(\tau) d\tau - \sum_{k=0}^{n+1} a_{j,n+1} D_*^\alpha y(t_j) \right| \le C_2 t_{n+1}^{\gamma_2} h^{\delta_2}, \tag{46}$$

with some $\gamma_1, \gamma_2 \ge 0$ and $\delta_1, \delta_2 \ge 0$. Then, for some suitably chosen $T \ge 0$, we have

$$\max_{0 \le j \le N} |y(t_j) - y_h(t_j)| = O(h^q), \tag{47}$$

with $q = \min\{\delta_1 + \alpha, \delta_2\}$ and $N = \lfloor T/h \rfloor$.

Theorem 1. *For the fractional initial problem (1), if $D_*^\alpha y(t) \in C^2[0,T]$ for some suitable T, then the convergent order of our algorithm with the predictor and corrector formulae (16) and (17) gives* $\max_{j=0,1,\cdots,n+1} |y(t_j) - y_h(t_j)| = O(h^{\min\{2,1+2\alpha\}})$.

Proof. Noting

$$\left| \int_{t_n}^{t_{n+1}} (t_{n+1}-\tau)^{\alpha-1} (g(\tau) - g(t_n)) d\tau \right| \le \|g'\|_\infty \frac{h^{\alpha+1}}{\alpha(\alpha+1)}, \tag{48}$$

and applying lemmas 2, 4, 6, we have the above result.

Theorem 2. *When $\alpha > 1$, if $\partial^2 f(\tau, y(\tau))/\partial^2\tau \in C[0,t)$ for some suitable t, then the local truncation error of our algorithm with the predictor and corrector formulae (26)-(27) ($\alpha \in (1,2)$) and (28)-(29) ($\alpha \in (2,\infty)$) is $O(h^3)$, and the convergent order is 2, i.e.,* $\max_{j=0,1,\cdots,n+1} |y(t_j) - y_h(t_j)| = O(h^2)$.

Proof. This proof will be used based on mathematical induction. In view of the given initial condition, the induction basis $j = 0$ is presupposed, it has convergent order 2. Now, assuming that the convergent order is 2 for $j = 0, 1, \cdots, k$, $k \leq n$, we have the local truncation error

$$\begin{aligned}
&\Bigg| y(t_{n+1}) - \Bigg\{ \sum_{k=1}^{\lceil\alpha\rceil-1} \frac{y_0^{(k)}}{k!}(t_{n+1}^k - t_n^k) + y(t_n) \\
&+ \frac{h^\alpha}{\Gamma(\alpha+2)}(\alpha f(t_n, y(t_n)) + f(t_{n+1}, y_h^P(t_{n+1}))) + \frac{h^\alpha}{\Gamma(\alpha+2)} \sum_{j=0}^{n} \widetilde{a}_{j,n} f(t_j, y_h(t_j)) \Bigg\} \Bigg| \\
&= \Bigg| \Bigg\{ \sum_{k=1}^{\lceil\alpha\rceil-1} \frac{y_0^{(k)}}{k!}(t_{n+1}^k - t_n^k) + y(t_n) + \frac{1}{\Gamma(\alpha)} \int_{t_n}^{t_{n+1}} (t_{n+1} - \tau)^{\alpha-1} f(\tau, y(\tau)) d\tau \\
&+ \frac{1}{\Gamma(\alpha)} \int_0^{t_n} ((t_{n+1} - \tau)^{\alpha-1} - (t_n - \tau)^{\alpha-1}) f(\tau, y(\tau)) d\tau \Bigg\} - \Bigg\{ \sum_{k=1}^{\lceil\alpha\rceil-1} \frac{y_0^{(k)}}{k!}(t_{n+1}^k - t_n^k) + y(t_n) \\
&+ \frac{h^\alpha}{\Gamma(\alpha+2)}(\alpha f(t_n, y(t_n)) + f(t_{n+1}, y_h^P(t_{n+1}))) + \frac{h^\alpha}{\Gamma(\alpha+2)} \sum_{j=0}^{n} \widetilde{a}_{j,n} f(t_j, y_h(t_j)) \Bigg\} \\
&= \Bigg| \Bigg\{ \frac{1}{\Gamma(\alpha)} \int_{t_n}^{t_{n+1}} (t_{n+1} - \tau)^{\alpha-1} f(\tau, y(\tau)) d\tau - \frac{h^\alpha}{\Gamma(\alpha+2)}(\alpha f(t_n, y(t_n)) + f(t_{n+1}, y(t_{n+1}))) \Bigg\} \\
&+ \frac{h^\alpha}{\Gamma(\alpha+2)}(f(t_{n+1}, y(t_{n+1})) - f(t_{n+1}, y_h^P(t_{n+1}))) \\
&+ \Bigg\{ \frac{1}{\Gamma(\alpha)} \int_0^{t_n} ((t_{n+1} - \tau)^{\alpha-1} - (t_n - \tau)^{\alpha-1}) f(\tau, y(\tau)) d\tau - \frac{h^\alpha}{\Gamma(\alpha+2)} \sum_{j=0}^{n} \widetilde{a}_{j,n} f(t_j, y(t_j)) \Bigg\} \\
&+ \Bigg\{ \frac{h^\alpha}{\Gamma(\alpha+2)} \sum_{j=0}^{n} \widetilde{a}_{j,n} f(t_j, y(t_j)) - \frac{h^\alpha}{\Gamma(\alpha+2)} \sum_{j=0}^{n} \widetilde{a}_{j,n} f(t_j, y_h(t_j)) \Bigg\} \Bigg|.
\end{aligned}$$

Then we have

$$\begin{aligned}
&\Bigg| y(t_{n+1}) - \Bigg\{ \sum_{k=1}^{\lceil\alpha\rceil-1} \frac{y_0^{(k)}}{k!}(t_{n+1}^k - t_n^k) + y(t_n) \\
&+ \frac{h^\alpha}{\Gamma(\alpha+2)}(\alpha f(t_n, y(t_n)) + f(t_{n+1}, y_h^P(t_{n+1}))) + \frac{h^\alpha}{\Gamma(\alpha+2)} \sum_{j=0}^{n} \widetilde{a}_{j,n} f(t_j, y_h(t_j)) \Bigg\} \Bigg| \\
&\leq C_1 h^{\alpha+2} + \frac{\alpha L}{\Gamma(\alpha+2)} h^{\alpha+\min\{\alpha+1,3\}} + \frac{\alpha L}{\Gamma(\alpha+2)} h^{\alpha+\min\{\alpha+2,3\}} + \left| \left(-\frac{1}{2} h^\alpha + (z_*)^{\alpha-1} h \right) \right| L h^2 \\
&\leq C h^3,
\end{aligned}$$

where $z_* \in (t_n, t_{n+1})$, Lemmas 4 and 5 in the above proof are used, and we also utilize the result $|y(t_{n+1}) - y_h^P(t_{n+1})| = O(h^{\min\{\alpha+1,3\}})$ which can be proved by using Lemmas 4 and 5 and the similar idea to above proof, its sketch proof is given as

$$\begin{aligned}
&\Bigg| y(t_{n+1}) - \Bigg\{ \sum_{k=1}^{\lceil\alpha\rceil-1} \frac{y_0^{(k)}}{k!}(t_{n+1}^k - t_n^k) + y(t_n) \\
&+ \frac{h^\alpha}{\Gamma(\alpha+1)} f(t_n, y(t_n)) + \frac{h^\alpha}{\Gamma(\alpha+2)} \sum_{j=0}^{n} \widetilde{a}_{j,n} f(t_j, y_h(t_j)) \Bigg\} \Bigg| \\
&= \Bigg| \Bigg\{ \sum_{k=1}^{\lceil\alpha\rceil-1} \frac{y_0^{(k)}}{k!}(t_{n+1}^k - t_n^k) + y(t_n) + \frac{1}{\Gamma(\alpha)} \int_{t_n}^{t_{n+1}} (t_{n+1}-\tau)^{\alpha-1} f(\tau, y(\tau)) d\tau \\
&+ \frac{1}{\Gamma(\alpha)} \int_0^{t_n} ((t_{n+1}-\tau)^{\alpha-1} - (t_n-\tau)^{\alpha-1}) f(\tau, y(\tau)) d\tau \Bigg\} - \Bigg\{ \sum_{k=1}^{\lceil\alpha\rceil-1} \frac{y_0^{(k)}}{k!}(t_{n+1}^k - t_n^k) + y(t_n) \\
&+ \frac{h^\alpha}{\Gamma(\alpha+1)} f(t_n, y(t_n)) + \frac{h^\alpha}{\Gamma(\alpha+2)} \sum_{j=0}^{n} \widetilde{a}_{j,n} f(t_j, y_h(t_j)) \Bigg\} \\
&= \Bigg| \Bigg\{ \frac{1}{\Gamma(\alpha)} \int_{t_n}^{t_{n+1}} (t_{n+1}-\tau)^{\alpha-1} f(\tau, y(\tau)) d\tau - \frac{h^\alpha}{\Gamma(\alpha+2)} f(t_n, y(t_n)) \Bigg\} \\
&+ \Bigg\{ \frac{1}{\Gamma(\alpha)} \int_0^{t_n} ((t_{n+1}-\tau)^{\alpha-1} - (t_n-\tau)^{\alpha-1}) f(\tau, y(\tau)) d\tau - \frac{h^\alpha}{\Gamma(\alpha+2)} \sum_{j=0}^{n} \widetilde{a}_{j,n} f(t_j, y_h(t_j)) \Bigg\} \\
&\le \Bigg| \frac{1}{\Gamma(\alpha)} \int_{t_n}^{t_{n+1}} (t_{n+1}-\tau)^{\alpha-1} (f(\tau, y(\tau)) - f(t_n, y(t_n))) d\tau \Bigg| \\
&+ \Bigg| \frac{1}{\Gamma(\alpha)} \int_0^{t_n} ((t_{n+1}-\tau)^{\alpha-1} - (t_n-\tau)^{\alpha-1}) f(\tau, y(\tau)) d\tau - \frac{h^\alpha}{\Gamma(\alpha+2)} \sum_{j=0}^{n} \widetilde{a}_{j,n} f(t_j, y_h(t_j)) \Bigg| \\
&\le \cdots \le C h^{\min\{\alpha+1,3\}}.
\end{aligned}$$

We have proved that the local truncation error of our algorithm (26)-(27) and (28)-(29) is $O(h^3)$ when $\alpha > 1$, so the convergent order is 2.

Lemma 7. *(Ford Simpson [6,Theorem 1]) The nested mesh scheme preserves the order of the underlying quadrature rule on which it is based.*

Because of Theorem 2, Lemma 7, and the analysis in above section, we have

Theorem 3. *When $\alpha > 1$, if $\partial^2 f(\tau, y(\tau))/\partial^2\tau \in C[0,t)$ for some suitable t, then the local truncation error of our algorithm with the predictor and corrector formulae (32)-(33) ($\alpha \in (1,2)$) is $O(h^3)$ and the convergent order is 2, i.e., $\max_{j=0,1,\cdots,n+1} |y(t_j) - y_h(t_j)| = O(h^2)$.*

The comparison of the local truncation error, convergent order and arithmetic complexity for different predictor-corrector schemes are presented in the following table. So far, our convergence results are obtained under the smoothness assumptions of $D_*^\alpha y(t)$ and f, which obviously depend on the smoothness properties of the solution $y(t)$ (Deng, 2010).

4. Numerical results

In this section we present two numerical examples to illustrate the performance of our proposed predictor-corrector schemes. We show the order of convergence in the the absolute

Schemes	Local truncation error	Convergent order	Arithmetic complexity
(11),(12)	$O(h^3)$	$O(h^{\min\{2,1+\alpha\}})$	$O(h^{-2})$
(16),(17)	$O(h^3)$	$O(h^{\min\{2,1+2\alpha\}})$	$O(h^{-2})$
(26)-(29)	$O(h^3)$	$O(h^2)$	$O(h^{-2})$
(32),(33)	$O(h^3)$	$O(h^2)$	$O(h^{-1}\log(h^{-1}))$

Table 1. Comparison of different predictor-corrector schemes.

error. And the convergence order is measured by

$$\text{Order} = \log_2\left(\frac{error(h)}{error(h/2)}\right)$$

where $error(h)$ is the absolute error $|y(t) - y_h(t)|$ with the step h.

Example 1. *Consider the following fractional differential equation*

$$D_*^\alpha y(t) = \frac{\Gamma(5)}{\Gamma(5-\alpha)} t^{4-\alpha} - y(t) + t^4, \quad \alpha \in (0,2), \tag{49}$$

with the initial conditions

$$y(0) = 0,\ \alpha \in (0,1), \tag{50}$$

or

$$y(0) = 0,\ y'(0) = 0,\ \alpha \in (1,2), \tag{51}$$

note that the exact solution to this problem is

$$y(t) = t^4. \tag{52}$$

The numerical results for the predictor-corrector scheme (16)-(17) at time $t = 1$ with different steps and different α are reported in Tables 2 and 3. Table 1 shows the numerical errors at time $t = 1$ between the exact solution and the numerical solution of the predictor-corrector schemes (16)-(17) for (49) with different $\alpha \in (0,1)$ for various step sizes. It is seen that the rate of convergence of the numerical results is of order $O(\tau^{\min\{2,1+2\alpha\}})$ for the predictor-corrector schemes (16)-(17). Tables 2 and 3 show the ratio of the error reduction as the grid is refined. The last row states the theoretical orders of convergence (abbreviated as TOC), which are the results we theoretically prove in the above theorems. From Tables 2 and 3, it can be noted that the numerical results are in excellent agreement with the theoretical ones given in Theorem 1.

Example 2. *Now, consider the following fractional differential equation (Diethelm et al., 2004)*

$$D_*^\alpha y(t) = \begin{cases} \frac{2}{\Gamma(3-\alpha)} t^{2-\alpha} - \frac{1}{\Gamma(2-\alpha)} t^{1-\alpha} - y(t) + t^2 - t, & \text{if} \quad 0 < \alpha < 1, \\ \\ \frac{2}{\Gamma(3-\alpha)} t^{2-\alpha} - y(t) + t^2 - t, & \text{if} \quad \alpha \geq 1. \end{cases} \tag{53}$$

with the initial conditions

$$y(0) = 0,\ \alpha \in (0,1), \tag{54}$$

or

$$y(0) = 0,\ y'(0) = -1,\ \alpha \in (1,2), \tag{55}$$

the exact solution to this problem is

$$y(t) = t^2 - t. \tag{56}$$

h	$\alpha = 0.1$	Order	$\alpha = 0.5$	Order	$\alpha = 0.9$	Order
1/10	3.64×10^{-1}	-	3.55×10^{-2}	-	1.07×10^{-2}	-
1/20	1.70×10^{-1}	1.10	8.79×10^{-3}	2.01	2.31×10^{-3}	2.21
1/40	7.13×10^{-2}	1.26	2.16×10^{-3}	2.03	5.21×10^{-4}	2.15
1/80	2.88×10^{-2}	1.31	5.31×10^{-4}	2.02	1.22×10^{-4}	2.09
1/160	1.15×10^{-2}	1.32	1.31×10^{-4}	2.02	2.94×10^{-5}	2.06
1/320	4.64×10^{-3}	1.31	3.24×10^{-5}	2.02	7.18×10^{-6}	2.03
1/640	1.88×10^{-3}	1.30	8.03×10^{-6}	2.01	1.77×10^{-6}	2.01
TOC		1.20		2.00		2.00

Table 2. Absolute errors and convergence orders for predictor-corrector schemes (16)-(17) at time $t = 1$ with different $0 < \alpha < 1$.

h	$\alpha = 1.25$	Order	$\alpha = 1.5$	Order	$\alpha = 1.85$	Order
1/10	8.48×10^{-3}	-	8.58×10^{-3}	-	9.04×10^{-3}	-
1/20	2.03×10^{-3}	2.06	2.12×10^{-3}	2.02	2.25×10^{-3}	2.00
1/40	5.00×10^{-4}	2.02	5.28×10^{-4}	2.00	5.63×10^{-4}	2.00
1/80	1.24×10^{-4}	2.01	1.32×10^{-4}	2.00	1.41×10^{-4}	2.00
1/160	3.10×10^{-5}	2.00	3.30×10^{-5}	2.00	3.52×10^{-5}	2.00
1/320	7.75×10^{-6}	2.00	8.24×10^{-6}	2.00	8.79×10^{-6}	2.00
1/640	1.94×10^{-6}	2.00	2.06×10^{-6}	2.00	2.20×10^{-6}	2.00
TOC		2.00		2.00		2.00

Table 3. Absolute errors and convergence orders for predictor-corrector schemes (16)-(17) at time $t = 1$ with different $1 < \alpha < 2$.

Table 4 shows the numerical errors at time $t = 1$ between the exact solution and the predictor-corrector schemes (16)-(17) for (53) with different $\alpha \in (0, 1)$ for various step sizes. Again we numerically verify that the rate of convergence for predictor-corrector schemes (16)-(17) is $O(\tau^{\min\{2,1+2\alpha\}})$. From Table 4, it can be seen that both the convergence orders

h	$\alpha = 0.1$	Order	$\alpha = 0.5$	Order	$\alpha = 1.25$	Order
1/10	1.04×10^{-1}	-	9.27×10^{-3}	-	7.96×10^{-3}	-
1/20	4.66×10^{-2}	1.16	2.29×10^{-3}	2.02	2.08×10^{-3}	1.94
1/40	1.87×10^{-2}	1.32	5.87×10^{-4}	1.96	5.53×10^{-4}	1.91
1/80	7.39×10^{-3}	1.34	1.56×10^{-4}	1.91	1.49×10^{-4}	1.89
1/160	2.94×10^{-3}	1.33	4.31×10^{-5}	1.86	4.07×10^{-5}	1.87
1/320	1.18×10^{-3}	1.31	1.23×10^{-5}	1.81	1.12×10^{-5}	1.86
TOC		1.20		2.00		2.00

Table 4. Absolute errors and convergence orders for predictor-corrector schemes (16)-(17) at time $t = 1$ with different α.

p	$\mathcal{T}$	Computing value	Absolute errors	Relative error (%)
1	1	2448.8	1.2	0.0489
2	4	2465.6	15.6	0.6367
2	3	2466.0	16.0	0.6530
2	2	2432.7	18.0	0.7347
3	4	2467.2	17.2	0.7020
3	3	2467.0	17.0	0.6939
3	2	2466.1	16.1	0.6571
4	4	2469.3	19.3	0.7878
4	3	2469.1	19.1	0.7796
4	2	2467.0	17.0	0.6939

Table 5. Error behavior versus the variation of p and $\mathcal{T}$ (the definition of p and $\mathcal{T}$ are given in (3)) at time $t = 50$ with exact (analytical) value $y(50) = 2450$, fractional order $\alpha = 1.5$, step length $h = 1/80$.

and the errors of the improved predictor-corrector scheme (16)-(17) are improved significantly compared with the Tables 3 and 4 given in (Diethelm et al., 2004).

Table 5 shows the computing values, the absolute numerical errors, and the relative numerical errors by using the scheme (32) and (33) to compute Example 2 for different values of p and $\mathcal{T}$ defined in (3). According to the numerical results we can see that the computing errors are generally acceptable in engineering when the computational cost is greatly minimized, especially the computing error is not sensitive to the value of p. On the other hand, this numerical example also illuminates that the algorithm is numerically stable.

5. Conclusions

We briefly review the numerical techniques for efficiently solving fractional ordinary differential equations. The possible improvements for the fractional predictor-corrector algorithm are presented. Even though the algorithm is designed for scalar fractional ordinary differential equation, it can be easily extended to the systems. In fact, based on this algorithm, we have simulated the dynamics of fractional systems (Deng, 2007c;d). In addition, the fractional predictor-corrector algorithm combining with the scheme of generating stochastic variables also works well for the stochastic fractional ordinary differential equations (Deng & Barkai, 2009).

6. Acknowledgments

This research was partly supported by the Program for New Century Excellent Talents in University under Grant No. NCET-09-0438, the National Natural Science Foundation of China under Grant No. 10801067, and the Fundamental Research Funds for the Central Universities under Grant No. lzujbky-2010-63 and No. lzujbky-2012-k26.

7. References

Bouchaud, J. & Georges, A. (1990). Anomalous diffusion in disordered media: Statistical mechanisms models and physical applications. *Phys. Rep.*, Vol. 195, No. 4-5, pp. 127-293.

Deng, W.H. (2007a). Numerical algorithm for the time fractional Fokker-Planck equation. *J. Comput. Phys.*, Vol. 227, No. 2, pp. 1510-1522.

Deng, W.H. (2007b). Short memory principle and a predictor-corrector approach for fractional differential equations. *J. Comput. Appl. Math.*, Vol. 206, No. 1, pp. 174-188.

Deng, W.H. (2007c). Generating 3-D scroll grid attractors of fractional differential systems via stair function. *Int. J. Bifurcation Chaos Appl. Sci. Eng.*, Vol. 17, No. 11, pp. 3965-3983.

Deng, W.H. (2007d). Generalized synchronization in fractional order systems. *Phys. Rev. E*, Vol. 75, 056201.

Deng, W.H. & Barkai, E. (2009). Ergodic properties of fractional Brownian-Langevin motion. *Phys. Rev. E*, Vol. 79, 011112.

Deng, W.H. (2010). Smoothness and stability of the solutions for nonlinear fractional differential equations. *Nonl. Anal.:TMA*, Vol. 72, No. 2, pp. 1768-1777.

Diethelm, K. & Ford, N.J. (2002). Analysis of fractional differential equations. *Nonl. Anal.:TMA*, Vol. 265, No. 2, pp. 229-248.

Diethelm, K., Ford, N.J. & Freed, A.D. (2002b). A predictor-corrector approach for the numerical solution of fractional differential equations. *Nonlinear Dynam.*, Vol. 29, No. 1-4, pp. 3-22.

Diethelm, K., Ford, N.J. & Freed, A.D. (2004). Detailed error analysis for a fractional Adams method. *Numer. Algorithms*, Vol. 36, No. 1, pp. 31-52.

Ford, N.J. & Simpson, A.C. (2001). The numerical solution of fractional differential equations: Speed versus accuracy. *Numer. Algorithms*, Vol. 26, No. 4, pp. 333-346.

Heymans, N., & Podlubny, I. (2005). Physical interpretation of initial conditions for fractional differential equations with Riemann-Liouville fractional derivatives. *Rheol. Acta*, Vol. 45, No. 5, pp. 765-771.

Hilfer, H. (2000). *Applications of Fractional Calculus in Physics*, World Scientific Press, Singapore.

Kirchner, J. W., Feng, X. & Neal, C. (2000). Fractal stream chemistry and its implications for contaminant transport in catchments. *Nature*, Vol. 403, No. 3, pp. 524-526.

Lubich, Ch. (1985). Fractional linear multistep methods for Abel-Volterra integral equations of the second kind. *Math. Comp.*, Vol. 45, No. 172, pp. 463-469.

Lubich, Ch. (1986). Discretized fractional calculus, *SIAM J. Appl. Math.*, Vol. 17, No. 3, pp. 704-719.

Metzler, R. & Klafter, J. (2000). The random walk's guide to anomalous diffusion: a fractional dynamics approach. *Phys. Rep.*, Vol. 339, No. 1, pp. 1-77.

Oldham, K.B. & Spanier, J. (1974). *The fractional calculus*, Academic Press, New York.

Podlubny, I. (1999). *Fractional differential equations*, Academic Press, San Diego.

Podlubny, I. (2002). Geometric and Physical interpretation of fractional integration and fractional differentiation. *Fract. Calc. Appl. Anal.*, Vol. 5, No. 4, pp. 367-386.

Shlesinger, M. F., Zaslavsky, G.M. & Klafter, J. (1993). Strange kinetics, *Nature*, Vol. 363 No. 6, pp. 31-37.

Samko, S.G., Kilbas A.A. & Marichev, O.I. (1993). *Fractional Integrals and Derivatives: Theory and Applications*, Gordon and Breach, London.

Zaslavsky, G.M. (2002). Chaos, fractional kinetic, and anomalous transport, *Phys. Rep.*, Vol. 371, No. 6, pp. 461-580.

Mainrdi, F. (2008). *Fractional calculus and waves in linear viscoelasticity: an introduction to mathematical models*. Imperial College Press, London.

16

Phenomenological Modelling of Cyclic Plasticity

Radim Halama[1], Josef Sedlák and Michal Šofer[1]
[1]*Centre of Excellence IT4Innovations*
Department of Mechanics of Materials
VŠB-Technical University of Ostrava
Czech Republic

1. Introduction

The stress-strain behaviour of metals under a cyclic loading is very miscellaneous and needs an individual approach for different metallic materials. There are many different models that have been developed for the case of cyclic plasticity. This chapter will address only so called phenomenological models, which are based purely on the observed behaviour of materials. The second section of this chapter describes the main experimental observations of cyclic plasticity for metals. Material models development for the correct description of particular phenomenon of cyclic plasticity is complicated by such effects as cyclic hardening/softening and cyclic creep (also called ratcheting). Effect of cyclic hardening/softening corresponds to hardening or softening of material response, more accurately to decreasing/increasing resistance to deformation of material subjected to cyclic loading. Some materials show very strong cyclic softening/hardening (stainless steels, copper, etc.), others less pronounced (medium carbon steels). The material can show cyclic hardening/softening behaviour during force controlled or strain controlled loading. On the contrary, the cyclic creep phenomenon can arise only under force controlled loading. The cyclic creep can be defined as accumulation of any plastic strain component with increasing number of cycles and can influence the fatigue life of mechanical parts due to the exhaustion of plastic ability of material earlier than the initiation of fatigue crack caused by low-cycle fatigue is started.

The third section of this chapter deals with the cyclic plasticity models included in the most popular Finite Element packages (Ansys, Abaqus, MSC.Nastran/Marc). A particular attention is paid to the calibration of classical nonlinear kinematic hardening models. Stress-strain behaviour of materials may be significantly different for proportional and non proportional loading, i.e. loading which leads to the rotation of principal stresses. In case of stainless steels an additional hardening occurs under non proportional loading. This additional hardening is investigated mostly under tension/torsion loading using the circular, elliptical, cross, star and other loading path shapes. Classical cyclic plasticity models implemented in the commercial Finite Element software are not able to describe well the non proportional hardening and correct prediction of multiaxial ratcheting is also problematic. This problem can be solved by implementation of more complex cyclic plasticity model to a FE code. As a conclusion there are briefly summarized phenomenological modelling theories of ratcheting. The main attention is focused on the most progressive group of cyclic plasticity models with a one yield surface only. The

AbdelKarim-Ohno model is also described, which gives very good prediction of ratcheting under uniaxial as well as multiaxial loading.

Comparison of the AbdelKarim-Ohno model and classical models is presented through simulations in the fourth section. Numerical analyses were performed for various uniaxial and multiaxial loading cases of specimen made from the R7T wheel steel. It is shown that classical models can also get sufficient ratcheting prediction when are correctly calibrated.

2. Experimental facts

Good understanding the nature of particular effects of cyclic plasticity plays a key role in the phenomenological modelling. Main findings from this chapter will be useful for a reader in the field of understanding the calibration of cyclic plasticity models and for correct numerical analysis results evaluation.

2.1 Bauschinger's effect

Bauschinger's effect is a basic and well known phenomenon of cyclic plasticity. It describes the fact that due to uniaxial loading of a specimen above yield limit in one direction the limit of elasticity in the opposite direction is reduced. As an example can serve the stress-strain curve corresponding to the first cycle of strain controlled low cycle fatigue test of the steel ST52 (see Fig.1). If the yield limit is marked as σ_Y, then the material during unloading from maximal axial stress state σ_1 behaves elastically up to the point, where the difference between maximal and immediate stress $\sigma_1 - \sigma_2$ is equal to the double of yield limit $2\sigma_Y$.

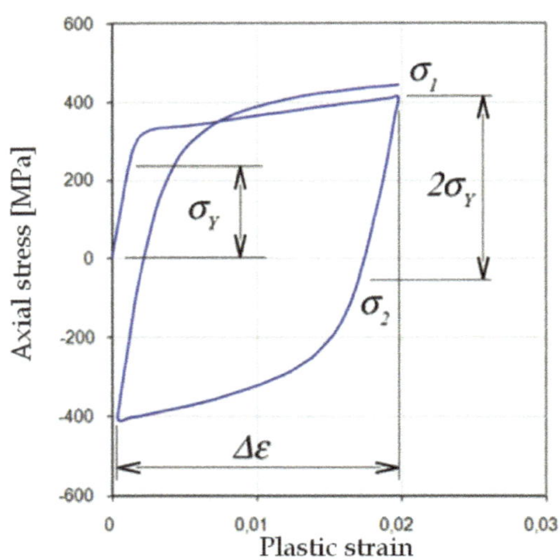

Fig. 1. Presentation of Bauschinger's effect.

2.2 Cyclic hardening/softening

Results of the micro structural changes in the beginning stage of cyclic loading are changes of physical properties and stress response in the material. Cyclic softening/hardening effect relates to softening/hardening of material response or decreasing/increasing of resistance against material deformation under cyclic loading. Its intensity usually decrease with number of cycles until the saturated state is reached. During uniaxial cyclic loading, the condition is characterized by closed hysteresis loop. Transient responses in initial cycles caused by cyclic hardening/softening under plastic strain control and stress control are shown at the Fig.2.

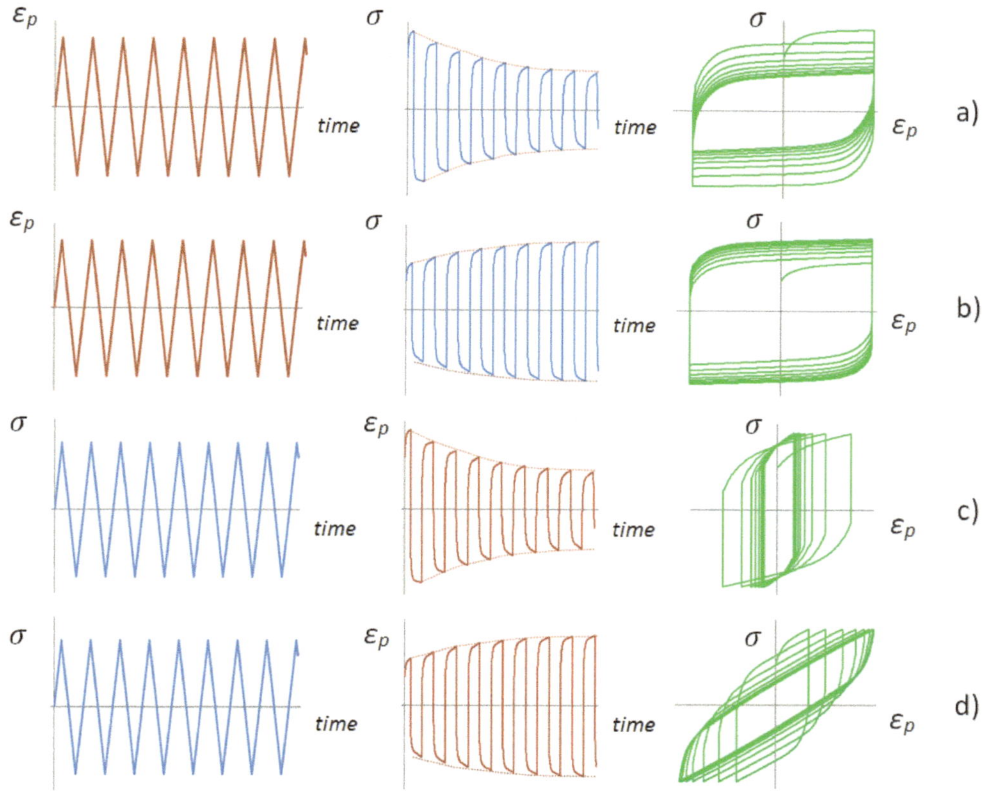

Fig. 2. Uniaxial fatigue test material response: Cyclic softening a) and cyclic hardening b) under plastic strain controlled loading and cyclic hardening c) and cyclic softening d) under stress controlled loading.

Some materials show very strong cyclic softening/hardening (stainless steels, cooper, etc.) some less obvious (structural steels). There can be also notable cyclic hardening in certain cycles range and in the remaining lifetime cyclical softening. Properties of cyclic hardening/softening don't depend only on material microstructure, but also on loading amplitude or more generally on previous strain history. Such transient behaviour of material makes accurate stress-strain modelling more difficult. There is very often mentioned possibility of transient stress-strain behaviour estimation according to its strength limit and yield limit ratio, but also very simple hypothesis is used, claiming that hard material cyclically softens whereas soft material cyclically hardens.

From upper peaks of several hysteresis loops corresponding to half lifetime is possible to obtain cyclic strain curve (Fig.3), which is often used in engineering computations.

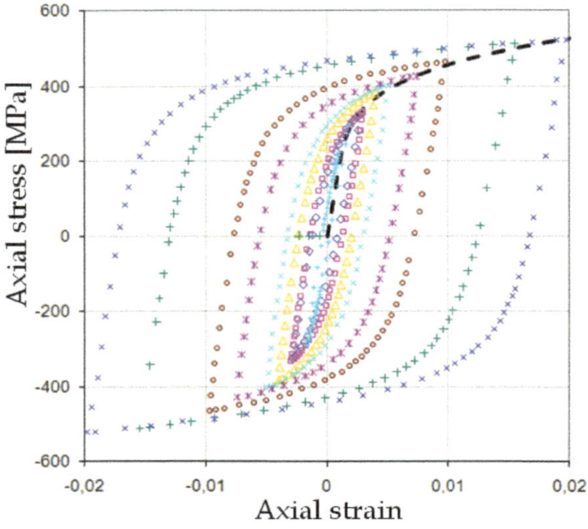

Fig. 3. Cyclic stress-strain curve of ST52 steel.

2.3 Non-masing behaviour

A material obeys Masing behaviour when the upper branches of hysteresis loops with different strain ranges after alignment in lower peaks overlap. More accurately, in the ideal case, single solid curve is created. From microscopic point of view Masing behaviour indicates stable microstructure in fatigue process. Most steel materials haven't Masing behaviour. Some engineering materials show Masing behaviour under certain testing conditions (Jiang & Zhang, 2008). As can be seen from the Fig.4, where the upper branches of hysteresis loops of the investigated material are displayed, the non-Masing behaviour is dependent on the amplitude of plastic strain ε_{ap}.

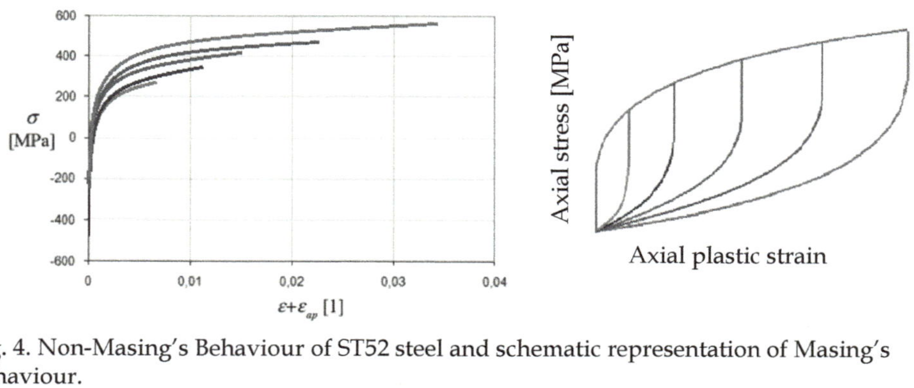

Fig. 4. Non-Masing's Behaviour of ST52 steel and schematic representation of Masing's Behaviour.

2.4 Non-proportional hardening

The Figure 5 illustrates the basic types of loading in the stress space. The tension-compression and simple shear belongs to the category of proportional loading, because there is no change of principal stress directions. This group also includes multi-axialloading in which the stress tensor components change proportionally. Non-proportional loading can be therefore defined as a loading that does not meet the specified condition, and is generally characterized by the loading path in the form of curve (Fig. 5).

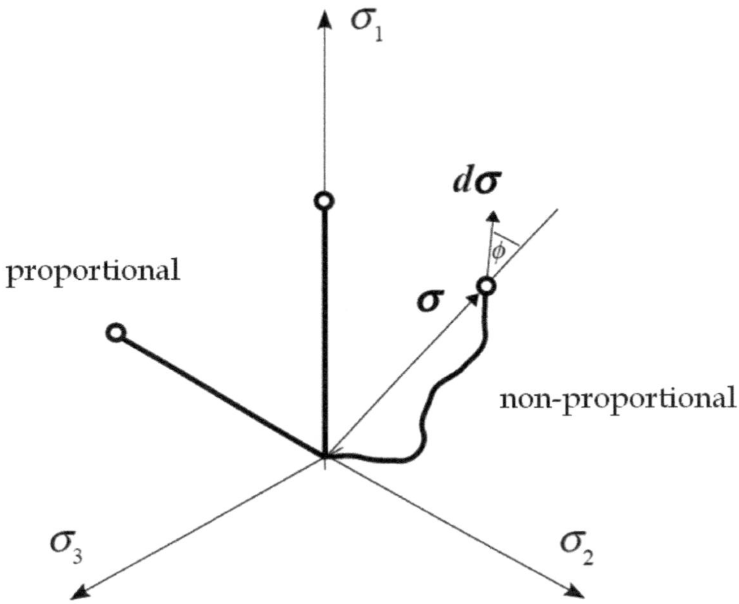

Fig. 5. Loading paths for non-proportional and proportional loading.

Conception of non-proportional hardening represents material hardening as a result of non-proportional loading. Most often it is investigated under tension-compression/torsion loading. Generally, the non-proportional hardening depends on material and shape of loading path. Thereafter we can express stress amplitude

$$\sigma_a(\Phi)=(1+\alpha\Phi)\ \sigma_a^p\ , \tag{1}$$

where σ_a^p is the equivalent stress amplitude under proportional loading, whereas the influence of loading path shape in a cycle is involved in the non proportional parameter Φ and the material parameter of additional hardening is define as

$$\alpha= \sigma_a^n\ /\ \sigma_a^p - 1 \tag{2}$$

where the quantity σ_a^n is the maximum value of von Mises equivalent stresses under non-proportional deformation (circular path). The equivalent stress amplitude is the radius of the minimum circle that circumscribes the loading path in the deviatoric stress space, see Fig.6.

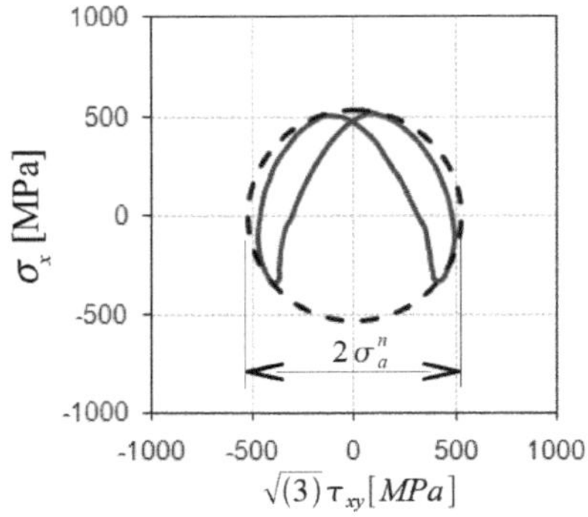

Fig. 6. Definition of equivalent stress amplitude under non-proportional loading.

Non-proportional hardening of FCC alloys pertinents to the stacking fault energy. For strain controlled 90° out-of-phase loading (circular path) it was found out, that the material parameter of non proportional strain hardening is higher for materials with lower value of the stacking fault energy (Doquet & Clavel, 1996).

2.5 Ratcheting

In an uniaxial test under load controll with non-zero mean stress σ_m the accumulation of axial plastic strain can occur cycle by cycle. This effect is called cyclic creep or ratcheting, see Fig.7. The uniaxial ratcheting is characterised by an open hysteresis loop and it is a result of different nonlinear behaviour of the material in tension and compression. The accumulation of plastic strain in initial cycles depends on the cyclic hardening/softening behaviour.

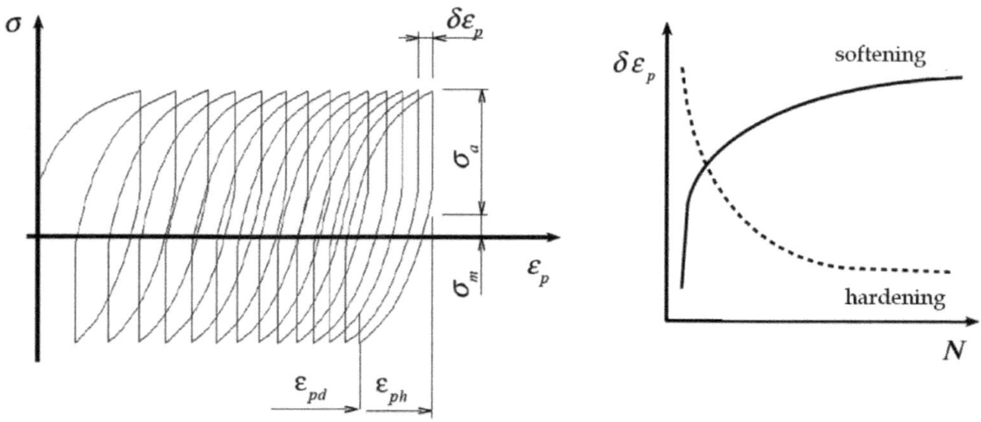

Fig. 7. Scheme of uniaxial ratcheting and influence of hardening/softening behaviour.

Generally, the ratcheting effect can be described as an accumulation of any component of strain tensor with increasing number of cycles. From the practical point of view the research of ratcheting, which occurs under multiaxial stress states, is also very important. There has been investigated mainly the multiaxial ratcheting under combined tension-compression/torsion loading or the biaxial ratcheting caused by internal/external pressure with simultaneous cyclic tension-compression, bending or torsion. The ratcheting strain corresponds to the stress component with non-zero mean stress. The typical example is thin-walled tube subjected to internal (external) pressure and cyclic axial tension (Fig.8c,d). For pure symmetrical bending case (a) it was experimentally observed, that the cross-section becomes more and more oval with increasing number of cycles. This process is then strengthened, when the external pressure is applied too (b).

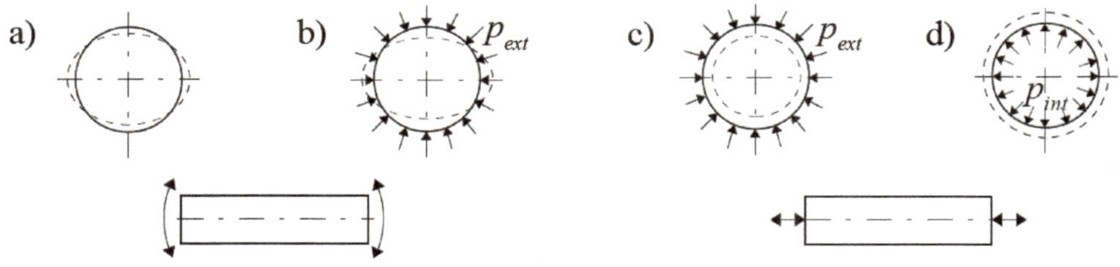

Fig. 8. Ratcheting of a piping component due to a) pure bending, b) bending and external pressure, c) external pressure and push-pull and d) internal pressure and push-pull.

As a next sample results of the fatigue test realised under tension - compression and torsion can serve (Fig.9). The test simulates ratcheting of shear strain, which occurs in surface layer subjected to rolling/sliding contact loading and was proposed by (McDowell, 1995). In the both axes force control was used. For measuring the axial and shear strain during the fatigue test two strain gauges rosette HBM RY3x3/120 were glued to the specimen. The pulsating torque leads to the accumulation of shear strain in the direction of applied torsional moment. The tested material R7T steel becomes almost elastic in initial cycles and then shows significant softening behaviour. After two hundred of loading cycles steady state is reached and the ratcheting rate is constant.

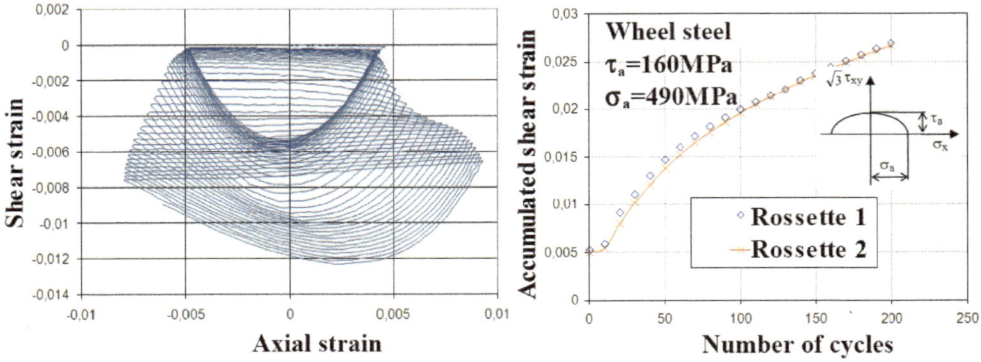

Fig. 9. Ratcheting of shear strain in the McDowell´s tension/torsion test.

A lot of rail and wheel steels show the decreasing ratcheting rate with the increasing number of cycles, which complicates accurate modelling of ratcheting effect. Ratcheting makes also life prediction of fatigue crack initiation difficult as well because the material could fail due to the fatigue or to the accumulation of a critical unidirectional plastic strain (ratcheting failure).

2.6 Other effects in cyclic plasticity

From the theory of elasticity and strength it is well known that the yield locus of ductile materials can be described by an ellipse in the diagram shear stress - normal stress. However, through experiments carried out under uniaxial loading was found (Williams & Svensson, 1971), that if the specimen is loaded by torsion prior to tensile test, then the yield locus (yield surface) has deformed shape. The anisotropy is usually neglected in cyclic plasticity modelling. All of the reported effects of cyclic plasticity are dependent on temperature. With increasing temperature is also strengthened the influence of strain rate on the material response.

3. Constitutive modelling

Basically, cyclic plasticity models can be devided into these groups:

Overlay models (Besseling, 1958)
Single surface models (Armstrong&Frederick, 1966)
Multisurface models (Mroz, 1967)
Two-surface models (Dafalias&Popov, 1976)
Endochronicmodels (Valanis, 1971)
Models with yield surface distortion (Kurtyka, 1988)

Due to its wide popularity and robustness we focus only to the group of models with a single yield surface based on various evolution equations for internal variables.

3.1 Basics of incremental theory of plasticity

The elastoplasticity theory is based on the observations found in the case of uniaxial loading (Fig.10). The rate-independent material's behaviour model includes the additive rule, i.e. the total strain tensor

$$\boldsymbol{\varepsilon} = \boldsymbol{\varepsilon}^e + \boldsymbol{\varepsilon}^p \tag{3}$$

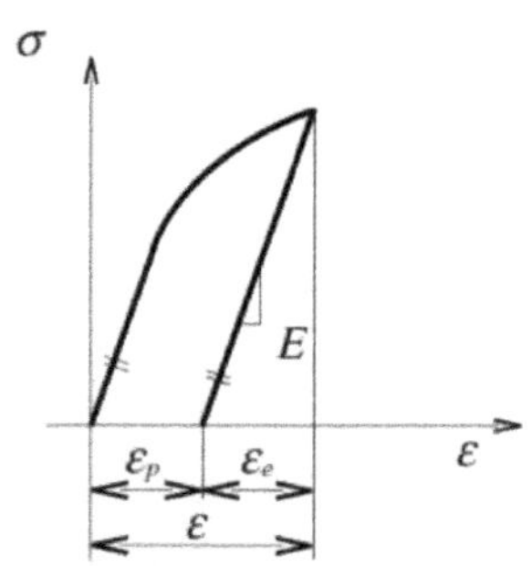

Fig. 10. Decomposition of total strain under uniaxial loading.

is composed of the plastic strain tensor $\boldsymbol{\varepsilon}^p$ and the elastic strain tensor $\boldsymbol{\varepsilon}^e$. The second consideration is that stresses and elastic strains are subjected to Hook's law

$$\boldsymbol{\sigma} = \mathbf{D}^e : \boldsymbol{\varepsilon}^e = \mathbf{D}^e : \left(\boldsymbol{\varepsilon} - \boldsymbol{\varepsilon}^p \right), \tag{4}$$

where $\mathbf{D}^e$ is the elastic stiffness matrix and the symbol ":" is contraction, i.e. using Einstein summation convention $d_{ij}=B_{ijkl}c_{kl}$.

In the uniaxial case, the development of irreversible deformation occurs due to crossing the yield limit σ_Y. Under multiaxial stress state it is necessary to consider an appropriate yield condition. For metallic materials the von Mises condition is mostly used

$$f = \sqrt{\frac{3}{2}(\mathbf{s}-\mathbf{a}):(\mathbf{s}-\mathbf{a})} - Y = 0\,, \quad Y = \sigma_Y + R \tag{5}$$

where $\mathbf{s}$ is the deviatoric part of stress tensor $\boldsymbol{\sigma}$, $\mathbf{a}$ is the deviatoric part of back-stress $\boldsymbol{\alpha}$, which states the centre position for the yield surface with the initial size σ_Y and R is the isotropic internal variable. The contraction operation ":" in (5) can be expressed again in terms of Einstein summation convention $d=b_{ij}c_{ij}$. Now it is necessary to answer the question: When happens a change of plastic strain increment? If the point representing the current stress state lies on the yield surface it can be supposed that this point do not leave the yield surface, the so called consistency condition $\dot{f} = 0$ must be valid. In case of active loading

$$f = 0, \quad \dot{f} = 0 \text{ and } \frac{\partial f}{\partial \boldsymbol{\sigma}} : d\sigma \geq 0\,, \tag{6}$$

the plastic deformation development is directed by the associated plastic flow rule

$$d\boldsymbol{\varepsilon}^p = \sqrt{\frac{3}{2}} d\lambda \frac{\partial f}{\partial \boldsymbol{\sigma}}\,, \quad \frac{\partial f}{\partial \boldsymbol{\sigma}} = \sqrt{\frac{3}{2}} \frac{\mathbf{s}-\mathbf{a}}{Y} = \mathbf{n}\,, \tag{7}$$

where the plastic multiplier $d\lambda$ in (7) corresponds to the equivalent plastic strain increment

$$dp = \sqrt{\frac{2}{3} d\boldsymbol{\varepsilon}^p : d\boldsymbol{\varepsilon}^p}\,. \tag{8}$$

In this concept of single yield surface, the kinematic hardening rule

$$d\mathbf{a} = g\left(\mathbf{a}, \dot{\boldsymbol{\varepsilon}}^p, d\dot{\boldsymbol{\varepsilon}}^p, dp, \mathbf{n}, etc.\right) \tag{9}$$

and the isotropic hardening rule

$$dY = h\left(R, dp, \boldsymbol{\sigma}, \mathbf{a}, \dot{\boldsymbol{\varepsilon}}^p, etc.\right) \tag{10}$$

play an essential role for the robustness of stress-strain model response. When the both hardening rules are used we speak about mixed hardening. Transient effects from initial cycles (cyclical hardening/softening), non-proportional hardening, ratcheting and other effects of cyclic plasticity can be described by a suitable superposition of kinematic and isotropic hardening rules.

In the case of cyclic loading a kinematic hardening rule should always be included in the plasticity model, otherwise the Bauschinger's effect cannot be correctly described.

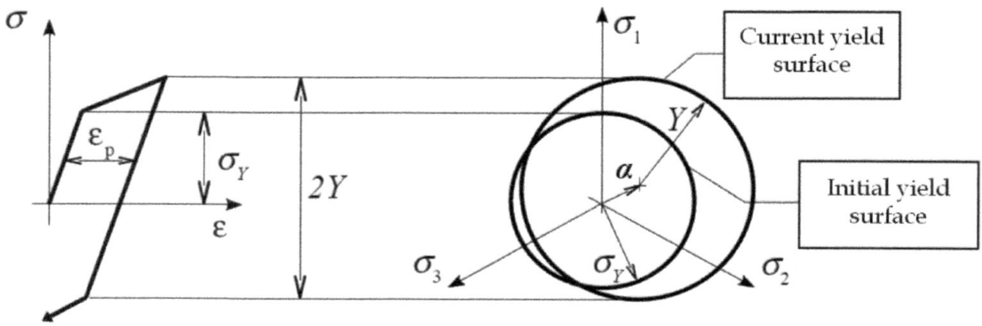

Fig. 11.Von Mises yield function with mixed hardening in the deviatoric plane.

The kinematic hardening rule is important in terms of the needs of capturing the cyclical response of the material. Description of the classical kinematic hardening rules and the resulting cyclic plasticity models is contained in next four subsections. Their availability in selected commercial software based on finite element method is given in Table 1.

Kinematic hardening	Ansys 13	Abaqus	MSC.Marc	MSC. Nastran
Bilinear	x (Prager)	x	x (Ziegler)	x (Ziegler)
Multilinear	x (Besseling)	-	-	-
Armstrong-Frederick	x	x	x	x
Chaboche	x (M_{max}=5)	x (M_{max}=3)	-	-

Table 1.Occurrence of cyclic plasticity models in some popular FE software.

3.2 Bilinear and multilinear kinematic hardening models

There are two bilinear kinematic hardening rules coded in the most popular FE software, suggested by Prager (1953) and Ziegler (1959). The models predict the same response for von Mises material (Ottosen&Ristinmaa, 2005) and for uniaxial loading their response is bilinear. The models show no ratcheting under uniaxial loading and tend to plastic shakedown for a biaxial history of loading. The nonlinearity in stress-strain behaviour can be introduced by a multisurface model, when each surface represents a constant work hardening modulus in the stress space (Mroz, 1967).

Besseling in 1958 introduced a multilinear overlay model, which has a physical meaning and does not use any notion of surfaces. The Besseling model predicts plastic shakedown for uniaxial loading independently of mean stress value. Unfortunately, the mean stress relaxation effect cannot be described too.

3.3 Armstrong-Frederick kinematic hardening model

The important work, leading to the introduction of nonlinearity in the kinematic hardening rule, was the research report of Armstrong and Frederick (1966). In their model the memory term is added to the Prager rule

$$d\boldsymbol{\alpha} = \frac{2}{3} C d\boldsymbol{\varepsilon}_p - \gamma \boldsymbol{\alpha} dp \tag{11}$$

where C and γ are material parameters. Their physical meaning will be explained for push-pull loading. The quantity dp is an increment of accumulated plastic strain, which is expressed as follows

$$dp = \sqrt{\frac{2}{3} d\boldsymbol{\varepsilon}_p : d\boldsymbol{\varepsilon}_p} \tag{12}$$

Considering initially isotropic homogenous material, Von-Misses condition can be again used as follows

$$f = \sqrt{\frac{3}{2}(\boldsymbol{s} - \boldsymbol{a}) : (\boldsymbol{s} - \boldsymbol{a})} - \sigma_Y \tag{13}$$

where $\boldsymbol{a}$ is a deviator of backstress $\boldsymbol{\alpha}$ and $\boldsymbol{s}$ is deviator of stress tensor $\boldsymbol{\sigma}$.

For the uniaxial loading case, the von Mises yield condition becomes to the simpler form

$$f = |\sigma - \alpha| - \sigma_Y = 0 \tag{14}$$

Similarly we can modify nonlinear kinematic hardening rule if we will consider only deviatoric part of the equation (11) taking into account plastic incompressibility.

Then the nonlinear kinematic hardening rule leads to the differential equation

$$d\alpha = C d\varepsilon_p - \gamma \alpha |d\varepsilon_p| \tag{15}$$

Now, we can use a multiplier $\psi = \pm 1$ to dispose of the absolute value

$$d\alpha = C d\varepsilon_p - \gamma \alpha |d\varepsilon_p| = C d\varepsilon_p - \gamma \alpha \psi d\varepsilon_p = (C - \gamma \alpha \psi) d\varepsilon_p \tag{16}$$

separate variables

$$\int_{\alpha_0}^{\alpha} \frac{d\alpha}{C - \gamma \alpha \psi} = \int_{\varepsilon_0}^{\varepsilon} d\varepsilon_p \tag{17}$$

and integrate to get the equation for backstress evolution

$$\alpha = \psi \frac{C}{\gamma} + \left(\alpha_0 - \psi \frac{C}{\gamma} \right) e^{-\psi\gamma\left(\varepsilon_p - \varepsilon_{p0}\right)} \tag{18}$$

Therefore, the relation for stress is given by yield condition

$$\sigma = \alpha + \psi \sigma_Y \tag{19}$$

For tension $\psi = 1$ and considering zeros initial values of plastic strain and backstress this equation is given

$$\sigma = \sigma_Y + \frac{C}{\gamma}\left(1 - e^{-\gamma\varepsilon_p}\right) \tag{20}$$

Now, we can investigate the limit values of the nonlinear function and its first derivation to get a concept about influence of parameters C and γ on stress - strain response of the Armstrong-Frederick model

$$\lim_{\varepsilon_p \to 0} Ce^{-\gamma\varepsilon_p} = C \tag{21}$$

$$\lim_{\varepsilon_p \to \infty} \sigma_Y + \frac{C}{\gamma}\left(1 - e^{-\gamma\varepsilon_p}\right) = \sigma_Y + \frac{C}{\gamma} \tag{22}$$

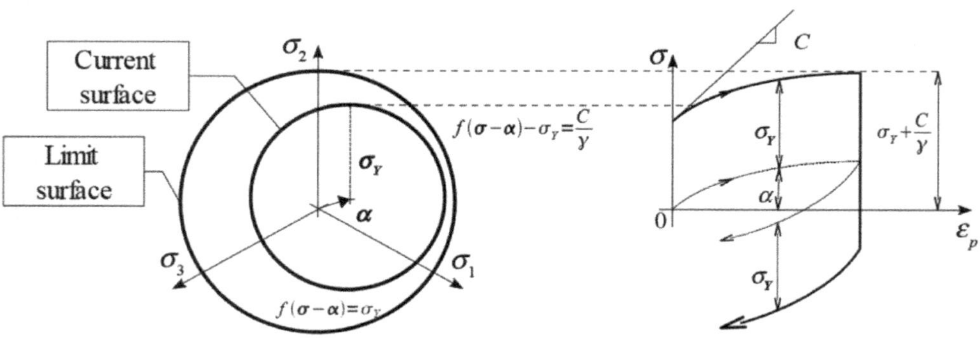

Fig. 12. Properties of the nonlinear kinematic hardening model of Armstrong/Frederick.

Described nonlinear kinematic hardening model allows to correctly capture Bauschinger effect and even behavior by nonsymmetrical loading in tension-compression. The large advantage of Armstrong-Frederick model is its easy implementation and the mentioned nonlinear behavior of the model. On the other hand, the model can not descibe the hysteresis loop shape precisely.

For the case of cyclic loading the parameters σ_Y, C and γ should be estimated from the cyclic strain curve. It is possible to determine the equation corresponding to the cyclic curve of Armstrong-Frederick model by application of equation (18) for the upper branch and the bottom branch of hysteresis loop. For tension $\psi = 1$ is valid and we have

$$\alpha_{\max} = \frac{C}{\gamma} + \left(\alpha_{\min} - \frac{C}{\gamma} \right) e^{-\gamma(\varepsilon_p - \varepsilon_{ap})} \tag{24}$$

For the compression ($\psi = -1$) similarly

$$\alpha_{\min} = -\frac{C}{\gamma} + \left(\alpha_{\max} + \frac{C}{\gamma} \right) e^{\gamma(\varepsilon_p - \varepsilon_{ap})} \tag{25}$$

After substitution of (24) to the equation (25) we get

$$\sigma_a = \sigma_Y + \frac{C}{\gamma} \tanh\left(\gamma \varepsilon_{ap} \right) \tag{26}$$

where tanh(x) is the hyperbolic tangent function and σ_a, ε_{ap} are stress amplitude and plastic strain amplitude respectively.

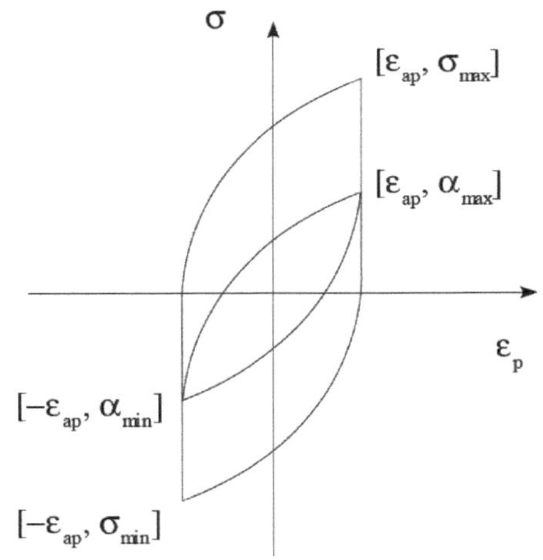

Fig. 13. Initial conditions for the backstress and plastic strain.

For a proper ratcheting description, the condition of equality for computed and experimentally stated ratcheting strain rate for opened stabilized hysteresis loop (Fig.7) should be satisfied

$$\delta \varepsilon_{pFEM} = \delta \varepsilon_{pEXP} \tag{27}$$

According to (Chaboche & Lemaitre, 1990), for Armstrong-Frederick kinematic hardening rule the plastic strain increment per cycle can be written as follows (with absence of isotropic hardening)

$$\delta\varepsilon_{pFEM} = \frac{1}{\gamma}\cdot \ln\left[\frac{\left(\frac{C}{\gamma}\right)^2 - \left(\sigma_{\min} + \sigma_Y\right)^2}{\left(\frac{C}{\gamma}\right)^2 - \left(\sigma_{\max} + \sigma_Y\right)^2}\right] \tag{28}$$

where $\delta\varepsilon_p$ is measured between upper peaks of two hysteresis loops as can be seen in Fig.7.

3.4 Chaboche kinematic hardening model

Very important improvement was the proposal of nonlinear kinematic hardening model by Chaboche (1979), which eliminated Armstrong-Frederick model disadvantages by creating a backstress through superposition of M parts

$$a = \sum_{i=1}^{M} a^{(i)} \tag{29}$$

whereas for each part the evolution equation of Armstrong and Frederick is used

$$da^{(i)} = \frac{2}{3} C_i d\varepsilon_p - \gamma_i a^{(i)} dp \tag{30}$$

where C_i and γ_i are material parameters.

Due to the usage of Armstrong-Frederick evolution law we can directly write the expression for static strain curve

$$\sigma = \psi\sigma_Y + \sum_{i=1}^{M} \psi\frac{C_i}{\gamma_i} + \left(\alpha_0^{(i)} - \psi\frac{C_i}{\gamma_i}\right) e^{-\psi\gamma_i(\varepsilon_p - \varepsilon_{p0})} \tag{31}$$

and for cyclic strain curve

$$\sigma_a = \sigma_Y + \sum_{i=1}^{M} \frac{C_i}{\gamma_i} \tanh\left(\gamma_i \varepsilon_{ap}\right) \tag{32}$$

The quality of cyclic strain curve description is adequate in the case of Chaboche model with the three evolution parts.

Thanks to the similar properties of functions tanh(x) and 1-exp(-x), including its derivatives, it is possible to use the same approach for parameter estimation from the static even cyclic strain curve. Parameters should be determined for example by a nonlinear least-squares method. It is useful to consider Prager's rule for the last backstress part ($\gamma_M = 0$). The parameter influence ratcheting and mean stress relaxation effects. Therefore, we can use this approximation function for cyclic and static strain curves respectively

$$\sigma_a = \sigma_Y + \sum_{i=1}^{M-1} \frac{C_i}{\gamma_i} \tanh\left(\gamma_i \varepsilon_{ap}\right) + C_M \varepsilon_{ap} \tag{33}$$

$$\sigma_a = \sigma_Y + \sum_{i=1}^{M-1} \frac{C_i}{\gamma_i}\left(1 - e^{-\gamma_i \varepsilon_p}\right) + C_M \varepsilon_p \tag{34}$$

When the cyclic strain curve of the investigated material is not available, it is possible to use for the calibration of the model also large saturated hysteresis loop. Based on the relationship (31), considering tension ($\psi = 1$) and these initial values (see Fig15)

$$\alpha_0^{(i)} = -\frac{C_i}{\gamma_i}, \varepsilon_{p0} = -\varepsilon_{ap} \tag{35}$$

we can get for the upper branch of the hysteresis loop this expression

$$\sigma = \sigma_Y + \sum_{i=1}^{2} \frac{C_i}{\gamma_i}\left(1 - 2e^{(-\gamma_i \varepsilon_p - (-\varepsilon_{ap}))}\right) + C_3 \varepsilon_p \tag{36}$$

In the Chaboche model the parameter γ_M influences ratcheting (provided that the last backstress part has the lowest value of the parameter γ_i) and is chosen to be small (up to γ_M=10). For the case of γ_M= 0 ratcheting cannot occur. However, after particular number of cycles the stabilized hysteresis loop will be formed (the Chaboche model tends to plastic shakedown) as it is clear from the graph at the Fig.16. For many materials such behavior does not correspond with reality and during numerical modeling, constant deformation increment can be achieved with aim of suitable choice of parameter γ_M. We can also provide the relation

$$\gamma_M = \frac{\delta\varepsilon_p \cdot C_M}{2 \cdot \sigma_m \cdot \Delta\varepsilon_p}, \tag{37}$$

published elsewhere (Chaboche and Nouailhas 1989).

Thus, with suitable choice of γ_M we get very good model for uniaxial ratcheting prediction (ratcheting with steady state only). In case of non-proportional loading the Chaboche model with three backstresses (M=3) considered in Fig.14 and Fig.15 drastically over predicts ratcheting as has been shown by other researchers (Bari & Hassan, 2000).

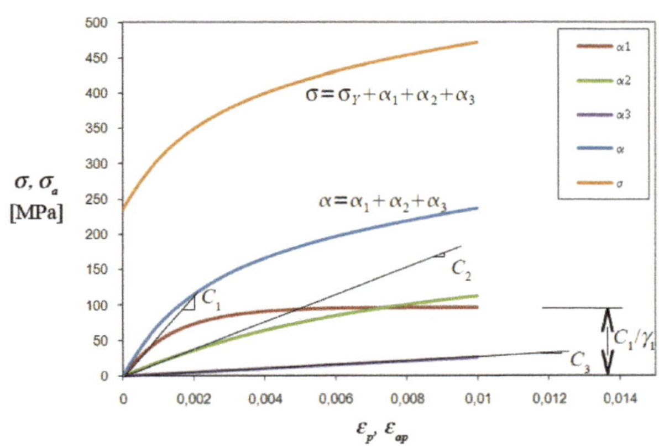

Fig. 14. Properties of constants of Chaboche nonlinear kinematic hardening model (case M=3).

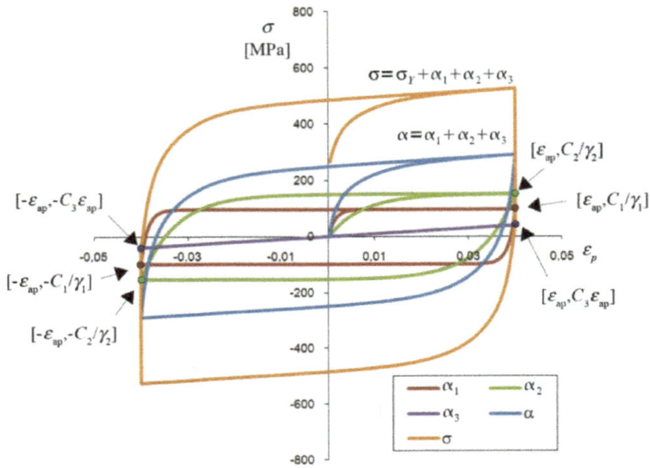

Fig. 15. Scheme for use of the hysteresis loop to identify parameters of Chaboche model.

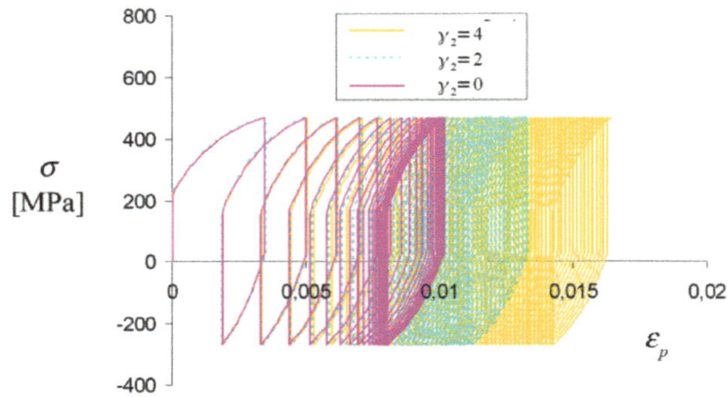

Fig. 16. Influence of parameter γ_2 on ratcheting response of the Chaboche model (M=2).

3.5 Mixed hardening models

Most of materials exhibit Masing and cyclic softening/hardening behaviour, which can be described by superposition of isotropic hardening to a kinematic hardening rule. In this case the size of yield surface Y is expressed with aim of initial value of σ_Y and isotropic variable R, which is usually dependent on the accumulated plastic deformation.

If we would like to describe cyclic softening/hardening, it is convenient to use simple evolutionary equation, leading to nonlinear isotropic hardening rule

$$dR = b(R_\infty - R)dp \tag{38}$$

where R_∞, b are material parameters and dp is an increment of accumulated plastic strain. The value of constant b, which determines the rate of stabilization of the hysteresis loop in case of loading with constant strain amplitude (Fig.17), can be determined directly from following equation

$$R = R_\infty(1 - e^{-b \cdot p}) \tag{39}$$

which follows from integration of equation (39) under following assumption: Change of variable R from zero to R_∞ and p from zero to p. Other possibilities how to identify the constant b are briefly described in (Chaboche & Lemaitre, 1990).

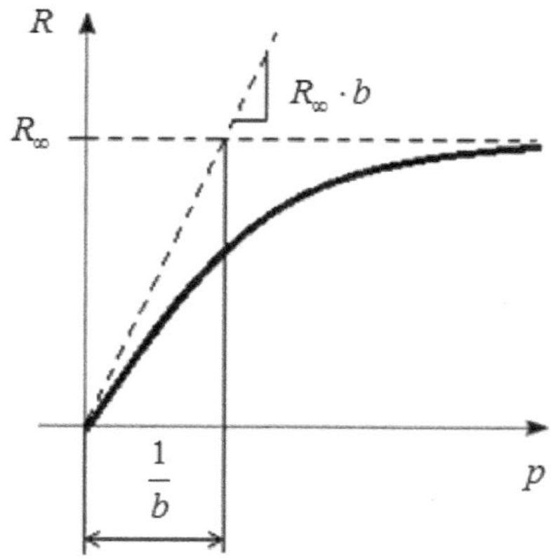

Fig. 17. Evolution of isotropic internal variable R for the nonlinear isotropic hardening rule.

The material parameter R_∞ can be determined, for example, by comparison of static and cyclic strain curve of particular material.

3.6 Other cyclic plasticity models

After Chaboche (1979) there were designed many evolution equations for better ratcheting prediction in the category of nonlinear kinematic hardening rules, but mostly based on the Chaboche superposition of several backstress parts. Because of the large number of theories we choose for their presentation form of table, which contains links to the original publications (Table 2). Presented group of cyclic plasticity models of Chaboche type, which considers the backstress to be defined by M parts

$$d\mathbf{a} = \sum_{i=1}^{M} d\mathbf{a}^{(i)} \tag{40}$$

can be generalized considering the evolution equation in the form

$$d\mathbf{a}^{(i)} = \frac{2}{3} C_i d\boldsymbol{\varepsilon}^p - \gamma_i \mathbf{a}^{(i)} dp^{(i)} \tag{41}$$

where C_i, γ_i are material parameters, $d\boldsymbol{\varepsilon}^p$ is plastic strain increment and $dp^{(i)}$ is the increment of accumulated plastic strain causing dynamic recovery of $\alpha^{(i)}$.

Authors do not guarantee completeness of the set of theories. There are also some new models and approaches. Very progressive are so called models with yield surface distortion, for example Vincent (2004), which are able to model the anisotropy induced by previous plastic deformation (see chapter 2.6).

Proposed modification	Authors
1. $\boldsymbol{a}^{(i)}dp^{(i)}=\left(\boldsymbol{a}^{(i)}:\frac{\partial f}{\partial\boldsymbol{\sigma}}\right)\frac{\partial f}{\partial\boldsymbol{\sigma}}dp,$	Burlet-Cailletaud 1986
2. $dp^{(i)}=dp$ for $i=1,2,3$ $dp^{(i)}=\langle 1-\frac{\bar{a}_i}{f(\alpha_i)}\rangle dp$ for $i=4$, $f(\alpha_i)=\sqrt{\frac{3}{2}\boldsymbol{a}^{(i)}:\boldsymbol{a}^{(i)}}$	Chaboche 1991, 1994
3. $dp^{(i)}=H(f_i)\langle d\boldsymbol{\varepsilon}_p:\frac{\boldsymbol{a}^{(i)}}{\bar{a}_i}\rangle,$ where $\bar{a}_i=\sqrt{\frac{3}{2}\boldsymbol{a}^{(i)}:\boldsymbol{a}^{(i)}}$, $f_i=\bar{a}_i^2-(\frac{C_i}{\gamma_i})^2$ and $H(f_i)$ denotes Heavyside step function.	Ohno-Wang 1993
4. $dp^{(i)}=\left(\frac{\bar{a}_i}{C_i/\gamma_i}\right)^{m_i}\langle d\boldsymbol{\varepsilon}_p:\frac{\boldsymbol{a}^{(i)}}{\bar{a}_i}\rangle,$	Ohno-Wang 1993
5. $dp^{(i)}=\left(\frac{\bar{a}_i}{C_i/\gamma_i}\right)^{m_i}\langle d\boldsymbol{\varepsilon}_p:\frac{\boldsymbol{a}^{(i)}}{\bar{a}_i}\rangle,\; m_i=A_i\langle\frac{d\boldsymbol{\varepsilon}_p}{dp}:\frac{\boldsymbol{a}^{(i)}}{\bar{a}_i}\rangle^{B_i}$	McDowell 1995
6. $dp^{(i)}=\left(\frac{\bar{a}_i}{C_i/\gamma_i}\right)^{m_i}dp,$ $m_i=A_{0i}\left(2-\frac{d\boldsymbol{\varepsilon}_p}{dp}:\frac{\boldsymbol{a}^{(i)}}{\bar{a}_i}\right),\; A_{0i}=Q_i\left(1+a_X e^{b_X R_M}\right),$	Jiang-Sehitoglu 1996
7. $dp^{(i)}=\left[\mu_i+H(f_i)\left\langle\frac{\partial f}{\partial\boldsymbol{\sigma}}:\frac{\boldsymbol{a}^{(i)}}{C_i/\gamma_i}-\mu_i\right\rangle\right]dp,$	Abdel-Karim / Ohno 2000
8. $\boldsymbol{a}^{(i)}dp^{(i)}=\left[\delta'\boldsymbol{a}^{(i)}+(1-\delta')\left(\boldsymbol{a}^{(i)}:\frac{\partial f}{\partial\boldsymbol{\sigma}}\right)\frac{\partial f}{\partial\boldsymbol{\sigma}}\right]dp,$ (pro i=1,2,3) $\boldsymbol{a}^{(i)}dp^{(i)}=$ $=\left[\delta'\boldsymbol{a}^{(i)}+(1-\delta')\left(\boldsymbol{a}^{(i)}:\frac{\partial f}{\partial\boldsymbol{\sigma}}\right)\frac{\partial f}{\partial\boldsymbol{\sigma}}\right]\langle 1-\frac{\bar{a}_i}{f(\alpha_i)}\rangle dp,$ (pro i=4)	Bari a Hassan 2002
9. $\boldsymbol{a}^{(i)}dp^{(i)}=\left(\frac{\bar{a}_i}{C_i/\gamma_i}\right)^{\chi_i}dp-\frac{\boldsymbol{a}^{(i)}}{r_i}\dot{r}_i,\quad \dot{r}_i=b(r_i^T-r_i)\dot{p},$ $r_i^T=r_i^0\left[1+\frac{a_i^1}{(1+b_1p)^2}+\frac{a_i^2}{(1+b_2p)^2}+\frac{a_i^3}{(1+b_3p)^2}\right],$ $\chi_i=\left\langle\chi_i^0+(\chi_i^0+1)(c_i^\chi-1)\left(1-\left\|\frac{\partial f}{\partial\boldsymbol{\sigma}}:\frac{\boldsymbol{a}^{(i)}}{\bar{a}_i}\right\|\right)\right\rangle,\quad \chi_i^0=Q_i\left(1+\frac{a_X}{(1+b_X R_\varepsilon)^2}\right)$	Döring 2003
10. $\boldsymbol{a}^{(i)}dp^{(i)}=$ $=\left(\frac{\bar{a}_i}{C_i/\gamma_i}\right)^{m_i}\left[\delta'\boldsymbol{a}^{(i)}+(1-\delta')\left(\boldsymbol{a}^{(i)}:\frac{\partial f}{\partial\boldsymbol{\sigma}}\right)\frac{\partial f}{\partial\boldsymbol{\sigma}}\right]\langle d\boldsymbol{\varepsilon}_p:\frac{\boldsymbol{a}^{(i)}}{\bar{a}_i}\rangle,$ $d\delta'=\beta(d\delta'_\infty-d\delta')dp.$	Chen-Jiao 2004
11. $dp^{(i)}=\left[\mu_i+H(f_i)\left\langle\frac{\partial f}{\partial\boldsymbol{\sigma}}:\frac{\boldsymbol{a}^{(i)}}{C_i/\gamma_i}-\mu_i\right\rangle\right]dp,$ $\mu_i=\mu=\frac{\mu_0}{(1+a\Phi)},\;\Phi=1-\sqrt{\frac{\boldsymbol{s}:\dot{\boldsymbol{s}}}{\bar{s}:\dot{\bar{s}}}}.$	Kang-Gao-Yang 2004
12. $dp^{(i)}=\langle\frac{\partial f}{\partial\boldsymbol{\sigma}}:\frac{\boldsymbol{a}^{(i)}}{\bar{a}_i}\rangle^{\chi_i}\left(\frac{\bar{a}_i}{C_i/\gamma_i}\right)^{m_i}\langle d\boldsymbol{\varepsilon}_p:\frac{\boldsymbol{a}^{(i)}}{\bar{a}_i}\rangle$	Chen et al. 2005
13. $\boldsymbol{a}^{(i)}dp^{(i)}=H(f_i)\langle d\boldsymbol{\varepsilon}_p:\frac{\boldsymbol{a}^{(i)}}{\bar{a}_i}\rangle\boldsymbol{a}^{(i)}+\xi_i(\bar{a}_i)^{m-1}\boldsymbol{a}^{(i)},$	Yaguchi - Takahashi 2005
14. $dp^{(i)}=\left[\mu_i+H(f_i)\left\langle\frac{\partial f}{\partial\boldsymbol{\sigma}}:\frac{\boldsymbol{a}^{(i)}}{C_i/\gamma_i}-\mu_i\right\rangle\right]dp,\quad \mu_i=\eta\left\langle\frac{\partial f}{\partial\boldsymbol{\sigma}}:\frac{\boldsymbol{a}^{(i)}}{\bar{a}_i}\right\rangle^{\chi},$ $d\eta=d\eta_1+d\eta_2,\; d\eta_1=\omega_1(\eta_{\infty 1}-\eta_1)dp,\; d\eta_2=\omega_2(\eta_{\infty 2}-\eta_2)dp,$ $\chi=\chi_\infty+(\chi_0-\chi_\infty)e^{-\omega_\chi p},$	Halama 2007

Table 2. Overview of some kinematic hardening rules.

It is obvious that many theories differ very little. For correct description of ratcheting in proportional and non-proportional loading more and more authors introduced a non-proportional parameter, which enables simultaneous correct description of uniaxial and multiaxial ratcheting (Chen & Jiao 2004; Chen, 2005; Halama, 2008). Significant improvements of prediction capability can be reached by using memory surfaces (Jiang & Sehitoglu, 1996 and Döring, 2003). Presented models should be compared in terms of nonlinearity, established for each backstress in the case of uniaxial loading. The Fig.18 compares four basic hardening rules.

Common values of parameters are considered for models Ohno-Wang-II and AbdelKarim-Ohno in the Fig.18. Both models lead to the Ohno-Wang model I at a certain choice of parameters affecting ratcheting. In the case of the AbdelKarim-Ohno model this occurs when $\mu_i = 0$ for all i, see Fig.19. Ohno-Wang model II corresponds to the Ohno-Wang model I, if $m_i = \infty$ for all i. Thus, the nonlinearity in the Fig.18 is weak for common parameters (Ohno-Wang II: m_i>>1, AbdelKarim-Ohno: μ_i<0.2) and we can therefore use the same procedure for estimation of the parameters C_i, γ_i ($i = 1, ..., M$) as in the case of multilinear Ohno-Wang model I. For determination of these material parameters we can use again either cyclic strain curve of the material (Ohno-Wang, 1993), (AbdelKarim-Ohno, 2000), or a stabilised hysteresis loop (Bari & Hassan, 2000).

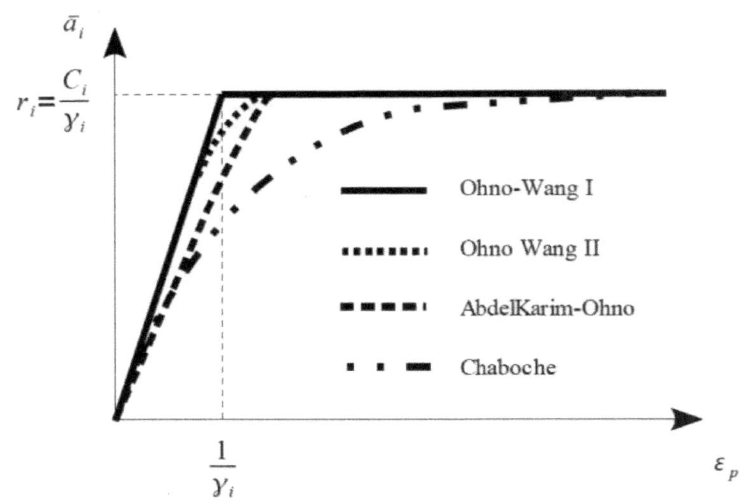

Fig. 18. Nonlinearity introduced in four basic cyclic plasticity models.

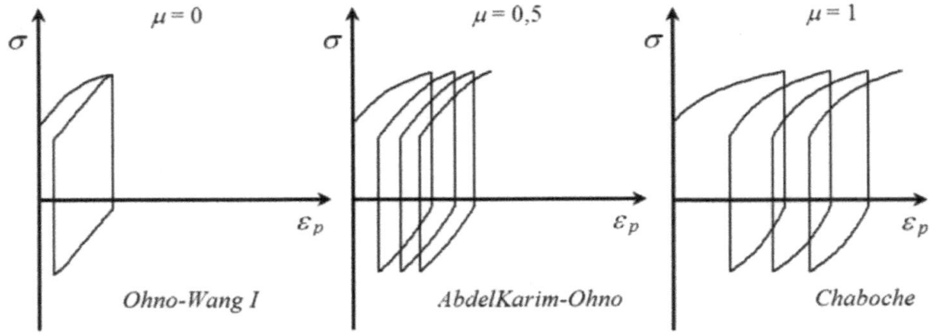

Fig. 19. Influence of ratcheting parameter on the response of AbdelKarim-Ohno model.

As shown in Fig.19, appropriate choice of parameter gives desired ratcheting rate. Considering the only one parameter for ratcheting $\mu_i = \mu$ and its evolution by equation

$$d\mu = \omega(\mu_\infty - \mu)\,dp \tag{42}$$

transient effects in initial cycles can be described too.

4. Ratcheting simulations for a wheel steel

There was realized a set of low-cycle fatigue tests of specimen made from R7T wheel steel at the Czech Technical University in Prague. The specimens were subjected to tension-compression and tension/torsion on the test machine MTS 858 MiniBionix. All tests were force controlled. More detailed description of experiments was reported elsewhere (Halama, 2009). Four cases considered for simulations in this book are shown in Fig.20.

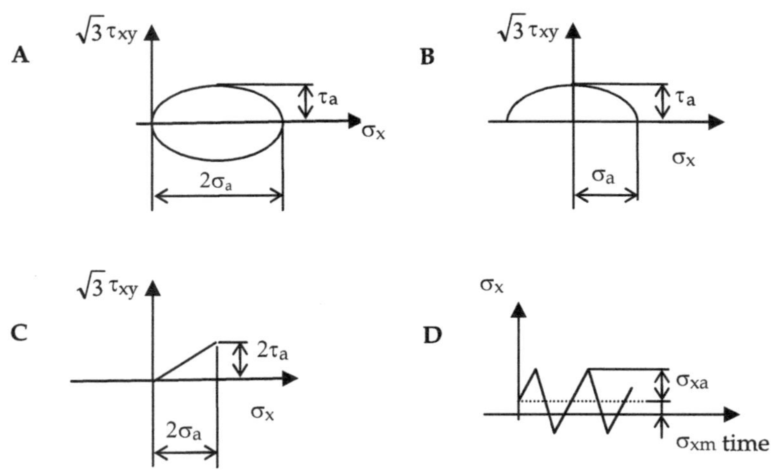

Fig. 20. The scheme of four realized loading paths.

For a description of the stress strain behaviour the AbdelKarim-Ohno model and two classical models of cyclic plasticity were chosen - Armstrong-Frederick model and Chaboche model with two backstress parts (M=2). The AbdelKarim-Ohno model was implemented to the ANSYS program via user subroutine. Material parameters used in simulations are listed in the tables below, except elastic constants (Young modulus E=180000MPa and Poisson´s ratio ν=0.3).

Plasticitymodel	Material parameters
Armstrong-Frederick and Voce rule (AF)	σ_Y=500 MPa, C=108939, γ=2.5 R_∞=-250MPa, b=30
Chabocheand Voce rule (CHAB)	σ_Y=500 MPa, C_1=264156, γ_1=873, C_2=20973, γ_2=1 R_∞=-320MPa, b=30
AbdelKarim-Ohno (AKO)	$\sigma_Y = 200MPa$ $C_{1-6} = 310600,130770,36290,32420,12940,18350MPa$ $\gamma_{1-6} = 5884,2020,980,520,255,3$ $\mu_0 = 0.5, \omega = 0.5, \mu_\infty = 0.14$

Table 3. Material parameters of used cyclic plasticity models.

Cyclic plasticity models were calibrated using saturated hysteresis loop from the test with strain range of 1.5% (Fig.21) and a uniaxial ratcheting test. The calibration procedure used for AbdelKarim-Ohno model was described in the paper Halama (2008). The results of ratcheting prediction gained from simulations of the low cycle fatigue test with nonzero mean stress (case D: 500 cycles with σ_m = 40MPa and σ_a = 500MPa) are shown for all three material models in Fig. 22.

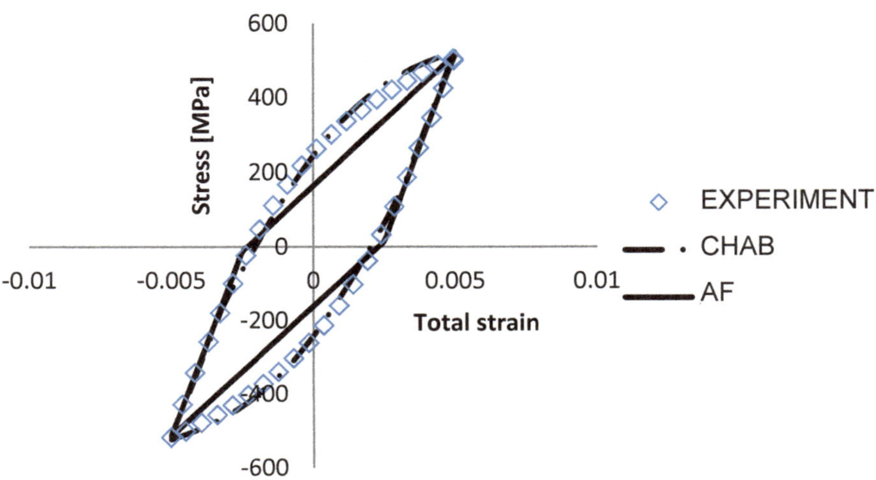

Fig. 21. Saturated uniaxial hysteresis loop and its prediction by AF and CHAB models.

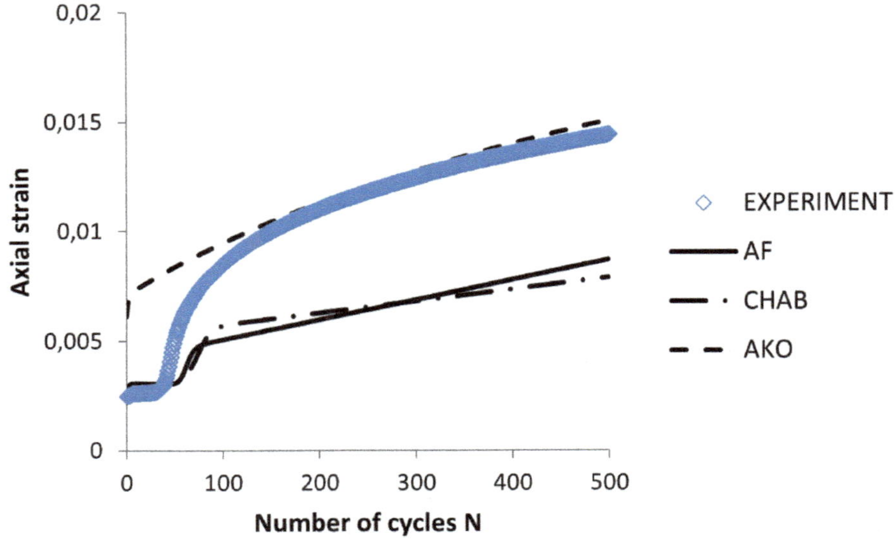

Fig. 22. Comparison of uniaxial ratcheting predictions with experiment (case D).

The results of multiaxial ratcheting predictions corresponding to simulations of the low cycle fatigue test performed under tension/torsion non-proportional (case A: 150 cycles with σ_a = 125MPa and τ_a = 300MPa) and proportional loading (case C: 100 cycles with σ_a = 225MPa and τ_a = 65MPa) are displayed in Fig. 23.

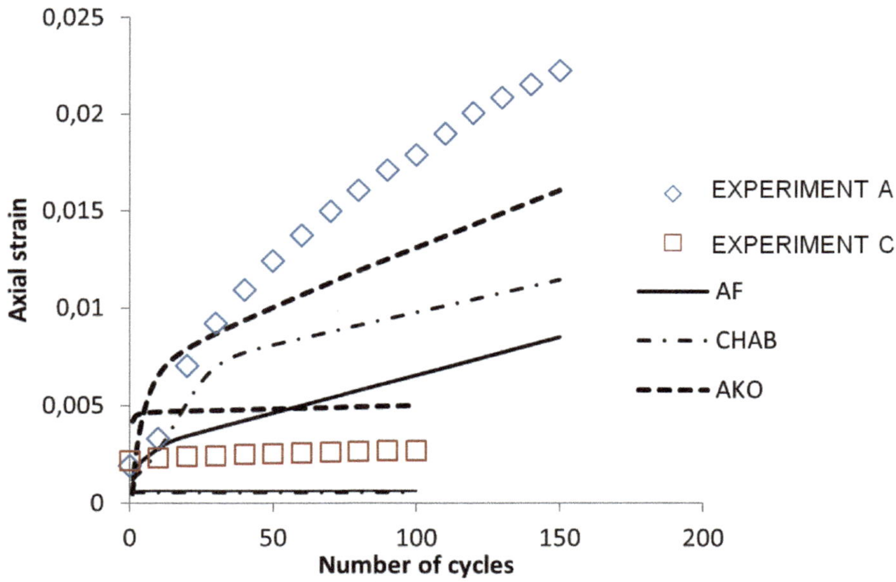

Fig. 23. Comparison of multiaxial ratcheting predictions with experiment (cases A and C).

The models CHAB and AF contain the nonlinear isotropic hardening rule (39), which enables to describe cyclic softening in initial cycles, see Fig.22. On the other hand, ratcheting prediction is better in the case of AbdelKarim-Ohno model. The same conclusion we have for simulations of the last loading case (case B: 500 cycles with σ_a = 490MPa and τ_a = 170MPa, 250 cycles with σ_a = 490MPa and τ_a = 115MPa, 250 cycles with σ_a = 490MPa and τ_a = 215MPa) as can be seen at the Fig. 24.

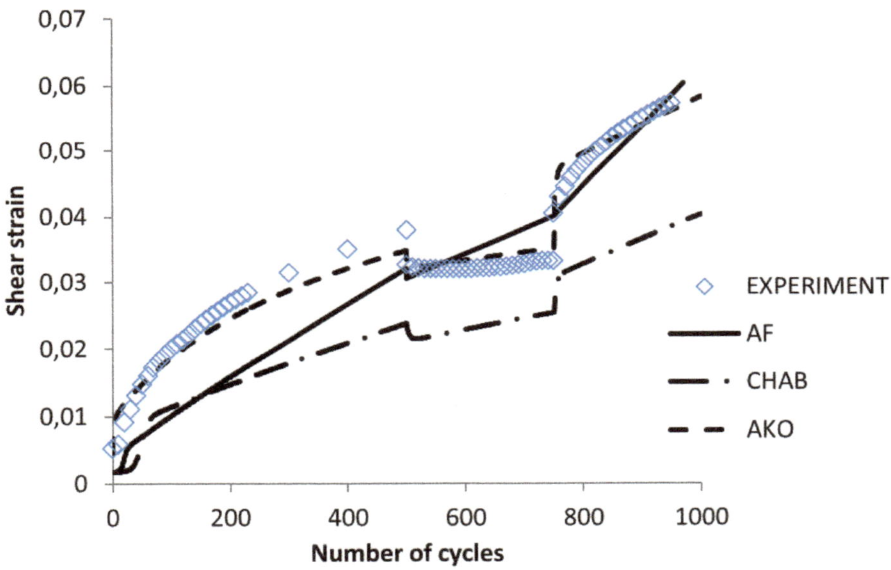

Fig. 24. Comparison of multiaxial ratcheting predictions with experiment (case B).

5. Conclusion

Background of the particular effects in cyclic plasticity of metals explained in the second section makes possible to understand well described incremental theory of plasticity and main features of cyclic plasticity models of Chaboche type. There have been shown interesting results of fatigue test simulations with emphasis on cyclic creep (ratcheting) prediction. It can be concluded from the results of simulations of the section 4 that used combined hardening model of Chaboche with two backstress parts can fairly well predicts the trend of accumulation of plastic deformation (ratcheting) for uniaxial and multiaxial loading cases, even under non-proportional loading, in comparison with the experimental observations of the R7T wheel steel. Indeed, the AbdelKarim-Ohno model gives better prediction of ratcheting for all cases than Armstrong-Frederick and Chaboche model.

6. Acknowledgment

This chapter has been elaborated in the framework of the IT4Innovations Centre of Excellence project, reg. no. CZ.1.05/1.1.00/02.0070 supported by Operational Programme 'Research and Development for Innovations' funded by Structural Funds of the European Union and state budget of the Czech Republic. Authors would like to appreciate the assistance of BONATRANS GROUP a.s. Bohumín company during the preparation and realization of experiments.

7. References

Abdel-Karim, M. & Ohno, N. (2000). Kinematic Hardening Model Suitable for Ratchetting with Steady-State, *International Journal of Plasticity* 16, p. 225-240.

Armstrong, P.J. & Frederick, C.O. (1966). A Mathematical Representation of the Multiaxial Bauschinger Effect, *G.E.G.B. Report RD/B/N*, 731.

Bari, S. & Hassan, T. (2000). Anatomy of Coupled Constitutive Models for Ratcheting Simulations. *International Journal of Plasticity* 16, p. 381-409.

Bari, S. & Hassan, T. (2001). Kinematic Hardening Rules in Uncoupled Modeling for Multiaxial Ratcheting Simulation. *International Journal of Plasticity* 17, p. 885-905.

Burlet, H. & Cailletaud, G. (1987). Modelling of cyclic plasticity in Finite element codes. In: Proc. of 2nd Conference on Constitutive Laws for Engineering Materials: Theory and Applications, Elsevier, New York, p. 1157-1164.

Besseling, J.F. (1958). A Theory of Elastic, Plastic, and Creep Deformations of an Initially Isotropic Material Showing Anisotropic Strain-Hardening, Creep Recovery, and Secondary Creep. *Journal of Applied Mechanics*, vol. 25, p. 529-536.

Chaboche, J. L. & Lemaitre, J. (1990). *Mechanics of Solid Materials.* Cambridge University Press, Cambridge, ISBN 0-521-47758-1

Chaboche, J.L. & Dang Van, K. & Cordier, G. (1979). Modelization of The Strain Memory Effect on The Cyclic Hardening of 316 Stainless Steel, In: *5th International Conference on Structural Mechanics in Reactor Technology, Division L11/3*, Berlin, 13.-17. August 1979, Ed. Jaeger A and Boley B A. Berlin: Bundesanstalt für Material prüfung, p.1-10.

Chaboche, J. L. & Nouailhas, D. (1989). *Constitutive Modeling of Ratchetting Effects -Part I: Experimental Facts and Properties of the Classical Models,* Journal of Engineering Materials and Technology, Vol. 111, p. 384-416.

Chaboche, J.L. (1991). On some modifications of kinematic hardening to improve the description of ratcheting effects. *International Journal of Plasticity* 7, p. 661-678.

Chaboche, J.L. (1994). Modelling of ratchetting: evaluation of various approaches. *European Journal of Mechanics, A/Solids* 13, p. 501-518.

Chen, X. & Jiao, R. (2004). Modified kinematic hardening rule for multiaxial ratchetting prediction. *International Journal of Plasticity* 20, p. 871–98.

Chen, X. & Jiao, R. & Kim, K.S. (2005). On the Ohno-Wang Kinematic Hardening Rules for Multiaxial Ratchetting Modeling of Medium Carbon Steel, *International Journal of Plasticity* 21, p. 161–184.

Dafalias, Z.F. & Popov, E.P. (1976). Plastic Internal Variables Formalism of Cyclic Plasticity, *Journal of Applied Mechanics* 43, p. 645–650.

Doquet, V. & Clavel, M. (1996). Stacking-fault energy and cyclic hardening of FCC solid solutions under multiaxial non-proportional loadings. In: Pineau, A., Cailletaud, G., Lindley, T.C. (Eds.), *Multiaxial Fatigue and Design*, ESIS 21. Mechanical Engineering Publication, London, pp. 43–60.

Döring, R. & Hoffmeyer, J. & Seeger, T. & Vormwald M. (2003). A plasticity model for calculating stress-strain sequences under multiaxial nonproportional cyclic loading. Computl Mater Sci, 2003, Vol. 28, p. 587–96.

Halama, R. (2008). A Modification of AbdelKarim-Ohno Model for Ratcheting Simulations. *Technical Gazette* 15 (3), p. 3-9. ISSN 1330-3651

Halama, R. & Fojtík, F. & Brumek, J. & Fusek, M. (2009). Ratcheting measurement during fatigue testing. In: *Proceedings of the 11th International Conference Applied Mechanics 2009*, April 6–8, Slovakia, pp. 43–44.

Jiang, S. & Sehitoglu, H. (1996). Modeling of Cyclic Ratchetting Plasticity, Part I: Development of Constitutive Relations, Part II: Comparison of Model Simulations With Experiment. *Journal of Applied Mechanics* 63, p. 720-733.

Jiang, Y. & Zhang, J. (2008). Benchmark experiments and characteristic cyclic plastic deformation behavior. *International Journal of Plasticity*, 2008, Vol. 24, p. 1481–1515.

Kang, G.Z. & Gao, Q. & Yang, X.J. (2004). Uniaxial and multiaxial ratcheting of SS304 stainless steel at room temperature: experiments and viscoplastic constitutive model. *Int J Non-linear Mech* 39, p. 843–57.

Kurtyka, T. (1988). Parameter identification of a distortional model of subsequent yield surfaces. *Arch. Mech.* 40 (4), p. 433-454.

McDowell, D.L. (1995). Stress state dependence of cyclic ratchetting behavior of two rail steels, *International Journal of Plasticity* 11 (4) 397–421.

Mroz, Z. (1967). On the Description of Anisotropic Work-Hardening, *Journal of the Mechanics and Physics of Solids* 15, p. 163–175.

Ottosen, N.S. & Ristinmaa, M. (2005). *The Mechanics of Constitutive Modeling.* Elsevier Ltd. Sweden. ISBN 978-0-08-044606-6

Ohno, N. & Wang, J.D. (1993). Kinematic Hardening Rules with Critical State of Dynamic Recovery, Part I: Formulation and Basic Features for Ratchetting Behavior, *International Journal of Plasticity* 9, p. 375-390.

Prager, W. (1956). A New Method of Analysing Stresses and Strains in Work Hardening Plastic Solids. *Journal of Applied Mechanics* 23, p. 493-496.

Valanis, K. C. (1971) A theory of viscoplasticity without a yield surface. Part I: General theory. *Arch. Of Mechs.* 23, p. 217.

Vincent, L. & Calloch, S. & Marquis, D. (2004). A general cyclic plasticity model taking intoaccount yield surface distortion for multiaxial ratchetting. *International Journal of Plasticity* 20, p.1817–1850.

Williams, J.F. & Svensson, N. L. (1971). Effect of torsional prestrain on the yield locus of 1100-F aluminium. *Journal of Strain Analysis*, Vol. 6, p. 263.

Yaguchi, M. & Takahashi, Y. (2005). Ratcheting of viscoplastic material with cyclic softening, part 2: Application of constitutive models. *International Journal of Plasticity* 21, p. 835–860.

Ziegler, H. (1959). A modification of Prager's hardening rule. *Quart. Appl. Math.* 17, p. 55-65.

17

Biorthogonal Decomposition for Wide-Area Wave Motion Monitoring Using Statistical Models

P. Esquivel[1], D. Cabuto[1], V. Sanchez[2] and F. Chan[2]
[1]Technological Institute of Tepic, Electrical and Electronics Engineering Division, Nayarit,
[2]University of Quintana Roo, Sciences and Engineering Division, Quintana Roo
México

1. Introduction

Characterization of spatial and temporal changes in the dynamic pattern that arise when a wide-area system is subjected to a perturbation becomes a significant problem of great theoretical and practical importance. The computation time required to solve large analytical models might become prohibitive for practical systems. Thus, to reduce the complexity of the problem, several simplifications have been commonly used which may result in a poor characterization of global system behaviour. Therefore, a great deal of attention has been paid to identify and to characterize oscillatory activity in large interconnected systems through use of wide-area monitoring schemes such as global positioning systems (GPS) based in multiple phasor measurements units (PMUs) (Messina, et al., 2010). When simultaneously measured responses throughout an interconnected system are available, modal behaviour should be extracted using correlation techniques rather that individual analysis of the system response. This provides a global picture on the system behaviour and enables statistical characterization of the observed phenomena. The problem of selecting the most significant modes is of considerable interest and it has been studied intensively for several researchers (Esquivel & Messina, 2008; Hannachi, et al., 2007; Hasselmann, 1988; Holmes, et al., 1996; Kwasniok, 1996, 2007). Statistical models have been widely used in many engineering and science applications for the analysis of space-time varying system response from measured data (Aubry, et al., 1990; Dankowicz, et al., 1996; Delsole, 2001;Lezama et al., 2009; Messina, et al., 2010, 2011; Spletzer, et al., 2010); i.e., unsteady fluid flow (Terradas, et al., 2004), turbulence (Hannachi, et al. 2007; Leonardi, et al., 2002; Susanto, et al., 1997; Toh, 1987), optimal control (Wallaschek, 1988), structural dynamics (Feeny & Kappagantu, 1998; Han & Feeny, 2003; Holmes, et al., 1996; Marrifield & Guza,1990; Oey, 2007), heat transfer (Barnett, 1983; Kaihatu, et al., 1997) and system identification have been reported (Esquivel, et al., 2009; Feeny, 2008; Hasselmann, 1988; Horel, 1984; Kwasniok, 1996, 2007). These methodologies use statistical techniques such as, empirical orthogonal function (EOF) (Esquivel & Messina, 2008), principal interaction pattern (PIP) (Achatz, et al., 1995), principal oscillation pattern (POP) (Hasselmann, 1988), optimal persistent pattern (OPP) (DelSol, 2001), and the canonical correlation analysis (CCA) (Kwasniok, 2007) that capture various forms of spatio-temporal variability. Among other approaches, empirical orthogonal functions (EOFs) have been used since the mid-1970s for the identification of space-time dynamic systems. More recently, these techniques have gained wide popularity in applications related to wide-area data analysis and reduced-order modelling of various physical processes or models (Messina, et al. 2010; Spletzer, et al., 2010). Underlying issues of these techniques, such as the estimation and localization of propagating and standing features that may be associated with observed or measured data and their applications to space-time varying processes do not seem to be recognized or, at least, they have not been reported. This fact motivates the derivation of a model based on statistical techniques to identify the behaviour of multivariate processes such as the seismic wave propagation components that surge during an earthquake which involve variability over both space and time. These processes may contain moving patterns, travelling waves of different spatial scales and temporal frequencies that are proposed to identify in our study using complex EOF analysis.

2. Theoretical fundamentals of empirical orthogonal functions

The conventional analysis of empirical orthogonal functions is primarily a method of compressing of time and space variability of a data set into the lower possible number of spatial patterns. Each one of these patterns is composed of standing modes of variability and modulated by a time function. The conventional formulation of EOF analysis involves a set of optimal basis which is forced to approach the original field with modes at infinite frequency. In this section is shown that this requirement reduces the ability in the conventional method to characterize the travelling and standing features in dynamical systems because the spatial variation of the original field are combined with the temporal variations. As such, conventional-EOF analysis detects only standing wave components, not travelling wave components. The key point to observe is that real-EOF analysis cannot deal with propagating features and it only uses spatial correlation of the data set, it is necessary to use both spatial and time information in order to identify such features (Esquivel, 2009). In this chapter, we extend the conventional empirical orthogonal function analysis to the study and detection of propagating features in nonlinear patters such as seismic wave propagation components that surge during an earthquake recorded from wide-area monitoring schemes such as GPS-based in multiple PMUs, most of the notation used in this text is standard, vectorial quantities are denoted by boldface letters and scalar quantities by italic letters; others symbols used in the text are too defined. Unlike the real case, complex EOF analysis allows compressing the data into the lowest possible number of spatial patters, each one composed of modes of variability, which may be either travelling or standing modes. The technique allows us to explicitly describe and localize standing and propagating oscillations to the leading seismic wave as a number of complex empirical modes.

In this section, we provide a spatio-temporal decomposition based in the use of time synchronized measured data recorded from multiple phasor measurement units (PMUs) in dynamical systems to cope with increasing complexity of information in the use of wide-area monitoring schemes. The methodology is proposed to identify and extract dynamically independent spatio-temporal patterns using a biorthogonal decomposition based in the complex EOF analysis and the separability of complex correlation functions considered from a statistical perspective (Aubry, et al., 1990; Dankowicz, et al., 1996; Spletzer, et al., 2010). This approach provides an efficient and accurate way to compute standing and propagating features of general nonstationary processes identifying important information for the analysis of dynamical phenomena such as seismic wave components recorded from earthquakes. Moreover, this may lead to greater understanding of the oscillatory activity in interconnected systems. The method allows the introduction of several measures that define moving features in space-time varying fields as: spatial amplitude and phase function, temporal amplitude and phase function, spatial and temporal energy, wave number, angular frequency and average phase speed (Barnett, 1983; Esquivel & Messina, 2008; Susanto, et al. 1997; Terradas, et al., 2004; Hannachi, et al., 2007). The method developed is general and could be applied without loss of generality to measured or simulated data. As an illustrative case, the method is applied to a synthetic example; additionally, data recorded from GPS-based multiple phasor measurements units from a real event of seismic wave components recorded during a submarine earthquake are used to study the practical applicability of the method to characterize spatio-temporal behaviour in wide-area systems.

2.1 Theoretical development

Empirical orthogonal function (EOF) analysis is a procedure for extracting a basis for a modal decomposition from an ensemble of signals in multidimensional measurements. A very appealing property of the basis is its optimality. Among all possible decompositions of a random field, the EOF analysis is the most efficient in the sense that for a given number of modes, the projection on the subspace used for modelling the random field will on average contain the most energy possible. Although EOF analysis has been regularly applied to non-linear problems (Marrifield & Guza, 1990; Susanto, et al., 1997; Toh, 1987; Kaihatu, et al., 1997), it is essential to underline that it is a linear technique and that it is optimal only with respect to other linear representations. Empirical orthogonal function analysis, also known as proper orthogonal decomposition (POD) and Karhunen-Loève transform was introduced by (Kosambi, 1943). It is also worth pointing out that EOF analysis is closely related to principal component analysis (PCA) introduced by (Hotelling, 1933). For a detailed historical review of POD or PCA, the reader is referred to (Barnett, 1983; Hasselmann, 1988; Hostelling, 1933; Horel, 1984; Kosambi, 1943; Toh, 1987).

Let

$$\boldsymbol{u}(x,t) \tag{1}$$

be a zero mean random field on a domain Ω. In practice, the field is sampled at a finite number of pints in time. Then, at time t_k, the system displays a snapshot $\boldsymbol{u}(x,t_k)$ which is a continuous function of x in Ω. The aim of the EOF analysis is to find the most persistent structure among the ensemble of N snapshots. More precisely, assume that $\mathbf{X}(x_j,t_k)$, j=1,...,n and k=1,...,N denotes a sequence of observations on some domain $x\epsilon\Omega$ where x is a vector of spatial variables, and t_k is the time at which the observations are made. The method of EOF analysis, both spatial and time-dependent, is a specification of the general theory of expansion of random functions (random fields or random processes) in a series of some deterministic (nonrandom) functions with random uncorrelated coefficients (Feeny & Kappagantu, 1998). The essential idea of the proper orthogonal decomposition is to generate an optimal basis, $\varphi(x)$, for the representation of an ensemble of data collected from measurements or numerical simulations of a dynamic system as is shown in Fig. 1.

Given an ensemble of measured data, the data set can be written as the $N\times n$-dimension matrix

$$\mathbf{X}(x_j,t_k)=\begin{bmatrix} u(x_1,t_1) & \cdots & u(x_n,t_1) \\ \vdots & \ddots & \vdots \\ u(x_1,t_N) & \cdots & u(x_n,t_N) \end{bmatrix} \tag{2}$$

where typically $n\neq N$, so $\mathbf{X}$ is generally rectangular (Messina, et al., 2010). The technique yields an orthogonal basis for linear, infinite-dimensional Hilbert space $\boldsymbol{L}^2([0,1])$, that maximizes the averaged projection of the response matrix for the representation of the ensemble of data that is fully orthogonal, and it is assumed to be normalized, i.e.,

$$\max_{\varphi_j(x)\in L^2([0,1])} \frac{\left\langle |(\mathbf{X}(x,t),\boldsymbol{\varphi}_j(x))|^2\right\rangle}{\left\|\boldsymbol{\varphi}_j(x)\right\|^2} \text{ subject to } \left\|\boldsymbol{\varphi}_j(x)\right\|^2=1 \tag{3}$$

where $|.|$ denotes the modulus, $\|.\|$ is the L^2-norm and, $\langle . \rangle$ implies the use of an average operation (Holmes, et al., 1996). The corresponding functional for the constrained variational problem is solved and reduced to:

$$\int_0^1\left[\int_0^1\left\langle \boldsymbol{u}(x)\boldsymbol{u}^*(x')\right\rangle \boldsymbol{\varphi}(x')dx'-\lambda\boldsymbol{\varphi}(x)\right]^*\boldsymbol{\psi}^*(x)dx=0 \tag{4}$$

where the (*) denotes the conjugate transpose (sometimes denoted as Hermitian, H), and the (') denotes transpose vector. Thus, if $\boldsymbol{\psi}^*(x)=0$, the optimal basis are given by the eigenfunctions $\boldsymbol{\varphi}_j(x)$ of the integral equation,

$$\int_0^1\left\langle \boldsymbol{u}(x)\boldsymbol{u}^*(x')\right\rangle \boldsymbol{\varphi}(x')dx'=\lambda\boldsymbol{\varphi}(x) \tag{5}$$

whose kernel is the averaged autocorrelation function $\left\langle \boldsymbol{u}(x)\boldsymbol{u}^*(x)\right\rangle \triangleq \mathbf{C}(x,x')$. Under this assumption, the integral (5) can be written as

$$\mathbf{C}\boldsymbol{\varphi}(x)=\lambda\boldsymbol{\varphi}(x) \tag{6}$$

where the resulting autocorrelation matrix **C**, is real, symmetric, positive and semi-definite matrix. Therefore, the optimization problem can be recast as the problem of finding the largest eigenvectors, $\varphi(x)$, of the equation (6), called empirical orthogonal functions (EOFs); its corresponding eigenvalues are real, nonnegative, and ordered so that $\lambda_1\geq\lambda_2\geq,\ldots,\geq\lambda_j\geq 0$. This method, also called conventional EOF analysis, cannot be used to detect propagation features due to the assumption that each field is represented as a spatial fixed pattern of behaviour and lack of phase information, becoming prohibitive to practical applications.

Now, if we assume that $\boldsymbol{\psi}^*(x)\neq 0$, then (4) can be rewritten as

$$\int_0^1\int_0^1 \boldsymbol{\varphi}^*(x')\langle u(x)u^*(x')\rangle \boldsymbol{\psi}^*(x)dx'dx = \int_0^1 \boldsymbol{\varphi}^*(x)\lambda\boldsymbol{\psi}^*(x)dx \tag{7}$$

such that, the inner product $\left(\boldsymbol{\varphi}^*(x)\mathbf{C}\boldsymbol{\psi}^*(x)\right) \neq 0$, with orthogonal eigenvectors $\boldsymbol{\varphi}(x)$, $\boldsymbol{\psi}(x)$, i.e.,

$$\boldsymbol{\varphi}_i^T\boldsymbol{\varphi}_j = \begin{cases} 0, & i \neq j \\ \delta(\boldsymbol{\varphi}), & i = j \end{cases} \text{ and, } \boldsymbol{\psi}_i^T\boldsymbol{\psi}_j = \begin{cases} 0, & i \neq j \\ \delta(\boldsymbol{\psi}), & i = j \end{cases} \tag{8}$$

From (4) it can be seen that if there exists an arbitrary variation (spatial), $\boldsymbol{\psi}^*(x) \neq 0$, then the original field can be reconstructed using two optimal orthogonal basis given from (7). Based in this notion, an efficient technique to find the optimal basis using complex EOF analysis (CEOFs) is proposed (Esquivel, 2009).

Our proposed methodology based in EOF analysis and the Hilbert transform is developed to be applied for representations of complex data fields in a biorthogonal decomposition illustrating the phenomenon of spatial and temporal variability in interconnected systems. This method consists first in extend each real field data to the complex world using the Hilbert transform to provide the phase information; and second, the EOF analysis is developed to the complex data field for the detection and localization of propagation features into dynamical systems.

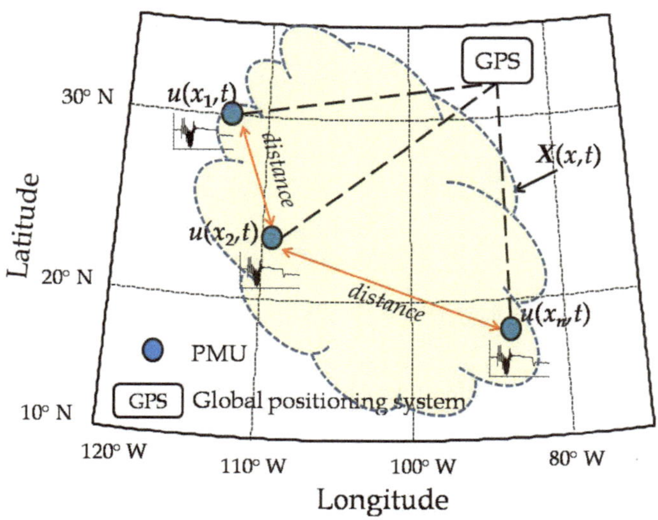

Fig. 1. Global positioning system (GPS) and PMU data in terms of a time-space varying field.

2.2 Complex data fields

Conventional EOF analysis of real-data fields is commonly carried out under the assumption that each field can be represented as a spatially fixed pattern of behaviour. This method, however, cannot be used for detection of propagating features because of the lack of phase information (Esquivel & Messina, 2008). To fully utilize the data, a technique is necessary unknowing the nonstationarity of the time-series data.

Let $\boldsymbol{u}(x_j,t_k)$ be a space-time varying scalar field representing a time series recorded from a wide-area distribution system, where x_j, j=1,...,n is a set of spatial variables on a space Ω, and t_k, k=1,...,N is the time at which the observations are made. Provide $\boldsymbol{u}(x,t)$ is simple and square integrable, it has a Fourier representation of the form

$$\boldsymbol{u}(x_j,t)=\sum_{m=1}^{\infty}[a_{j(m)}(\omega)\cos(m\omega t)+b_{j(m)}(\omega)\sin(m\omega t)] \tag{9}$$

where $a_{j(m)}(\omega)$ and $b_{j(m)}(\omega)$ are the Fourier coefficients defined as

$$\begin{aligned} a_{j(m)} &= \frac{1}{\pi}\int_{-\pi}^{\pi} u(x_j,t)\cos\left(m\omega t\right)d\omega \\ b_{j(m)} &= \frac{1}{\pi}\int_{-\pi}^{\pi} u(x_j,t)\sin\left(m\omega t\right)d\omega \end{aligned} \tag{10}$$

This allows the description of travelling waves propagating throughout the system. Equation (9) can be rewritten in the form

$$\boldsymbol{u}_c(x_j,t)=\sum_{m=1}^{\infty}c_{j(m)}(\omega)e^{-im\omega t} \tag{11}$$

where $\boldsymbol{c}_{j(m)}(\omega)= a_{j(m)}(\omega)+ib_{j(m)}(\omega)$, $i=\sqrt{-1}$ is the unit complex number. Expanding (11) and collecting terms gives

$$\begin{aligned} \boldsymbol{u}_c(x_j,t) &= \sum_{m=1}^{\infty}\left\{[a_{j(m)}(\omega)\cos(m\omega t)+b_{j(m)}(\omega)\sin(m\omega t)]\right\} \\ &\quad + i\sum_{m=1}^{\infty}\left\{[b_{j(m)}(\omega)\cos(m\omega t)-a_{j(m)}(\omega)\sin(m\omega t)]\right\} \\ &= \boldsymbol{u}(x_j,t)+i\,\boldsymbol{u}_H(x_j,t) \end{aligned} \tag{12}$$

where the real part of $\boldsymbol{u}_c(x_j,t)$ is given by (9) and the imaginary part is the Hilbert transform of $\boldsymbol{u}(x_j,t)$. In formal terms, the Hilbert transform of a continuous time series $\boldsymbol{u}(x_j,t)$ is defined by the convolution

$$\boldsymbol{u}_H(x_j,t)=\frac{1}{\pi}\int_{-\infty}^{\infty}\frac{u(y)}{t-y}dy \tag{13}$$

where the integral is taken to mean the Cauchy principal value. The most well-known classical methods for computing the Hilbert transform are derived from the Fourier transform. However, this transform has a global character and hence, it is not suitable for the characterization of local signal parameters. Alternatives for local implementation of the Hilbert transformation which are based on local properties are developed and tested in this analysis (Hannchi, et al., 2007; Lezama, et al., 2009; Terradas, et al., 2004, Barnett, 1983).

For discretely sampled data, the Hilbert transform can be derived in the time domain by applying a rectangular rule to (13). It can be shown that

$$u_H(x_j,t) \approx \frac{2}{\pi}\sum_{k=-\infty}^{\infty}\frac{u(t+(2k+1)\tau)}{2k+1} \tag{14}$$

where τ is the step size. When (13) is applied to a discrete time series $u(x_j,t)$, k=0,±1,..., we get

$$\begin{aligned} u_H(x_j,t_k) &= \frac{2}{\pi}\sum_{k=-\infty}^{\infty}\frac{u(t+2k+1)}{2k+1} \\ &= \frac{2}{\pi}\sum_{k\geq 0}\frac{1}{2k+1}[u(t+2k+1)-u(t-2k-1)] \end{aligned} \tag{15}$$

In previous formulations, the Hilbert transform was estimated by truncating the series (15). This truncation was approximated using a convolution filter as

$$u_H(x_j,t_k) = \sum_{\ell=-L}^{L} u(x_j,t_k-\ell)h(\ell)\,, \quad L=\infty \tag{16}$$

where h is a convolution filter with an unit amplitude response and 90° phase shift. In this research, it has been found that a simple filter that has the desired properties of approximate unit amplitude response and $\pi/2$ phase shift is given by

$$h(l) = \begin{cases} \dfrac{2}{\pi l}\sin^2(\pi l/2)\,, & l \neq 0 \\ 0\,, & l = 0 \end{cases} \tag{17}$$

where $-L \leq l \leq L$. We omit the calculations.

As $L\to\infty$, equation (16) yields an exact Hilbert transform. This represents a filtering operation upon $u(x_j,t)$ in which the amplitude of each Fourier spectral component remains unchanged while its phase is advantaged by $\pi/2$. In (Hannachi, et al., 2007) has been found that $7 \leq L \leq 25$ provides adequate values for the filter response.

In what follows, we discuss the extension of the conventional EOF analysis using the above approach to compute standing and propagating features of general nonstationary processes where the eigenvectors of the covariance matrix are complex and it can be expressed alternatively as a magnitude and phase pair.

3. Complex empirical orthogonal function analysis

The method of complex EOF analysis is an optimal technique of biorthogonal decomposition to find a spatial and temporal basis that spans an ensemble of data collected from experiments or numerical simulations. The method essentially decomposes a fluctuating field into a weighted linear sum of spatial orthogonal modes and temporal orthogonal modes such that the projection onto the first few modes is optimal.

Drawing on the above approach, an efficient formulation to compute a complex expansion for the data set has been derived.

Assume that $\mathbf{X}(x,t)$ is augmented by their imaginary components to form a complex data matrix such as (Esquivel, 2009)

$$\mathbf{X}_c(x,t) = \mathbf{X}_R(x,t) + i\mathbf{X}_I(x,t) \tag{18}$$

where the subscripts c, R and I indicate the complex, real and imaginary vectors respectively. Implicit in the model is the assumption that $\mathbf{X}_c$ can be represented as

$$\mathbf{X}_c = \|\mathbf{X}_c\|[cos(\boldsymbol{\theta}_{X_c} t) + isin(\boldsymbol{\theta}_{X_c} t)] \tag{19}$$

where $\|\mathbf{X}_c\|$ and $\boldsymbol{\theta}_{X_c}$ are the magnitude and phase of $\mathbf{X}_c$. Under this assumption, the complex autocorrelation matrix becomes,

$$\mathbf{C} = \frac{1}{N}\mathbf{X}_c^H \mathbf{X}_c \tag{20}$$

where it is straightforward to show that the autocorrelation matrix ,**C**, for the case complex data can be written in the form $\mathbf{C}=\mathbf{C}_R+i\mathbf{C}_I$ which the real part is a symmetrical matrix, (i.e., $\mathbf{C}_R = \mathbf{C}_R^T$) and the imaginary part is a skew-symmetric matrix (i.e., $\mathbf{C}_I^T = -\mathbf{C}_I$). If the size of $\mathbf{C}_I$ is odd, then the determinant of $\mathbf{C}_I$ will always be zero. Because the symmetrical matrix is a particular case of the Hermitian matrix, then all its eigenvectors are real. Furthermore, the eigenvalues of the skew-symmetric matrix are all imaginary pure and, it is a normal matrix; its eigenvectors are complex conjugate.

From (20), It can be easily verified that

$$\begin{aligned}
\mathbf{X}_c^H \mathbf{X}_c &= \|\mathbf{X}_c^T\| \|\mathbf{X}_c\| [cos(\boldsymbol{\theta}_{\mathbf{X}_c^T} t) - i\sin(\boldsymbol{\theta}_{\mathbf{X}_c^T} t)][cos(\boldsymbol{\theta}_{X_c} t) + i\sin(\boldsymbol{\theta}_{X_c} t)] \\
&= \|\mathbf{X}_c^T\| \|\mathbf{X}_c\| \{[cos(\boldsymbol{\theta}_{\mathbf{X}_c^T} t) cos(\boldsymbol{\theta}_{X_c} t) + \sin(\boldsymbol{\theta}_{\mathbf{X}_c^T} t)\sin(\boldsymbol{\theta}_{X_c} t)] \\
&\quad + i[cos(\boldsymbol{\theta}_{\mathbf{X}_c^T} t)\sin(\boldsymbol{\theta}_{X_c} t) - \sin(\boldsymbol{\theta}_{\mathbf{X}_c^T} t) cos(\boldsymbol{\theta}_{X_c} t)]\}
\end{aligned} \tag{21}$$

From the decomposition given in (21) can be seen that the imaginary part is zero when the time is in phase with the extremum of the cosine or sine, that is, the sum of the two components is zero; at this time instant both are symmetrical matrices (Feeny, 2008). The imaginary part of (21) measures the degree of asymmetry when the sum of both matrices is different from zero; this is used to define the existence of arbitrary variations into the space, $\boldsymbol{\psi}^*(x)\neq 0$; this feature is used to define the existence of travelling wave components in the space-time varying fields and to determine leading seismic wave propagation components.

From the decomposition for the complex autocorrelation matrix (20), the optimal basis for the proposed spatio-temporal decomposition is defined by the eigenfunctions $\boldsymbol{\varphi}_R(x)$ and $\boldsymbol{\varphi}_I(x)$ for the real and imaginary part respectively. A test to split the spatial-temporal covariance functions is given by (Wallaschek, 1988; Fuentes, 2006).

Once the spatial eigenvectors associated with real and imaginary part of (20) are computed, the original field can be approximated by a spatio-temporal model. Assuming that this model is composed of standing and travelling wave components, the space-time varying field can be written as

$$\mathbf{X}(x,t) = \mathbf{X}_{swc}(x,t) + \mathbf{X}_{twc}(x,t) \tag{22}$$

where $\mathbf{X}_{swc}$ and $\mathbf{X}_{twc}$ denotes the standing and travelling wave components respectively. Therefore, the associated approximation for the complex data field (19) in terms of a truncated sum of dominant modes (EOFs basis) p and q, is defined as

$$\mathbf{X}_c(x,t) = \sum_{j=1}^{p} \mathbf{A}_{R(j)}(t)\boldsymbol{\varphi}_{R(j)}^{\mathrm{H}}(x) + i\sum_{j=1}^{q} \mathbf{A}_{I(j)}(t)\boldsymbol{\varphi}_{I(j)}^{\mathrm{H}}(x) \tag{23}$$

where the time-dependent complex coefficients associated with each eigenfuntion, $\mathbf{A}_{R(j)}(t)$ and $\mathbf{A}_{I(j)}(t)$ are obtained as the projection of the basis $\boldsymbol{\varphi}_{R(j)}(x)$ and $\boldsymbol{\varphi}_{I(j)}(x)$ respectively into complex field $\mathbf{X}_c$ of the form

$$\begin{aligned} \mathbf{A}_{R(j)}(t) &= \mathbf{X}_c \boldsymbol{\varphi}_{R(j)}(x) \\ \mathbf{A}_{I(j)}(t) &= \mathbf{X}_c \boldsymbol{\varphi}_{I(j)}(x) \end{aligned} \tag{24}$$

These complex coefficients are conveniently split into their amplitude and phase, therefore, from the complex model (23), the ensemble of data can be expressed as

$$\mathbf{X}_c(x,t) = \sum_{j=1}^{p} \mathbf{R}_{R(j)}(t)\mathbf{S}_{R(j)}(x)e^{i(\boldsymbol{\theta}_{R(j)}(t)+\boldsymbol{\Phi}_{R(j)}(x))} + \sum_{j=1}^{q} \mathbf{R}_{I(j)}(t)\mathbf{S}_{I(j)}(x)e^{i(\boldsymbol{\theta}_{I(j)}(t)+\boldsymbol{\Phi}_{I(j)}(x)+\pi)} \tag{25}$$

where $\mathbf{R}(t)$ and $\mathbf{S}(x)$ are the temporal and spatial amplitude functions associated with the wave decomposition respectively and, $\boldsymbol{\theta}(t)$ and $\boldsymbol{\Phi}(x)$ are the temporal and spatial phase function.

Now, four measurements that define moving features in $\boldsymbol{u}(x,t)$ can then be defined:

1. Spatial distribution of variability of each eigenmode
2. Relative phase fluctuation
3. Temporal variability in magnitude
4. Variability of the phase of a particular oscillation

The succeeding sections describe the properties of these representations to assess and to extract swing oscillations patterns and modal characteristics directly from recorded data in wide-area dynamical systems. It is shown that the proposed method can be used to predict the correct spatial location in the modal distribution of seismic wave.

3.1 Spatial amplitude function, S(*x*)

This function shows the spatial distribution of variability associated with each eigenmode. The spatial amplitude functions in the proposed model (25) are defined as (Hannchi, et al.,

2007; Marrifield & Guza, 1990; Susanto, et al., 1997; Terradas, et al., 2004; Toh, 1987; Barnett, 1983).,

$$\begin{aligned} \mathbf{S}_{R(j)}(x) &= \sqrt{\boldsymbol{\varphi}_{R(j)}^{\mathrm{H}}(x)\boldsymbol{\varphi}_{R(j)}(x)} \\ \mathbf{S}_{I(j)}(x) &= \sqrt{\boldsymbol{\varphi}_{I(j)}^{\mathrm{H}}(x)\boldsymbol{\varphi}_{I(j)}(x)} \end{aligned} \tag{26}$$

3.2 Spatial phase function, $\Phi(x)$

This function shows the relative phase fluctuation among various spatial locations where $u(x,t)$ is defined, it is given by

$$\begin{aligned} \boldsymbol{\Phi}_{R(j)}(x) &= \tan^{-1}\left\{\frac{im[\boldsymbol{\varphi}_{R(j)}(x)]}{re[\boldsymbol{\varphi}_{R(j)}(x)}\right\} \\ \boldsymbol{\Phi}_{I(j)}(x) &= \tan^{-1}\left\{\frac{im[\boldsymbol{\varphi}_{I(j)}(x)]}{re[\boldsymbol{\varphi}_{I(j)}(x)}\right\} \end{aligned} \tag{27}$$

3.3 Temporal amplitude function, $\mathrm{R}(t)$

This function gives a measure of the temporal variability in the magnitude of the modal structure in the original field. Similar to the description of the spatial amplitude function, the temporal amplitude function is defined as

$$\begin{aligned} \mathbf{R}_{R(j)}(t) &= \sqrt{\mathbf{A}_{R(j)}^{\mathrm{H}}(t)\mathbf{A}_{R(j)}(t)} \\ \mathbf{R}_{I(j)}(t) &= \sqrt{\mathbf{A}_{I(j)}^{\mathrm{H}}(t)\mathbf{A}_{I(j)}(t)} \end{aligned} \tag{28}$$

3.4 Temporal phase function, $\theta(t)$

This function shows the temporal variation of the phase associated with the magnitude of the modal structure of $u(x,t)$. It is given by

$$\begin{aligned} \boldsymbol{\theta}_{R(j)}(t) &= \tan^{-1}\left\{\frac{im[\mathbf{A}_{R(j)}(t)]}{re[\mathbf{A}_{R(j)}(t)}\right\} \\ \boldsymbol{\theta}_{I(j)}(t) &= \tan^{-1}\left\{\frac{im[\mathbf{A}_{I(j)}(t)]}{re[\mathbf{A}_{I(j)}(t)}\right\} \end{aligned} \tag{29}$$

Equations (26-29) provide a complete characterization of any propagating effects and periodicity in the original data field which might be obscured by standard cross-spectral analysis. These equations give a measure of the space-time distribution and can be used to identify the dominant modes and their phase relationships. Furthermore, for each dominant mode of interest, a mode shape can be computed by using the spatial part of (23). This method effectively decomposes the data into spatial and temporal modes.

4. Analysis of propagating features in space-time varying fields

In this section, we turn our attention to the analysis of spatial and temporal behaviour of propagating features in space-time varying fields.

4.1 Space-time biorthogonal decomposition

In order to investigate travelling and standing features into a space-time varying field, the real physical field is reconstructed by taking the real part of the complex model given in (25), so, its wave form is given by (Esquivel & Messina, 2008; Esquivel, 2009)

$$\mathbf{X}(x,t)=\sum_{j=1}^{p}\mathbf{R}_{R(j)}(t)\mathbf{S}_{R(j)}(x)\cos(\boldsymbol{\omega}_{R(j)}t)+\sum_{j=1}^{q}\mathbf{R}_{I(j)}(t)\mathbf{S}_{I(j)}(x)\cos(\boldsymbol{\omega}_{I(j)}t+\mathbf{K}_{I(j)}x+\pi) \tag{30}$$

where $\mathbf{K}(x)$ is the wave number, and $\boldsymbol{\omega}_R(t)$, $\boldsymbol{\omega}_I(t)$ represent the angular frequency of the real and imaginary wave components, respectively. The wave number is only defined for travelling waves and its components in terms of the complex representation (25) are given by: $\mathbf{K}=d(\boldsymbol{\Phi})/dx$, with physical units of $rad.m^{-1}$, and $\boldsymbol{\omega}=d(\boldsymbol{\theta})/dt$, in *rad/s*. The relationship between complex modes and the wave motion is given from average phase speeds $c_{R(j)}$, $c_{I(j)}$ obtained by using the relation $\mathbf{c}=\boldsymbol{\omega}/\mathbf{K}$, in m/s.

From (30), it can be seen that the term associated with the j-th travelling wave component can be expressed as

$$\begin{aligned}&\mathbf{R}_{I(j)}\mathbf{S}_{I(j)}\cos(\boldsymbol{\omega}_{I(j)}t+\mathbf{K}_{I(j)}x+\pi)=\\&\mathbf{R}_{I(j)}\mathbf{S}_{I(j)}\{\sin(\boldsymbol{\omega}_{I(j)}t)\sin(K_{I(j)}x)-\cos(\boldsymbol{\omega}_{I(j)}t)\cos(\mathbf{K}_{I(j)}x)\}\\&\qquad=-\mathbf{R}_{I(j)}\mathbf{S}_{I(j)}\cos(\boldsymbol{\omega}_{I(j)}t+\mathbf{K}_{I(j)}x)\end{aligned} \tag{31}$$

where we can see that the travelling wave components are also identified as the sum of two intermodulated standing wave components with negative sign. To obtain the decomposition of the original data field in its pure standing wave components, it is necessary to compute the difference with the pure travelling wave components as

$$\mathbf{X}_{swc}(x,t)=\mathbf{X}(x,t)-\mathbf{X}_{twc}(x,t)=\sum_{j=1}^{p}\boldsymbol{R}_{R(j)}(t)\boldsymbol{S}_{R(j)}(x)\cos\left(\boldsymbol{\omega}_{R(j)}t\right) \tag{32}$$

with

$$\mathbf{X}_{twc}(x,t)=\sum_{j=1}^{q}\mathbf{R}_{I(j)}(t)\mathbf{S}_{I(j)}(x)\cos(\boldsymbol{\omega}_{I(j)}t+\mathbf{K}_{I(j)}x+\pi) \tag{33}$$

where $\mathbf{X}_{swc}$ and $\mathbf{X}_{twc}$ represent the decomposition of the original field given by the pure standing and travelling wave components respectively. Furthermore, the damping factor of each mode is given by its amplitude.

From the modal decomposition given in (32-33), the statistical modes are also called orthogonal temporal and spatial modes respectively. Based in the proposed model, a practical criterion for choosing the relevant modes is given in the next section.

4.2 Approximation order and energy distribution in the space-time varying modes

The relationship between spatial and temporal behaviour in space-time varying fields can be obtained by noting that the spatio-temporal information can be mapping into a space and time grid, i.e., each component $\boldsymbol{u}(x,t)$ of the space-time varying field is represented by the field value at time t and spatial position x. Based in the proposed biorthogonal method, the analysis is used to determine the spatial and temporal energy distribution in the space-time varying field, a criterion for choosing the number of relevant modes from proposed model is given by the energy percentage contained in the p and q dominant modes of the form

$$\%E(p,q) = \frac{\sum_{j=1}^{p} \lambda_{(j)swc} + \sum_{j=1}^{q} \lambda_{(j)twc}}{\|\mathbf{X}\|_F^2} \times 100 = 99\% \tag{34}$$
$$= \text{subject to} \operatorname{argmin}\{E(p,q) : E(p,q) \geq E_0\}$$

where $\|.\|_F^2$ denotes the Frobenius norm, E_0 is an appropriate energy level, and $0 \leq \%E(p,q) \leq 100$ is the percentage of energy that is captured by the optimal basis. By neglecting modes corresponding to the small eigenvalues a reduced-order model can be constructed (Esquivel, 2009; Messina, et al. 2010).

We note from (34) that $E = \|\mathbf{X}\|_F^2$; so the spatial-temporal energy distribution can be computed by

$$\%E_{swc} = \frac{\|\mathbf{X}_{swc}\|_F^2}{\|\mathbf{X}\|_F^2} \times 100 \tag{35}$$

which is associated with the temporal energy distribution, and

$$\%E_{twc} = \frac{\|\mathbf{X}_{twc}\|_F^2}{\|\mathbf{X}\|_F^2} \times 100 \tag{36}$$

is associated with the spatial energy distribution. Figure 2 shows a conceptual representation of spatial and temporal variability illustrating the energy distribution in a

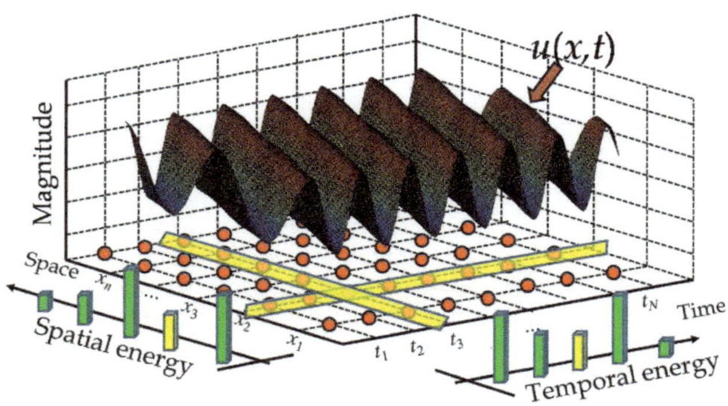

Fig. 2. Three-dimensional view of energy distribution of a time-space varying field.

space-time varying field. In an effort to better understand the mechanism of wave propagating using complex EOF analysis, in the next section is presented a first example as illustrative case to determine the theoretical fundamentals from the complex EOF analysis.

5. Motivating examples: Modeling of propagating wave using the covariance matrix

As illustrative case, in this section we consider a first example of wave propagation to study the modelling of propagating wave using the complex EOF analysis (Marrifield, 1990).

For simplicity, we consider a nondispersive plane wave propagating at phase speed *c*, and wavenumber as $k=\omega/c$, past an array of sensors at positions *j* given by

$$\boldsymbol{u}_j(t)=\sum_{\omega}[\alpha(\omega)\cos(kx_j-\omega t)+\beta(\omega)\sin(kx_j-\omega t)] \tag{37}$$

Expanding (37) and using identities

$$\begin{aligned} a_j(\omega)&=\alpha(\omega)\cos(kx_j)+\beta(\omega)\sin(kx_j)\\ b_j(\omega)&=\alpha(\omega)\sin(kx_j)-\beta(\omega)\cos(kx_j)\end{aligned} \tag{38}$$

we can rewritten (37) as

$$\boldsymbol{u}_j(t)=\sum_{\omega}\left\{a_j(\omega)\cos(\omega t)+b_j(\omega)\sin(\omega t)\right\} \tag{39}$$

where, for this example, $\boldsymbol{u}_j(t)$ is a white, band-limited signal given by

$$\alpha^2(\omega)+\beta^2(\omega)=\begin{cases}\mathrm{A}^2, & if \quad \omega_1\le\omega\le\omega_2\\ 0\;, & if \quad \omega>\omega_2,\omega<\omega_1\end{cases} \tag{40}$$

To obtain phase information between stations, a complex representation of (39) is invoked. Its complex covariance matrix $\mathbf{C}_{jk}=\left\langle \boldsymbol{u}_j^*(t)\boldsymbol{u}_k(t)\right\rangle_t$, where $\langle\cdots\rangle_t$ denotes time averaging and the asterisk complex conjugation, is given as

$$\begin{aligned}\mathbf{C}_{jk}&=\sum_{\omega}\left\{[a_j(\omega)-ib_j(\omega)]e^{i\omega t}\right\}*\sum_{\omega}\left\{[a_k(\omega)+ib_k(\omega)]e^{-i\omega t}\right\}\\ &=\sum_{\omega}\left\{[a_j(\omega)-ib_j(\omega)]*[a_k(\omega)+ib_k(\omega)]\right\},\\ &=\sum_{\omega}\left\{[a_j(\omega)a_k(\omega)+b_j(\omega)b_k(\omega)]+i[a_j(\omega)b_k(\omega)-b_j(\omega)a_k(\omega)]\right\}\end{aligned} \tag{41}$$

which, simplifying and using condition given in (40), $\mathbf{C}_{jk}$ can be rewritten as

$$\mathbf{C}_{jk}=\mathrm{A}^2\sum_{\omega}\left\{\cos(kx_j-kx_k)-i\sin(kx_j-kx_k)\right\}=\mathrm{A}^2\sum_{\omega=\omega_1}^{\omega_2}e^{-ik(\omega)\Delta x_{jk}} \tag{42}$$

where $\Delta x_{jk} = x_j - x_k$. Replacing the summatory of (42) with an integral, yields

$$\mathbf{C}_{jk} = \frac{A^2}{\Delta\omega}\int_{\omega_1}^{\omega_2} e^{-ik(\omega)\Delta x_{jk}} d\omega\,, \quad \text{to} \quad \Delta\omega = \frac{2\pi}{T} \tag{43}$$

Integrating (43) by parts and after some algebra, we can show that

$$\mathbf{C}_{jk} = A^2 M \sin c\left(\frac{\Delta k \Delta x_{jk}}{2}\right) e^{-i\bar{k}\Delta x_{jk}} \tag{44}$$

with

$$\begin{aligned} \Delta k &= k(\omega_2) - k(\omega_1) \\ \bar{k} &= \frac{k(\omega_2) + k(\omega_1)}{2} \\ M &= \frac{T(\omega_2 - \omega_1)}{2\pi} \end{aligned} \tag{45}$$

where Δk is the wave number bandwith, Δx is the array length and M is the frequency.

Equation (44) illustrates some important properties of **C**. General algebraic expression in order to computating the eigenvalues and eigenfunctions of **C** for an arbitrary number of sensors (n), are very difficult to determine and which are not purposed here. For the case of two sensors, the above model can be reduced to

$$\mathbf{C}_{jk} = \begin{bmatrix} A^2 M & A^2 M \sin c\left(\frac{\Delta k (x_1 - x_2)}{2}\right) e^{-i\bar{k}(x_1 - x_2)} \\ A^2 M \sin c\left(\frac{\Delta k (x_2 - x_1)}{2}\right) e^{-i\bar{k}(x_2 - x_1)} & A^2 M \end{bmatrix}, j,k = 1,2 \tag{46}$$

From the relation above is followed that the eigenvalues of **C** are given by $\det[\lambda \mathbf{I} - \mathbf{C}_{j,k}]$, i.e., it is easy to see further that

$$\lambda_{1,2} = A^2 M \pm A^2 M \left|\sin c\left(\frac{\Delta k \Delta x}{2}\right)\right| = A^2 M\left[1 \pm \left|\sin c\left(\frac{\Delta k \Delta x}{2}\right)\right|\right] \tag{47}$$

It then follows that the eigenvectors of **C** defined as $[\lambda \mathbf{I} - \mathbf{C}_{j,k}]\mathbf{B} = 0$, are given by

$$\begin{bmatrix} A^2 M \sin c\left(\frac{\Delta k \Delta x}{2}\right) & -A^2 M \sin c\left(\frac{\Delta k \Delta x}{2}\right) e^{-i\bar{k}\Delta x} \\ -A^2 M \sin c\left(\frac{\Delta k \Delta x}{2}\right) e^{i\bar{k}\Delta x} & A^2 M \sin c\left(\frac{\Delta k \Delta x}{2}\right) \end{bmatrix} \begin{bmatrix} B_{1,1} \\ B_{2,1} \end{bmatrix} = \begin{bmatrix} 0 \\ 0 \end{bmatrix} \tag{48}$$

and

$$\begin{bmatrix} -A^2 M \operatorname{sin}c\left(\frac{\Delta k \Delta x}{2}\right) & -A^2 M \operatorname{sin}c\left(\frac{\Delta k \Delta x}{2}\right) e^{-i\bar{k}\Delta x} \\ -A^2 M \operatorname{sin}c\left(\frac{\Delta k \Delta x}{2}\right) e^{i\bar{k}\Delta x} & -A^2 M \operatorname{sin}c\left(\frac{\Delta k \Delta x}{2}\right) \end{bmatrix} \begin{bmatrix} B_{1,2} \\ B_{2,2} \end{bmatrix} = \begin{bmatrix} 0 \\ 0 \end{bmatrix} \tag{49}$$

which can be simplified to

$$\begin{bmatrix} 1 & -e^{-i\bar{k}\Delta x} \\ 0 & 0 \end{bmatrix} \begin{bmatrix} B_{1,1} \\ B_{2,1} \end{bmatrix} = \begin{bmatrix} 0 \\ 0 \end{bmatrix} \quad \textit{and,} \quad \begin{bmatrix} 1 & e^{-i\bar{k}\Delta x} \\ 0 & 0 \end{bmatrix} \begin{bmatrix} B_{1,2} \\ B_{2,2} \end{bmatrix} = \begin{bmatrix} 0 \\ 0 \end{bmatrix} \tag{50}$$

where

$$\begin{aligned} & B_{1,1} = B_{2,1} e^{-i\bar{k}\Delta x}, \quad \textit{if} \quad B_{2,1} = T, \quad B_{1,1} = T e^{-i\bar{k}\Delta x} \\ & \begin{bmatrix} B_{1,1} \\ B_{2,1} \end{bmatrix} = \begin{bmatrix} e^{-i\bar{k}x_1} \\ e^{-i\bar{k}x_2} \end{bmatrix} T, \quad \text{with} \quad T = e^{-i\bar{k}x_2} \end{aligned} \tag{51}$$

and

$$\begin{aligned} & B_{1,2} = -B_{2,2} e^{-i\bar{k}\Delta x}, \quad \text{if} \quad B_{2,2} = -T, \quad B_{1,2} = T e^{-i\bar{k}\Delta x} \\ & \begin{bmatrix} B_{1,2} \\ B_{2,2} \end{bmatrix} = \begin{bmatrix} e^{-i\bar{k}x_1} \\ -e^{-i\bar{k}x_2} \end{bmatrix} T, \quad \text{with} \quad T = e^{-i\bar{k}x_2} \end{aligned} \tag{52}$$

The following observations can be made from the analysis:

1. As can be seen from (51) and (52), the method yields complex conjugate eigenvectors and an average wave number, $\bar{k}$.
2. The value $\bar{k}x_j$ gives the mode shape; this can be used for detection of wave propagating into original field and can be useful to identify the dominant stations involved in the propagating wave of dynamical oscillations. We remark that the performance of the complex EOF analysis as measured by the percentage of variance given in (47) depends on the spread in wave number relative to the array size, as the parameter $\Delta k \Delta x$ decreases, more of the variance is contained in the lowest complex EOF modes.
3. A point of particular interest is that, as a standard technique for describing coherent variability in spatial data, a relatively wide number bandwidth $[\Delta k \Delta x > 0(2\pi)]$ results from (47).

The development given in this section indicates that modal spatial patterns from a time domain complex EOF analysis may be computed in a straightforward manner. In the next section, data obtained from GPS-based multiple phasor measurements units from a real event of seismic wave of an earthquake are used to study the practical applicability of the method to characterize spatio-temporal behaviour in wide-area systems. Additionally, we discuss the practical computation of mode shape identification in relation to the proposed decomposition from measurements data that can be used to identify coherence groups in vast wide-area interconnected systems where the propagating wave are given.

6. Complex EOF analysis to wide-area system oscillatory dynamics

This section examines the application of the proposed technique to assess oscillations patterns in dynamical systems. Attention is focused on the identification of critical modes and the associated areas involved in the oscillations. In order to test the ability of the method to analyze complex oscillations, we use data recorded from time-synchronized measurements. The data were obtained from the Geophysical Institute of the National Autonomous University of México. A brief description of the data is given below.

At local time 15:36:14.730, October 9, 1995 a submarine earthquake was occurred near the Mexican coast (Colima-Jalisco); this earthquake was recorded by sixteen stations of phasor measurement units (PMUs) over a 225 s window sampled with time interval of 0.005 s during its propagating that was felt over much of Jalisco and parts of Colima. We examined evidence of seismic wave arrival times of the earthquake in PMUs based in global positing system (GPS). For simplicity, in Table 1 is given the description of the locations of each station. This earthquake was located at a depth of 5km about (18.740°N, 104.670°W). Figure 3 shows with a geographical diagram the PMUs locations and the location of the event.

		Location		
PMUs	Station	Latitude N	Longitude W	Altitude (msnm)
1	Ciudad Guzman	19.6°	103.4°	1507
2	Santa Rosa corona centro	20.912°	103.708°	770
3	Santa Rosa margen izquierda	20.912°	103.708°	780
4	Ciudad Granja	20.672°	103.398°	1680
5	Jardines del sur	20.648°	103.366°	1583
6	Arcos	20.671°	103.362°	1585
7	Obras publicas Zapopan	20.699°	103.361°	1561
8	Miravalle	20.633°	103.342°	1610
9	Rotonda	20.673°	103.34°	1542
10	San Rafael	20.654°	103.311°	1560
11	Planetario	20.717°	103.308°	1543
12	Tonala	20.641°	103.279°	1660
13	CICEJ superficie	20.6°	103.2°	1575
14	CICEJ pozo 9m	20.6°	103.2°	1566
15	CICEJ pozo 35m	20.6°	103.2°	1540
16	Oblatos	20.6°	103.2°	1580
	Earthquake	18.740°	104.670°	5km (depth)

Table 1. Description and location of stations of phasor measurements units.

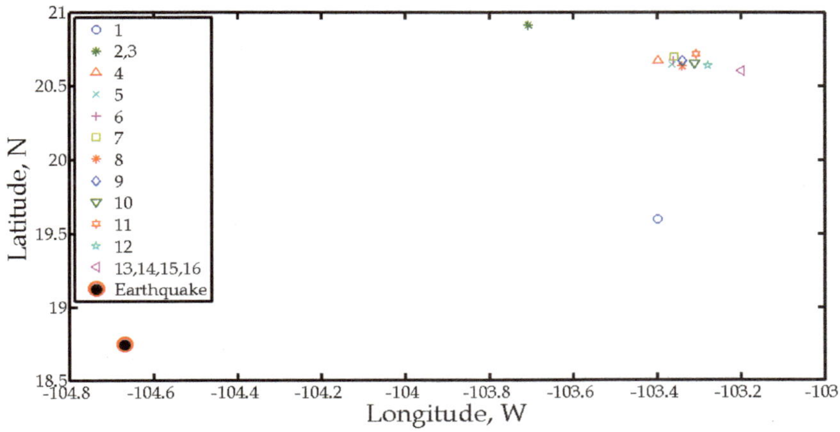

Fig. 3. Schematic showing the location of stations of the PMUs.

During the time interval 15:36:14.730-15:40:42 the earthquake experiment severe fluctuations where its seismic wave components such as frequency and amplitude were felt. Figures 4,5 and 6 give an extraction from PMUs measurements of this event showing the observed oscillations in the selected stations, where for simplicity, the seismic wave features are selected in longitude, latitude and altitude components. As a first step towards the development of the proposed methodology, the observed records are placed in a data matrix representing equally spaced measurements in sixteen different geographical locations. For our simulation, 45000 snapshots are available. Each time series is then augmented with an imaginary component by the Hilbert analysis to provide phase information and the corresponding birthogonal decomposition is applied to the dataset. System measurements in Figs. 4,5 and 6 demonstrate significant variability suggesting a nonstationary process in both space and time. Furthermore, in these figures are shown the associated mode to the travelling wave components based in the proposed method of birthogonal decomposition. The results clearly show the seismic wave decomposition, it is evident that the travelling wave mode in longitude and latitude is quite prominent at CICEJ and Oblatos stations, while that in the Ciudad Guzman and Santa Rosa stations are more stronger in altitude. A point of particular interest is the agreement between the results from the proposed model and the real behaviour of the space-time variability presented during the seismic wave. In (Ortiz, et al., 1998) was analyzed the tsunami data generated by the Colima-Jalisco earthquake, where the results of the tsunami arrival time are consistent with the presented in this analysis.

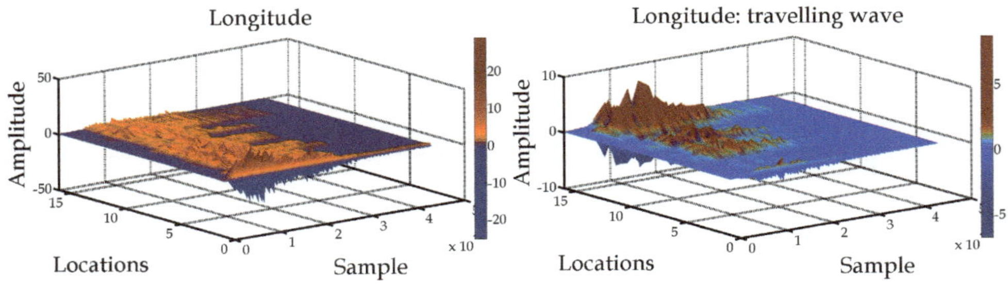

Fig. 4. Seismic fluctuating components in longitude and the leading mode showing spatio-temporal variability in the location of stations.

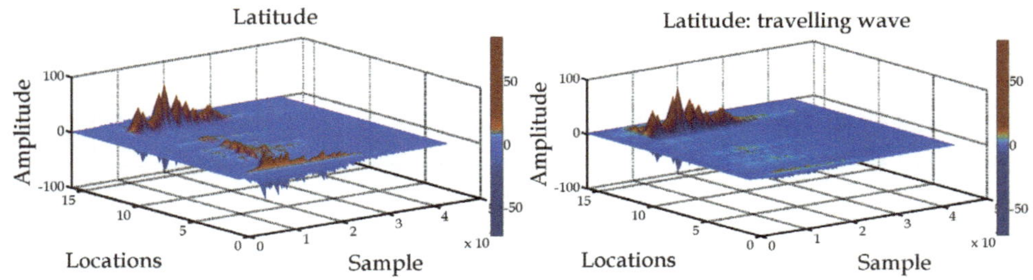

Fig. 5. Seismic fluctuating components in latitude and the leading mode showing spatio-temporal variability in the location of stations.

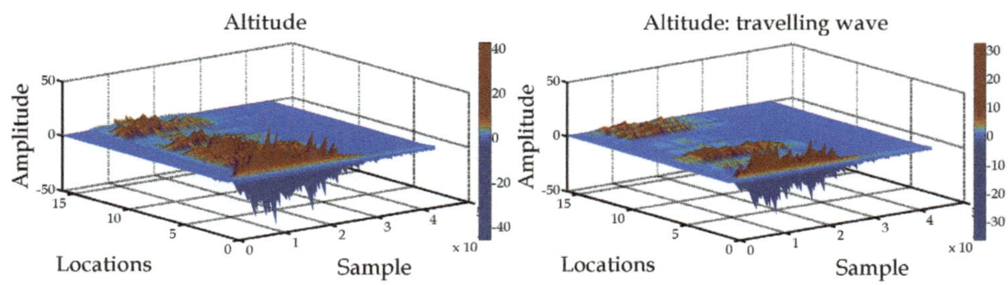

Fig. 6. Seismic fluctuating components in altitude and the leading mode showing spatio-temporal variability in the location of stations.

Additional insight into the frequency variability of the seismic oscillations can be obtained from the analysis of instantaneous frequency. Recognizing that the instantaneous frequency is the derivative of the temporal phase function given from the proposed model (30), the instantaneous frequency is estimated for each mode of concern. However, other approach can be used to characterize the spectral behaviour that requires other analytical formulations (Ortiz, et al., 2000).

The study focuses on the travelling wave mode which is the mode that captures most of the variability in the seismic wave. Figure 7 gives the spectrogram of the travelling wave modes associated to the longitude, latitude and altitude for the interval of interest in this study.

From this figure is evident that the earthquake was feeling with fluctuating components after 10 s since it was occurred.

Spectral analysis results for the leading travelling wave shows that the main power is concentred in oscillations with frequencies about 4.8, 5.2 and 4.0 Hz, to the longitude, latitude and altitude components which are associated with the major time interval of the seismic wave.

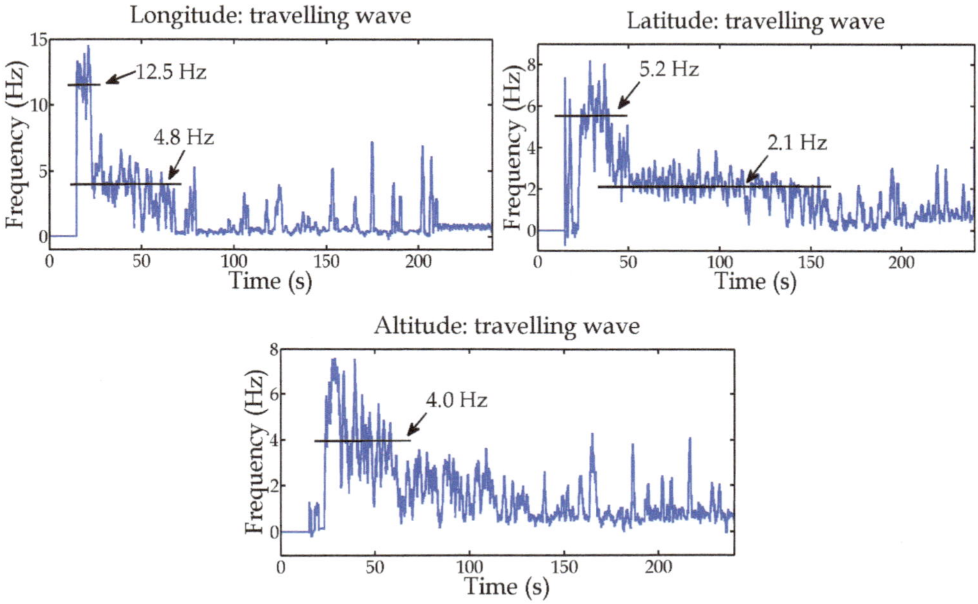

Fig. 7. Spectrograms to the seismic wave from the longitude, latitude and altitude components using the travelling wave modes.

One of the most attractive features of the proposed technique is its ability to detect changes in the mode shape properties of critical modes arising from systems. Changes in the mode shape may indicate changes in topology of the dynamic systems and may be useful for the design of special protection systems. This is a problem that has been recently addressed using spectral correlation analysis (Wallaschek, 1988).

Using the spatial phase and amplitude (the mode shape), the phase relationship between key system locations can be determined. In this analysis, we display the complex values as a vector with the length of its arrow proportional to eigenvector magnitude and direction equal to the eigenvector phase. Figure 8 shows the mode shape for the three travelling wave computed from the longitude, latitude and altitude components for the seismic wave, this information is useful to identify the dominant stations involved in the oscillations. Simulation results to the mode shape clearly show that the CICEJ (13,14,15) stations are more stronger evident at the longitude components; CICEJ (13,14) stations in latitude, and finally the Ciudad Guzman, Los Arcos, CICEJ (1,6,13,14) stations in altitude.

These results are in general agreement with the shown in Figs. 4,5 and 6 from the observed oscillations giving validity to the results. The new results provide clarification on the exact phase relationship between key stations as a function of space.

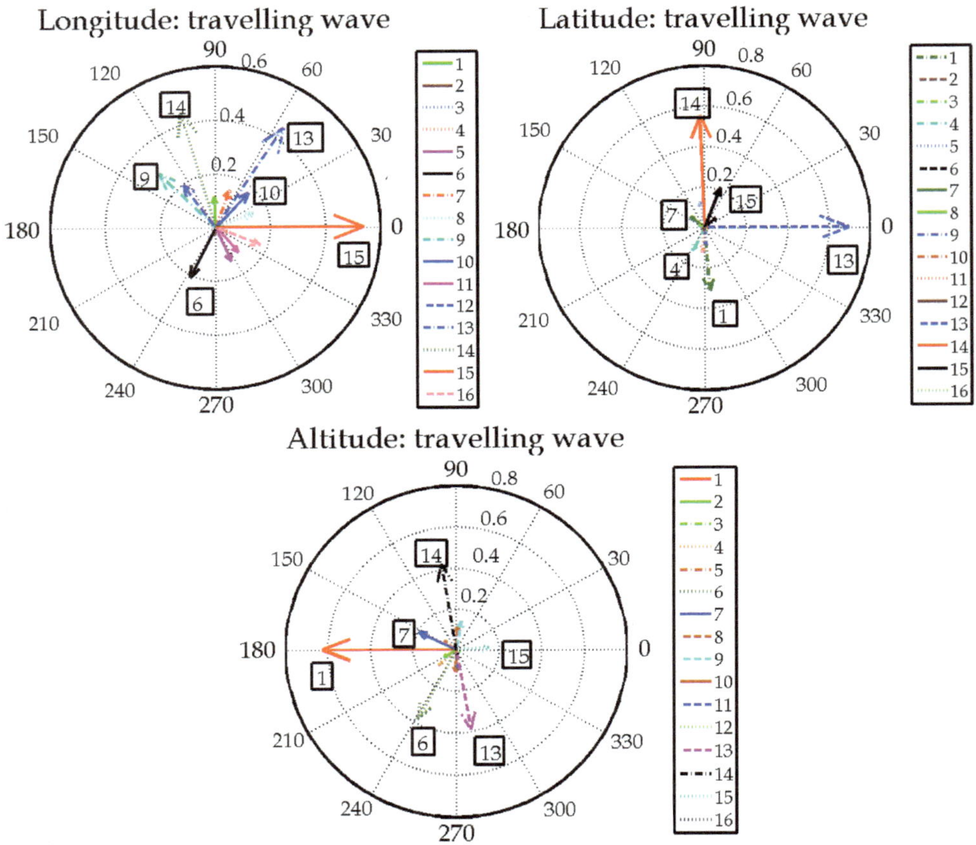

Fig. 8. Mode shape fluctuating in longitude, latitude and amplitude of the leading mode showing the phase relationship between stations.

7. Conclusion

Approaches for detection of propagation features in space-time varying system measurements through its travelling and standing components are proposed.

The conceptual framework developed provides bases for the analysis, detection, and simplification of seismic wave components through use of wide-area monitoring schemes such as global positioning systems (GPS) based in multiple phasor measurements units (PMUs) for interconnected systems, and enables the simultaneous study of synchronized measurements. The main advantage of the approach is its ability to compress the variability of large data sets into the fewest possible number of spatial and temporal modes. The technique is especially attractive, because, it does not require previous notion about the behaviour associated with abrupt changes in system topology or operating conditions.

Complex empirical orthogonal function analysis is shown to be a useful method to identify standing and travelling patterns in wide-area system measurements. In the use of information in interconnected systems, spatio-temporal analysis of wide-area time-synchronized measurements shows that transient oscillations may manifest highly complex phenomena, including nonstatonary behaviour. Numerical results show that the proposed method can provide accurate estimation of nonstationary effects, modal frequency, mode shapes, and time instants of intermittent transient responses. This information is important to determine strategies for wide-area monitoring and special protection systems.

The main contributions in this chapter are based in the estimation of propagating and standing features in space-time varying processes using statistical techniques to identify oscillatory activity in interconnected systems through the use of wide-area monitoring schemes in interconnected systems.

The proposed technique is based on the complex correlation structure from space-time varying fields, which can treat both, spatial and temporal information; this provides a global picture on the system behaviour to characterize oscillatory dynamics. Its significant drawbacks are associated to treat with their space-time scales. These include geographical distribution and the time interval to the modal extraction using measured data. For some applications may be desirable to have these data at very high space-time resolution that allow the study of processes close to inertial frequency, then a technique of space-time interpolation can be used. Wide-area monitoring may prove invaluable in interconnected system dynamic studies by giving a quick assessment of the damping and frequency content of dominant system modes after a critical contingency. The alternative technique based on space-time dependent complex EOF analysis of measured data is proposed to resolve the localized nature of transient processes and to extract dominant temporal and spatial information.

8. Acknowledgment

This work was supported under a scholarship granted by CONACYT México no. 172551 to CINVESTAV Guadalajara. We thank to the Geophysical Institute of the National Autonomous University of México for their support and facility to use the data presented in this work.

9. References

Achatz, U.; Schmitz, G. & Greisiger, K.-M. (1995). Principal Interaction Patterns in Baroclinic Wave Life Cycles, *Journal of the Atmospheric Sciences*, Vol.52, No.18, (September 1995), pp. 3201-3213, ISSN 15200469

Aubry, N.; Guyonnet, R. & Lima, R. (1990). Spatiotemporal Analysis of Complex Signals: Theory and Applications, *Journal of Statistical Physics*, Vol.64, No.3-4, (January 1991), pp. 683-738, ISSN 00224715

Barnett, T. P. (1983). Interaction of the Monsoon and Pacific Trade Wind Systems at Interannual Time Scales. Part I: The Equatorial Zone, *Monthly Weather Review*, Vol.111, No.4, (April 1983), pp. 756-773, ISSN 15200493

Dankowicz, H.; Holmes, P.; Berkooz, G. & Elezgaray, J. (1996). Local Models of Spatio-Temporally Complex Fields, *Physica D*, Vol.90, No.1996, (August 1995), pp. 387-407, ISSN 01672789

DelSole, T. (2001). Optimally Persistent Patterns in Time-Varying Fields, *Journal of Atmospheric Sciences*,Vol.58, No.11, (June 2001), pp. 1341-1356, ISSN 207053106

Esquivel, P. (2009). Wide-Area Wave Motion Analysis Using Complex Empirical Orthogonal Functions, *International Conference on Electrical Engineering, Computing Science and Automatic Control (ICEEE)*, ISBN 978-1-4244-4688-9, Toluca, Estado de México, México, November 10-13, 2009

Esquivel, P.; Barocio, E.; Andrade, M. A. & Lezama, F. (2009). Complex Empirical Orthogonal Function Analysis of Power-System Oscillatory Dynamics, In: *Inter-area Oscillations in Power Systems: A Nonlinear and Nonstationary Perspective*, A. R. Messina, Springer, 159-187, ISBN 9780387895291, New York, United States

Esquivel, P. & Messina, A. R. (2008). Complex Empirical Orthogonal Function Analysis of Wide-Area System Dynamics, *IEEE Power and Energy Society General Meeting Scheduled (PES)*, ISBN 978-1-4244-1905-0, Pittsburgh, Pennsylvania, United States, July 20-24, 2008

Feeny, B. F. (2008). A Complex Orthogonal Decomposition for Wave Motion Analysis, *Journal of Sound and Vibration*, Vol.310, No.1-2, (November 2008), pp. 77-90, ISSN 0022460X

Feeny, B. F. & Kappagantu, R. (1998). On the Physical Interpretation of Proper Orthogonal Modes in Vibrations, *Journal of Sound and Vibration*, Vol.211, No.4, (Abril 1999), pp. 607-616, ISSN 0022460X

Fuentes, M. (2006). Testing for Separability of Spatial-Temporal Covariance Functions, *Journal of Statistical Planning and Inference*, Vol.136, No.2, (July 2004), pp. 447-466, ISNN 03783758

Han, S. & Feeny B. F. (2003). Application of Proper Orthogonal Decomposition to Structural Vibration Analysis, *Mechanical Systems and Signals Processing*, Vol.17, No.5, (September 2003), pp. 989-1001, ISSN 08883270

Hannachi, A.; Jolliffe, I. T. & Stephenson, D. B. (2007). Empirical Orthogonal Functions and Related Techniques in Atmospheric Science: A Review, *International Journal of Climatology*, Vol.27, No.9, (July 2007), pp. 1119-1152, ISSN 08998418

Horel, J. D. (1984). Complex Principal Component Analysis: Theory and Examples, *Journal of Climate and Applied Meteorology*, Vol.23, No.12, (September 1984), pp. 1660-1673, ISSN 07333021

Hasselmann, K. (1988). PIP and POPs: the Reduction of Complex Dynamical Systems Using Principal Interaction and Oscillation Patterns, *Journal of Geophysical Research*, Vol.93, No.D9, (September 1988), pp. 11015-11021, ISSN 01480227

Holmes, P.; Lumley, J. L. & Berkooz, G. (1996). *Turbulence, Coherent Strustures, Dynamical Systems and Symmetry*, Cambridge University Press, ISBN 0521634199, New York, United States

Hostelling, H. (1933). Analysis of a Complex of Statistical Variables into Principal Components, *Journal of Educational Psychology*, Vol.24, No.6, (September 1933), pp. 417-441, ISSN 19392176

Kaihatu, J. M.; Handler, R. A.; Marmorino, G. O. & Shay, L. K. (1997). Empirical Orthogonal Function Analysis of Ocean Surface Currents Using Complex and Real-Vector Methods, Journal of Atmospheric and Oceanic Technology, Vol.15, No.1, (September 1997), pp. 927-941, ISSN 395295004

Kosambi, D. D. (1943). Statistics in Function Space, *Journal of Indian Mathematical Society*, Vol.7, No.46, (June 1943) pp. 76-88, ISSN 00195839

Kwasniok, F. (1996). The reduction of Complex Dynamical Systems Using Principal Interaction Patterns, *Physica D: Nonlinear Phenomena*, Vol.92, No.1-2, (April 1996), pp. 28-60, ISSN 01672789

Kwasniok, F. (2007). Reduced Atmospheric Models Using Dynamically Motivated Basis Functions, *Journal of the Atmospheric Sciences*, Vol.64, No.10, (October 2007), pp. 3452-3474, ISSN 00224928

Marrifield, M. A. & Guza, R. T. (1990). Detecting Propagating Signals with Complex Empirical Orthogonal Functions: A Cautionary Note, *Journal of Physical Oceanography*, Vol.20, No.10, (February 1990), pp. 1628-1633, ISSN 15200485

Messina, A. R.; Esquivel, P. & Lezama, F. (2010). Time-Dependent Statistical Analysis of Wide-Area Time-Synchronized Data, *Journal of Mathematical Problems in Engineering (MPE)*, Vol.2010, No.2010, (Abril 2010), pp. 1-17, ISSN 1024123X

Messina, A. R.; Esquivel, P. & Lezama, F. (2011). Wide-Area PMU Data Monitoring Using Spatio-Temporal Statistical Models, *2011 IEEE/PES Power Systems Conferences and Exposition*, ISBN 978-1-61284-789-4, Phoenix, Arizona, United States, March 20-23, 2011

Leonardi, A. P.; Morey, L. S. & O'Brien, J. J. (2002). Interannual Variability in the Eastern Subtropical North Pacific Ocean, *Journal of Physical Oceanography*, Vol.32, No.1, (June 2002), pp. 1824-1837, ISSN 15200485

Lezama, F.; Rios, A. L.; Esquivel, P. & Messina, A. R. (2009). A Hilbert-Huang Based Approach for On-Line Extraction of Modal Behavior from PMU Data, *IEEE PES 41st North American Power Symposium (NAPS)*, ISBN 978-1-4244-4428-1, Starkville, Mississippi, United States, October 4-6, 2009

Oey, L.-Y. (2007). Loop Current and Deep Eddies, *Journal of Physical Oceanography*, Vol.38, No.1, (Oct0ber 2007), pp. 1426-1449, ISSN 11753818

Ortiz, M.; Gomez-Reyes, E. & Velez-Muños, S. (2000). A Fast Preliminary Estimation Model for Transoceanic Tsunami Propagation, *Geophysical International*, Vol.39, No.3, (June 2000), pp. 207-220, ISSN 00167169

Ortiz, M; Singh, S. K.; Pacheco, J. & Kostoglodov, V. (1998). Rupture Length of the October 9, 1995 Colima-Jalisco Earthquake (M_W) Estimated from Tsunami Data, *Geophysical Research Letters*, Vol.25, No.15, (August 1998), pp. 2857-2860, ISSN 00948534

Spletzer, M.; Raman, A. & Reifenberger, R. (2010). Spatio-Temporal Dynamics of Microcantilevers Tapping on Samples Observed Under an Atomic Force Microscope Integrated with a Scanning Laser Doppler Vibrometer: Applications to Proper Orthogonal Decomposition and Model Reduction, Journal of Micromechanics and Microengineering, Vol.20, No.8, (August 2010), pp. 1088-1098, ISSN 09601317

Susanto, R. D.; Zheng, Q. & Yan, X. H. (1997). Complex Singular Value Decomposition Analysis of Equatorial Waves in the Pacific Observed by TOPEX/Poseidon Altimeter, *Journal of Atmospheric and Oceanic Technology*, Vol. 15, No.1, (July 1997), pp. 764-774, ISSN 197163501

Terradas, J.; Oliver, R. & Ballester, J. L. (2004). Application of Statistical Techniques to the Analysis of Solar Coronal Oscillations, *The Astrophysical Journal*, Vol.614, No.1, (October 2004), pp. 435-447, ISSN 0004637X

Toh, S. (1987). Statistical Model with Localized Structures Describing the Spatial-Temporal Chaos of Kuramoto-Sivashinky Equation, *Journal of the Physical Society of Japan*, Vol.56, No.3, (March 1987), pp. 949-962, ISSN 00319015

Wallaschek, J. (1988). Integral Covariance Analysis for Random Vibrations of Linear Continuous Mechanical Systems, *Dynamical and Stability of Systems*, Vol.3, No.1-2, (January 1988), pp. 99-107, ISNN 14565390

Permissions

The contributors of this book come from diverse backgrounds, making this book a truly international effort. This book will bring forth new frontiers with its revolutionizing research information and detailed analysis of the nascent developments around the world.

We would like to thank Peep Miidla, for lending his expertise to make the book truly unique. He has played a crucial role in the development of this book. Without his invaluable contribution this book wouldn't have been possible. He has made vital efforts to compile up to date information on the varied aspects of this subject to make this book a valuable addition to the collection of many professionals and students.

This book was conceptualized with the vision of imparting up-to-date information and advanced data in this field. To ensure the same, a matchless editorial board was set up. Every individual on the board went through rigorous rounds of assessment to prove their worth. After which they invested a large part of their time researching and compiling the most relevant data for our readers. Conferences and sessions were held from time to time between the editorial board and the contributing authors to present the data in the most comprehensible form. The editorial team has worked tirelessly to provide valuable and valid information to help people across the globe.

Every chapter published in this book has been scrutinized by our experts. Their significance has been extensively debated. The topics covered herein carry significant findings which will fuel the growth of the discipline. They may even be implemented as practical applications or may be referred to as a beginning point for another development. Chapters in this book were first published by InTech; hereby published with permission under the Creative Commons Attribution License or equivalent.

The editorial board has been involved in producing this book since its inception. They have spent rigorous hours researching and exploring the diverse topics which have resulted in the successful publishing of this book. They have passed on their knowledge of decades through this book. To expedite this challenging task, the publisher supported the team at every step. A small team of assistant editors was also appointed to further simplify the editing procedure and attain best results for the readers.

Our editorial team has been hand-picked from every corner of the world. Their multi-ethnicity adds dynamic inputs to the discussions which result in innovative outcomes. These outcomes are then further discussed with the researchers and contributors who give their valuable feedback and opinion regarding the same. The feedback is then collaborated with the researches and they are edited in a comprehensive manner to aid the understanding of the subject.

Apart from the editorial board, the designing team has also invested a significant amount of their time in understanding the subject and creating the most relevant covers. They scrutinized every image to scout for the most suitable representation of the subject and create an appropriate cover for the book.

The publishing team has been involved in this book since its early stages. They were actively engaged in every process, be it collecting the data, connecting with the contributors or procuring relevant information. The team has been an ardent support to the editorial, designing and production team. Their endless efforts to recruit the best for this project, has resulted in the accomplishment of this book. They are a veteran in the field of academics and their pool of knowledge is as vast as their experience in printing. Their expertise and guidance has proved useful at every step. Their uncompromising quality standards have made this book an exceptional effort. Their encouragement from time to time has been an inspiration for everyone.

The publisher and the editorial board hope that this book will prove to be a valuable piece of knowledge for researchers, students, practitioners and scholars across the globe.

List of Contributors

María Isabel Lamas Galdo and Carlos G. Rodríguez Vidal
Universidade da Coruña, Spain

Rekha R. Rao, Lisa A. Mondy, David R. Noble, Matthew M. Hopkins and Carlton F. Brooks
Sandia National Laboratories, USA

Thomas A. Baer
Procter and Gamble Company, USA

A. Georgoulas, P. Angelidis, K. Kopasakis and N. Kotsovinos
Democritus University of Thrace, Xanthi, Greece

Leonid Bazyma
National Aerospace University "Kharkov Aviation Institute", Ukraine

Vasyl Rashkovan
National Polytechnic Institute, Mexico

Vladimir Golovanevskiy
Western Australian School of Mines, Curtin University, Australia

Surendra Adhikari and Shawn J. Marshall
Department of Geography, University of Calgary, Canada

Seyed Mahmood Kashefipour
Department of Hydraulic Structures, Shahid Chamran University, Ahwaz, Khuzestan, Iran

Ali Roshanfekr
Department of Civil and Resource Engineering, Dalhousie University, Halifax, NS, Canada

A.K. Ferouani, B. Liani and M. Lemerini
Laboratory of Theoretical Physics, University Abou Bakr Belkaid, Tlemcen, Algéria

Olga Lavrova and Viktor Polevikov
Belarusian State University, Belarus

Lutz Tobiska
Otto-von-Guericke University, Germany

Philip May
Elcometer Ltd., UK

Erping Zhou
University of Bolton, UK

M'hemed Rachek and Tarik Merzouki
University Mouloud Mammeri of Tizi-Ouzou, Algeria

Marcin Hojny and Miroslaw Glowacki
AGH University of Science and Technology, Krakow, Poland

Nadia Bhuiyan
Concordia University, Canada

Marwan Al-Heib
Ineris - Ecole des Mines de Nancy, Parc de Saurupt, France

Miroslaw Glowacki
AGH University of Science and Technology, Poland

Weihua Deng and Can Li
School of Mathematics and Statistics, Lanzhou University, Lanzhou, People's Republic of China

Radim Halama and Michal Šofer
Centre of Excellence IT4Innovations, Department of Mechanics of Materials, VŠB-Technical University of Ostrava, Czech Republic

Josef Sedlák

P. Esquivel and D. Cabuto
Technological Institute of Tepic, Electrical and Electronics Engineering Division, Nayarit, México

V. Sanchez and F. Chan
University of Quintana Roo, Sciences and Engineering Division, Quintana Roo, México